Basic Concepts in Relativity and Early Quantum Theory

SECOND EDITION

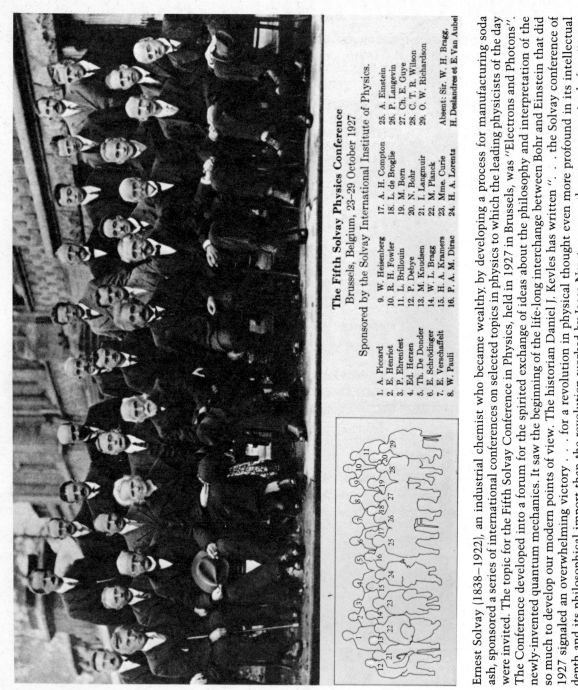

The Fifth Solvay Physics Conference
Brussels, Belgium, 23–29 October 1927
Sponsored by the Solvay International Institute of Physics.

1. A. Piccard
2. E. Henriot
3. P. Ehrenfest
4. Ed. Herzen
5. Th. De Donder
6. E. Schrödinger
7. E. Verschaffelt
8. W. Pauli
9. W. Heisenberg
10. R. H. Fowler
11. L. Brillouin
12. P. Debye
13. M. Knudsen
14. W. L. Bragg
15. H. A. Kramers
16. P. A. M. Dirac
17. A. H. Compton
18. L. de Broglie
19. M. Born
20. N. Bohr
21. I. Langmuir
22. M. Planck
23. Mme. Curie
24. H. A. Lorentz
25. A. Einstein
26. P. Langevin
27. Ch. E. Guye
28. C. T. R. Wilson
29. O. W. Richardson

Absent: Sir. W. H. Bragg,
H. Deslandres et E. Van Aubel

Ernest Solvay (1838–1922), an industrial chemist who became wealthy by developing a process for manufacturing soda ash, sponsored a series of international conferences on selected topics in physics to which the leading physicists of the day were invited. The topic for the Fifth Solvay Conference in Physics, held in 1927 in Brussels, was "Electrons and Photons". The Conference developed into a forum for the spirited exchange of ideas about the philosophy and interpretation of the newly-invented quantum mechanics. It saw the beginning of the life-long interchange between Bohr and Einstein that did so much to develop our modern points of view. The historian Daniel J. Kevles has written ". . . the Solvay conference of 1927 signaled an overwhelming victory . . . for a revolution in physical thought even more profound in its intellectual depth and its philosophical import than the revolution worked by Isaac Newton more than two centuries before."*

Except for J. J. Thomson and Ernest Rutherford, all of the physicists prominently mentioned in the pages that follow were among the 33 conferees. In due course 19 Nobel Prizes were awarded to the members of this group, with Marie Curie receiving two, one in Physics and one in Chemistry.

* *The Physicists* (Knopf, New York, 1978), Chapter XI.

Basic Concepts in Relativity and Early Quantum Theory

SECOND EDITION

Robert Resnick
Rensselaer Polytechnic Institute

David Halliday
University of Pittsburgh

JOHN WILEY & SONS

New York   Chichester   Brisbane   Toronto   Singapore

Library of Congress Cataloging in Publication Data:

Resnick, Robert, 1923–
    Basic concepts in relativity and early quantum theory.

    Includes bibliographical references and index.
    1. Relativity (Physics)   2. Quantum theory.
I. Halliday, David, 1916–   .   II. Title.
QC173.55.R47   1985   530.1'1   84-22211
ISBN 0-471-88813-3
ISBN 0-471-88858-3 (pbk.)

Printed in the United States of America

10 9 8 7 6 5 4 3 2

Preface

Since the publication of the first edition of this book,[*] the need for such a flexible supplement to introductory classical physics courses and to beginning quantum physics courses has increased. We have taken advantage of the experience gained in the classroom with this material and the upsurge in interest in relativity and early quantum theory to prepare a significantly improved version of the text. Amongst the changes we have made are these:

(a) In relativity we have included and evaluated the more recent experimental tests, given a fuller treatment of the spacetime interval, and expanded on the significant Kennedy-Thorndike experiment. In addition, we have included three supplementary topics at the end of the book on spacetime diagrams, the twin paradox, and general relativity, that are intended to serve as optional supplementary material for instructors seeking greater breadth or depth of treatment.

(b) In early quantum theory we have treated the quantum theory of specific heats, which played a crucial historical role in gaining acceptance of the quantization idea, included material on synchrotron radiation, and modernized the treatment of the electron microscope. At the very end we have added a preview of quantum mechanics. We also present six supplementary topics, intended for optional use, giving brief treatments of a detailed or advanced nature on selected items discussed in the main text.

(c) The thought questions and problems have been expanded into a large, rich, and varied set. The number has been nearly doubled (from 215 to 420 problems and from 150 to 240 questions, plus 40 questions and problems in the supplementary topics), the more difficult ones have been starred, and a set of problems for handheld programmable calculators (designated with a C) has been included.

(d) There has been a fifty percent increase in worked-out examples in the text—now numbering seventy—the selection being based on what the teaching experience with the first edition indicated was the most effective way to enhance the student's understanding.

(e) We have made modest additions of relevant and interesting material of historical, philosophical, or biographical nature, reflecting the great expansion in research and publications of the past decade concerning the history and origins of relativity and early quantum theory.

(f) Pedagogic features of the text have been enhanced in many other ways. We now give titles to all the worked examples and all the problems, use relevant chapter-opening quotes, and have improved the physical layout of the text. All the figures have been redone for greater clarity and additions of photographs and figures have been made where helpful and interesting to the student. References, which are cited to encourage students to read

[*] Robert Resnick, *Basic Concepts in Relativity and Early Quantum Theory*, 1st ed., (Wiley, New York, 1972).

original or popular sources, have been updated and the use of summary tables has been increased. Answers to the problems and tables of useful data are also provided.

There are two principal ways this material has been used. One way has been to use it as the conclusion of an introductory course in classical physics, bringing the subject matter right up to modern quantum physics. As such, this text can be regarded as a supplement to 'Physics, 3d Edition' by the same authors. Another way has been to use it as the beginning of a course in quantum physics, wherein typically the treatment of relativity and early quantum theory has been thin, if present at all. However, there are and have been still other effective ways to use this text. For a semester course in modern physics, one could use this text together with a survey of atomic, solid state, and nuclear physics, such as is provided in the concluding five chapters of 'Fundamentals of Physics, Extended Version, 2d Edition' by the same authors. The material has also proven useful on its own in a variety of settings, such as in minicourses and in summer courses for teachers.

Adding to its utility is the flexibility of the text. For example, the supplementary topics are optional. The first chapter on relativity has occasionally been left for a reading assignment and Chapter 4—which emphasizes the idea of quantization—can be skipped without a serious lack of continuity. The experienced instructor will find still other ways in which he or she can expand or contract the time needed for a coherent presentation of the subject. We should point out that some marginal material has been cut out from the first edition and that most of the increased length of the second edition resides either in its optional material or the pedagogically enhanced areas, such as worked examples, questions and problems, figures, photographs, and the like.

We acknowledge with pleasure our debt to the professional staff at John Wiley & Sons, organized with great effectiveness by Robert McConnin, Physics Editor, for its outstanding cooperation and assistance. We also wish to thank Edward Derringh (Wentworth Institute of Technology) for many useful comments and for valuable assistance with the problem sets.

Robert Resnick

Rensselaer Polytechnic Institute
Troy, New York 12181

David Halliday

3 Clement Road
Hanover, New Hampshire 03755

January 1, 1985

Contents

Introduction to Chapters 1 to 3   1

CHAPTER 1

THE EXPERIMENTAL BACKGROUND OF THE THEORY OF SPECIAL RELATIVITY   3

1-1 Introduction   3
1-2 Galilean Transformations   5
1-3 Newtonian Relativity   8
1-4 Electromagnetism and Newtonian Relativity   12
1-5 Attempts to Locate the Absolute Frame—The
     Michelson-Morley Experiment   13
1-6 Attempts to "Save the Ether"—The Lorentz-Fitzgerald
     Contraction   19
1-7 Attempts to "Save the Ether"—The Ether-Drag Hypothesis   21
1-8 Attempts to Modify Electrodynamics   25
1-9 The Postulates of Special Relativity Theory   26
1-10 Einstein and the Origin of Relativity Theory   29

CHAPTER 2

RELATIVISTIC KINEMATICS   39

2-1 The Relativity of Simultaneity   39
2-2 Derivation of the Lorentz Transformation Equations   44
2-3 Some Consequences of the Lorentz Transformation Equations   50
2-4 The Lorentz Equations—A More Physical Look   58
2-5 The Observer in Relativity   65
2-6 The Relativistic Addition of Velocities   66
2-7 Aberration and Doppler Effect in Relativity   70
2-8 The Common Sense of Special Relativity   76

CHAPTER 3

RELATIVISTIC DYNAMICS   93

3-1 Mechanics and Relativity   93
3-2 The Need to Redefine Momentum   93
3-3 Relativistic Momentum   95
3-4 The Relativistic Force Law and the Dynamics of a Single Particle   98
3-5 Some Experimental Results   105
3-6 The Equivalence of Mass and Energy   109
3-7 Relativity and Electromagnetism   115

Introduction to Chapters 4 to 7   125

CHAPTER 4

THE QUANTIZATION OF ENERGY   127

4-1 Introduction   127
4-2 Thermal Radiation   127
4-3 The Theory of Cavity Radiation   131
4-4 The Rayleigh-Jeans Radiation Law   134
4-5 The Quantization of Energy   136
4-6 Energy Quantization and the Heat Capacities of Solids   140
4-7 The Franck-Hertz Experiment   147
4-8 Epilogue   151

CHAPTER 5

THE PARTICLE NATURE OF RADIATION   161

5-1 Introduction   161
5-2 The Photoelectric Effect   161
5-3 Einstein's Quantum Theory of the Photoelectric Effect   165
5-4 The Compton Effect   170
5-5 Pair Production   176
5-6 Photons Generated by Accelerating Charges   180
5-7 Photon Production by Pair Annihilation   184

CHAPTER 6

THE WAVE NATURE OF MATTER AND THE UNCERTAINTY PRINCIPLE   195

6-1 Matter Waves   195
6-2 Testing the de Broglie Hypothesis   198
6-3 The Electron Microscope   204
6-4 The Wave–Particle Duality   212
6-5 The Uncertainty Principle   218
6-6 A Derivation of the Uncertainty Principle   222
6-7 Interpretation of the Uncertainty Principle   222

CHAPTER 7

EARLY QUANTUM THEORY OF THE ATOM   231

7-1 J. J. Thomson's Model of the Atom   231
7-2 The Nuclear Atom   233
7-3 The Stability of the Nuclear Atom   238
7-4 The Spectra of Atoms   239

7-5   The Bohr Atom   241
7-6   The Bohr Semiclassical Planetary Model of the One-Electron
       Atom   247
7-7   The Quantization of Angular Momentum   250
7-8   Correction for the Nuclear Mass   252
7-9   Bohr Theory—The High-Water Mark   256
7-10   Quantum Mechanics—A Preview   257

SUPPLEMENTARY TOPIC A
THE GEOMETRIC REPRESENTATION OF SPACETIME   **269**

SUPPLEMENTARY TOPIC B
THE TWIN PARADOX   **281**

SUPPLEMENTARY TOPIC C
*THE PRINCIPLE OF EQUIVALENCE AND THE GENERAL
THEORY OF RELATIVITY*   **291**

SUPPLEMENTARY TOPIC D
THE RADIANCY AND THE ENERGY DENSITY   **305**

SUPPLEMENTARY TOPIC E
EINSTEIN'S DERIVATION OF PLANCK'S RADIATION LAW   **307**

SUPPLEMENTARY TOPIC F
DEBYE'S THEORY OF THE HEAT CAPACITY OF SOLIDS   **311**

SUPPLEMENTARY TOPIC G
PHASE SPEED AND GROUP SPEED   **315**

SUPPLEMENTARY TOPIC H
MATTER WAVES—REFRACTION AT THE CRYSTAL SURFACE   **319**

SUPPLEMENTARY TOPIC I
RUTHERFORD SCATTERING   **321**

ANSWERS TO PROBLEMS   327

APPENDIXES   333

INDEX   335

Introduction to Chapters 1 to 3

Modern physics can be defined as physics that requires relativity theory or quantum theory for its interpretation. These theories emerged in the early decades of the twentieth century as the classical theories encountered increasing difficulty in explaining experimental observations.

In this first section of the book (Chapters 1 to 3), we examine the experimental background to relativity, the development of the special theory of relativity, and the experimental confirmation of relativistic predictions. We shall see that classical mechanics breaks down in the region of very high speeds and that relativistic mechanics is a generalization that includes the classical laws as a special case. Gradually the student will develop a physical feeling for the principles of relativity. The point of view and the results that emerge from relativistic considerations prove to be useful and necessary in many areas of modern physics, including atomic, nuclear, and particle physics as well as the physics of the solid state.

CHAPTER 1

the experimental background of the theory of special relativity

> *. . . whenever energy is transmitted from one body to another in time, there must be a medium or substance in which the energy exists after it leaves one body and before it reaches the other. . . .*
>
> J. C. Maxwell (1873)

> *. . . I came to the opinion quite some time ago that Fresnel's idea, hypothesizing a motionless ether, is on the right path.*
>
> H. A. Lorentz (1895)

> *The introduction of a "luminiferous ether" will prove to be superfluous inasmuch as the view here developed will not require an "absolute stationary space". . . .*
>
> A. Einstein (1905)

1-1 INTRODUCTION

To send a signal from one point to another as fast as possible, we use a beam of light or some other electromagnetic radiation such as a radio wave. *No faster method of signaling has ever been discovered.* This experimental fact suggests that the speed of light in free space ($c = 3.00 \times 10^8$ m/s)* is an appropriate limiting reference speed with respect to which other speeds, such as the speeds of particles or of mechanical waves, can be compared.

In the macroscopic world of our ordinary daily experiences, the speed u of moving objects or mechanical waves with respect to any observer is always much less than c. For example, a satellite circling the earth may move at 16,000 mi/h with respect to the earth; here $u/c = 0.000024$. Sound waves in air at room temperature travel at 343 m/s through the air, so that $u/c = 0.0000011$. It was in this ever-present, but limited, macroscopic environment that our ideas about

---

* In October 1983, the speed of light was adopted as a defined standard and assigned the value of (exactly) 2.99792458×10^8 m/s.

space and time were first formulated and in which Newton developed his system of mechanics.

In the microscopic world, however, it is possible to find particles whose speeds are quite close to that of light. For an electron accelerated through a 10-million-volt potential difference, a value reasonably easy to obtain, we find that $u/c = 0.9988$. We cannot be certain without direct experimental test that Newtonian mechanics can be safely extrapolated from the ordinary region of low speeds ($u/c \ll 1$) in which it was developed to this high-speed region ($u/c \rightarrow 1$). Experiment shows, in fact, that Newtonian mechanics does *not* predict the correct answers when it is applied to such fast particles. Indeed, in Newtonian mechanics there is no limit in principle to the speed attainable by a particle, so that the speed of light c plays no special role at all. For example, if the energy of the 10-MeV electron above is increased by a factor of four (to 40 MeV), experiment shows that the speed is not doubled to $1.9976c$, as we might expect from the Newtonian relation $K = \frac{1}{2}m_e v^2$, but remains below c; it increases only from $0.9988c$ to $0.9999c$, a change of 0.11 percent. Or, if the 10-MeV electron moves at right angles to a magnetic field of 2.0 T, the measured radius of curvature of its path is not 0.53 cm (as may be computed from the classical relation $r = m_e v/qB$) but, instead, 1.7 cm. Hence, no matter how well Newtonian mechanics may work at low speeds, it fails badly as $u/c \rightarrow 1$.

In 1905 Albert Einstein published his special theory of relativity. Although motivated by a desire to gain deeper insight into the nature of electromagnetism, Einstein, in his theory, extended and generalized Newtonian mechanics as well. He correctly predicted the results of mechanical experiments over the complete range of speeds from $u/c = 0$ to $u/c \rightarrow 1$. Newtonian mechanics was revealed to be an important special case of a more general theory. In developing this theory of relativity, Einstein critically examined the procedures used to measure length and time intervals. These procedures require the use of light signals and, in fact, an assumption about the way light is propagated is one of the two central hypotheses on which the theory is based. His theory resulted in a completely new view of the nature of space and time.

The connection between mechanics and electromagnetism is not surprising because light, which (as we shall see) plays a basic role in making the fundamental space and time measurements that underlie mechanics, is an electromagnetic phenomenon. However, our low-speed Newtonian environment is so much a part of our daily life that almost everyone has some conceptual difficulty in understanding Einstein's ideas of space-time when he or she first studies them. Einstein may have put his finger on the difficulty when he said: "Common sense is that layer of prejudices laid down in the mind prior to the age of eighteen." Indeed, it has been said that every great theory begins as a heresy and ends as a prejudice. The ideas of motion of Galileo and Newton may very well have passed through such a history already. More than three-quarters of a century of experimentation and application has removed special relativity theory from the heresy stage and put it on a sound conceptual and practical basis. Furthermore, we shall show that a careful analysis of the basic assumptions of Einstein and of Newton makes it clear that the assumptions of Einstein are really much more reasonable than those of Newton.

In the following pages, we shall develop the experimental basis for the ideas of special relativity theory. Because, in retrospect, we found that Newtonian mechanics fails when applied to high-speed particles, it seems wise to begin by examining the foundations of Newtonian mechanics. Perhaps, in this way, we can find clues as to how it might be generalized to yield correct results at high speeds while still maintaining its excellent agreement with experiment at low speeds.

1-2 GALILEAN TRANSFORMATIONS

Let us begin by considering a physical *event*. An event is something that happens independent of the reference frame we might use to describe it. For concreteness, we can imagine the event to be a collision of two particles or the flashing of a tiny light source. The event happens at a point in space and at an instant in time. We specify an event by four (space-time) measurements in a particular frame of reference, say, the position numbers *x*, *y*, *z* and the time *t*. For example, the collision of two particles may occur at $x = 1$ m, $y = 4$ m, $z = 11$ m, and at time $t = 7$ s in one frame of reference (for example, a laboratory on earth), so that the four numbers (1, 4, 11, 7) specify the event in that reference frame. The same event observed from a different reference frame (for example, an airplane flying overhead) would also be specified by four numbers, although the numbers might be different than those in the laboratory frame. Thus, if we are to describe events, our first step is to establish a frame of reference.

We define an *inertial system* as a frame of reference in which the law of inertia—Newton's first law—holds. In such a system, which we may also describe as an *unaccelerated* system, a body that is acted on by zero net external force will move with a constant velocity. Newton assumed that a frame of reference fixed with respect to the stars is an inertial system. A rocketship drifting in outer space, without spinning and with its engines cut off, provides an ideal inertial system. Any frame moving at constant velocity with respect to such a system is also an inertial frame. However, frames accelerating with respect to such a system are *not* inertial.

In practice, we can often neglect the small (acceleration) effects due to the rotation and the orbital motion of the earth and to solar motion. Thus, we may regard any set of axes fixed on the earth as forming (approximately) an inertial coordinate system. Likewise, any set of axes moving at uniform velocity with respect to the earth, as in a train, ship, or airplane, will be (nearly) inertial, because motion at uniform velocity does not introduce acceleration. However, a system of axes that accelerates with respect to the earth, such as one fixed to a spinning merry-go-round or to an accelerating car, is *not* an inertial system. A particle acted on by zero net external force will not move in a straight line with constant speed according to an observer in such a noninertial system.

The special theory of relativity, which we consider here, deals only with the description of events by observers in inertial reference frames. The objects whose motions we study may be accelerating with respect to such frames, but the frames themselves are unaccelerated. The general theory of relativity, presented by Einstein in 1917, concerns itself with all frames of reference, including noninertial ones. See Supplementary Topic C for a brief discussion of general relativity theory.

Consider now an inertial frame *S* and another inertial frame *S'* that moves at a constant velocity **v** with respect to *S*, as shown in Fig. 1-1. For convenience, we choose the three sets of axes to be parallel and allow their relative motion to be along the common *x-x'* axis. We can easily generalize to arbitrary orientations and relative velocity of the frames later, but the physical principles involved are not affected by the particular simple choice we make at present. Note also that we can just as well regard *S* as moving with velocity −**v** with respect to *S'* as we can regard *S'* as moving with velocity **v** with respect to *S*.

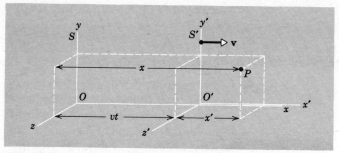

Figure 1-1. Two inertial frames with a common x-x' axis and with the y-y' and z-z' axes parallel. As seen from frame S, frame S' is moving in the positive x direction at speed v. Similarly, as seen from frame S', frame S is moving in the negative x' direction at this same speed. Point P suggests an *event*, whose space-time coordinates may be measured by each observer. The origins O and O' coincide at time $t = 0$, $t' = 0$.

Let an event occur at point P, whose space and time coordinates are measured in each inertial frame. An observer attached to S specifies by means of meter sticks and clocks, for instance, the location and time of occurrence of this event, ascribing space coordinates x, y, and z and time t to it. An observer attached to S', using his or her measuring instruments, specifies the *same* event by space-time coordinates x', y', z', and t'. The coordinates x, y, z will give the position of P relative to the origin O as measured by observer S, and t will be the time of occurrence of P that observer S records with his or her clocks. The coordinates x', y', and z' likewise refer the position of P to the origin O' and the time of P, t', to the clocks of inertial observer S'.

We now ask what the relationship is between the measurements x, y, z, t and x', y', z', t'. The two inertial observers use meter sticks, which have been compared and calibrated against one another, and clocks, which have been synchronized and calibrated against one another. The classical procedure, which we look at more critically later, is to assume thereby that length intervals and time intervals are absolute, that is, that they are the same for all inertial observers of the same events. For example, if meter sticks are of the same length when compared at rest with respect to one another, it is implicitly assumed that they are of the same length when compared in relative motion to one another. Similarly, if clocks are calibrated and synchronized when at rest, it is assumed that their readings and rates will agree thereafter, even if they are put in relative motion with respect to one another. These are examples of the "common sense" assumptions of classical theory.

We can state these results explicitly, as follows. For simplicity, let us say that the clocks of each observer read zero at the instant that the origins O and O' of the frames S and S', which are in relative motion, coincide. Then the *Galilean coordinate transformations*, which relate the measurements x, y, z, t to x', y', z', t', are

$$
\begin{aligned}
x' &= x - vt \\
y' &= y \\
z' &= z.
\end{aligned}
\qquad\qquad (1\text{-}1a)
$$

These equations agree with our classical intuition, the basis of which is easily seen from Fig. 1-1. It is assumed that time can be defined independently of any particular frame of reference. This is an implicit assumption of classical physics, which is expressed in the transformation equations by the absence of a transformation for t. We can make this assumption of the universal nature of time explicit by adding to the Galilean transformations the equation

$$t' = t. \tag{1-1b}$$

It follows at once from Eqs. 1-1a and 1-1b that the time interval between occurrence of two given events, say, P and Q, is the same for each observer, that is,

$$t'_P - t'_Q = t_P - t_Q, \tag{1-2a}$$

and that the distance, or space interval, between two points, say, A and B, measured at a given instant, is the same for each observer, that is,

$$x'_B - x'_A = x_B - x_A. \tag{1-2b}$$

We have assumed for simplicity that points A and B both lie on the x axis, or on a line parallel to that axis.

EXAMPLE 1.

A Galilean Result. Derive the classical space interval result, Eq. 1-2b.

Let A and B be the end points of a rod, for example, which is at rest in the S frame, parallel to the common x-x' axis. Then, the primed observer, for whom the rod is moving with velocity $-\mathbf{v}$, will measure the endpoint locations as x'_B and x'_A, whereas the unprimed observer locates them at x_B and x_A. Using the Galilean transformations, however, we find that

$$x'_B = x_B - vt_B \quad \text{and} \quad x'_A = x_A - vt_A,$$

so that

$$x'_B - x'_A = x_B - x_A - v(t_B - t_A).$$

Since the two end points, A and B, are measured at the same instant, we must put $t_A = t_B$, and we obtain

$$x'_B - x'_A = x_B - x_A,$$

as found above.

Or, we can imagine the rod to be at rest in the primed frame, and moving therefore with velocity $\mathbf{v}$ with respect to the unprimed observer. Then the Galilean transformations, which can be written equivalently as

$$\begin{aligned} x &= x' + vt \\ y &= y' \\ z &= z' \\ t &= t', \end{aligned} \tag{1-3}$$

give us $x_B = x'_B + vt'_B$ and $x_A = x'_A + vt'_A$ and, with $t'_A = t'_B$, we once again obtain $x_B - x_A = x'_B - x'_A$.

Notice carefully that two measurements (the end points x'_A, x'_B or x_A, x_B) are made for each observer and that we assumed they were made at the *same time* ($t_A = t_B$, or $t'_A = t'_B$). The assumption that the measurements are made at the same time—that is, simultaneously—is a crucial part of our definition of the length of the moving rod. Surely we should not measure the locations of the end points at different times to get the length of the moving rod; it would be like measuring the location of the tail of a swimming fish at one instant and of if its head at another instant in order to determine its length (see Fig. 1-2).

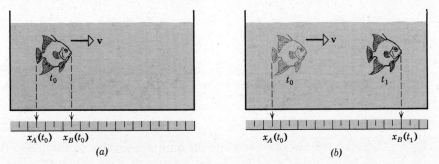

Figure 1-2. To measure the length of a swimming fish, one must mark the positions of its head and tail simultaneously (a), rather than at arbitrary times (b).

The time-interval and space-interval measurements made above are absolutes according to the Galilean transformation; that is, they are the same for all inertial observers, the relative velocity **v** of the frames being arbitrary and not entering into the results. When we add to this result the assumption of classical physics that the mass of a body is a constant, independent of its motion with respect to an observer, then we can conclude that classical mechanics and the Galilean transformations imply that length, mass, and time—the three basic quantities in mechanics—are all independent of the relative motion of the measurer (or observer).

1-3 NEWTONIAN RELATIVITY

How do the measurements of different inertial observers compare with regard to velocities and accelerations of objects? The position of a particle in motion is a function of time, so we can express particle velocity and acceleration in terms of time derivatives of position. We need only carry out successive time differentiations of the Galilean transformations. The velocity transformation follows at once. Starting from

$$x' = x - vt,$$

differentiation with respect to t gives

$$\frac{dx'}{dt} = \frac{dx}{dt} - v.$$

But, because $t = t'$, the operation d/dt is identical to the operation d/dt', so

$$\frac{dx'}{dt} = \frac{dx'}{dt'}.$$

Therefore,

$$\frac{dx'}{dt'} = \frac{dx}{dt} - v.$$

Similarly,

$$\frac{dy'}{dt'} = \frac{dy}{dt}$$

and

$$\frac{dz'}{dt'} = \frac{dz}{dt}.$$

However, $dx'/dt' = u'_x$, the x component of the velocity measured in S', and $dx/dt = u_x$, the x component of the velocity measured in S, and so on, so that we have simply the *classical velocity addition theorem:*

$$\begin{aligned} u'_x &= u_x - v \\ u'_y &= u_y \\ u'_z &= u_z. \end{aligned} \qquad (1\text{-}4)$$

Clearly, in the more general case in which **v**, the relative velocity of the frames, has components along all three axes, we would obtain the more general (vector) result

$$\mathbf{u}' = \mathbf{u} - \mathbf{v}. \qquad (1\text{-}5)$$

You have already encountered many examples of this. (See *Physics*, Part I, Sec. 4-6.*) Thus, the velocity of an airplane with respect to the air (**u'**) equals the

---

* We shall give occasional review references, in this format, to the authors' introductory physics text; see Ref. 1 at the end of the chapter. Other references will be given in the form [3], [4], and so forth.

velocity of the plane with respect to the ground (**u**) minus the velocity of the air with respect to the ground (**v**).

EXAMPLE 2.

Adding Velocities Classically (I). A passenger walks forward along the aisle of a train at a speed of 3.5 km/h as the train moves along a straight track at a constant speed of 92.5 km/h with respect to the ground. What is the passenger's speed with respect to the ground?

Let us choose the train to be the primed frame so that

$u'_x = 3.5$ km/h. The primed frame moves forward with respect to the ground (unprimed frame) at a speed $v = 92.5$ km/h. Hence, the passenger's speed with respect to ground is

$$u_x = u'_x + v = 3.5 \text{ km/h} + 92.5 \text{ km/h} = 96.0 \text{ km/h}.$$

EXAMPLE 3.

Adding Velocities Classically (II). Two electrons are ejected in opposite directions from radioactive atoms in a sample of radioactive material at rest in the laboratory. Each electron has a speed 0.67c as measured by a laboratory observer. What is the speed of one electron as measured from the other, according to the classical velocity addition theorem?

Here, we may regard one electron as the S frame, the laboratory as the S' frame, and the other electron as the object whose speed in the S frame is sought (see Fig. 1-3). In the S' frame, the other electron's speed is 0.67c, moving in the positive x' direction, say, and the speed of the S frame (one electron) is 0.67c, moving in the negative x' direction. Thus, $u'_x = +0.67c$ and $v = +0.67c$, so that the other electron's speed with respect to the S frame is

$$u_x = u'_x + v = +0.67c + 0.67c = +1.34c,$$

according to the classical velocity addition theorem.

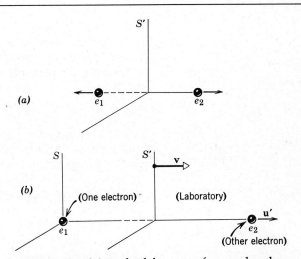

Figure 1-3. (*a*) In the laboratory frame, the electrons are observed to move in opposite directions at the same speed. (*b*) In the rest frame, S, of one electron, the laboratory moves at a velocity **v**. In the laboratory frame, S', the second electron has a velocity denoted by **u**′. What is the velocity of this second electron as seen by the first?

To obtain the acceleration transformation we merely differentiate the velocity relations (Eq. 1-4). Proceeding as before, we obtain

$$\frac{d}{dt}(u'_x) = \frac{d}{dt}(u_x - v),$$

or, recalling again that $t = t'$,

$$\frac{du'_x}{dt'} = \frac{du_x}{dt}, \qquad v \text{ being a constant,}$$

$$\frac{du'_y}{dt'} = \frac{du_y}{dt},$$

and

$$\frac{du'_z}{dt'} = \frac{du_z}{dt}.$$

That is, $a_x' = a_x$, $a_y' = a_y$, and $a_z' = a_z$. Hence, $\mathbf{a}' = \mathbf{a}$. The measured components of acceleration of a particle are unaffected by the uniform relative velocity of the reference frames. The same result follows directly from two successive differentiations of Eqs. 1-1a and applies generally when $\mathbf{v}$ has an arbitrary direction, as long as $\mathbf{v}$ = constant.

We have seen that different velocities are assigned to a particle by different observers when the observers are in relative motion. These velocities always differ by the relative velocity of the two observers, which in the case of inertial observers is a constant velocity. It follows then that when the particle velocity changes, the *change* will be the same for both observers. Thus, they each measure the same *acceleration* for the particle. The acceleration of a particle is the same in *all* reference frames that move relative to one another with constant velocity; that is,

$$\mathbf{a}' = \mathbf{a}. \tag{1-6}$$

In classical physics the *mass* is also unaffected by the motion of the reference frame. Hence, the product $m\mathbf{a}$ will be the same for all inertial observers. If $\mathbf{F} = m\mathbf{a}$ is taken as the definition of force, then obviously each observer obtains the same measure for each force. With $\mathbf{F} = m\mathbf{a}$ and $\mathbf{F}' = m\mathbf{a}'$, it follows from Eq. 1-6 that $\mathbf{F} = \mathbf{F}'$. *Newton's laws of motion and the equations of motion of a particle would be exactly the same in all inertial systems.* Since, in classical mechanics, the conservation principles—such as those for energy, linear momentum, and angular momentum—all can be shown to be consequences of Newton's laws, it follows that *the laws of mechanics are the same in all inertial frames.* In more formal language, we say that the laws of mechanics are *invariant* under a Galilean transformation. Let us make sure that we understand just what this paragraph says before we draw some important conclusions from it.

Although different inertial observers will record different velocities for the same particle, and hence different momenta and kinetic energies, they will agree that momentum is conserved in a collision or is not conserved, that mechanical energy is conserved or is not conserved, and so forth. A tennis ball on the court of a moving ocean liner will have a different velocity to a passenger than it has to an observer on shore, and the billiard balls on the table in a home will have different velocities to the player than they have to an observer on a passing train. But, whatever the values of the particle's or system's momentum or mechanical energy may be, when one observer finds that they do not change in an interaction, the other observer will find the same thing. Although the numbers assigned to such things as velocity, momentum, and kinetic energy may be different for different inertial observers, the laws of mechanics (for example, Newton's laws and the conservation principles) will be the same in all inertial systems. This is illustrated in the following example.

EXAMPLE 4.

A Collision, Viewed by Two Observers. A particle of mass $m_1 = 3$ kg, moving at a velocity of $u_1 = +4$ m/s along the x axis of frame S, approaches a second particle of mass $m_2 = 1$ kg, moving at a velocity $u_2 = -3$ m/s along this axis. After a head-on collision, we find that m_2 has a velocity $U_2 = +3$ m/s along the x axis.

(*a*) Calculate the expected velocity U_1 of m_1, after the collision.

We use the law of conservation of momentum.

Before the collision the momentum of the system of two particles is

$$P = m_1 u_1 + m_2 u_2 = (3 \text{ kg})(+4 \text{ m/s}) + (1 \text{ kg})(-3 \text{ m/s})$$
$$= +9 \text{ kg} \cdot \text{m/s}.$$

After the collision the momentum of the system,

$$P = m_1 U_1 + m_2 U_2,$$

is also +9 kg·m/s, so that

$$+9 \text{ kg·m/s} = (3 \text{ kg})(U_1) + (1 \text{ kg})(+3 \text{ m/s})$$
or $\qquad U_1 = +2 \text{ m/s}, \qquad$ along the x axis.

(*b*) Discuss the collision as seen by observer S' who has a velocity **v** of +2 m/s relative to S along the x axis.

The four velocities measured by S' can be calculated from the Galilean velocity transformation equation (Eq. 1-5), $\mathbf{u'} = \mathbf{u} - \mathbf{v}$, from which we get

$$u_1' = u_1 - v = +4 \text{ m/s} - 2 \text{ m/s} = 2 \text{ m/s},$$
$$u_2' = u_2 - v = -3 \text{ m/s} - 2 \text{ m/s} = -5 \text{ m/s},$$
$$U_1' = U_1 - v = +2 \text{ m/s} - 2 \text{ m/s} = 0,$$
$$U_2' = U_2 - v = +3 \text{ m/s} - 2 \text{ m/s} = 1 \text{ m/s}.$$

The system momentum in S' is

$$P' = m_1 u_1' + m_2 u_2' = (3 \text{ kg})(2 \text{ m/s}) + (1 \text{ kg})(-5 \text{ m/s})$$
$$= +1 \text{ kg·m/s}$$

before the collision, and

$$P' = m_1 U_1' + m_2 U_2' = (3 \text{ kg})(0) + (1 \text{ kg})(1 \text{ m/s})$$
$$= +1 \text{ kg·m/s}$$

after the collision.

Hence, although the velocities and momenta have different numerical values in the two frames, S and S', when momentum is conserved in S it is also conserved in S'.

An important consequence of the above discussion is that *no mechanical experiments carried out entirely in one inertial frame can tell the observer what the motion of that frame is with respect to any other inertial frame.* The billiard player in a closed boxcar of a train moving uniformly along a straight track cannot tell from the behavior of the balls what the motion of the train is with respect to ground. The tennis player in an enclosed court on an ocean liner moving with uniform velocity (in a calm sea) cannot tell from her game what the motion of the boat is with respect to the water. No matter what the relative motion may be (perhaps none), so long as it is constant, the results will be identical. Of course, we *can* tell what the *relative* velocity of two frames may be by comparing the data the different observers take on the very same event—but then we have not deduced the relative velocity from observations *confined to a single frame.*

Furthermore, there is no way at all of determining the *absolute* velocity of an inertial reference frame from our mechanical experiments. No inertial frame is preferred over any other, for the laws of mechanics are the same in all. Hence, there is no physically definable absolute rest frame. We say that all inertial frames are equivalent as far as mechanics is concerned. The person riding the train cannot tell absolutely whether he alone is moving, or the earth alone is moving past him, or if some combination of motions is involved. Indeed, would you say that you on earth are at rest, that you are moving 30 km/s (the speed of the earth in its orbit about the sun), or that your speed is much greater still (for instance, the sun's speed in its orbit about the galactic center)? Actually, no mechanical experiment can be performed that will detect an absolute velocity through empty space. This result, that we can speak only of the *relative* velocity of one frame with respect to another, and not of an absolute velocity of a frame, is sometimes called *Newtonian relativity.*

Transformation laws, in general, will change many quantities but will leave some others unchanged. These unchanged quantities are called *invariants* of the transformation. In the Galilean transformation laws for the relation between observations made in different inertial frames of reference, for example, acceleration is an invariant and—more important—so are Newton's laws of motion. A statement of what the invariant quantities are is called a relativity principle; it says that for such quantities the reference frames are equivalent to one another, no one having an absolute or privileged status relative to the others. Newton expressed his relativity principle as follows: "The motions of bodies included in a given space are the same amongst themselves, whether that space is at rest or moves uniformly forward in a straight line."

1-4 *ELECTROMAGNETISM AND NEWTONIAN RELATIVITY*

Let us now consider the situation from the electrodynamic point of view. That is, we inquire now whether the laws of physics other than those of mechanics (such as the laws of electromagnetism) are invariant under a Galilean transformation. If so, then the (Newtonian) relativity principle would hold not only for mechanics but for all of physics. That is, no inertial frame would be preferred over any other, and *no type of experiment* in physics, not merely mechanical ones, carried out in a single frame would enable us to determine the velocity of our frame relative to any other frame. There would then be no preferred, or absolute, reference frame.

To see at once that the electromagnetic situation is different from the mechanical one, as far as the Galilean transformations are concerned, consider a pulse of light (that is, an electromagnetic pulse) traveling to the right with respect to the medium through which it is propagated at a speed c. A "medium" of light propagation, given the name "ether," was considered necessary, for when the mechanical view of physics dominated physicists' thinking (late nineteenth century and early twentieth century), it was not really accepted that an electromagnetic disturbance could be propagated in empty space. Sound waves, for example, require a medium for propagation. For simplicity, we may regard the "ether" frame, S, as an inertial one in which the speed of light c $(=1/\sqrt{\mu_0\varepsilon_0})$ has its internationally agreed-upon value; see the footnote on page 3. In a frame S' moving at a constant speed v with respect to this ether frame, an observer would measure a different speed for the light pulse, ranging from $c + v$ to $c - v$ depending on the direction of relative motion, according to the Galilean velocity transformation (Eq. 1-5).

Hence, the speed of light is certainly *not* invariant under a Galilean transformation. If these transformations really do apply to optical or electromagnetic phenomena, then there is one inertial system, and only one, in which the measured speed of light is exactly c; that is, there is a unique inertial system in which the so-called ether is at rest. We would then have a physical way of identifying an absolute (or rest) frame and of determining by optical experiments carried out in some other frame what the relative velocity of that frame is with respect to the absolute one.

A more formal way of saying this is as follows. Maxwell's equations of electromagnetism, from which we deduce the electromagnetic wave equation for example, contain the constant $c = 1/\sqrt{\mu_0\varepsilon_0}$, which is identified as the velocity of propagation of a plane wave in vacuum (see *Physics*, Part II, Sec. 41-8.) But such a velocity cannot be the same for observers in different inertial frames, according to the Galilean transformations, so Maxwell's equations and therefore electromagnetic effects will probably not be the same for different inertial observers. But if we accept both the Galilean transformations and Maxwell's equations as basically correct, then it automatically follows that there exists a unique privileged frame of reference (the "ether" frame) in which Maxwell's equations are valid and in which light is propagated at a speed $c = 1/\sqrt{\mu_0\varepsilon_0}$.

The situation then seems to be as follows.[*] The fact that the Galilean relativity principle *does* apply to the Newtonian laws of mechanics but *not* to Maxwell's laws of electromagnetism requires us to choose the correct consequences from among the following possibilities.

---

[*] The treatment here follows closely that of Ref. 2.

1. *There is an ether.* A relativity principle exists for mechanics, but *not* for electrodynamics; in electrodynamics there *is* a preferred inertial frame, that is, the ether frame. Should this alternative be correct, the Galilean transformations would apply and we would be able to locate the ether frame experimentally.

2. *Maxwell was wrong.* A relativity principle exists *both* for mechanics and for electrodynamics, but the laws of electrodynamics as given by Maxwell are *not* correct. If this alternative were correct, we ought to be able to perform experiments that show deviations from Maxwell's electrodynamics and reformulate the electromagnetic laws. The Galilean transformations would apply here also.

3. *Newton was wrong.* A relativity principle exists *both* for mechanics and for electrodynamics, but the laws of mechanics as given by Newton are *not* correct. If this alternative is the correct one, we should be able to perform experiments that show deviations from Newtonian mechanics and reformulate the mechanical laws. In that event, the correct transformation laws would not be the Galilean ones (for they are inconsistent with the invariance of Maxwell's equations) but some other ones that are consistent with classical electromagnetism and the new mechanics.

We have already indicated (Section 1-1) that Newtonian mechanics breaks down at high speeds, so you will not be surprised to learn that alternative 3, leading to Einsteinian relativity, is the correct one. In the following sections, we shall look at the experimental bases for rejecting alternatives 1 and 2, as a fruitful prelude to finding the new relativity principle and transformation laws of alternative 3.

1-5 ATTEMPTS TO LOCATE THE ABSOLUTE FRAME— THE MICHELSON-MORLEY EXPERIMENT

The obvious experiment would be one in which we measure the speed of light in a variety of inertial systems, noting whether the measured speed is different in different systems, and if so, noting especially whether there is evidence for a single unique system—the "ether" frame—in which the speed of light is c, the value predicted from electromagnetic theory. A. A. Michelson in 1881 and Michelson and E. W. Morley in 1887 carried out such an experiment [3]. To understand the setting better, let us look a bit further into the "ether" concept.

When we say that the speed of sound in dry air at 0°C is 331.4 m/s, we have in mind an observer, and a corresponding reference system, fixed in the air mass through which the sound wave is moving. The speed of sound for observers moving with respect to this air mass is correctly given by the usual Galilean velocity transformation Eq. 1-5. However, when we say that the speed of light in a vacuum is 2.99792458×10^8 m/s, it is not at all clear what reference system is implied. A reference system fixed in the medium of propagation of light presents difficulties because, in contrast to sound, no medium seems to exist. However, it seemed inconceivable to nineteenth-century physicists that light and other electromagnetic waves, in contrast to all other kinds of waves, could be propagated without a medium. It seemed to be a logical step to postulate such a medium, called the ether, even though it was necessary to assume unusual properties for it, such as zero density and perfect transparency, to account for its undetectability.

The ether was assumed to fill all space and to be the medium with respect to which the speed c applies. It followed then that an observer moving through the ether with velocity **v** would measure a velocity **c'** for a light beam, where **c'** = **c** − **v**. It was this result that the Michelson-Morley experiment was designed to test.

If an ether exists, the spinning and revolving earth should be moving through it. An observer on earth would sense an "ether wind," whose velocity is **v** relative to the earth. If we were to assume that v is equal to the earth's orbital speed about the sun, about 30 km/s, then $v/c \cong 10^{-4}$. Optical experiments, which were accurate to the first order in v/c, were not able to detect the absolute motion of the earth through the ether, but Fresnel (and later Lorentz) showed how this result could be interpreted in terms of an ether theory. This interpretation had difficulties, however, so that the issue was not really resolved satisfactorily with first-order experiments (experiments whose results depend only on the first power of the ratio v/c). It was generally agreed that an unambiguous test of the ether hypothesis would require an experiment that measured the "second-order" effect, that is, one that measured $(v/c)^2$. The first-order effect is not large to begin with ($v/c = 10^{-4}$, an effect of one part in 10,000) but the second-order effect is really very small ($v^2/c^2 = 10^{-8}$, an effect of one part in 100 million).

It was A. A. Michelson who invented the optical interferometer whose remarkable sensitivity made such an experiment possible. Michelson first performed the experiment in 1881, and then—in 1887, in collaboration with E. W. Morley—carried out the more precise version of the investigation that was destined to raise troublesome questions for classical physics and that opened the way to a more receptive consideration of relativity theory. For his invention of the interferometer and his many optical experiments, Michelson was awarded the Nobel Prize in Physics in 1907, the first American to be so honored.

Let us now describe the Michelson-Morley experiment. The Michelson interferometer (Fig. 1-4) is fixed on the earth. If we imagine the "ether" to be fixed with respect to the sun, then the earth (and interferometer) moves through the

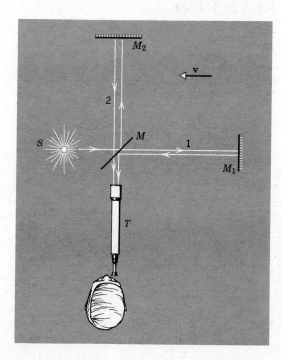

Figure 1-4. A simplified version of the Michelson interferometer showing how the beam from the source S is split into two beams by the partially silvered mirror M. The beams are reflected by mirrors M_1 and M_2, returning to the partially silvered mirror. The beams are then transmitted to the telescope T, where they interfere, giving rise to a fringe pattern. In this figure, **v** *is the presumed velocity of the ether with respect to the interferometer.*

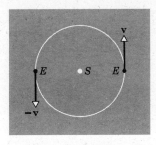

Figure 1-5. The earth E moves at an orbital speed of 30 km/s along its nearly circular orbit about the sun S, reversing the direction of its velocity every six months.

ether at a speed of 30 km/s, in different directions in different seasons (Fig. 1-5). For the moment, neglect the earth's spinning motion. The beam of light (plane waves, or parallel rays) from the laboratory source S (fixed with respect to the instrument) is split by the partially silvered mirror M into two coherent beams, beam 1 being transmitted through M and beam 2 being reflected off M. Beam 1 is reflected back to M by mirror M_1 and beam 2 by mirror M_2. Then the returning beam 1 is partially reflected and the returning beam 2 is partially transmitted by M back to a telescope at T, where they interfere. The interference is constructive or destructive depending on the phase difference of the beams. The partially silvered mirror surface M is inclined at 45° to the beam directions. If M_1 and M_2 are very nearly (but not quite) at right angles, we shall observe a fringe system in the telescope (Fig. 1-6) consisting of nearly parallel lines, much as we get from a thin wedge of air between two glass plates.

Let us compute the phase difference between the beams 1 and 2. This difference can arise from two causes, the different path lengths traveled, l_1 and l_2, and the different speeds of travel with respect to the instrument because of the "ether wind" v. The second cause, for the moment, is the crucial one. The different speeds are much like the different cross-stream and up-and-down-stream speeds with respect to shore of a swimmer in a moving stream. The time for beam 1 to travel from M to M_1 and back is

$$t_1 = \frac{l_1}{c-v} + \frac{l_1}{c+v} = l_1\left(\frac{2c}{c^2-v^2}\right) = \frac{2l_1}{c}\left(\frac{1}{1-v^2/c^2}\right),$$

for the light, whose speed is c in the ether, has an "upstream" speed of $c - v$ with respect to the apparatus and a "downstream" speed of $c + v$. The path of beam 2, traveling from M to M_2 and back, is a cross-stream path through the ether, as

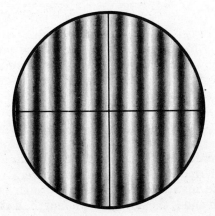

Figure 1-6. A typical fringe system seen through the telescope T when M_1 and M_2 are not quite at right angles.

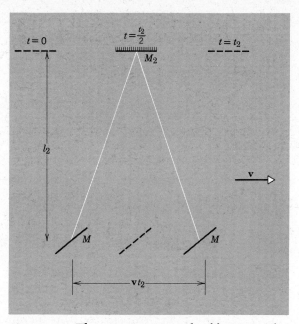

Figure 1-7. The cross-stream path of beam 2. The mirrors move through the "ether" at a speed v, the light moving through the "ether" at speed c. Reflection from the moving mirror automatically gives the cross-stream path. In this figure, **v** *is the presumed velocity of the interferometer with respect to the "ether."*

shown in Fig. 1-7, enabling the beam to return to the (advancing) mirror M. The transit time is given by

$$2\left[l_2{}^2 + \left(\frac{vt_2}{2}\right)^2\right]^{1/2} = ct_2$$

or

$$t_2 = \frac{2l_2}{\sqrt{c^2 - v^2}} = \frac{2l_2}{c}\frac{1}{\sqrt{1 - v^2/c^2}}.$$

The calculation of t_2 is made in the ether frame, that of t_1 in the frame of the apparatus. Because time is an absolute in classical physics, this is perfectly acceptable classically. Note that both effects are second-order ones $(v^2/c^2 \cong 10^{-8})$ and are in the same direction (they *increase* the transit time over the case $v = 0$). The difference in transit times is

$$\Delta t = t_2 - t_1 = \frac{2}{c}\left(\frac{l_2}{\sqrt{1 - \beta^2}} - \frac{l_1}{1 - \beta^2}\right) \tag{1-7a}$$

in which, for convenience, we have replaced the dimensionless ratio v/c by the symbol β, often referred to as the *speed parameter*.

Suppose that the instrument is rotated through 90°, thereby making l_1 the cross-stream length and l_2 the downstream length. If the corresponding times are now designated by primes, the same analysis as above gives the transit-time difference as

$$\Delta t' = t_2' - t_1' = \frac{2}{c}\left(\frac{l_2}{1 - \beta^2} - \frac{l_1}{\sqrt{1 - \beta^2}}\right). \qquad (1\text{-}7b)$$

Hence, *the rotation changes the differences* by

$$\Delta t' - \Delta t = \frac{2}{c}\left(\frac{l_2 + l_1}{1 - \beta^2} - \frac{l_2 + l_1}{\sqrt{1 - \beta^2}}\right).$$

Using the binomial expansion* and dropping terms higher than the second-order, we find

$$\Delta t' - \Delta t \cong \frac{2}{c}(l_1 + l_2)\left(1 + \beta^2 - 1 - \frac{1}{2}\beta^2\right) = \left(\frac{l_1 + l_2}{c}\right)\beta^2.$$

Therefore, the rotation should cause a shift in the fringe pattern, since it changes the phase relationship between beams 1 and 2.

If the optical path difference between the beams changes by one wavelength, for example, there will be a shift of one fringe across the crosshairs of the viewing telescope. Let ΔN represent the number of fringes moving past the crosshairs as the pattern shifts. Then, if light of wavelength λ is used, so that the period of one vibration is $T = 1/\nu = \lambda/c$,

$$\Delta N = \frac{\Delta t' - \Delta t}{T} \cong \frac{l_1 + l_2}{cT}\beta^2 = \frac{(l_1 + l_2)\beta^2}{\lambda}. \qquad (1\text{-}8)$$

Michelson and Morley were able to obtain an optical path length, $l_1 + l_2$, of about 22 m. In their experiment the arms were of (nearly) equal length, that is, $l_1 = l_2 = l$, so that $\Delta N = 2l\beta^2/\lambda$. If we choose $\lambda = 5.5 \times 10^{-7}$ m and $\beta(=v/c) = 10^{-4}$, we obtain, from Eq. 1-8,

$$\Delta N = \frac{(22\text{ m})(10^{-4})^2}{5.5 \times 10^{-7}\text{ m}} = 0.4,$$

or a shift of four-tenths a fringe!

Michelson and Morley mounted the interferometer on a massive stone slab for stability and floated the apparatus in mercury so that it could be rotated smoothly about a central pin. In order to make the light path as long as possible, mirrors were arranged on the slab to reflect the beams back and forth through eight round trips. The fringes were observed under a continuous rotation of the apparatus, and a shift as small as $\frac{1}{100}$ of a fringe definitely could have been detected (see Fig. 1-8). Observations were made day and night (as the earth spins about its axis) and during all seasons of the year (as the earth revolves about the sun), but the expected fringe shift was not observed. Indeed, the experimental conclusion was that *there was no fringe shift at all.*

---

* In these cases the binomial expansion (see Appendix 5) gives

$$\frac{1}{1 - \beta^2} = (1 - \beta^2)^{-1} = 1 + \beta^2 + \beta^4 + \beta^6 + \cdots,$$

and

$$\frac{1}{(1 - \beta^2)^{1/2}} = (1 - \beta^2)^{-1/2} = 1 + \frac{1}{2}\beta^2 + \frac{3}{8}\beta^4 + \frac{5}{16}\beta^6 + \cdots.$$

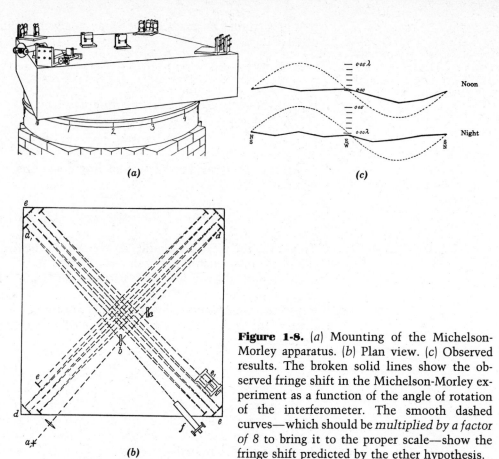

Figure 1-8. (*a*) Mounting of the Michelson-Morley apparatus. (*b*) Plan view. (*c*) Observed results. The broken solid lines show the observed fringe shift in the Michelson-Morley experiment as a function of the angle of rotation of the interferometer. The smooth dashed curves—which should be *multiplied by a factor of 8* to bring it to the proper scale—show the fringe shift predicted by the ether hypothesis.

This null result ($\Delta N = 0$) was such a blow to the ether hypothesis that the experiment was repeated by many workers over a 50-year period. The null result was amply confirmed (see Table 1-1) and provided a great stimulus to theoretical and experimental investigation. In 1958 Cedarholm, Bland, Havens, and Townes [4] carried out an "ether-wind" experiment using microwaves in which they showed that if there is an ether and the earth is moving through it, the earth's speed with respect to the ether would have to be less than $\frac{1}{1000}$ of the earth's orbital speed. This is an improvement of 50 in precision over the best experiment of the Michelson-Morley type. The null result is well established.

Note that the Michelson-Morley experiment depends essentially on the 90° rotation of the interferometer, that is, on interchanging the roles of l_1 and l_2, as the apparatus moves with a speed v through an "ether." In predicting an expected fringe shift, we took **v** to be the earth's velocity with respect to an ether fixed with the sun. However, the solar system itself might be in motion with respect to the hypothetical ether. Actually, the experimental results themselves determine the earth's speed with respect to an ether, if indeed there is one, and these results give $v = 0$. Now, if at some time the velocity were zero in such an ether, no fringe shift would be expected, of course. But the velocity cannot always be zero, since the velocity of the apparatus is *changing* from day to night (as the earth rotates on its axis) and from season to season (as the earth revolves about the sun). Therefore, the experiment does not depend solely on an "absolute" velocity of the earth

Table 1-1
EARLY TRIALS OF THE MICHELSON-MORLEY EXPERIMENT[a]

Observer	Year	Place	l in Meters	Fringe Shift Predicted by Ether Theory	Upper Limit of Observed Fringe Shift
Michelson	1881	Potsdam	1.2	0.04	0.02
Michelson and Morley	1887	Cleveland	11.0	0.40	0.01
Morley and Miller	1902–1904	Cleveland	32.2	1.13	0.015
Miller	1921	Mt. Wilson	32.0	1.12	0.08
Miller	1923–1924	Cleveland	32.0	1.12	0.030
Miller (sunlight)	1924	Cleveland	32.0	1.12	0.014
Tomaschek (starlight)	1924	Heidelberg	8.6	0.3	0.02
Miller	1925–1926	Mt. Wilson	32.0	1.12	0.088
Kennedy	1926	Pasadena and Mt. Wilson	2.0	0.07	0.002
Illingworth	1927	Pasadena	2.0	0.07	0.0004
Piccard and Stahel	1927	Mt. Rigi	2.8	0.13	0.006
Michelson et al.	1929	Mt. Wilson	25.9	0.9	0.010
Joos	1930	Jena	21.0	0.75	0.002

[a] From R. S. Shankland, S. W. McCuskey, F. C. Leone, and G. Kuerti, *Rev. Mod. Phys.*, **27**, 167 (1955).

through an ether, but also depends on the changing velocity of the earth with respect to the "ether." Such a changing motion through the "ether" would be easily detected and measured by the precision experiments, if there were an ether frame. The null result seems to rule out an ether (absolute) frame.

At the end of the last century the ether concept was so deeply ingrained and universally accepted that the null result of the Michelson-Morley experiment came as a real blow. Scientists, anxious to "save the ether," advanced several hypotheses designed to explain this null result without abandoning belief in the ether. We explore these hypotheses in succeeding sections, as a compelling prelude to the need for a totally new approach.

1-6 ATTEMPTS TO "SAVE THE ETHER"— THE LORENTZ-FITZGERALD CONTRACTION

A few years after Michelson and Morley reported the null result of their ether drift experiment, H. A. Lorentz and G. F. Fitzgerald [5] independently proposed a way to account for that result without abandoning the concept of the ether as a preferred reference frame. Although this so-called *Lorentz-Fitzgerald contraction* hypothesis can indeed account for the null Michelson-Morley result, we shall see that it fails to account for the equally crucial null result of a second ether-drift experiment, performed in 1932 by R. J. Kennedy and E. M. Thorndike. We now examine each of these experiments in relationship to the contraction hypothesis.

The Michelson-Morley Experiment   The hypothesis advanced by Lorentz and by Fitzgerald was that all bodies are <u>contracted</u> in the direction of motion relative to the stationary ether by a factor $\sqrt{1 - v^2/c^2}$. Again, let the ratio v/c be represented by the symbol β, so that this factor may be written as $\sqrt{1 - \beta^2}$. Now, if l^0 represents the length of a body at rest with respect to the ether (its rest length) and l its length when in motion with respect to the ether, then in the Michelson-Morley experiment

$$l_1 = l_1^0 \sqrt{1 - \beta^2} \quad \text{and} \quad l_2 = l_2^0.$$

This last result follows from the fact that in the hypothesis it was assumed that lengths at right angles to the motion are unaffected by the motion. Then (see Eq 1-7*a*)

$$\Delta t = \frac{2}{c} \frac{1}{\sqrt{1 - \beta^2}} (l_2^0 - l_1^0). \tag{1-9a}$$

On 90° rotation we have

$$l_1 = l_1^0 \quad \text{and} \quad l_2 = l_2^0 \sqrt{1 - \beta^2}$$

so that (see Eq. 1-7*b*)

$$\Delta t' = \frac{2}{c} \frac{1}{\sqrt{1 - \beta^2}} (l_2^0 - l_1^0). \tag{1-9b}$$

Hence, no fringe shift should be expected on rotation of the interferometer, for $\Delta t' - \Delta t = 0$.

Lorentz was able to account for such a contraction in terms of his electron theory of matter, but the theory was elaborate and somewhat contrived, and other results predicted by it could not be found experimentally.

The Kennedy-Thorndike Experiment   Kennedy and Thorndike carried out an ether-drift experiment using an interferometer with these characteristics:

1. The arms, instead of being (nearly) equal in length as in the Michelson-Morley experiment, were deliberately made as different in length as the coherence of their light source permitted. In their case the length difference was 16 cm.
2. Their interferometer was not free to rotate but was rigidly fixed in the laboratory. The observations consisted of watching for a fringe shift as a function of time as the days and months went by. The instrument had to be sufficiently rigid and stable that its dimensions did not change during these long observation periods.

The theory of the Kennedy-Thorndike experiment is based on the fact that as the earth spins on its axis and orbits the sun, the presumed "ether wind" should change periodically, both in magnitude and direction. Let us suppose that there really is an ether wind and that during a certain interval its magnitude changes from v to v'. The predicted fringe shift, accurate to second-order terms in v/c and *taking the Lorentz-Fitzgerald contraction hypothesis fully into account*, is given by

$$\Delta N = \frac{l_1^0 - l_2^0}{\lambda} \left(\frac{v^2 - v'^2}{c^2} \right). \tag{1-10}$$

After monitoring their instrument continuously for intervals ranging from a few days to a month and extending over a period of almost a year, Kennedy and

Thorndike concluded that there simply was no observable fringe shift; although the quantity $(v^2 - v'^2)/c^2$ should have changed appreciably during their observation period, the experimental finding was simply that $\Delta N = 0$, in spite of the prediction of Eq. 1-10.

The contraction hypothesis thus fails completely to account for the experimental results. Having failed here, we have no logical basis for invoking the contraction hypothesis to account for the null result of the Michelson-Morley experiment. We must discard this hypothesis completely and account for the null results of both of these important ether-drift experiments from an entirely new point of view, that of the special theory of relativity.

1-7 ATTEMPTS TO "SAVE THE ETHER"— THE ETHER-DRAG HYPOTHESIS

Another idea advanced to retain the notion of an ether was that of "ether drag." This hypothesis assumed that the ether frame was attached to all bodies of finite mass, that is, dragged along with such bodies. The assumption of such a "local" ether would automatically give a null result in the Michelson-Morley experiment. Its attraction lay in the fact that it did not require modification of classical mechanics or electromagnetism. However, there were two well-established effects that contradicted the ether-drag hypothesis: stellar aberration and the Fizeau convection experiment. Let us consider these effects now, since we must explain them eventually by whatever theory we finally accept.

Stellar Aberration   The aberration of light was first reported by the British astronomer James Bradley [6] in 1727. He observed that (with respect to astronomical coordinates fixed with respect to the earth) the stars appear to move in circles, the angular diameter of these circular orbits being about 41 seconds of arc. This can be understood as follows. Imagine that a star is directly overhead so that a telescope would have to be pointed straight up to see it if the earth were at rest in the ether. That is (see Fig. 1-9a), the rays of light coming from the star would proceed straight down the telescope tube. Now, imagine that the earth is moving to the right through the ether with a speed v. In order for the rays to pass down the telescope tube without hitting its sides—that is, in order to see the star—we would have to tilt the telescope as shown in Fig. 1-9b. The light proceeds straight down in the ether (as before) but, during the time Δt that the light travels the vertical distance $l = c \, \Delta t$ from the objective lens to the eyepiece, the telescope has moved a distance $v \, \Delta t$ to the right. The eyepiece, at the time the ray leaves the telescope, is on the same vertical line as the objective lens was at the time the ray entered the telescope. From the point of view of the telescope, the ray travels along the axis from objective lens to eyepiece. The angle of tilt of the telescope, α, is given by

$$\tan \alpha = \frac{v \, \Delta t}{c \, \Delta t} = \frac{v}{c}. \tag{1-11}$$

It was known that the earth goes around the sun at a speed of about 30 km/s, so that with $c = 3.00 \times 10^5$ km/s, we obtain an angle $\alpha = 20.5''$ of arc. This is a very small angle, being about half the angular diameter of the planet Jupiter at its mean distance from the earth. The earth's motion is nearly circular, so the direction of aberration reverses every six months, the telescope axis tracing out a cone of aberration during the year (Fig. 1-9c). The angular diameter of the cone, or of

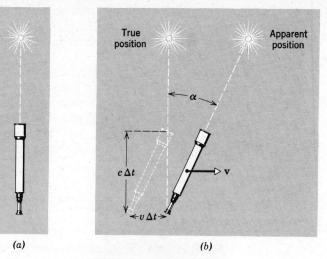

(a)   *(b)*

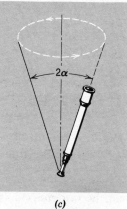

(c)

Figure 1-9. *(a)* The star and telescope have no relative motion (that is, both are at rest in the ether); the star is directly overhead. *(b)* The telescope now moves to the right at speed *v* through the ether; it must be tilted at an angle α (greatly exaggerated in the drawing) from the vertical to see the star, whose apparent position now differs from its true position. ("True" means with respect to the sun, that is, with respect to an earth that has no motion relative to the sun.) *(c)* A cone of abberation of angular diameter 2α is swept out by the telescope axis during the year.

the observed circular path of the star, would then be $2\alpha = 41''$ of arc, in excellent agreement with the observations.

The important thing we conclude from this agreement is that the ether is *not* dragged around with the earth. If it were, the ether would be at rest with respect to the earth, the telescope would not have to be tilted, and there would be no aberration at all. That is, the ether would be moving (with the earth) to the right with speed *v* in Fig. 1-9*b*, so there would be no need to correct for the earth's motion through the ether; the light ray would be swept along with the ether just as a wind carries a sound wave with it. Hence, *if* there is an ether, it is *not* dragged along by the earth but, instead, the earth moves freely through it. Therefore, we cannot explain the Michelson-Morley result by means of an ether-drag hypothesis.

The Fizeau Convection Experiment One way to investigate whether or not the ether (presuming it exists) is dragged along by bodies moving through it is to make a direct laboratory test. Assume that a transparent body (water filling a long pipe, say) is at rest in the laboratory. The speed at which light is propagated through it is given by c/n, where n is the index of refraction of water. Suppose now that the water is caused to move through the pipe at speed v_w with respect to the laboratory and that a beam of light is sent along the axis of the pipe, either parallel or antiparallel to the direction of the flow of the water. What will now be the measured laboratory speed of the light going through the (moving) water?

If the moving transparent medium drags the ether with it *totally*, the measured speed of light should be simply $(c/n \pm v_w)$, in which the choice of sign depends on whether the light beam is sent in the direction of motion (+) or opposite to that direction (−).

In 1817 (long before the Michelson-Morley experiment), the French physicist J. A. Fresnel predicted that light would be only *partially* dragged along by a moving medium and derived an exact formula for the effect on the basis of an ether hypothesis. The effect was confirmed experimentally by Fizeau in 1851. The setup of the Fizeau experiment is shown diagrammatically in Fig. 1-10. Light from the source S falls on a partially silvered mirror M, which splits the beam into two parts. One part is transmitted to mirror M_1 and proceeds in a counter-

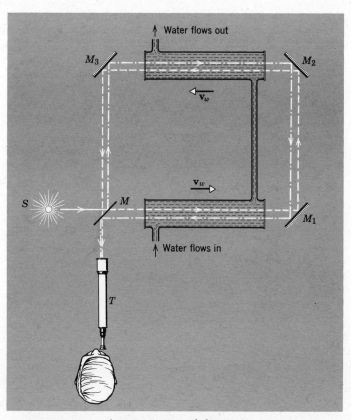

Figure 1-10. Schematic view of the Fizeau experiment.

clockwise sense back to M, after reflections at M_1, M_2, and M_3. The other part is reflected to M_3 and proceeds in a clockwise sense back to M, after reflections at M_3, M_2, M_1. At M, part of the returning first beam is transmitted and part of the returning second beam is reflected to the telescope T. Interference fringes, representing optical path differences of the beams, will be seen in the telescope. Water flows through the tubes (which have flat glass end sections) as shown, so that one light beam always travels in the direction of flow and the other always travels opposite to the direction of flow. The flow of water can be reversed, of course, but outside the tubes conditions remain the same for each beam.

Let the apparatus be our S frame. In this laboratory frame the velocity of light in still water is c/n and the velocity of the water is v_w. Does the flow of water, the medium through which the light passes, affect the velocity of light measured in the laboratory? According to Fresnel, the answer is yes. The velocity of light, v, in a body of refractive index, n, moving with a velocity v_w relative to the observer (that is, to the frame of reference S in which the free-space velocity of light would be c) is given by Fresnel as

$$v = \frac{c}{n} \pm v_w \left(1 - \frac{1}{n^2}\right). \tag{1-12}$$

The factor $(1 - 1/n^2)$ is called the Fresnel drag coefficient. The speed of light is changed from the value c/n because of the motion of the medium but, because the factor is less than unity, the change (increase or decrease) of speed is less than the speed v_w of the medium—hence the term "drag." For yellow sodium light in water, for example, the speed increase (or decrease) is $0.565\, v_w$. Notice that for $n = 1$ ("a moving vacuum"), Eq. 1-12 reduces plausibly to $v = c$.

This result can be understood by regarding the light as being carried along both by the refractive medium and by the ether that permeates it. Then, with the ether at rest and the refractive medium moving through the ether, the light will act to the rest observer as though only a part of the velocity of the medium were added to it. The result can be derived directly from electromagnetic theory.

In Fizeau's experiment, the water flowed through the tubes at a speed of about 7 m/s. Fringe shifts were observed from the zero flow speed to flow speeds of 7 m/s, and on reversing the direction of flow. Fizeau's measurements confirmed the Fresnel prediction. The experiment was repeated by Michelson and Morley in 1886 and by P. Zeeman and others after 1914 under conditions allowing much greater precision, again confirming the Fresnel drag coefficient.

EXAMPLE 5.

Fizeau's Ether-Drag Experiment. In Fizeau's experiment, the approximate values of the parameters were as follows: $l = 1.5$ m, $n = 1.33$, $\lambda = 5.3 \times 10^{-7}$ m, and $v_w = 7.1$ m/s. A shift of 0.23 fringe was observed from the case $v_w = 0$. Calculate the drag coefficient and compare it with the predicted value.

Let d represent the drag coefficient. The time for beam 1 to traverse the water is then

$$t_1 = \frac{2l}{(c/n) - v_w d}$$

and for beam 2,

$$t_2 = \frac{2l}{(c/n) + v_w d}.$$

Hence,

$$\Delta t = t_1 - t_2 = \frac{4l v_w d}{(c/n)^2 - v_w^2 d^2} \cong \frac{4l n^2 v_w d}{c^2}.$$

The period of vibration of the light is $T = \lambda/c$ so that

$$\Delta N = \frac{\Delta t}{T} \cong \frac{4l n^2 v_w d}{\lambda c},$$

and, with the values above, we obtain

$$d = \frac{\lambda c \Delta N}{4 l n^2 v_w}$$

$$= \frac{(5.3 \times 10^{-7} \text{ m}) (3.0 \times 10^8 \text{ m/s})(0.23)}{(4)(1.5 \text{ m}) (1.33)^2 (7.1 \text{ m/s})}$$

$$= 0.49.$$

The Fresnel prediction (see Eq. 1-12) is

$$d = 1 - \frac{1}{n^2} = 1 - \frac{1}{(1.33)^2} = 0.43,$$

the measured value being in rough agreement. Some 35 years later Michelson and Morley repeated the experiment with improved techniques and found a value of 0.43 ± 0.02 for d, in much better agreement with Fresnel's prediction.

If the ether were *totally* dragged along with the water, the speed of light in the laboratory frame would have been, as we have seen, $(c/n + v_w)$ in one tube and $(c/n - v_w)$ in the other tube. Instead, the Fizeau experiment is interpreted most simply in terms of only a *partial* drag by the moving medium. Indeed, the aberration experiment, when done with a telescope filled with water (see Question 21) leads to exactly the same result and interpretation. Hence, the hypothesis that moving bodies are permeated by or closely surrounded by a stagnant localized volume of ether is contradicted by the facts.

At this stage you may well point out that, even though the ether-drag hypothesis is not supported by the facts, Eq. 1-12, derived by Fresnel on the basis of an ether theory, *is* confirmed by experiment. Does not this fact alone amount to support for an ether theory of some kind? In the light of today's understanding, the answer to this question is "no." We shall show in Example 10 in Chapter 2 that Eq. 1-12 follows in a very direct and simple way from the theory of relativity; there is no need whatever to invoke an ether concept in order to derive it.

There appears to be no acceptable experimental basis, then, for the idea of an ether, that is, for a preferred frame of reference. This is true whether we choose to regard the ether as stationary or as dragged along. We must now face the alternative that a principle of relativity is valid in electrodynamics as well as in mechanics. If this is so, then either electrodynamics must be modified, so that it is consistent with the classical relativity principle, or else we need a new relativity principle that is consistent with electrodynamics, in which case classical mechanics will need to be modified.

1-8 ATTEMPTS TO MODIFY ELECTRODYNAMICS

Now let us consider attempts to modify the laws of electromagnetism, with the hope of gaining an understanding of the way light is propagated in free space and in transparent bodies. These laws, in the form of Maxwell's equations (see *Physics*, Part II, Sec. 41-8) predict the existence of electromagnetic waves that travel through free space with a speed c $(= 1/\sqrt{\mu_0 \varepsilon_0} = 3.00 \times 10^8 \text{ m/s})$. A central question is this: With respect to what reference frame does this speed apply? The arguments of the preceding three sections disposed of the ether as such a frame.

In 1908 Walter Ritz discarded the ether concept entirely and modified Maxwell's equations in such a way that the predicted speed c was identified with a reference frame attached to the *source* of the emitted radiation. The various theories that are based on this assumption are called *emission* theories. Common to them all is the hypothesis that the velocity of light is c relative to the original source and that this velocity is independent of the state of motion of the medium transmitting the light. This would automatically explain the null result of the Michelson-Morley experiment, in which the source is rigidly connected to the interferometer proper and moves with it. The various emission theories differ only in their predictions of the way light is reflected from a moving mirror, a detail that need not involve us here.

It turns out that none of the emission theories can pass the crucial test of experiment. All are contradicted by early experiments using extraterrestrial light sources, such as de Sitter's (1913) observations on double stars and a Michelson-Morley experiment using starlight or sunlight. More recently, measurements of the speed of the radiation emitted from fast-moving energetic particles emerging from laboratory accelerators are even more direct in contradicting emission theories.

Two stars that move in orbits about their common center of mass are called double stars. Imagine the orbits to be circles. Now assume that the velocity of the light by which we see them through the empty space is equal to $c + v_s$, where v_s is the component of the velocity of the source relative to the observer, at the time the light is emitted, along the line from the source to the observer. Then, the time for light to reach the earth from the approaching star would be smaller than that from the receding star. As a consequence, the circular orbits of double stars should appear to be eccentric as seen from earth. But measurements show no such eccentricities in the orbits of double stars observed from earth.

The results are consistent with the assumption that the velocity of the light is independent of the velocity of the source. De Sitter's conclusion was that, if the velocity of light is not equal to c but is equal to $c + kv_s$, then k experimentally must be less than 2×10^{-3} (see, however, Ref. 7). In 1977 Brecher [8], observing x-rays emitted by binary stars, claims to have reduced this limit to 2×10^{-9}.

More direct experiments [9] using fast-moving terrestrial sources confirm the conclusion that the velocity of electromagnetic radiation is independent of the velocity of the source. In such an experiment (1964), measurements were made of the speed of electromagnetic radiation emitted from the decay of rapidly moving π^0 mesons produced in the CERN synchrotron. The mesons had energies greater than 6 GeV ($v_s = 0.99975c$), and the speed of the γ radiation emitted from these fast-moving sources was measured absolutely by timing over a known distance. The result corresponded to a value of k equal to $(-3 \pm 13) \times 10^{-5}$.

The Michelson-Morley experiment, using an extraterrestrial source, has been performed by R. Tomaschek, who used starlight, and D. C. Miller, who used sunlight. If the source velocity (due to rotational and translational motions relative to the interferometer) affects the velocity of light, we should observe complicated fringe-pattern changes. No such effects were observed in either of the experiments.

We saw earlier that an ether hypothesis is untenable. Now we are forced by experiment to conclude further that the laws of electrodynamics are correct and do not need modification. The speed of light (that is, electromagnetic radiation) is the same in all inertial systems, independent of the relative motion of source and observer. Hence, a relativity principle, applicable both to mechanics *and* to electromagnetism, is operating. Clearly, it cannot be the Galilean principle, since that required the speed of light to depend on the relative motion of source and observer. We conclude that the Galilean transformations must be replaced and, therefore, the basic laws of mechanics, which were consistent with those transformations, need to be modified so that they will be consistent with the new transformations.

1-9 *THE POSTULATES OF SPECIAL RELATIVITY THEORY*

In 1905, before many of the experiments we have discussed were actually performed, Albert Einstein (1879–1955), provided a solution to the dilemma facing physics. In his paper, "On the Electrodynamics of Moving Bodies" [10], Einstein wrote:

> . . . no properties of observed facts correspond to a concept of absolute rest; . . . for all coordinate systems for which the mechanical equations hold, the equivalent electrodynamical and optical equations hold also. . . . In the following we make these assumptions (which we shall subsequently call the Principle of Relativity) and introduce the further assumption—an assumption which is at the first sight quite irreconcilable with the former one—that light is propagated in vacant space, with a velocity *c* which is independent of the nature of motion of the emitting body. These two assumptions are quite sufficient to give us a simple and consistent theory of electrodynamics of moving bodies on the basis of the Maxwellian theory for bodies at rest.

We can rephrase these assumptions of Einstein as follows.

1. *The laws of physics are the same in all inertial systems. No preferred inertial system exists* (the principle of relativity).
2. *The speed of light in free space has the same value c in all inertial systems* (the principle of the constancy of the speed of light).

Einstein's relativity principle goes beyond the Newtonian relativity principle, which dealt only with the laws of mechanics, to include *all* the laws of physics. It states that it is impossible by means of *any* physical measurements to designate an inertial system as intrinsically stationary or moving; we can speak only of the *relative* motion of two systems. Hence, no physical experiment of *any* kind made entirely *within* an inertial system can tell the observer what the motion of his system is with respect to any other inertial system.

Although the relativity principle seems quite acceptable to us (and, in retrospect, even compelling), there are conceptual difficulties for most of us in accepting Einstein's second postulate. Consider that a source emits a short pulse of light, whose spherical wavefronts sweep past three observers, *A*, *B*, and *C*. Let *A* be at rest with respect to the light source; let *B* be rushing toward the source at great speed; finally, let *C* be rushing away from the source, also at great speed. The principle of the constancy of the speed of light tells us that *all three observers will measure the same speed for the light pulse!* This is in flat contradiction to the predictions of the Galilean velocity transformation equations (Eq. 1-5). It is *not* the way sound pulses or baseballs behave. It is a tribute to Einstein that he first realized clearly that—conceptual difficulties aside—this *is* the way that light behaves.

The entire special theory of relativity is derived directly from the two assumptions stated above. Their simplicity, boldness, and generality are characteristic of Einstein's genius. The success of his theory, as indeed of any theory, can be judged only by comparison with experiment. It was able not only to explain all the existing experimental results but it also predicted new effects that were confirmed by later experiments. No experimental objection to Einstein's special theory of relativity has yet been found.

In Table 1-2 we list the seven theories proposed at various times and compare their predictions of the results of 13 crucial experiments, old and new. Notice that only the special theory of relativity is in agreement with *all* the experiments listed. We have already commented on the successes and failures of the ether and emission theories with most of the light-propagation experiments, and it remains for us to show how special relativity accounts for their results. In addition,

Table 1-2
EXPERIMENTAL BASIS FOR THE THEORY OF SPECIAL RELATIVITY[a]

	Theory	Light Propagation Experiments							Experiments from Other Fields					
		Aberration	Fizeau Convection Coefficient	Michelson-Morley	Kennedy-Thorndike	Moving Sources and Mirrors	De Sitter Spectroscopic Binaries	Michelson-Morley, Using Sunlight	Variation of Mass with Velocity	General Mass–Energy Equivalence	Radiation from Moving Charges	Meson Decay at High Velocity	Trouton-Noble	Unipolar Induction, Using Permanent Magnet
Ether Theories	Stationary Ether, No Contraction	√	√	×	×	√	√	×	×		√		×	×
	Stationary Ether, Lorentz Contraction	√	√	√	×	√	√	√	√		√		√	×
	Ether Attached to Ponderable Bodies	×	×	√	√	√	√	√	×				√	
Emission Theories	Original Source	√	√	√	√	√	×	×			×			
	Ballistic	√		√	√	×	×	×			×			
	New Source	√		√	√	×	×	√			×			
Special Theory of Relativity		√	√	√	√	√	√	√	√	√	√	√	√	√

Legend. √, the theory agrees with experimental results.
     ×, the theory disagrees with experimental results.
     ▢, the theory is not applicable to the experiment.

[a] From Ref. 2.

several experiments from other fields—some suggested by the predictions of relativity and in flat contradiction to Newtonian mechanics—remain to be examined. What emerges from this comparative preview is the compelling *experimental* basis of special relativity theory. It alone is in accord with the real world of experimental physics.

    As is often true in the aftermath of a great new theory, it seemed obvious to many in retrospect that the old ideas had to be wrong. For example, Herman Bondi has written:

> The special theory of relativity is a necessary consequence of any assertion that the unity of physics is essential, for it would be intolerable for all inertial systems to be equivalent from a dynamical point of view yet distinguishable by optical measurements. It now seems almost incredible that the possibility of such a discrimination was taken for granted in the nineteenth century, but at the same time it was not easy to see what was more important—the universal validity of the Newtonian principle of relativity or the absolute nature of time.

It was his preoccupation with the nature of time that led Einstein to his revolutionary proposals. We shall see later how important a clear picture of the concept of time was to the development of relativity theory. However, the program of the theory, in terms of our discussions in this chapter, should now be clear. First, we must obtain equations of transformation between two uniformly moving (inertial) systems that will keep the velocity of light constant. Second, we must examine the laws of physics to check whether or not they keep the same form (that is, are invariant) under this transformation. Those laws that are not invariant will need to be generalized so as to obey the principle of relativity.

The new equations of transformation obtained in this way by Einstein are known for historical reasons as a Lorentz transformation. We have seen (Section 1-3) that Newton's equation of motion is invariant under a Galilean transformation, which we now know to be incorrect. It is likely then that Newton's laws—and perhaps other commonly accepted laws of physics—will not be invariant under a Lorentz transformation. In that case, they must be generalized. We expect the generalization to be such that the new laws will reduce to the old ones for velocities much less than that of light, for in that range both the Galilean transformation and Newton's laws are at least approximately correct.

1-10 EINSTEIN AND THE ORIGIN OF RELATIVITY THEORY

It is so fascinating a subject that one is hard-pressed to cut short a discussion of Albert Einstein, the person. Common misconceptions of the man, who quite properly symbolized for his generation the very height of intellect, might be shattered by such truths as these: Einstein's parents feared for a while that he might be mentally retarded, for he learned to speak much later than customary; one of his teachers said to him, "You will never amount to anything, Einstein," in despair at his daydreaming and his negative attitude toward formal instruction; he failed to get a high-school diploma and, with no job prospects, at the age of 15 he loafed like a "model dropout"; Einstein's first attempt to gain admission to a polytechnic institute ended when he failed to pass an entrance examination; after gaining admittance he cut most of the lectures and, borrowing a friend's class notes, he crammed intensively for two months before the final examinations. He later said of this ". . . after I had passed the final examination, I found the consideration of any scientific problem distasteful to me for an entire year." It was not until two years after his graduation that he got a steady job, as a patent examiner in the Swiss Patent Office at Berne; Einstein was very interested in technical apparatus and instruments, but—finding that he could complete a day's work in three or four hours—he secretly worked there, as well as in his free time, on the problems in physics that puzzled him. And so it goes.*

The facts above are surprising only when considered in isolation, of course, Einstein simply could not accept the conformity required of him, whether in educational, religious, military, or governmental institutions. He was an avid reader who pursued his own intellectual interests, had a great curiosity about nature, and was a genuine "free thinker" and independent spirit. As Martin Klein points out, what is really surprising about Einstein's early life is that none of his "elders" recognized his genius.

But such matters aside, let us look now at Einstein's early work. It is appropriate to quote here from Martin Klein [13].

> In his spare time during those seven years at Berne, the young patent examiner wrought a series of scientific miracles; no weaker word is adequate. He did nothing less than to lay out the main lines along which twentieth-century theoretical physics has developed. A very brief list will have to suffice. He began by working out the subject of statistical mechanics quite independently and without knowing of the work of J. Willard Gibbs. He also took this subject seriously in a way that neither Gibbs nor Boltzman had ever done, since he used it to give the theoretical basis for a final proof of the atomic nature of matter. His reflections on the problems of the Maxwell-Lorentz electrodynamics led him to create the special theory of relativity. Before he left Berne he had formulated the principle of equivalence and was struggling with the problems of gravitation which he later solved with the general theory of relativity. And, as if these were not enough, Einstein introduced another new idea into physics, one that even he described as "very revolutionary," the idea that light consists of particles of energy. Following a line of reasoning related to but quite distinct from Planck's, Einstein not only introduced the light quantum hypothesis, but proceeded almost at once to explore its implications for phenomena as diverse as photochemistry and the temperature dependence of the specific heat of solids.
>
> What is more, Einstein did all this completely on his own, with no academic connections whatsoever, and with essentially no contact with the elders of his profession.

The discussion thus far emphasizes Einstein's independence of other contemporary workers in physics. Also characteristic of his work is the fact that he always made specific predictions of possible experiments to verify his theories. In 1905, at intervals of less than eight weeks, Einstein sent to the *Annalen der Physik* three history-making papers. The first paper, on the quantum theory of light, included an explanation of the photoelectric effect. The suggested experiments, which gave the proof of the validity of Einstein's equations, were successfully carried out by Robert A. Millikan nine years later! The second paper, on statistical aspects of molecular theory, included a theoretical analysis of Brownian movement. Einstein wrote later of this: "My major aim in this was to find facts which would guarantee as much as possible the existence of atoms of definite size. In the midst of this I discovered that, according to atomistic theory, there would have to be a movement of suspended microscopic particles open to observation, without knowing that observations concerning the Brownian motion were already long familiar." Einstein's specific predictions were confirmed in detail in 1908 by Jean Perrin. The third paper [10], on special relativity, included applications to electrodynamics such as the relativistic mass of a moving body, all subsequently confirmed by experiment.

As for the origins of his special theory of relativity, Einstein was probably convinced, independently of the experimental background that we have presented, that Maxwell's equations had to have exactly the same form in all inertial

---

* See Refs. 11–19 for some rewarding articles and books about Einstein.

frames of reference. He may have reached such a conclusion intuitively in his youth while meditating on how a light wave would look to an observer who was following it at speed c. The constancy of the velocity of light seemed to him indisputable, and his 1905 paper begins with the precise electrodynamical experimental situation described in a text on Maxwell's theory that he was studying at the time. This highlights another characteristic of Einstein's work, which suggests why his approach to a problem was usually not that of the mainstream: namely, his attempt to restrict hypotheses to the smallest number possible and to the most general kind. For example, Lorentz, who never really accepted Einstein's relativity, used a great many *ad hoc* hypotheses to arrive at the same transformations in 1904 as Einstein did in 1905 (and as Voigt did in 1887); furthermore, Lorentz had assumed these equations *a priori* in order to obtain the invariance of Maxwell's equations in free space. Einstein, on the other hand, *derived* them from the simplest and most general postulates—the two fundamental principles of special relativity. And he was guided by his solution to the problem that had occupied his thinking since he was 16 years old: the nature of time. Lorentz and Poincaré, two of the most eminent mathematical physicists of the time, had accepted Newton's universal time $(t = t')$, whereas Einstein abandoned that notion.

Newton, even more than many succeeding generations of scientists, was aware of the fundamental difficulties inherent in his formulation of mechanics, based as it was on the concepts of absolute space and absolute time. Einstein expressed a deep admiration for Newton's method and approach and can be regarded as bringing many of the same basic attitudes to bear on his analysis of the problem. In his Autobiographical Notes, after critically examining Newtonian mechanics, Einstein writes:

> Enough of this. Newton, forgive me; you found the only way which, in your age, was just about possible for a man of highest thought and creative power. The concepts, which you created, are even today still guiding our thinking in physics, although we now know that they will have to be replaced by others farther removed from the sphere of immediate experience, if we aim at a profounder understanding of relationships.

It seems altogether fitting that Einstein should have extended the range of Newton's relativity principle, generalized Newton's laws of motion, and later incorporated Newton's law of gravitation into his space-time scheme. In subsequent chapters we shall see how this was accomplished.

questions _____

1. Quasars (*quasi*-stell*ar* objects) are the most intrinsically luminous objects in the universe. Many of them fluctuate in brightness, often on a time scale of a day or so. How can the rapidity of these brightness changes be used to estimate an upper limit to the size of these objects? (*Hint:* Separated points cannot change in a coordinated way unless information is sent from one to the other.)

2. How would you test a proposed reference frame to find out whether or not it is an inertial frame?

3. Give several examples of reference frames that are not inertial frames.

4. If you were confined to the hold of an ocean liner on a calm sea, how could you test whether the ship was moving with a constant velocity? Was accelerating?

5. Give examples in which effects associated with the earth's rotation are significant enough in practice to rule out a laboratory frame as being a good enough approximation to an inertial frame.

6. How does the concept of simultaneity enter into the measurement of the length of a body?

7. Could a mechanical experiment be performed in a given reference frame that would reveal information about the *acceleration* of that frame relative to an inertial frame?

8. In an inelastic collision, the amount of thermal energy (internal mechanical kinetic energy) developed is independent of the inertial reference frame of the observer. Explain why, in words.

9. Why is it necessary to rotate the interferometer in the Michelson-Morley experiment?

10. You are in deep space, in command of a spaceship equipped with all the resources of a modern optics laboratory, including a Michelson interferometer. Taking full advantage of the maneuverability of your ship, how would you go about testing for the presence or absence of a presumed luminiferous ether?

11. Describe an acoustic Michelson-Morley experiment, in analogy with the optical one. What could you measure with such a device? Would you expect a null result, similar to that obtained in the optical case?

12. A simple way to test for the presence or absence of an "ether wind" would be to make one-way, rather than round-trip, measurements of the speed of light. That is, we could measure the speed along a straight line, first in one direction and then in the other. Explain how the speed of the ether wind could be deduced from such data. Is the experiment practical?

13. Does the Lorentz-Fitzgerald contraction hypothesis contradict the classical notion of a rigid body?

14. Could the Lorentz-Fitzgerald contraction hypothesis have been disproved by simply measuring the length of a rod at different orientations to the presumed ether wind?

15. What is the relation between the apparent angular position of the star in Fig. 1-9 and the apparent direction of motion of raindrops falling on the vertical windshield of a bus?

16. Does the fact that stellar aberration is observable contradict the principle of the relativity of uniform motion (that is, does it determine an absolute velocity)? How, in this regard, does it differ from the Michelson-Morley experiment?

17. In the text we assumed that the star whose stellar aberration we were measuring was directly overhead. How would the situation change if the star were close to the horizon?

18. Even if the speed of light were infinite, the star in Fig. 1-9 would appear to move in a small circle during the course of a year, due to the different directions from which it is viewed from different parts of the earth's orbit. This effect is called *parallax*. How can effects due to parallax be separated from those due to aberration?

19. If the earth's motion, instead of being nearly circular about the sun, were uniformly along a straight line through the "ether," could an aberration experiment measure its speed?

20. How can we use aberration observations to refute the Ptolemaic, or earth-centered, model of the solar system?

21. If the "ether" were dragged along with water, what would be the expected result of the aberration experiment when done with a telescope filled with water? (Such an experiment was done by Sir George Airy in 1871. The results were the same as without water.)

22. Describe in your own words the essential difference between the Michelson-Morley and the Kennedy-Thorndike experiments.

23. Borrowing two phrases from Herman Bondi, we can catch the spirit of Einstein's two postulates by labeling them: (1) the principle of "the irrelevance of velocity" and (2) the principle of "the uniqueness of light." In what senses are velocity irrelevant and light unique in these two statements?

24. A beam from a laser falls at right angles on a plane mirror and rebounds from it. What is the speed of the reflected beam if the mirror is (*a*) fixed in the laboratory? (*b*) Moving directly toward the laser with speed *v*?

25. Comment on the assertion that if one accepts Einstein's principle of the constancy of the speed of light, then there is no reason to interpret the null result of the Michelson-Morley experiment as evidence against an ether hypothesis.

26. What boxes in Table 1-2 have been accounted for in this chapter?

27. Show that the two postulates of the special theory of relativity account satisfactorily for the results of (*a*) the Michelson-Morley experiment, (*b*) the Kennedy-Thorndike experiment, (*c*) the stellar aberration observations, (*d*) the binary star observations, and (*e*) the CERN experiment described in Section 1-8.

28. The speed of light in a vacuum is a true constant of nature, independent of the wavelength of the light or the choice of an (inertial) reference frame. Is there any sense, then, in which Einstein's second postulate can be viewed as contained within the scope of his first postulate?

29. Discuss the problem that young Einstein grappled with; that is, what would be the appearance of an electromagnetic wave to a person running along with it at speed *c*?

30. Can a particle move through a (transparent) medium at a speed greater than the speed of light in that medium? (See *Physics*, Part II, Sec. 43-4.)

*problems**

1. **Some lengths and some times.** (a) How many meters are there in a light-day? (b) How many kilometers in a light-microsecond? (c) How many years in a light-parsec? (d) How many seconds in a light-fermi? A *parsec* (= 3.09×10^{16} m) is a length unit much used in astronomy. A *fermi* (= 10^{-15} m) is a length unit much used in nuclear and particle physics.

2. **Some speeds.** What fraction of the speed of light does each of the following speeds represent? (a) A typical rate of continental drift (1 inch per year, say). (b) A typical drift speed for electrons in a current-carrying conductor (0.5 mm/s, say). (c) A typical highway speed limit of 55 mi/h. (d) The root-mean-square speed of a hydrogen molecule at room temperature. (e) A supersonic plane flying at Mach 2.5 (= 1200 km/h, say). (f) The escape speed of a projectile from the surface of the earth. (g) The speed of the earth in its orbit around the sun. (h) The recession speed of a distant quasar (3.0×10^4 km/s, say).

3. **Some units for c—useful and otherwise.** Express the speed of light in (a) cm/ns, (b) ft/μs, (c) ly/y, (d) $J^{1/2} \cdot kg^{-1/2}$, (e) $MeV^{1/2} \cdot u^{-1/2}$, and (f) $m \cdot F^{-1/2} \cdot H^{-1/2}$. (Hint: Consider the relations $E = mc^2$ and $c = 1/\sqrt{\mu_0 \varepsilon_0}$. In the above "ly" stands for light year and "u" for atomic mass unit.)

4. **Win, place, and show.** Electrons emerging from the 2-mile-long Stanford Electron Accelerator may have a kinetic energy of 20 GeV, corresponding to an electron speed that is less than the speed of light by only 9.78 cm/s. Consider a three-way race over a straight 10-km course between the electron beam in a vacuum, a light beam in a vacuum, and a light beam in air. (a) List the order of finish. (b) Give the time interval between first and second place and (c) between second and third place. Take the index of refraction of air to be 1.00029.

5. **"... this goodly frame, the earth ..."** (Hamlet). For some purposes the earth cannot be taken as a totally "goodly" inertial frame because its motion is accelerated. Calculate the accelerations associated with (a) the earth's rotation about its axis (assume an equatorial point), (b) the earth's orbital motion about the sun and (c) the orbital motion of the solar system about the galactic center. The sun is 10 kpc from the galactic center and orbits around it at 300 km/s. One parsec = 1 pc = 3.09×10^{16} m.

6. **Electrons also fall.** Quite apart from effects due to the earth's rotational and orbital motions, a laboratory frame is not strictly an inertial frame because a particle placed at rest there will not, in general, remain at rest; it will fall under gravity. Often, however, events happen so quickly that we can ignore free fall and treat the frame as inertial.

*The more difficult problems have been starred ★.

Consider, for example, a 1.0-MeV electron (for which $v = 0.992c$) projected horizontally into a laboratory test chamber and moving through a distance of 20 cm. (a) How long would it take, and (b) how far would the electron fall during this interval? What can you conclude about the suitability of the laboratory as an inertial frame in this case?

7. **The Galilean transformation generalized.** Write the Galilean coordinate transformation equations (see Eqs. 1-1a and 1-1b) for the case of an arbitrary direction for the relative velocity **v** of one frame with respect to the other. Consider that the corresponding axes of the two frames remain parallel. (*Hint:* Let **v** have components v_x, v_y, and v_z.)

8. **Momentum is conserved for all. ...** An observer on the ground watches a collision between two particles whose masses are m_1 and m_2 and finds, by measurement, that momentum is conserved. Use the classical velocity addition theorem (Eq. 1-5) to show that an observer on a moving train will also find that momentum is conserved in this collision.

9. **(... but only if mass is conserved!)** Repeat Problem 8 under the assumption that a transfer of mass from one particle to the other takes place during the collision, the initial masses being m_1 and m_2 and the final masses being m'_1 and m'_2. Again, assume that the ground observer finds, by measurement, that momentum is conserved. Show that the train observer will also find that momentum is conserved *only* if mass is also conserved, that is, if

$$m_1 + m_2 = m'_1 + m'_2.$$

★ 10. **The invariance of "elastic."** A collision between two particles in which kinetic energy is conserved is described as *elastic*. Show, using the Galilean velocity transformation equations, that if a collision is found to be elastic in one inertial reference frame, it will also be found to be elastic in all other such frames. Could this result have been predicted from the conservation of energy principle?

11. **The work–energy theorem holds in all inertial frames.** Observer G is on the ground and observer T is on a train moving with uniform velocity **v** with respect to the ground. Each observes that a particle of mass m, initially at rest with respect to the train, is acted on by a constant force **F** applied to it in the forward direction for a time t. (a) Show that the two observers will find, for the work done on the particle by force **F**,

$$W_T = \tfrac{1}{2}ma^2t^2 \quad \text{and} \quad W_G = \tfrac{1}{2}ma^2t^2 + mvat,$$

in which a is the common acceleration of the particle. (b) Show that ΔK_T and ΔK_G, the changes in kinetic energy calculated by each observer, are also given by these same two expressions. Thus the work–energy theorem ($W = \Delta K$) is valid in all inertial reference frames.

12. More work–energy. (a) In Problem 11, evaluate W_T and W_G for $m = 1.0$ kg, $a = 0.20g$, $t = 5.0$ s, and $v = 25$ m/s. (b) Explain the fact that the two observers find that different amounts of work are done by the same force in terms of the different distances through which the observers measure that force to act during the time t. (c) Find also the initial and final kinetic energies for the particle, as reported by each observer. (d) Verify that the work–energy theorem is valid in each frame. (e) Explain the fact that the observers find different final kinetic energies for the particle in terms of the work that the particle could do in being brought to rest relative to each observer's frame.

13. Work, friction, and thermal energy. Suppose that, in Problem 11, the "particle of mass m" is a block sliding on the floor of the train and that a frictional force acts between the block and the floor. Assume that the external applied force acts for the same time t and is of such a magnitude that the block experiences the same acceleration a as in that problem. (a) Show that the amount of thermal energy developed by the action of the frictional

force is the same for each observer. (*Hint:* Work done against friction depends on the *relative* motion of the surfaces.) (b) Calculate this thermal energy, assuming that the frictional force is equal to 0.10 mg and that, as in Problem 12, $a = 0.20g$, $m = 1.0$ kg, and $t = 5.0$ s.

14. Steve and Sally watch an elastic collision. (a) Observer Steve watches a head-on elastic collision, in which one of the colliding particles in initially at rest in his reference frame. The masses and the initial velocities are shown in the table below. Fill in the blanks in Steve's column in this table. Verify numerically that momentum and kinetic energy are indeed conserved. (b) Observer Sally, moving past Steve at 2.5 m/s along the collision line in the same direction as m_1, also sees this event. Using the Galilean transformation equations, fill in the blanks in *her* column and, again, verify that momentum and kinetic energy are conserved.

15. Steve and Sally watch an inelastic collision. The two students of Problem 14 now observe a second head-on collision between two particles. In this case, however, it is *not* assumed that kinetic energy is conserved; instead, the final velocity of the particle of mass m_2 is given. Fill in the

Table 1-3
PROBLEM 14

	Symbol[a]	Steve's Data	Sally's Data
Particle Masses, kg	M_1	0.107	0.107
	M_2	0.345	0.345
Particle Velocities, m/s	u_1	+3.25	
	u_2	0.00	
	u_1'		
	u_2'		
Total System Momentum, kg·m/s	P		
	P'		
	$P' - P$		
Total System Kinetic Energy, J	K		
	K'		
	$K' - K$	0.000	

[a] Primed quantities refer to values after the collision.

Legend: ☐ Given data.

Table 1-4
PROBLEM 15

	Symbol[a]	Steve's Data	Sally's Data
Particle Masses, kg	M_1	0.107	0.107
	M_2	0.345	0.345
Particle Velocities, m/s	u_1	+3.25	
	u_2	0.00	
	u_1'		
	u_2'	+1.25	
Total System Momentum, kg·m/s	P		
	P'		
	$P' - P$		
Total System Kinetic Energy, J	K		
	K'		
	$K' - K$		

[a] Primed quantities refer to values after the collision.

Legend: ☐ Given data.

blanks, (a) for Steve and (b) for Sally, as before. (c) Do Steve and Sally still agree that momentum is conserved? Do they agree that kinetic energy is *not* conserved? Do they find the same value for the decrease in kinetic energy? Where has this energy gone?

16. Maxwell versus Galileo. Starting from Maxwell's equations, it is possible to derive a *wave equation* whose solutions represent electromagnetic waves. For the special case of a plane wave traveling parallel to the x axis in the direction of increasing x, the wave equation proves to be

$$\frac{\partial^2 E}{\partial x^2} = \frac{1}{c^2} \frac{\partial^2 E}{\partial t^2},$$

in which $E(x, t)$ is the wave amplitude. Show that this equation does not retain its form when written in terms of the coordinates (x', t') appropriate to another reference frame; That is, show that this equation is *not* invariant under a Galilean transformation. *Hint:* Use the chain rule in which, if $f = f(x, t)$, then

$$\frac{\partial f}{\partial x} = \frac{\partial f}{\partial x'}\left(\frac{\partial x'}{\partial x}\right) + \frac{\partial f}{\partial t'}\left(\frac{\partial t'}{\partial x}\right).$$

17. The "binumeral explosion" (one student's preference). (a) Use the binomial expansion theorem to verify each of the expansions displayed in the footnote on page 17. (b) Assuming that $\beta = 0.30$, what percent error is made in the first of these expansions if only two terms are kept? Only three terms? (c) Answer these same questions for the second of these expansions.

★ **18. Michelson-Morley, generalized.** Figure 1-11 shows a Michelson interferometer (compare Fig. 1-4) with a presumed ether wind whose uniform velocity **v** makes an angle ϕ with the arm of length l_1. (a) Show that the round-trip travel times for the two arms are

$$t_1 = \frac{2l_1}{c} \frac{\sqrt{1 - \beta^2 \sin^2 \phi}}{1 - \beta^2} \quad \text{and} \quad t_2 = \frac{2l_2}{c} \frac{\sqrt{1 - \beta^2 \cos^2 \phi}}{1 - \beta^2}.$$

(b) Let the interferometer be rotated through 90°, starting from the position shown in the figure. Show that the predicted fringe shift is given by

$$\Delta N = \frac{\beta^2(l_1 + l_2)}{\lambda} \cos 2\phi.$$

(c) Show that the three expressions given in (a) and (b) reduce to those given in Section 1-5 for $\phi = 0$ or 90° and, further, that the predictions for $\phi = 45°$ are reasonable.

19. Michelson-Morley with a real wind. A pilot plans to fly due east from A to B and back again. If u is his airspeed and if l is the distance between A and B, it is clear that his round-trip time t_0—if there is no wind—will be $2l/u$. (a) Suppose, however, that a steady wind of speed v blows from the east (or from the west). Show that the round-trip travel time will now be

$$t_1 = \frac{t_0}{1 - (v/u)^2}.$$

(b) If the wind is from the north (or from the south), show that the expected round-trip travel time is

$$t_2 = \frac{t_0}{\sqrt{1 - (v/u)^2}}.$$

(c) Note that these two travel times are *not* equal. In the Michelson-Morley experiment, however, experiment seems to show that (for arms of equal length) the travel times for light *are* equal; otherwise these experimenters would have found a fringe shift when they rotated the interferometer. What is the essential difference between the two situations?

20. A Michelson-Morley derby. Two light planes, whose cruising airspeeds are 100 knots (= 115 mi/h), each complete a round trip over a 40-km course laid out on the ground. The two courses are at right angles, and if there is no wind the race will be a tie. Suppose, however, that there is a steady 15-knot wind parallel to one of the tracks. (a) What heading with respect to her ground track must the cross-wind pilot adopt? (b) Which pilot will win and by how much?

21. Not much of a contraction. In the Michelson-Morley experiment, by about how many atomic diameters would the appropriate arm of the interferometer have to shrink according to the Lorentz-Fitzgerald contraction hypothesis? The actual arm length of the interferometer was 2.8 m, but (see Fig. 1-8) the beam was reflected back

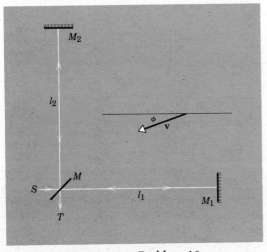

Figure 1-11. Problem 18.

and forth four times on each arm. Take 0.1 nm as a typical atomic diameter.

22. Still not much! If the Lorentz-Fitzgerald contraction hypothesis were true, by how much would the diameter of the earth be contracted by virtue of its orbital motion?

23. A neat little result. Suppose that a rod in the form of a circular cylinder is immersed in a presumed ether wind, the wind velocity being at right angles to the axis of the rod. Show that the effect of the Lorentz-Fitzgerald contraction would be to change the cross section of the rod from a circle to an ellipse. Show further that the eccentricity of this ellipse is simply β.

★ **24. The incredible shrinking rod.** Figure 1-12 shows a rod placed at an angle ϕ to the direction of an ether wind of velocity **v**. (a) Show that, according to the Lorentz-Fitzgerald contraction hypothesis, the length of the rod is given by

$$l = l^0 \sqrt{1 - \beta^2 \cos^2 \phi},$$

in which $\beta = v/c$ and l^0 is the length of the rod when there is no ether wind. (b) Show further that according to the Lorentz-Fitzgerald hypothesis the round-trip travel time for a light signal sent back and forth along the rod is given, to order β^2, by

$$t = \frac{2l^0}{c}\left(1 + \frac{1}{2}\beta^2\right).$$

(*Hint:* Make use of the expression for t_1 displayed in Problem 18.) Note that the round-trip travel time is independent of ϕ. What is the relationship of this problem to the Michelson-Morley experiment? Discuss this problem from the point of view of Einstein's second postulate.

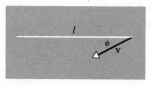

Figure 1-12. Problem 24.

★ **25. A fascinating "thought experiment."** Paul Ehrenfest (1880–1933) proposed the following thought experiment to illustrate the different behavior expected for light under the ether-wind hypothesis and under Einstein's second postulate.

Imagine yourself seated at the center of a spherical shell of radius 3×10^8 m, the inner surface being perfectly reflecting. A source at the center of the sphere emits a sharp pulse of light, which travels outward through the darkness with uniform intensity in all directions. What would you see during the 3-s interval following the pulse under the assumptions that (a) there is a steady ether wind

blowing through the sphere at 100 km/s and (b) that there is no ether and Einstein's second postulate holds. (c) Discuss the relationship of this thought experiment to the Michelson-Morley experiment.

26. Aberration and the rotating earth. In addition to revolving about the sun, the earth also rotates on its axis once a day. Find the largest aberration angle (tilt of the telescope) due to the earth's rotation alone for an observer at (a) the equator, (b) latitude 60°, and (c) the north pole.

27. Aberration and relativity. Consider the aberration arrangement of Fig. 1-9b. (a) At what speed does light pass along the telescope axis according to the ether hypothesis? (b) According to the special theory of relativity? (c) Show that, according to relativity theory, the classical aberration equation (see Eq. 1-11)

$$\tan \alpha = \frac{v}{c} \quad \text{(classical theory)}$$

must be replaced by

$$\sin \alpha = \frac{v}{c} \quad \text{(relativity theory)}$$

Thus the ether theory and the theory of relativity makes different predictions for the aberration of starlight. As the next problem shows, however, the predictions prove to be not as different as they seem.

28. Not a very sensitive test. Problem 27 displays the classical and the relativistic predictions for the aberration constant α. If α_c and α_r are the predictions of these two theories, find their fractional difference; that is, find $(\alpha_r - \alpha_c)/\alpha_r$. Assume that v, the earth's orbital speed, is 11 km/s, and take 3.00×10^8 m/s for the speed of light. (*Hint:* The predictions of the two theories are so close that the calculation of their difference may well be beyond the scope of your hand calculator! Use the series expansions:

$$\sin^{-1} x = x + \tfrac{1}{6}x^3 + \tfrac{3}{40}x^5 \ldots$$

and

$$\tan^{-1} x = x - \tfrac{1}{3}x^3 + \tfrac{1}{5}x^5 \ldots .).$$

29. Weighing the sun. (a) Show that M, the mass of the sun, is related to the aberration constant α (see Eq. 1-11) by

$$M = \frac{\alpha^2 c^2 R}{G},$$

in which R is the radius of the earth's orbit (assumed circular) and $G (= 6.67 \times 10^{-11}$ N·m²/kg²) is the universal gravitational constant. (*Hint:* Apply Newton's second law to the earth's motion around the sun.) (b) Calculate M, given that $\alpha = 20.5''$ and $R = 1.50 \times 10^{11}$ m.

Figure 1-13. Problem 32.

30. Dragging the ether. (*a*) In the Fizeau ether-drag experiment (Fig. 1-10), identify the frames *S* and *S'* and the relative velocity **v** that correspond to Fig. 1-1. (*b*) Show that in the Fresnel ether-drag formula (Eq. 1-12) $v \rightarrow v_w$ for very large values of *n*. How would you interpret this? (*c*) Under what circumstances will the Fresnel drag coefficient be zero? To what does this correspond physically?

31. A fast-moving fluid. According to the Fresnel ether-drag formula (Eq. 1-12), with what speed would the fluid have to move in order that the speed of light in the fluid equal the speed of light in a vacuum? Assume an index of refraction of $n = 1.33$.

32. Twinkle, twinkle, double star. . . . Consider one star in a binary system moving in uniform circular motion with speed *v*. Consider two positions: (I) the star is moving *away* from the earth along the line connecting them, and (II) the star is moving *toward* the earth along the line connecting them (see Fig. 1-13). Let the period of the star's motion be *T* and its distance from earth be *l*. Assume that *l* is large enough that positions I and II are a half-orbit apart. (*a*) Show that the star would appear to go from position (I) to position (II) in a time $T/2 - 2lv/(c^2 - v^2)$ and from position (II) to position (I) in a time $T/2 + 2lv/(c^2 - v^2)$, assuming that the emission theories are correct. (*b*) Show that the star would appear to be at both positions I and II at the same time if $T/2 = 2lv/(c^2 - v^2)$.

33. Three emission theories (all wrong!) Emission theories differ in their predictions of what the speed of light will be on reflection from a moving mirror. (*a*) The *original-source* theory assumes that the speed remains *c* rela-

tive to the source. (*b*) The *ballistic* theory assumes that the speed becomes *c* relative to the mirror. (*c*) The *new-source* theory assumes that the speed becomes *c* relative to the mirror image of the source. Figure 1-14 shows a

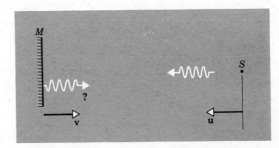

Figure 1-14. Problems 33 and 34.

source of light *S* moving to the left with speed *u* in the laboratory frame and a mirror *M* moving to the right with speed *v*. What is the speed of the reflected light beam according to each of the above three theories? (Compare Problem 34.)

34. Moving sources, moving mirrors, and Einstein's second postulate. According to Einstein's second postulate, what would be the measured speed of a light pulse (both before and after reflection from the moving mirror of Fig. 1-14) as viewed from (*a*) the laboratory reference frame, (*b*) a frame attached to the source, and (*c*) a frame attached to the mirror? Compare the simplicity of your answers with those predicted by the various emission theories described in Problem 33.

references

1. Robert Resnick and David Halliday, *Physics*, Part I, 3rd ed., (Wiley, New York, 1977) and David Halliday and Robert Resnick, *Physics*, Part II, 3rd ed. (Wiley, New York, 1978).

2. Wolfgang K. H. Panofsky and Melba Phillips, *Classical Electricity and Magnetism*, 2nd ed., (Addison–Wesley, Reading, Mass., 1962), chap. 14.

3. R. S. Shankland, "The Michelson-Morley Experiment," *Am. J. Phys.* **32**, 16 (1964). A detailed account of this experiment and a clear portrayal of the centrality of

the ether concept in prerelativity days. Also, a long extract from the Michelson-Morley paper is given in W. F. Magie, *Source Book in Physics* (McGraw-Hill, New York, 1935), p. 369.

4. J. P. Cedarholm, G. L. Bland, B. L. Havens, and C. H. Townes, "New Experimental Tests of Special Relativity," *Phys. Rev. Lett.*, **1**, 342 (1958).

5. Sir Edmund Whittaker, "G. F. Fitzgerald," *Scientific American* (November 1953). An account of Fitzgerald's life and the contraction hypothesis.

6. Albert Stewart, "The Discovery of Stellar Aberration," *Scientific American* (March 1964). An interesting and detailed description of Bradley's work.

7. J. G. Fox, "Evidence Against Emission Theories," *Am. J. Phys.*, **33**, 1 (1965). A critical analysis of the de Sitter and other experiments. See also J. G. Fox, "Constancy of the Velocity of Light," *Am. J. Phys.*, **35**, 967 (1967).

8. Kenneth Brecher, "Is the Speed of Light Independent of the Velocity of the Source?" *Phys. Rev. Lett.*, **39**, 1051 (1977).

9. See, for example, T. Alvager, F. J. M. Farley, J. Kjellman, and I. Wallen, "Test of the Second Postulate of Special Relativity in the GeV Region," *Phys. Lett.*, **12**, 260 (1964).

10. A full translation of this 1905 paper by Einstein, along with translations of other significant papers by Einstein and by others, appears in *The Principle of Relativity* (Dover, New York, 1952).

11. Gerald Holton, "On the Orgins of the Special Theory of Relativity," *Am. J. Phys.* **28**, 627 (1960).

12. R. S. Shankland, "Conversations with Einstein," *Am. J. Phys.*, **31**, 47 (1963).

13. Martin J. Klein, "Einstein and Some Civilized Discontents," *Phys. Today*, **18**, 38 (1965).

14. Gerald Holton, "Influences on Einstein's Early Work," *Am.* Scholar (Winter 1967–68).

15. Gerald Holton, "Einstein and the 'Crucial' Experiment," *Am. J. Phys.*, **37**, 968 (1969).

16. Jeremy Bernstein, *Einstein* (Viking, New York, 1973).

17. A. P. French, Ed., *Einstein, a Centenary Volume,* (Harvard University Press, Cambridge, Mass., 1979). Authorized by the International Commission on Physics Education in celebration of the centenary of Einstein's birth.

18. Robert Resnick, "Misconceptions About Einstein, His Work and His Views," *J. Chem. Educ.*, **57**, 854 (1980).

19. Abraham Pais, '*Subtle is the Lord . . .* '—*The Science and Life of Albert Einstein* (Oxford University Press, New York, 1982).

relativistic kinematics

We will raise this conjecture (whose intent will from now on be referred to as the "Principle of Relativity") to the status of a postulate, and also introduce another postulate, which is only apparently irreconcilable with the former: light is always propagated in empty space with a definite velocity c *which is independent of the state of motion of the emitting body.*

Albert Einstein (1905)

2-1 THE RELATIVITY OF SIMULTANEITY

In a paper entitled "Conversations with Albert Einstein," R. S. Shankland* writes: "I asked Professor Einstein how long he had worked on the Special Theory of Relativity before 1905. He told me that he had started at age 16 and worked for ten years; first as a student when, of course, he could spend only part-time on it, but the problem was always with him. He abandoned many fruitless attempts, 'until at last it came to me that time was suspect!'" What was it about time that Einstein questioned? It was the assumption, often made unconsciously and certainly not stressed, that there exists a universal time that is the same for all observers. Indeed, it was only to bring out this assumption explicitly that we included the equation $t = t'$ in the Galilean transformation equations (Eq. 1-1). In prerelativistic discussions, the assumption was there implicitly by the absence of a transformation equation for t in the Galilean equations. That the same time scale applied to all inertial frames of reference was a basic premise of Newtonian mechanics.

In order to set up a universal time scale, we must be able to give meaning, independent of a frame of reference, to statements such as "Events A and B occurred at the same time." Einstein pointed out that when we say that a train arrived at 7 o'clock, this means that the exact pointing of the clock hand to 7 and the arrival of the train at the clock were simultaneous. We certainly shall not have a universal time scale if different inertial observers disagree as to whether two events are simultaneous. Let us first try to set up an unambiguous time scale in a single frame of reference; then we can set up time scales in exactly the same way in all inertial frames and compare what different observers have to say about the sequence of two events, A and B.

Suppose that the events occur at the same place in one particular frame of reference. We can have a clock at that place which registers the time of occurrence of each event. If the reading is the same for each event, we can logically regard the events as simultaneous. But what if the two events occur at *different* locations? Imagine now that there is a clock at the positions of each event—the clock at A being of the same nature as that at B, of course. These clocks can

* See Ref. 12 in Chapter 1.

record the time of occurrence of the events but, before we can compare their readings, we must be sure that they are synchronized.

Some "obvious" methods of synchronizing clocks turn out to be erroneous. For example, we can set the two clocks so that they always read the same time as seen by the observer at the site of event A (observer A). This means that whenever A looks at the B clock, it reads the same to him as his clock. The defect here is that if observer B uses the same criterion (that is, that the clocks are synchronized if they always read the same time to *him*), he will find that the clocks are *not* synchronized if A says that they *are*. The reason is that this method neglects the fact that it takes time for light to travel from B to A and vice versa. The student should be able to show that, if the distance between the clocks is L, one observer will see the other clock lag his by $2L/c$ when the other observer claims that they are synchronous. We certainly cannot have observers in the same reference frame disagree on whether clocks are synchronized or not, so we reject this method.

An apparent way out of this difficulty is simply to set the two clocks to read the same time and then move them to the positions where the events occur. (In principle, we need clocks everywhere in our reference frame to record the time of occurrence of events, but once we know how to synchronize two clocks we can, one by one, synchronize all the clocks.) The difficulty here is that we do not know ahead of time, and therefore cannot assume, that the motion of the clocks (which may have different velocities, accelerations, and path lengths in being moved into position) will not affect their readings or time-keeping ability. Even in classical physics, the motion can affect the rate at which clocks run. For example, a pendulum clock in free fall will not run at all!

Hence, the logical thing to do is to put our clocks into position and synchronize them by means of signals. If we had a method of transmitting signals with infinite speed, there would be no complications. The signals would go from clock A to clock B to clock C, and so on, in zero time. We could use such a signal to set all clocks at the same time reading. But no signal known has this property. All known signals require a finite time to travel some distance, the time increasing with the distance traveled. The best signal to choose would be one whose speed depends on as few factors as possible. We choose electromagnetic waves because they do not require a material medium for transmission and their speed in vacuum does not depend on their wavelength, amplitude, or direction of propagation. Furthermore, their propagation speed is the highest known and—most important for finding a universal method of synchronization—experiment shows their speed to be the same for all inertial observers.

Now we must account for the finite time of transmission of the signal and our clocks can be synchronized. To do this let us imagine an observer with a light source that can be turned on and off (for example, a flashbulb) at each clock, A and B. Let the measured distance between the clocks (and observers) be L. The agreed-upon procedure for synchronization then is that A will turn on his light source when his clock reads $t = 0$ and observer B will set his clock to $t = L/c$ the instant he receives the signal. This accounts for the transmission time and synchronizes the clocks in a consistent way. For example, if B turns on his light source at some later time t by his clock, the signal will arrive at A at a time $t + L/c$, which is just what A's clock will read when A receives the signal.

A method equivalent to the above is to put a light source at the exact midpoint of the straight line connecting A and B and instruct each observer to put his clock at $t = 0$ when the turned-on light signal reaches him. The light will take an equal amount of time to reach A and B from the midpoint, so this procedure does indeed synchronize the clocks.

Now that we have a procedure for synchronizing clocks in one reference frame, we can judge the time order of events in that frame. The time of an event is measured by the clock whose location coincides with that of the event. Events occurring at two different places in that frame must be called *simultaneous* when the clocks at the respective places record the same time for them. Suppose that one inertial observer does find that two separated events are simultaneous. Will these same events be measured as simultaneous by an observer on another inertial frame that is moving with speed v with respect to the first? (Remember, each observer uses an identical procedure to synchronize the clocks in his reference frame.) If not, simultaneity is not independent of the frame of reference used to describe events. Instead of being absolute, simultaneity would be a relative concept. Indeed, this is exactly what we find to be true, in direct contradiction to the classical assumption.

We can understand the relativity of simultaneity by a simple example in which, in fact, clocks play no role. It will also be clear from this example in what a direct and compelling way the relativity of simultaneity follows from Einstein's second postulate, namely, that the speed of light is the same for all inertial observers. Let there be two inertial reference frames S' and S, having a relative velocity v. Each observer is provided with a measuring rod. The observers note that two lightning bolts strike each frame, hitting and leaving permanent marks.* Assume that afterwards, by using their measuring rods, each inertial observer finds that he was located exactly at the midpoint of the marks that were left on his reference frame. In Fig. 2-1a, these marks are left at A and B on the S frame and at A' and B' on the S' frame, the observers being at O and O'. Because each observer knows he was at the midpoint of the marks left by these events, he will conclude that the events were simultaneous if the light signals from them arrive simultaneously at his position (see the definitions of simultaneity given earlier). If, on the other hand, one signal arrives before the other, he will conclude that one event preceded the other.

Many different possibilities exist in principle as to what the measurements might show. Let us suppose, for the sake of argument, that the S observer finds that the lightning bolts struck simultaneously. Will the S' observer also find these events to be simultaneous? In Fig. 2-1 we take the point of view of the S observer and see the S' frame moving, say, to the right. At the instant the lightning struck at A and A', these two points coincide, and at the instant the lightning struck at B and B' those two points coincide. The S observer found these two events to occur at the same instant, so at that instant O and O' must coincide also for him. However, *the light signals from the events take a finite time to reach O and during this time O' travels to the right* (Figs. 2-1b to 2-1d). Hence, the signal from event BB' arrives at O' (Fig. 2-1b) before it gets to O (Fig. 2-1c), whereas the signal from event AA' arrives at O (Fig. 2-1c) before it gets to O' (Fig. 2-1d). Consistent with our starting assumption, the S observer finds the events to be simultaneous (both signals arrive at O at the same instant). The S' observer, however, finds that event BB' precedes event AA' in time; they are *not* simultaneous to him. Since O' is to the right of O at all times after the strokes occur, the right wavefront *must* pass O' before it reaches O. Thus the two wavefronts

---

* The essential point is to have light sources that leave marks. Note that, because the speed of light is the same in all inertial frames, only a single expanding wavefront originates at each pair of marks.

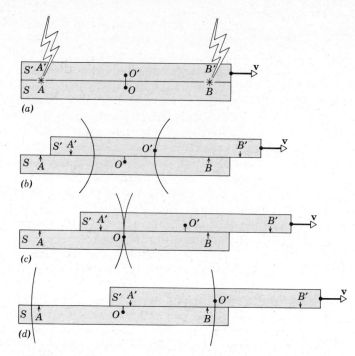

Figure 2-1. *The point of view of the S observer*, the *S'* frame moving to the right. (*a*) Light waves leave *AA'* and *BB'*. Successive drawings correspond to the assumption that event *AA'* and event *BB'* are simultaneous in the *S* frame. (*b*) The right wavefront reaches *O'*. (*c*) Both wavefronts reach *O*. (*d*) The left wavefront reaches *O'*.

cannot reach *O'* simultaneously, and observer *S'* *must* judge the two events to be nonsimultaneous.

Thus we see that two separated events which are judged to be simultaneous in one reference frame will, in general, not be so judged by an observer in a different frame. Note in what a direct and simple way this concept of the relativity of simultaneity follows from the principle of the constancy of the speed of light. If the speed of light were not the same for the two observers in Fig. 2-1, the entire argument we have just given would fall apart.

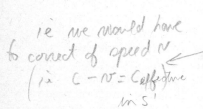

Now we could have supposed, just as well, that the lightning bolts struck so that the *S'* observer found them to be simultaneous. In that case the light signals reach *O'* simultaneously, rather than *O*. We show this in Fig. 2-2, where now we take the point of view of *S'*. The *S* frame moves to the left relative to the *S'* observer. But, in this case, the signals do not reach *O* simultaneously; the signal from event *AA'* reaches *O* before that from event *BB'*. Here the *S'* observer finds the events to be simultaneous but the *S* observer finds that event *AA'* precedes event *BB'*.

Hence, *neither* frame is preferred and the situation is perfectly reciprocal. Simultaneity is genuinely a relative concept, not an absolute one. Indeed, the two figures become indistinguishable if you turn one of them upside down. Neither observer can assert absolutely that he is at rest. Instead, each observer correctly states only that the other one is moving relative to him and that the signals travel with (the same) finite speed *c* relative to him. It should be clear that if we had an infinitely fast signal, then simultaneity *would* be an absolute concept; for the

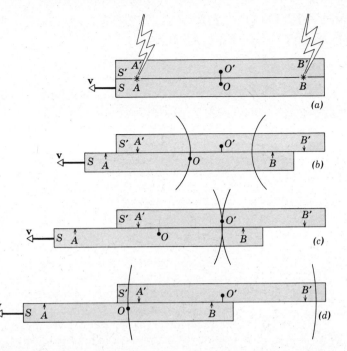

Figure 2-2. *The point of view of the S' observer,* the *S* frame moving to the left. (*a*) Light waves leave *AA'* and *BB'*. Successive drawings correspond to the assumption that event *AA'* and event *BB'* are simultaneous in the *S'* frame. (*b*) The left wavefront reaches *O*. (*c*) Both wavefronts reach *O'*. (*d*) The right wavefront reaches *O*.

frames would not move at all relative to one another in the (zero) time it would take the signal to reach the observers.

Note that the time order of two events *at the same place* can be absolutely determined. It is in the case that two events are *separated* in space that simultaneity is a relative concept. In our arguments, we have shown that *if* one observer finds the events to be simultaneous, *then* the other one will find them not to be simultaneous. Of course, it could also happen that neither observer finds the events to be simultaneous, but then they would disagree either on the time order of the events or on the time interval elapsing between the events, or both.

Some other conclusions suggest themselves from the relativity of simultaneity. To measure the length of an object (a rod, say) means to locate its end points simultaneously. Because simultaneity is a relative concept, length measurements will also depend on the reference frame and be relative. Furthermore, we find that the rates at which clocks run also depend on the reference frame. We can illustrate this as follows. Consider two clocks, one on a train and one on the ground, and assume that at the moment they pass one another (that is, the instant that they are coincident), they read the same time (that is, the hands of the clocks are in identical positions). Now, if the clocks continue to agree, we can say that they go at the same rate. But, when they are a great distance apart, we know from the preceding discussion that their hands cannot have identical positions simultaneously as measured both by the ground observer and the train observer. Hence, time interval measurements are also relative; that is, they depend on the reference frame of the observer.

2-2 DERIVATION OF THE LORENTZ TRANSFORMATION EQUATIONS

We have seen that the Galilean transformation equations must be replaced by new ones consistent with experiment. Here we shall derive these new equations, using the postulates of special relativity theory. To show the consistency of the theory with the discussion of the previous section, we shall then derive all the special features of the new transformation equations again from the more physical approach of the measurement processes discussed there.

We observe an event in one inertial reference frame S and characterize its location and time by recording the coordinates x, y, z, t of the event. In a second inertial frame S', this *same event* is recorded by the observer there as the space-time coordinates x', y', z', t'. We now seek the functional relationships $x' = x'(x, y, z, t)$, $y' = y'(x, y, z, t)$, $z' = z'(x, y, z, t)$, and $t' = t'(x, y, z, t)$. That is, we want the equations of transformation which relate one observer's space-time coordinates of an event with the other observer's coordinates of the same event.

We shall use the fundamental postulates of relativity theory and, in addition, the assumption that space and time are homogeneous. This homogeneity assumption (which can be paraphrased by saying that all points in space and time are equivalent) means, for example, that the results of a measurement of a length or time interval between two specific events should not depend on where or when the interval happens to be in our reference frame. We shall illustrate its application shortly.

We can simplify the algebra by choosing the relative velocity of the S and S' frames to be along a common x-x' axis and by keeping corresponding planes parallel (see Fig. 1-1). This does not impose any fundamental restrictions on our results, for space is isotropic (that is, has the same properties in all directions), a result contained in the homogeneity assumption. Also, at the instant the origins O and O' coincide, we let the clocks there read $t = 0$ and $t' = 0$, respectively. Now, as explained below, the homogeneity assumption requires that transformation equations must be linear (that is, they involve only the first power in the variables), so that the most general form they can take is

$$
\begin{aligned}
x' &= a_{11}x + a_{12}y + a_{13}z + a_{14}t \\
y' &= a_{21}x + a_{22}y + a_{23}z + a_{24}t \\
z' &= a_{31}x + a_{32}y + a_{33}z + a_{34}t \\
t' &= a_{41}x + a_{42}y + a_{43}z + a_{44}t.
\end{aligned}
\tag{2-1}
$$

Here, the subscripted coefficients are constants that we must determine to obtain the exact transformation equations. Notice that we do not exclude the possible dependence of space and time coordinates on one another.

If the equations were not linear, we would violate the homogeneity assumption. For example, suppose that x' depended on the square of x, that is, as $x' = a_{11}x^2$. Then the distance between two points in the primed frame would be related to the location of these points in the unprimed frame by $x'_2 - x'_1 = a_{11}(x_2^2 - x_1^2)$. Suppose now that a rod of unit length in S had its end points at $x_2 = 2$ and $x_1 = 1$; then $x'_2 - x'_1 = 3a_{11}$. If, instead, the same rod happens to be located at $x_2 = 5$ and $x_1 = 4$, we would obtain $x'_2 - x'_1 = 9a_{11}$. That is, the measured length of the rod would depend on where it is in space. Likewise, we can reject any dependence on t that is not linear, for the time interval between two events should not depend on the numerical setting of the hands of the observer's clock. The relationships must be linear then in order not to give the choice of origin of our space-time coordinates (or some other point) a physical preference over all other points.

Now, regarding the 16 coefficients in Eq. 2-1, it is expected that their values will depend on the relative velocity v of the two inertial frames. For example, if $v = 0$, then the two frames coincide at all times and we expect $a_{11} = a_{22} = a_{33} = a_{44} = 1$, all other coefficients being zero. More generally, if v is small compared to c, the coefficients should lead to the (classical) Galilean transformation equations. We seek to find the coefficients for *any* value of v, that is, as functions of v.

How then do we determine the values of these 16 coefficients? Basically, we use the postulates of relativity, namely, (1) the principle of relativity—that no preferred inertial system exists, the laws of physics being the same in all inertial systems—and (2) the principle of the constancy of the speed of light—that the speed of light in free space has the same value c in all inertial systems. Let us proceed.

With no relative motion of the frames in the y or z direction, we might expect $y' = y$ and $z' = z$. This result does indeed follow directly from arguments using the relativity postulate (see Sec. 2-2 of Ref. 1 for a proof), so that $a_{22} = a_{33} = 1$ and $a_{21} = a_{23} = a_{24} = a_{31} = a_{32} = a_{34} = 0$ and eight of the coefficients are thereby determined. Therefore, our two middle transformation equations become

$$y' = y \quad \text{and} \quad z' = z. \tag{2-2}$$

There remain transformation equations for x' and t', namely,

$$x' = a_{11}x + a_{12}y + a_{13}z + a_{14}t$$

and

$$t' = a_{41}x + a_{42}y + a_{43}z + a_{44}t.$$

Let us look first at the t' equation. For reasons of symmetry, we assume that t' does not depend on y and z. Otherwise, clocks placed symmetrically in the y-z plane (such as at $+y$, $-y$ or $+z$, $-z$) about the x axis would appear to disagree as observed from S', which could contradict the isotropy of space. Hence, $a_{42} = a_{43} = 0$. As for the x' equation, we know that a point having $x' = 0$ appears to move in the direction of the positive x axis with speed v, so that the statement $x' = 0$ must be identical to the statement $x = vt$. Putting $x' = 0$ and $x = vt$ in the x' equation above then yields

$$0 = a_{11}(vt) + a_{12}y + a_{13}z + a_{14}t$$

or

$$0 = (a_{11}v + a_{14})t + a_{12}y + a_{13}z.$$

Because t, y, and z are all independent variables, the only way this last equation can be satisfied for all values of these variables is for the coefficients of the variables to vanish. Thus we must have

$$a_{14} = -a_{11}v$$

and

$$a_{12} = a_{13} = 0.$$

Our four transformation equations are thus reduced to

$$\begin{aligned} x' &= a_{11}(x - vt) \\ y' &= y \\ z' &= z \\ t' &= a_{41}x + a_{44}t. \end{aligned} \tag{2-3}$$

There remains the task of determining the three coefficients a_{11}, a_{41}, and a_{44}. To do this, we use the principle of the constancy of the velocity of light. Let us assume that at the time $t = 0$ a spherical electromagnetic wave leaves the origin of S, which coincides with the origin of S' at that moment. The wave propagates with a speed c in all directions in each inertial frame. Its progress, then, is described by the equation of a sphere whose radius expands with time at the same rate c in terms of *either* the primed or unprimed set of coordinates. That is,

$$x^2 + y^2 + z^2 = c^2t^2 \tag{2-4}$$

or

$$x'^2 + y'^2 + z'^2 = c^2t'^2. \tag{2-5}$$

If now we substitute the transformation equations (Eqs. 2-3) into Eq. 2-5, we get

$$a_{11}^2(x - vt)^2 + y^2 + z^2 = c^2(a_{41}x + a_{44}t)^2.$$

Rearranging the terms gives us

$$(a_{11}^2 - c^2a_{41}^2)x^2 + y^2 + z^2 - 2(va_{11}^2 + c^2a_{41}a_{44})xt = (c^2a_{44}^2 - v^2a_{11}^2)t^2.$$

In order for this expression to agree with Eq. 2-4, which represents the same thing, we must have

$$a_{11}^2 - c^2a_{41}^2 = 1,$$
$$va_{11}^2 + c^2a_{41}a_{44} = 0,$$
and $$c^2a_{44}^2 - v^2a_{11}^2 = c^2.$$

Here we have three equations in three unknowns, whose solution (as you can verify by substitution into the three equations above) is

$$a_{11} = \frac{1}{\sqrt{1 - v^2/c^2}},$$
$$a_{41} = -\frac{(v/c^2)}{\sqrt{1 - v^2/c^2}}, \tag{2-6}$$
and $$a_{44} = \frac{1}{\sqrt{1 - v^2/c^2}}.$$

By substituting these values into Eqs. 2-3, we obtain, finally, the new sought-after transformation equations,

$$x' = \frac{x - vt}{\sqrt{1 - v^2/c^2}}$$
$$y' = y$$
$$z' = z \tag{2-7}$$
$$t' = \frac{t - (v/c^2)x}{\sqrt{1 - v^2/c^2}},$$

the so-called* *Lorentz transformation equations.*

Simple inspection of Eqs. 2-7 shows at once that serious difficulties arise if the relative speed v of the two observers is equal to or greater than the speed of light.

---

* Poincaré originally gave this name to the equations. Lorentz, in his classical theory of electrons, had proposed them before Einstein did. However, Lorentz took v to be the speed relative to an absolute ether frame and gave a different interpretation to the equations.

Under these circumstances the predicted quantities x' and t' are no longer finite real numbers. They are either infinitely great (for $v = c$) or imaginary* (for $v > c$). We conclude that c represents a limiting speed and that the speeds of material objects must always be less than this value, regardless of the reference frame from which the objects are observed. This conclusion is in complete agreement with experiment, no exceptions ever having been found.

Before probing the meaning of the Lorentz equations, we should put them to two necessary tests. First, if we were to exchange our frames of reference or—what amounts to the same thing—consider the given space-time coordinates of the event to be those observed in S' rather than in S, the only change allowed by the relativity principle is the physical one of a change in relative velocity from v to $-v$. That is, from S' the S frame moves to the left, whereas from S the S' frame moves to the right. When we solve Eqs. 2-7 for x, y, z, and t in terms of the primed coordinates (see Problem 5), we obtain

$$x = \frac{x' + vt'}{\sqrt{1 - v^2/c^2}},$$

$$y = y',$$

$$z = z', \qquad\qquad\qquad (2\text{-}8)$$

$$t = \frac{t' + (v/c^2)x'}{\sqrt{1 - v^2/c^2}},$$

which are identical in form with Eqs. 2-7 except that, as required, v changes to $-v$.

Another requirement is that for speeds small compared to c, that is, for $v/c \ll 1$, the Lorentz equations should reduce to the (approximately) correct Galilean transformation equations. A formal but useful way of imposing the requirement that the speed v is very much less than c, the speed of light, is to imagine a world in which the speed of light is infinitely great. In such a world, $v/c \ll 1$ would *always* be true, for all (finite) speeds. If, then, we let $c \to \infty$ in Eqs. 2-7, we obtain

$$x' = x - vt$$
$$y' = y$$
$$z' = z \qquad\qquad\qquad (2\text{-}9)$$
$$t' = t,$$

which are the classical Galilean transformation equations.

In dealing algebraically with the Lorentz equations, and with other equations that we shall encounter later, it simplifies matters greatly to introduce two new parameters, β and γ. Both are dimensionless, and both are simple functions of the relative speed v. The *speed parameter β* ($= v/c$) is simply the ratio of the relative speed v to the speed of light. The second parameter, γ, often called the *Lorentz factor*, is defined from

$$\gamma = \frac{1}{\sqrt{1 - v^2/c^2}} = \frac{1}{\sqrt{1 - \beta^2}}. \qquad\qquad (2\text{-}10)$$

Table 2-1 shows some selected values of β along with the corresponding values of γ computed from Eq. 2-10. Figure 2-3 is a plot of Eq. 2-10. In the classical low-speed limit, which is described by $\beta \ll 1$, we see that $\gamma \to 1$. In the high-speed limit, which is described by $\beta \to 1$, we see that $\gamma \to \infty$.

---

* In the mathematical sense, meaning a quantity containing $\sqrt{-1}$ as a factor.

Table 2-1
SOME VALUES OF γ

β	$\sqrt{1-\beta^2}$	γ	β	$\sqrt{1-\beta^2}$	γ
0.000	1.000	1.000	0.900	0.437	2.29
0.050	0.9987	1.0013	0.950	0.312	3.21
0.100	0.9950	1.0050	0.990	0.141	7.09
0.300	0.9542	1.048	0.9990	0.0446	22.4
0.600	0.8000	1.25	0.99990	0.0141	70.7

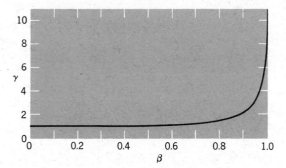

Figure 2-3. A plot of γ $(= 1/\sqrt{1-\beta^2})$ against β $(= v/c)$. See Eq. 2-10 and Table 2-1.

Table 2-2 displays the Lorentz transformation equations (Eqs. 2-7 and 2-8) with γ written in place of the quantity $1/\sqrt{1-v^2/c^2}$.

The Lorentz transformation equations displayed in Table 2-2 refer to a single event whose space-time coordinates are measured by two different inertial observers, S and S'. Commonly, however, we are confronted with pairs of events (event 1 and event 2, say) and our interest centers on the differences between their corresponding space-time coordinates rather than on the coordinates themselves. We are often more concerned, for example, with Δx $(= x_2 - x_1)$ and with Δt $(= t_2 - t_1)$ than with the separate quantities $x_2, x_1, t_2,$ and t_1. Table 2-3 shows the Lorentz transformation equations written in a difference form, suitable to such occasions.

Table 2-2
THE LORENTZ TRANSFORMATION EQUATIONS

1. $x = \gamma(x' + vt')$	1'. $x' = \gamma(x - vt)$
2. $y = y'$	2'. $y' = y$
3. $z = z'$	3'. $z' = z$
4. $t = \gamma\left(t' + \dfrac{vx'}{c^2}\right)$	4'. $t' = \gamma\left(t - \dfrac{vx}{c^2}\right)$

$$\gamma = \frac{1}{\sqrt{1-v^2/c^2}}$$

Table 2-3
THE LORENTZ
TRANSFORMATION EQUATIONS[a]

1. $\Delta x = \gamma(\Delta x' + v\,\Delta t')$	1'. $\Delta x' = \gamma(\Delta x - v\,\Delta t)$
2. $\Delta y = \Delta y'$	2'. $\Delta y' = \Delta y$
3. $\Delta z = \Delta z'$	3'. $\Delta z' = \Delta z$
4. $\Delta t = \gamma\left(\Delta t' + \dfrac{v\,\Delta x'}{c^2}\right)$	4'. $\Delta t' = \gamma\left(\Delta t - \dfrac{v\,\Delta x}{c^2}\right)$

[a] Written for event pairs, as difference equations.

EXAMPLE 1.

Two Observers View a Single Event. An observer in the S frame of Fig. 1-1 notes that an event occurs along the x-x' axis and records its space-time coordinates as $x = 2.00$ m and $t = 5.00$ ns $(= 5.00 \times 10^{-9}$ s$)$. The S' frame is moving with a speed v with respect to the S frame along their common axis. (a) If $v = 0.500c$, what space-time coordinates would the S' observer record for this event? Take the speed of light (c) to be 3.00×10^8 m/s, noting that this can be usefully written as 0.300 m/ns. (b) What space-time coordinates would observer S' record if the Galilean transformation equations held?

(a) For $\beta = 0.500$ we have, from Eq. 2-10,

$$\gamma = \frac{1}{\sqrt{1 - \beta^2}} = \frac{1}{\sqrt{1 - (0.50)^2}} = 1.16,$$

and, from Table 2-2 (Eqs. 1' and 4'),

$$x' = \gamma(x - vt)$$
$$= (1.16)[2.00\text{ m} - (0.500 \times 0.300\text{ m/ns})(5.00\text{ ns})]$$
$$= (1.16)(1.25\text{ m}) = 1.45\text{ m},$$

and

$$t' = \gamma\left(t - \frac{vx}{c^2}\right)$$

$$= \gamma\left(t - \frac{\beta x}{c}\right)$$

$$= (1.16)\left[5.00\text{ ns} - \frac{(0.500)(2.00\text{ m})}{(0.300\text{ m/ns})}\right]$$

$$= (1.16)(1.67\text{ ns}) = 1.94\text{ ns}.$$

(b) The Galilean transformation equations (Eq. 2-9) yield

$$x' = x - vt$$
$$= 2.00\text{ m} - (0.500 \times 0.300\text{ m/ns})(5.00\text{ ns}) = 1.25\text{ m}$$

and

$$t' = t = 5.00\text{ ns}.$$

Summarizing the results for easy comparison:

Observer S	Observer S'	
	Lorentz Equations	**Galilean Equations**
x 2.00 m	x' 1.45 m	(1.25 m)
t 5.00 ns	t' 1.94 ns	(5.00 ns)

We see that for such a large relative speed (half the speed of light!), the Lorentz transformation equations, which are correct for all speeds, predict results that differ substantially from the predictions of the Galilean equations. The Galilean predictions do not agree with the observations, and we have enclosed them in parentheses to remind us of that fact.

Recall that the length measurements (x and x') are from the appropriate origin (O or O') to the event. The time measurements (t and t') are the times that have elapsed since the moment the origins of the two reference frames passed each other. Each observer set his clocks to zero at that instant.

EXAMPLE 2.

The Galilean Equations Fail at High Speeds. At what relative speed will the Galilean and the Lorentz expressions for position x differ by 0.10 percent? By 1.0 percent? By 10 percent?

We can write the Galilean expression as $x'_G = x - vt$ and the Lorentz expression (see Table 2-2; Eq. 1') as $x'_L = \gamma(x - vt)$. We seek the value of v for which

$$\frac{x'_L - x'_G}{x'_L} = 0.10\% = 0.001.$$

We can write this as

$$\frac{x'_G}{x'_L} = 1 - 0.001 = 0.9990.$$

Substituting from above yields

$$\frac{(x - vt)}{\gamma(x - vt)} = 0.9990$$

or

$$\frac{1}{\gamma} = 0.9990.$$

From Eq. 2-10, the defining equation for γ, we then have

$$\frac{1}{\gamma} = \sqrt{1 - \beta^2} = 0.9990.$$

Solving for β yields

$$\beta = 0.045.$$

Thus

$$v = \beta c = (0.045)(3.00 \times 10^8 \text{ m/s}) = 1.4 \times 10^7 \text{ m/s}$$

is the answer we seek.

At this value of v (almost 5 percent of the speed of light) the Lorentz and the Galilean transformation equations differ in their predictions by only 0.1 percent. Small as this difference might appear, it still occurs at a speed (32 million mi/h!) much greater than any speed we encounter in the macroscopic world. We are quite safe in using the Galilean transformation equations when dealing with macroscopic objects such as baseballs or spaceships.

For a difference of 1 percent we find similarly that $\beta = 0.14$, or

$$v = \beta c = (0.14)(3.00 \times 10^8 \text{ m/s}) = 4.2 \times 10^7 \text{ m/s}.$$

For a difference of 10 percent we find $\beta = 0.44$ and $v = 1.3 \times 10^8$ m/s. This latter speed, which is 44 percent of the speed of light, may seem high—and indeed it is impossibly so for terrestrial macroscopic objects. However, as we shall see in later sections, it corresponds to an electron with a kinetic energy of only ~60 keV, a modest laboratory achievement. We shall encounter speeds much higher than this later in examples from microscopic physics. For example, electrons emerging from the 2-mile-long Stanford Linear Accelerator with a kinetic energy of 30 GeV ($= 3 \times 10^{10}$ eV) have $v = 0.99999999986c$. At these enormous energies the Galilean equations are hopelessly inadequate; the Lorentz equations, however, continue to give results that agree with experiment.

2-3 SOME CONSEQUENCES OF THE LORENTZ TRANSFORMATION EQUATIONS

The Lorentz transformation equations (Eqs. 2-7 and 2-8 and Table 2-2), derived rather formally in the last section from the relativity postulates, have some interesting consequences for length and time measurements. We shall look at them briefly in this section. In the next section we shall present a more physical interpretation of these equations and their consequences, relating them directly to the operations of physical measurement. Throughout the chapter we shall cite experiments that confirm these consequences.

Moving Rods Contract   The Lorentz transformation equations predict that: *When a body moves with a velocity v relative to the observer, its measured length is contracted in the direction of its motion by the factor $\sqrt{1 - v^2/c^2}$, whereas its dimensions perpendicular to the direction of motion are unaffected.* To prove the italicized statement, imagine a rod lying at rest along the x' axis of the S' frame. Its end points are measured to be at x_2' and x_1', so that its rest length is $x_2' - x_1'$ ($= \Delta x'$). What is the rod's length as measured by the S-frame observer, for whom the rod moves with a relative speed v? From the Lorentz equations (Table 2-3, Eq. 1'), we have

$$\Delta x' = \gamma(\Delta x - v \Delta t). \tag{2-11}$$

We can identify Δx as the length of the rod in the S frame if (and only if) the positions x_2 and x_1 of the end points of the rod are measured in this frame *at the same time.* Thus, with $\Delta t = 0$ in the above we obtain

$$\Delta x' = \gamma \Delta x$$

or (see Eq. 2-10)

$$\Delta x = \Delta x'(1/\gamma) = \Delta x' \sqrt{1 - \beta^2}. \tag{2-12a}$$

Putting $l \, (= \Delta x)$ for the length of the moving rod and $l_0 \, (= \Delta x')$ for its length at rest allows us to write

$$l = l_0 \sqrt{1 - \beta^2} \qquad \text{(length contraction)} \tag{2-12b}$$

so that the measured length l of the moving rod is contracted by the factor $\sqrt{1 - \beta^2}$ from its rest length l_0. As for the dimensions of the rod along y and z, perpendicular to the relative motion, it follows at once from the transformation equations $y' = y$ and $z' = z$ that these are measured to be the same by both observers.

Equation 2-12b predicts that a rod has its greatest length $(= l_0)$ when it is at rest (that is, when $\beta = 0$). Finally, our conclusions about the contraction of a moving rod are completely symmetrical, as required by the principle of relativity. We have seen above that a rod fixed in the S' frame is measured to be shorter by an observer in the S frame. It is equally true that a rod at rest in the S frame would be measured as shorter by an S' observer.

Moving Clocks Run Slow The Lorentz transformation equations also predict that *when a clock moves with a velocity v with respect to an observer, its rate is measured to have slowed down by a factor* $\sqrt{1 - (v/c)^2}$. To prove this, consider a clock C to be at rest at the position x'_0 in the S' frame, as shown in Fig. 2-4a. The figure shows two fiduciary marks on the clock face. Let t'_1 be the time at which the hand of the clock, as seen by the S' observer, coincides with the first of these marks, and let t'_2 be the time at which it coincides with the second.

The S-frame observer, on the other hand, records these same two events as occurring at times t_1 and t_2, as Fig. 2-4b shows. It is important to realize that observer S (like observer S' for that matter) has at his disposal a whole array of synchronized clocks, imagined to be there just for the purpose of assigning time coordinates to various events. S reads the times t_1 and t_2 from *two different* clocks

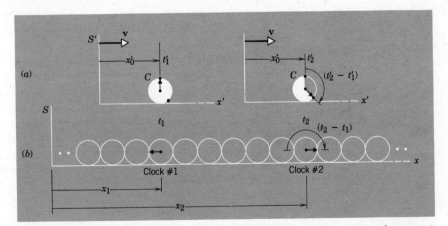

Figure 2-4. Clock C is fixed at position x'_0 in reference frame S'. Observer S sees this as a *single moving clock* and compares its readings with *two different stationary clocks* from the array of synchronized clocks that he has established in his own reference frame. As inspection of the clock hands shows, the interval $t_2 - t_1$ is greater than the interval $t'_2 - t'_1$. Observer S thus declares that the moving clock (C) is running slow by comparison with his own clocks.

in this array. These are the stationary S clocks that happen to coincide in position with the moving clock C both at the beginning and the end of the interval under consideration. The curved arrows in Fig. 2-4 identify the intervals $\Delta t'$ $(= t_1' - t_2')$ and Δt $(= t_1 - t_2)$ as measured by the S' observer and the S observer, respectively, and show clearly that the first of these intervals is shorter than the second.

From the Lorentz transformation equations (Table 2-3, Eq. 4) we have

$$\Delta t = \gamma(\Delta t' + v\,\Delta x'/c^2). \tag{2-13}$$

However, we have assumed that clock C is at rest in the S' frame, so that $\Delta x' = 0$. Putting this information into the above equation yields, for the time interval as measured by the S observer, $\Delta t = \gamma \Delta t' = \Delta t'/\sqrt{1 - \beta^2}$ which we can write as

$$t_2 - t_1 = \frac{t_2' - t_1'}{\sqrt{1 - \beta^2}}. \tag{2-14a}$$

Because $\sqrt{1 - \beta^2} < 1$ we see that the interval $(t_2 - t_1)$ is greater than the interval $(t_2' - t_1')$, as reflected by the shaded areas and the curved arrows on the clock faces in Fig. 2-4. A time interval measured on the single S' clock is recorded as a *longer* time interval by the S clocks. From the point of view of observer S, the moving S' clock appears slowed down, that is, it appears to run at a rate which is slow by the factor $\sqrt{1 - \beta^2}$. This result applies to all S' clocks observed from S, for in our proof the location x_0' was arbitrary.

In general, let us suppose that two events occur at a given place and let us represent by $\Delta\tau$ the time interval between them, measured on a clock at rest at that same place. Let Δt be the interval between these same two events, measured by an observer for whom the single clock is moving. We can identify $\Delta\tau$ with $(t_2' - t_1')$ in Eq. 2-14a and Δt with $(t_2 - t_1)$ and write

$$\Delta t = \frac{\Delta\tau}{\sqrt{1 - \beta^2}}. \qquad \text{(time dilation)} \tag{2-14b}$$

Equation 2-14b predicts that a clock runs the fastest (that is, $\Delta\tau$ will equal Δt rather than being smaller than Δt) when it is at rest (that is, when $\beta = 0$). Finally, our conclusions about clock rates are completely symmetrical, as required by the principle of relativity. We have seen that a clock fixed in the S' frame appears to run slow as seen by an observer in the S frame. It is equally true that a clock fixed in the S frame would appear to run slow to an S' observer.

It is common in relativity to speak of the frame in which the observed body is at rest as the *proper* frame. The length of a rod in such a frame is then called its *proper length* or, equivalently, its *rest length*. Likewise, the *proper time interval* is the time interval recorded by a clock attached to the observed body. The proper time interval can be thought of equivalently as the time interval between two events occurring at the same place in a given frame or the time interval measured by a single clock at one place. A nonproper time interval would be a time interval measured by two different clocks at two different places. Thus, in Eq. 2-14 (see also Fig. 2-4), the interval $\Delta\tau$ $[= (t_2' - t_1')]$ is a proper interval, being recorded by a single clock (C) fixed in the S' frame. The interval Δt $[= (t_2 - t_1)]$, on the other hand, is a nonproper interval, being recorded on two different clocks, separated by a distance $(x_2 - x_1)$ in the S frame. Later we shall define other proper quantities, such as *proper mass* (often called *rest mass*), and shall examine the usefulness of these concepts in relativity theory.

The Relativity of Clock Synchronization   A third consequence of the Lorentz transformation equations is this: Although clocks in a moving frame all appear to go at the same slow rate when observed from a stationary frame with respect to which the clocks move, *the moving clocks appear to differ from one another in their readings by a phase constant that depends on their location; that is, they appear to be unsynchronized.* This becomes evident at once from the transformation equation (see Table 2-2, Eq. 4)

$$t = \gamma \left(t' + \frac{vx'}{c^2} \right).$$

For consider an instant of time in the S frame, that is, a given value of t. Then, to satisfy this equation, $t' + vx'/c^2$ must have a definite fixed value. This means the greater is x' (that is, the farther away an S' clock is stationed on the x' axis), the smaller is t' (that is, the farther behind in time its reading appears to be). Hence, the moving clocks appear to be out of phase, or synchronization, with one another. We shall see in the next section that this is just another manifestation of the fact that two events that occur simultaneously in the S frame are not, in general, measured to be simultaneous in the S' frame, and vice versa.

The lack of synchronization of clocks in a moving reference frame, like the contraction of moving rods and the slowing down of moving clocks, is also reciprocal. If observer S declares that the clocks in her frame are synchronized but that those in the (moving) S' frame are not, observer S' can make a similar statement. He can, with equal validity, declare that the clocks in *his* reference frame are synchronized but that those in the (moving) S frame are not.

The Relativity of Simultaneity   In Section 2-1 we developed a physical argument (see Figs. 2-1 and 2-2) to show that observers in different reference frames cannot agree as to whether two events are simultaneous or not. Our argument there followed in a direct and logical way from the principle of the constancy of the speed of light. This concept of the relativity of simultaneity should also be contained in the Lorentz transformation equations, because they were derived from this same principle.

Consider two events, observed both by S and by S'. The time differences between these events, as reported by these two observers, are related by Eq. 2-13, which follows directly from the Lorentz transformation equations (Table 2-3, Eq. 4). Thus,

$$(t_2 - t_1) = \gamma(t_2' - t_1') + \left(\frac{\gamma v}{c^2} \right) (x_2' - x_1'). \tag{2-13}$$

We see at once that if S' finds the events to be simultaneous (that is, if $t_2' = t_1'$), then S will *not* find them so, *unless* S' also finds that the events occur in the same place, that is, unless $x_2' = x_1'$. Note also that, if the events occur at different places it is possible for S and S' to disagree even about the sequence of the two events. If $t_2' > t_1'$, for example, observer S' declares that event 1 (having the smaller time of occurrence) comes first. However, if x_1' is large enough (that is, if event 1 takes place far enough from the origin of the S' frame), the last term in Eq. 2-13 can be sufficiently negative to require that $t_1 > t_2$. This means that observer S will declare that event 2 comes first. All of this is in complete accord with what we have already learned in Section 2-1 by direct application of the principle of the constancy of the speed of light.

The Spacetime* Interval—An Invariant Quantity   In the limit of low speeds, where the Galilean transformation equations hold sufficiently well for all

---

* To stress the intimate interconnection between space and time in relativity theory it is common to treat "spacetime" as a single word, without a hyphen.

practical purposes, the length of a rod is a fixed quantity, having the same value for all inertial observers. The same thing is true for the time interval between any two events. Quantities that have the same value for all inertial observers are said be *invariant*. We see that under a Galilean transformation there are at least two invariant quantities, one involving space coordinates and one the time coordinate.

Under a Lorentz transformation, however, we have seen that the lengths of rods and the time intervals between events are precisely *not* invariant. Different observers obtain, by measurement, different numerical results. The question naturally arises: Is there any quantity involving the space and time coordinates of events that *is* invariant under a Lorentz transformation? The answer proves to be "yes."

Consider two events, viewed by observers S and S'. For simplicity, let us imagine that the events occur on the common $x - x'$ axis. We state without proof (see, however, Problem 41 and Supplementary Topic A) that the quantity

$$c^2(t_2 - t_1)^2 - (x_2 - x_1)^2,$$

which we can write as

$$(c\,\Delta t)^2 - (\Delta x)^2,$$

is invariant under a Lorentz transformation. By this we mean that although observer S' would find, by measurement, that $\Delta t' \neq \Delta t$ and that $\Delta x' \neq \Delta x$, he would also find that

$$(c\,\Delta t')^2 - (\Delta x')^2 = (c\,\Delta t)^2 - (\Delta x)^2. \tag{2-15}$$

It is not unexpected that in relativity theory any quantity found to be invariant would involve both space *and* time coordinates. The invariance of the quantity displayed above has been described in a remarkable *tour de force* by W. A. Shurcliff entitled *Special Relativity—A back-of-an-envelope summary in words of one syllable;* see page 55.

The invariant quantity displayed in Eq. 2-15 is called the (square of the) *spacetime interval* (or, more commonly, simply the *interval*) between the two events and is symbolized by $(\Delta s)^2$. Thus,

$$(\Delta s)^2 = (c\,\Delta t)^2 - (\Delta x)^2. \tag{2-16}$$

For any given pair of events, $(\Delta s)^2$ is the only measureable characteristic kinematic quantity for which all inertial observers would obtain the same numerical value. Inspection of Eq. 2-16 shows that, although the two terms on the right of that equation are always positive, $(\Delta s)^2$ itself can be positive, negative, or zero, depending on the relative magnitudes of those terms. Note also that if $(\Delta s)^2$ is positive, say, in any given frame, it will be positive in all frames; it is an invariant quantity.

For $(\Delta s)^2$ in Eq. 2-16 to be positive, the events must be such that $(c\,\Delta t)^2 > (\Delta x)^2$. It helps in visualizing such pairs of events to imagine them as spaced relatively close together along the x axis (Δx small) and/or separated by a relatively long time interval (Δt large). The spacetime interval associated with such event pairs is described as *timelike* because the term containing Δt predominates.

The *proper time* interval $\Delta \tau$ between any two events is defined as

$$\Delta \tau = \frac{\Delta s}{c} = \sqrt{(\Delta t)^2 - \left(\frac{\Delta x}{c}\right)^2}. \tag{2-17}$$

We see that $\Delta \tau$ is invariant because of the invariance of Δs. Equation 2-17 can be used to calculate the proper time interval between two events from measurements of Δt and Δx made in *any* inertial frame. The choice does not matter, because all observers will calculate the same result.

Special Relativity
Back-of-Envelope Summary
(in words of one syllable)

There is no one true frame (no test frame, no fixed frame) from which to judge time, space, mass, or speed. One frame is as good as the next. All of our great laws are the same in each frame. The speed of light, c, is the same in each frame, and there is no way to change this speed: no change in your speed, or in the speed of the light source, can change the speed at which light comes by you. No star, ship, man, rock, or speck of dust can reach the speed of light. This holds true for all things that can be slowed to a stop. None of these can quite match the speed of light.

If a car zooms by you at high speed, you find that its length is less (and its mass is more) than when it stands still in your frame. Also, its clock runs slow. Of course, the deal works both ways; the man in the car finds you to have more mass than when you and he are in the same frame.

If two guns (one to your right, one to your left) are fired, and you find they were fired at the same time, a man who went by at high speed may have found them not to have been fired at the same time. For him, the time span Δt may be large, not nil.

It turns out that time and space are in some sense joined. If two guns, one here and one there, are fired the time span Δt and the space span Δx are linked in a strange way. Take $\sqrt{c^2(\Delta t)^2 - (\Delta x)^2}$ and note that, though the Δt and Δx found from some frame A are not the same as the Δt and Δx found from some frame B, yet $\sqrt{c^2(\Delta t)^2 - (\Delta x)^2}$ is the same for each frame—in fact for all frames.

Personal correspondence from William A. Shurcliff, 19 Appleton Street, Cambridge, Massachusetts 02138.

The proper time interval $\Delta\tau$, however, will only be equal to the measured time interval Δt in a frame in which the two events occur at the same place, that is, in a frame in which $\Delta x = 0$; frame S' in Fig. 2-4 is such a frame. In all other frames, Eq. 2-17 shows that the measured time interval will be *greater than* the proper time interval, corresponding to the slowing down of moving clocks.

If the pair of events is not timelike, that is, if $(c\,\Delta t)^2$ in Eq. 2-16 is, in fact, *less than* $(\Delta x)^2$, then no frame can be found such that the events coincide in space; Δs, calculated from Eq. 2-16, is a mathematically imaginary quantity, and no physically meaningful proper time interval can be assigned using Eq. 2-17. In such cases the term $(\Delta x)^2$ in Eq. 2-16 dominates, and the pair of events is called *spacelike*. It helps in visualizing such pairs of events to imagine them as spaced relatively far apart (Δx large) and/or separated by a relatively short time interval (Δt small).

We can define a *proper distance* $\Delta\sigma$ for such events from Eq. 2-16. Thus,

$$\Delta\sigma = \sqrt{-(\Delta s)^2} = \sqrt{(\Delta x)^2 - (c\,\Delta t)^2}. \tag{2-18}$$

We see that $\Delta\sigma$ is also an invariant quantity, again because of the invariance of Δs. Equation 2-18 can be used to calculate the proper distance associated with any (spacelike) pair of events from measurements of Δx and Δt made in *any* inertial frame. Again, the choice does not matter, because all observers will get the same answer.

The proper distance $\Delta\sigma$, however, will only be equal to the measured distance Δx in a frame in which the measurements of the two endpoints of Δx are made simultaneously, that is, in a reference frame in which $\Delta t = 0$. Just as proper time has no physical meaning for spacelike pairs of events, proper distance has no physical meaning for timelike pairs of events; it proves impossible to find a reference frame in which timelike events occur simultaneously.

When $\Delta\tau$ is real the interval is called timelike; when $\Delta\sigma$ is real the interval is called spacelike. In the timelike region we can find a frame in which the two events occur at the same place, so that $\Delta\tau$ can be thought of as the time interval between the events in that frame. In the spacelike region we can find a frame in which the events are simultaneous, so that $\Delta\sigma$ can be thought of as the spatial interval between the events in that frame.

It is possible to find event pairs for which the two terms on the right side of Eq. 2-16 are exactly equal. Such a pair is neither timelike or spacelike but is, instead, identified as *lightlike*. The name derives from the fact that, in view of the equality just assumed, $\Delta x/\Delta t = c$. The significance of this is that if a light pulse leaves one event just as it occurs, it will arrive at the other event just as *it* occurs. As Eqs. 2-17 and 2-18 show, both the proper time $\Delta\tau$ and the proper distance $\Delta\sigma$ vanish for lightlike event pairs.

EXAMPLE 3.

Riding a Fast Electron. (a) An electron with a kinetic energy of 50 MeV ($= 5.0 \times 10^7$ eV), such as might be produced in a linear accelerator, can be shown (see Section 3-4) to have a speed parameter β of 0.999949. A beam of such electrons moves along the axis of an evacuated tube that is 10 m long, measured in a reference frame S fixed in the laboratory. Imagine a second frame S', attached to an electron in the beam and moving with it. How long would this tube seem to be to an observer in this frame? (b) Repeat the calculation for a 30-GeV ($= 3.0 \times 10^{10}$-eV) elec-

tron, such as might be produced in the Stanford Linear Accelerator. Such an electron can be shown to have a speed parameter β of 0.99999999986.

(a) In frame S' the electron is at rest and the tube is moving and thus is contracted in length according to Eq. 2-12, or

$$\Delta x' = \Delta x \sqrt{1 - \beta^2}$$
$$= (10 \text{ m}) \sqrt{1 - (0.999949)^2}$$
$$= 0.10 \text{ m} = 10 \text{ cm}.$$

(*b*) In attempting this calculation for a 30-GeV electron, a difficulty arises in that evaluating β^2 overloads the capacity of an ordinary hand calculator. For electrons in this extreme relativistic realm the quantity $(1 - \beta)$ is, in fact, both more significant and more manageable than β itself. To take advantage of this, let us put

$$\Delta x' = \Delta x \sqrt{1 - \beta^2}$$
$$= \Delta x \sqrt{(1 + \beta)(1 - \beta)}.$$

Now "$(1 + \beta)$" can, to an extremely good approximation, be replaced by "2," and $(1 - \beta)$ is $(1 - 0.99999999986) = 0.00000000014 = 1.4 \times 10^{-10}$. Thus,

$$\Delta x' = (10 \text{ m}) \sqrt{(2)(1.4 \times 10^{-10})}$$
$$= 1.7 \times 10^{-4} \text{ m} = 0.17 \text{ mm}.$$

Relativity theory is a very practical engineering matter in the design of high-energy accelerators. If it were not properly taken into account, those machines simply would not work. Purcell [16] has referred to those engineering ventures (particle accelerators, klystrons, high-voltage television tubes, electron microscopes, global navigation systems, and so forth) in which relativistic considerations play a role, as "high-gamma engineering."

EXAMPLE 4.

Finding the Proper Time Interval. Two events are viewed by an observer fixed in an inertial reference frame S. They occur along the x axis and are separated in space by Δx and in time by Δt. What is the proper time interval between these events? Consider three cases:

Event Pair	Δx	Δt
(*a*)	9.0×10^8 m	5.0 s
(*b*)	7.5×10^8 m	2.5 s
(*c*)	5.0×10^8 m	1.5 s

(*a*) Let us first calculate the proper time interval by finding the proper frame for these events. Recall that the proper time interval is the time interval recorded in a frame S' chosen so that both events in the pair occur at the same point when viewed from that frame. A (single) clock placed at that point reads intervals of proper time. Frame S' must move, as seen by S, at a speed v such that it covers the distance Δx (= 9.0×10^8 m) in a time Δt (= 5.0 s); in that way observer S' (and his clock) can be at both events. Thus

$$v = \frac{\Delta x}{\Delta t} = \frac{9.0 \times 10^8 \text{ m}}{5.0 \text{ s}} = 1.8 \times 10^8 \text{ m/s} = 0.60c.$$

The time interval $\Delta t'$ (= $\Delta \tau$) read by S' (on his single clock) will be related to the time interval Δt read by S (on two of his clocks, separated in space) by the time dilation formula. Thus, from Eq. 2-14*b* we have

$$\Delta \tau = \Delta t \sqrt{1 - \beta^2}$$
$$= (5.0 \text{ s}) \sqrt{1 - (0.60)^2}$$
$$= 4.0 \text{ s}.$$

As we have seen, it is not necessary to find the (proper) frame S' to calculate the proper time interval $\Delta \tau$. We can calculate it from measurements in *any* frame and will get the same answer because the proper time interval is an invariant quantity. From Eq. 2-17, then, using measurements in frame S, we have

$$\Delta \tau = \sqrt{(\Delta t)^2 - \left(\frac{\Delta x}{c}\right)^2}$$
$$= \sqrt{(5.0 \text{ s})^2 - \left(\frac{9.0 \times 10^8 \text{ m}}{3.0 \times 10^8 \text{ m/s}}\right)^2}$$
$$= 4.0 \text{ s},$$

in full agreement with the preceding direct calculation. Note that event pair (*a*), because it possesses a physically observable proper time, constitutes what we have called a *timelike* pair.

(*b*) For this event pair we have

$$v = \frac{\Delta x}{\Delta t} = \frac{7.5 \times 10^8 \text{ m}}{2.5 \text{ s}} = 3.0 \times 10^8 \text{ m/s} = c.$$

We thus see that the proper frame S' would have to move at the speed of light. The event pair is *lightlike* and the proper time interval, calculated by either of the approaches used in (*a*), is zero.

(*c*) For this event pair we have

$$v = \frac{\Delta x}{\Delta t} = \frac{5.0 \times 10^8 \text{ m}}{1.5 \text{ s}} = 3.3 \times 10^8 \text{ m/s} = 1.1c.$$

No reference frame can move so fast relative to another, so we conclude that a proper frame simply does not exist. There is no frame, that is, in which the two events would occur at the same place; they are separated in space for *all* inertial observers. Calculations of the proper time interval, carried out as in (*a*), would yield a mathematically imaginary result, devoid of physical meaning. We have called such event pairs *spacelike*.

EXAMPLE 5.

Two Observers View Two Events. In inertial system S an event occurs on the x axis at point A and then, 1.0 μs $(= 1.0 \times 10^{-6}$ s) later, an event occurs at point B farther out on the x axis. A and B are 600 m apart in frame S. (See Fig. 2-5.) (a) Does there exist another inertial frame S' in which the two events will be seen to occur simultaneously? If so, what are the magnitude and direction of the velocity of S' with respect to S? (b) What is the separation of events A and B in frame S'? Assume that S and S' are related as in Fig. 1-1. (c) What is the situation if the separation between the events in frame S is 100 m, all else remaining unchanged?

(a) From the Lorentz transformation equations (Table 2-3, Eq. 4'), we have

$$\Delta t' = \gamma(\Delta t - v \, \Delta x/c^2).$$

If the events are to be simultaneous in S', we must have $\Delta t' = 0$, which leads to

$$0 = \Delta t - \left(\frac{v}{c^2}\right) \Delta x$$

and thus to

$$v = \frac{c^2 \, \Delta t}{\Delta x}$$
$$= \frac{(3.0 \times 10^8 \text{ m/s})^2 (1.0 \times 10^{-6} \text{ s})}{600 \text{ m}}$$
$$= 1.5 \times 10^8 \text{ m/s} = 0.50c.$$

So, an observer in a system S' moving from A toward B at half the speed of light would record the events as simultaneous. We have seen that a pair of events for which such a frame can be found is described as *spacelike*. You can show, using the methods of Example 4, that it is not possible to assign a proper time interval to this pair of events.

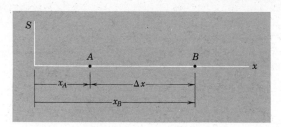

Figure 2-5. Example 5.

(b) Again, from the Lorentz transformation equations (Table 2-3; Eq. 1), we have

$$\Delta x = \gamma(\Delta x' + v \, \Delta t').$$

But the events are simultaneous in S', so $\Delta t' = 0$. Thus (see Eq. 2-10), the separation of the events in S' is

$$\Delta x' = \frac{\Delta x}{\gamma} = \Delta x \sqrt{1 - \beta^2}$$
$$= (600 \text{ m})\sqrt{1 - (0.50)^2} = 520 \text{ m}.$$

This is simply the familiar length contraction relationship (Eq. 2-12a), the length $AB \, (= \Delta x)$ being the length of a rod at rest in frame S.

(c) The relative velocity v, worked out as in (a), proves to be

$$v = \frac{c^2 \, \Delta t}{\Delta x} = \frac{(3.0 \times 10^8 \text{ m/s})^2 (1.0 \times 10^{-6} \text{ s})}{100 \text{ m}}$$
$$= 9.0 \times 10^8 \text{ m/s} = 3.0c,$$

which exceeds the speed of light. Thus there is no frame in which these events would be seen as simultaneous. They occur at different times for all inertial observers. Such events are called *timelike*. You can show, using the methods of Example 4, that it is possible to calculate a proper time interval for this pair of events.

2-4 THE LORENTZ EQUATIONS—A MORE PHYSICAL LOOK

Among the most important consequences of the Lorentz transformation equations are these: (1) Lengths perpendicular to the relative motion are measured to be the same in both frames; (2) the time interval indicated on a clock is measured to be longer by an observer for whom the clock is moving than by one at rest with respect to the clock; (3) lengths parallel to the relative motion are measured to be contracted compared to the rest lengths by the observer for whom the measured bodies are moving; and (4) two clocks, which are synchronized and separated in one inertial frame, are observed to be out of synchronism from another inertial frame. Here we rederive these features one at a time by thought experiments that focus on the measuring process.

Comparison of Lengths Perpendicular to the Relative Motion   Imagine two frames whose relative motion v is along a common x-x' axis. In each frame an observer has a stick extending up from the origin along her vertical (y and y') axis, which she measures to have a (rest) length of exactly 1 m, say. As these observers approach and pass each other, we wish to determine whether or not, when the origins coincide, the top ends of the sticks coincide. We can arrange to have the sticks mark each other permanently by a thin pointer at the very top of each (for example, a razor blade or a paintbrush bristle) as they pass one another. (We displace the sticks very slightly so that they will not collide, always keeping them parallel to the vertical axis.) Notice that the situation is perfectly symmetrical. Each observer claims that her stick is a meter long, each sees the other approach with the same speed v, and each claims that her stick is perpendicular to the relative motion. Furthermore, the two observers must agree on the result of the measurements because they agree on the simultaneity of the measurements (the measurements occur at the instant the origins coincide). After the sticks have passed, either each observer will find her pointer marked by the other's pointer, or else one observer will find a mark below her pointer, the other observer finding no mark. That is, either the sticks are found to have the same length by both observers, or else there is an absolute result, agreed on by both observers, that the same one stick is shorter than the other. That each observer finds the other stick to be the *same* length as hers follows at once from the contradiction any other result would indicate with the relativity principle. Suppose, for example, that observer S finds that the S' stick has left a mark (below her pointer) on her stick. She concludes that the S' stick is *shorter* than hers. This is an absolute result, for the S' observer will find no mark on her stick and will conclude *also* that her stick is shorter. If, instead, the mark was left on the S' stick, then *each* observer would conclude that the S stick is the *shorter* one. In either case, this would give us a physical basis for preferring one frame over another, for although all the conditions are symmetrical, the results would be unsymmetrical—a result that contradicts the principle of relativity. That is, the laws of physics would not be the same in each inertial frame. We would have a property for detecting absolute motion, in this case; a shrinking stick would mean absolute motion in one direction and a stretching stick would mean absolute motion in the other direction. Hence, to conform to the relativity postulate, we conclude that the length of a body (or space interval) transverse to the relative motion is measured to be the same by all inertial observers.

Comparison of Time-Interval Measurements   A simple thought experiment that reveals in a direct way the quantitative relation connecting the time interval between two events as measured from two different inertial frames is the following. Imagine a passenger sitting on a train that moves with uniform velocity v with respect to the ground. The experiment will consist of turning on a flashlight aimed at a mirror directly above on the ceiling and measuring the time it takes the light to travel up and be reflected back down to its starting point. The situation is illustrated in Fig. 2-6. The passenger, who has a wristwatch, sees the light ray follow a strictly vertical path (Fig. 2-6a) from A to B to C and times the event by her clock (watch). This interval $\Delta\tau$ is a proper time interval, measured by a single clock at one place, the departure and arrival of the light ray occurring at the same place in the passenger's (S') frame. Another observer, fixed to the ground (S) frame, sees the train and passenger move to the right during this interval. He will measure the time interval Δt from the readings on two clocks, stationary in his frame, one being at the position the experiment began (turning on of flashlight)

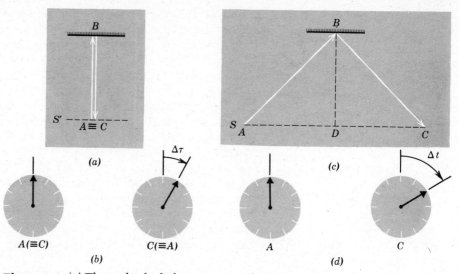

Figure 2-6. (a) The path of a light ray as seen by a passenger in the S' frame. B is a mirror on the ceiling. A and C are the *same* point, namely, the bulb of the flashlight, in this frame. (b) The readings of the passenger's clock at the start and at the end of the event, showing the time interval Δt' (= Δτ) on the (single) S' clock, stationary in this frame. (c) The path of a light ray as seen by a ground observer, in the S frame. A and C are *different locations* of the flashbulb at the start and at the end of the event, as the train moves to the right with speed v, in this frame. (d) Readings on the *two* (synchronized) *clocks*, stationary in the S frame and located at the start (A) of the event and the end (C) of the event, showing the time interval Δt.

and a second at the position the experiment ended (arrival of light to flashlight). Hence, he compares the reading of one moving clock (the passenger's watch) to the readings on two stationary clocks. For the S observer, the light ray follows the oblique path shown in Fig. 2-6c. Thus, the observer on the ground measures the light as traveling a greater distance than does the passenger (we have already seen that the transverse distance is the same for each observer). Because the speed of light is the same in both frames, the ground observer sees more time elapse between the departure and the return of the ray of light than does the passenger. He therefore concludes that the passenger's clock runs slow (see Fig. 2-6b and 2-6d). The quantitative result follows at once from the Pythagorean theorem, for

$$\Delta\tau = \frac{2BD}{c} \quad \text{and} \quad \Delta t = \frac{AB + BC}{c} = \frac{2AB}{c};$$

but

$$(BD)^2 = (AB)^2 - (AD)^2,$$

so that

$$\frac{\Delta\tau}{\Delta t} = \frac{BD}{AB} = \frac{\sqrt{(AB)^2 - (AD)^2}}{AB}$$

$$= \sqrt{1 - \left(\frac{AD}{AB}\right)^2} = \sqrt{1 - \frac{v^2}{c^2}}.$$

Here AD is the horizontal distance traveled at speed v during the time the light traveled with speed c along the hypotenuse AB. This result can be written as

$$\Delta t = \frac{\Delta\tau}{\sqrt{1 - \beta^2}}$$

and is identical to Eq. 2-14b, derived earlier in a more formal way.

Comparison of Lengths Parallel to the Relative Motion The simplest deduction of the length contraction uses the time dilation result just obtained and shows directly that length contraction is a necessary consequence of time dilation. Imagine, for example, that two different inertial observers, one sitting on a train moving through a station with uniform velocity v and the other at rest in the station, want to measure the length of the station's platform. The ground observer, for whom the platform is at rest, measures the length to be l_0 and claims that the passenger covered this distance in a time l_0/v. This time, Δt, is a nonproper time, for the events observed (passenger passes back end of platform, passenger passes front end of platform) occur at two different places in the ground (S) frame and are necessarily timed by two different clocks. The passenger, however, observes the platform approach and recede and finds the two events to occur at the same place in her (S') frame. That is, her clock (wristwatch) is located at each event as it occurs. She measures a proper time interval $\Delta\tau$, which, as we have just seen (Eq. 2-14b), is related to Δt by $\Delta\tau = \Delta t\sqrt{1 - \beta^2}$. But $\Delta t = l_0/v$, so that $\Delta\tau = l_0\sqrt{1 - \beta^2}/v$. The passenger claims that the platform moves with the same speed v relative to her so that she would measure the distance from back to front of the platform as $v\,\Delta\tau$. Hence, the length of the platform to her is $l = v\,\Delta\tau = l_0\sqrt{1 - \beta^2}$, which is precisely Eq. 2-12b, the length-contraction result. Thus, a body of rest length l_0 is measured to have a length $l_0\sqrt{1 - \beta^2}$ parallel to the relative motion in a frame in which the body moves with speed v.

The Phase Difference in the Synchronization of Clocks The Lorentz transformation equation for the time variable (see Table 2-2, Eq. 4) can be written as

$$t = \gamma\left(t' + \frac{vx'}{c^2}\right) = \gamma\left(t' + \frac{\beta x'}{c}\right).$$

Here we wish to give a physical interpretation of the $\beta x'/c$ term, which we call the *phase difference*. We shall synchronize two clocks in one frame and examine what an observer in another frame concludes about the process.

Imagine that we have two clocks, A and B, at rest in the S' frame. Their separation is L' in this frame. We set off a flashbulb, which is at the exact midpoint, and instruct two assistants, one at each clock, to set them to read $t' = 0$ when the light reaches them (see Fig. 2-7a). This is an agreed-upon procedure for synchronizing two separated clocks (see Section 2-1). We now look at this synchronization process as seen by an observer in the S frame, for whom the clocks A and B move to the right (see Fig. 2-7b) with speed v. The S observer has at her disposal her own fixed array of synchronized clocks, so she can assign times of occurrence to various events.

To the S observer, the separation of the two clocks will be $L'\sqrt{1 - \beta^2}$. She observes the following sequence of events. The flash goes off and leaves the midpoint traveling in all directions with a speed c. As the wavefront expands at the rate c, the clocks move to the right at the rate v. Clock A intercepts the flash first, before B, and the assistant at clock A sets his clock at $t' = 0$ (third picture in sequence). Hence, as far as the S observer is concerned, the assistant at A sets his clock to zero time *before* the assistant at B does, and the setting of the two primed clocks does not appear simultaneous to her. Here again we see the relativity of

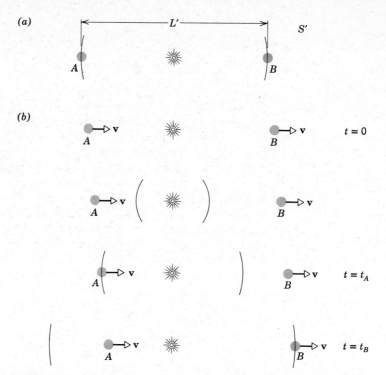

Figure 2-7. (a) A flash sent from the midpoint of clocks A and B, at rest in the S' frame a distance L' apart, arrives simultaneously at A and at B. (b) The sequence of events as seen from the S frame, in which the clocks are a distance L apart and move to the right with speed v.

simultaneity; that is, the clocks at rest in the primed frame are *not* synchronized according to the unprimed observer, who uses exactly the same procedure to synchronize her own clocks.

By how much do the two S' clocks differ in their readings according to the S observer? Let $t = 0$ be the time S sees the flash go off. Then, when the light pulse meets clock A, at $t = t_A$, we have

$$ct_A = (L'/2)\sqrt{1 - \beta^2} - vt_A$$

or

$$t_A = \frac{(L'/2)\sqrt{1 - \beta^2}}{c + v} = \left(\frac{L'}{2c}\right)\frac{\sqrt{1 - \beta^2}}{1 + \beta}.$$

That is, the distance the pulse travels to meet A is less than their initial separation by the distance A travels to the right during this time. When the light pulse later meets clock B (fourth picture in sequence), at $t = t_B$, we have

$$ct_B = \frac{L'}{2}\sqrt{1 - \beta^2} + vt_B$$

or

$$t_B = \frac{(L'/2)\sqrt{1 - \beta^2}}{c - v} = \left(\frac{L'}{2c}\right)\frac{\sqrt{1 - \beta^2}}{1 - \beta}.$$

The distance the pulse travels to meet B is greater than their initial separation by the distance B travels to the right during this time. As measured by the clocks in S, therefore, the time interval between the setting of the primed clocks (A and B) is

$$\Delta t = t_B - t_A$$
$$= \left(\frac{L'\sqrt{1 - \beta^2}}{2c}\right)\left(\frac{1}{1 - \beta} - \frac{1}{1 + \beta}\right)$$
$$= \frac{L'\beta}{c}\frac{1}{\sqrt{1 - \beta^2}}.$$

During this interval, however, S observes clock A to run slow by the factor $\sqrt{1 - \beta^2}$ (for "moving clocks run slow"), so to observer S it will read

$$\Delta t' = \Delta t\sqrt{1 - \beta^2} = \frac{L'\beta}{c}$$

when clock B is set to read $t' = 0$.

The result is that the S observer finds the S' clocks to be out of synchronization, with clock A reading *ahead* in time by an amount $L'\beta/c$ ($= x'\beta/c$). The greater the separation L' of the clocks in the primed frame, the further behind in time is the reading of the B clock as observed at a given instant from the unprimed frame. This is in exact agreement with the Lorentz transformation equation for the time.

Hence, all the features of the Lorentz transformation equations, which we derived in a formal way directly from the postulates of relativity in Section 2-2, can be derived more physically from the measurement processes, which were, of course, chosen originally to be consistent with those postulates.

EXAMPLE 6.

Simultaneity at Low Speeds. Why is the fact that simultaneity is not an absolute concept an unexpected result to the classical mind? It is because the speed of light has such a large value compared to ordinary speeds.

Consider these two cases, which are symmetrical in terms of an interchange of the space and time coordinates. *Case 1:* S' observes that two events occur at the same place but are separated in time; S will then declare that the two events occur in different places. *Case 2:* S' observes that two events occur at the same time but are separated in space; S will then declare that the two events occur at different times.

Case 1 is readily acceptable on the basis of daily experience. If a person (S') on a moving train winks and then—ten minutes later—winks again, these events occur at the same place on *his* reference frame (the train). A ground observer (S), however, would assert that these same events occur at different places in *his* reference system (the ground). Case 2, although true, cannot be easily supported on the basis of daily experience. Suppose that S', seated at the center of a moving railroad car, observes that two children, one at each end of the car, wink simultaneously. The ground observer S, watching the railroad car go by, would assert (if he could make precise enough measure-

ments) that the child in the back of the car winked a little before the child in the front of the car did. The fact that the speed of light is so high compared to the speeds of familiar large objects makes Case 2 less intuitively reasonable than Case 1, as we now show.

(*a*) In Case 1, assume that the time separation in S' is 10 min; what is the distance separation observed by S? (*b*) In Case 2, assume that the distance separation in S' is 25 m; what is the time separation observed by S? Take $v = 20.0$ m/s, which corresponds to 45 mi/h or $\beta = v/c = 6.7 \times 10^{-8}$.

(*a*) From Table 2-3, Eq. 1 we have

$$x_2 - x_1 = \frac{x_2' - x_1'}{\sqrt{1 - \beta^2}} + \frac{v(t_2' - t_1')}{\sqrt{1 - \beta^2}}.$$

We are given that $x_2' = x_1'$ and $t_2' - t_1' = 10$ min, so

$$x_2 - x_1 = \frac{(20.0 \text{ m/s})(600 \text{ s})}{\sqrt{1 - (6.7 \times 10^{-8})^2}} = 12000 \text{ m} = 12 \text{ km.}$$

This result is readily accepted. Because the denominator above is unity for all practical purposes, the result is even numerically what we would expect from the Galilean equations.

(b) From Table 2-3, Eq. 4 we have

$$t_2 - t_1 = \frac{t_2' - t_1'}{\sqrt{1 - \beta^2}} + \frac{(v/c^2)(x_2' - x_1')}{\sqrt{1 - \beta^2}}.$$

We are given that $t_2' = t_1'$ and that $x_2' - x_1' = 25$ m, so

$$t_2 - t_1 = \frac{[(20 \text{ m/s})/(3.0 \times 10^8 \text{ m/s})^2](25 \text{ m})}{\sqrt{1 - (6.7 \times 10^{-8})^2}}$$
$$= 5.6 \times 10^{-15} \text{ s}.$$

The result is *not* zero, a value that would have been expected by classical physics, but the time interval is so short that it would be very hard to show experimentally that it really was not zero.

If we compare the expressions for $x_2 - x_1$ and for $t_2 - t_1$ above, we see that, whereas v appears as a factor in the second term of the former, v/c^2 appears in the latter. Thus the relatively high value of c puts Case 1 within the bounds of familiar experience but puts Case 2 out of these bounds.

In the following example we consider the realm wherein relativistic effects are easily observable.

EXAMPLE 7

The Decay of Moving Pions. Among the particles of high-energy physics are charged pions, particles of mass between that of the electron and the proton and of positive or negative electronic charge. They can be produced by bombarding a suitable target in an accelerator with high-energy protons, the pions leaving the target with speeds close to that of light. It is found that the pions are radioactive and, when they are brought to rest, their half-life is measured to be 1.8×10^{-8} s. That is, half of the number present at any time have decayed 1.8×10^{-8} s later. A collimated pion beam, leaving the accelerator target at a speed of $0.99c$, is found to drop to half its original intensity 38 m from the target.

(a) Are these results consistent?

If we take the half-life to be $1.8 \text{ s} \times 10^{-8}$ s and the speed to be 2.97×10^8 m/s $(=0.99c)$, the distance traveled over which half the pions in the beam should decay is

$$d = vt = 2.97 \times 10^8 \text{ m/s} \times 1.8 \times 10^{-8} \text{ s} = 5.3 \text{ m}.$$

This appears to contradict the direct measurement of 38 m.

(b) Show how the time dilation accounts for the measurements.

If the relativistic effects did not exist, then the half-life would be measured to be the same for pions at rest and pions in motion (an assumption we made in part a). In relativity, however, the nonproper and proper half-lives are related by Eq. 2-14b, or

$$\Delta t = \frac{\Delta \tau}{\sqrt{1 - \beta^2}}.$$

The proper time in this case is 1.8×10^{-8} s, the time interval measured by a clock attached to the pion, that is, at one place in the rest frame of the pion. In the laboratory frame, however, the pions are moving at high speeds and the time interval there (a nonproper one) will be measured to be larger (moving clocks appear to run slow). The nonproper half-life, measured by two different clocks in the laboratory frame, would then be

$$\Delta t = \frac{1.8 \times 10^{-8} \text{ s}}{\sqrt{1 - (0.99)^2}} = 1.28 \times 10^{-7} \text{ s}.$$

This is the half-life appropriate to the laboratory reference frame. Pions that live this long, traveling at a speed $0.99c$, would cover a distance

$$d = 0.99c \times \Delta t = 2.97 \times 10^{-8} \text{ m/s} \times 1.28 \times 10^{-7} \text{ s}$$
$$= 38 \text{ m},$$

exactly as measured in the laboratory.

(c) Show how the length contraction accounts for the measurements.

In part a we used a length measurement (38 m) appropriate to the laboratory frame and a time measurement $(1.8 \times 10^{-8}$ s) appropriate to the pion frame and incorrectly combined them. In part b we used the length (38 m) and time $(1.28 \times 10^{-7}$ s) measurements appropriate to the laboratory frame. Here we use length and time measurements appropriate to the pion frame.

We already know the half-life in the pion frame, that is, the proper time 1.8×10^{-8} s. What is the distance covered by the pion beam during which its intensity falls to half its original value? If we were sitting on the pion, the laboratory distance of 38 m would appear much shorter to us because the laboratory moves at a speed $0.99c$ relative to us (the pion). In fact, we would measure the distance

$$d' = d \sqrt{1 - \beta^2} = 38 \sqrt{1 - (0.99)^2} \text{ m}.$$

The time elapsed in covering this distance is $d'/0.99c$ or

$$\Delta \tau = \frac{38 \text{ m} \sqrt{1 - (0.99)^2}}{0.99c} = 1.8 \times 10^{-8} \text{ s},$$

exactly the measured half-life in the pion frame.

Thus, depending on which frame we choose to make measurements in, this example illustrates the physical reality of either the time-dilation or the length-contraction predictions of relativity. Each pion carries its own clock, which determines the proper time τ of decay, but the decay time observed by a laboratory observer is much greater. Or, expressed equivalently, the moving pion sees the laboratory distances contracted and in its proper decay time can cover laboratory distances greater than those measured in its own frame.

Notice that in this region of $v \cong c$ the relativistic effects are large. There can be no doubt whether, in our example, the distance is 38 m or 5.3 m. If the proper time were applicable to the laboratory frame, the time $(1.28 \times 10^{-7} \text{ s})$ to travel 38 m would correspond to more than seven half-lives (that is, $1.28 \times 10^{-7} \text{ s}/1.8 \times 10^{-8} \text{ s} \cong 7$). Instead of the beam being reduced to half its original intensity, it would be reduced to $(1/2)^7$ or $1/128$ its original intensity in travelling 38 m. Such differences are very easily detectable. See Problems 23–27 and also [2] for other examples in which relativistic considerations are a central part of the problem at hand.

EXAMPLE 8.

Two Spaceships Pass Each Other. Two spaceships, each of proper length 100 m, pass near one another heading in opposite directions; see Fig. 2-8. An astronaut at the front of one ship (S) measures a time interval of 2.50×10^{-6} s for the second ship (S') to pass her. (a) What is the relative speed v of the two ships? (b) What time interval is measured on ship S for the front of ship S' to pass from the front to the back of S? In the figure the front and back ends of spaceship S are labeled A and B, respectively, those of spaceship S' being labeled A' and B'. Let AA' mean the coincidence of points A and A', AB' the coincidence of A and B', and so forth.

(a) The time interval between the occurrences of events AA' and AB', measured by a single clock at A, is a proper time interval and is given as 2.50×10^{-6} s $(= \Delta\tau)$. Each ship has a proper length of 100 m $(= L_0)$ so the space interval between A' and B' that the astronaut at A measures is the contracted length $L_0 \sqrt{1 - \beta^2}$, appropriate, as Eq. 2-12b shows, for an object of rest length L_0 moving at a speed v. Therefore,

$$v(= \beta c) = \frac{L_0 \sqrt{1 - \beta^2}}{\Delta\tau}$$

or

$$\left(\frac{\beta\Delta\tau c}{L_0}\right)^2 = 1 - \beta^2.$$

This can be written as

$$\begin{aligned}
\frac{1}{\beta^2} &= \left(\frac{\Delta\tau c}{L_0}\right)^2 + 1 \\
&= \left[\frac{(2.50 \times 10^{-6} \text{ s})(3.00 \times 10^8 \text{ m/s})}{100 \text{ m}}\right]^2 + 1 \\
&= 57.25,
\end{aligned}$$

which yields

$$\beta = \frac{1}{\sqrt{57.25}} = 0.132$$

and

$$v = \beta c = (0.132)(3.00 \times 10^8 \text{ m/s}) = 3.96 \times 10^7 \text{ m/s}.$$

(b) We want to find the time interval between events AA' and BA' measured by two clocks in spaceship S, one at A and one at B. This is a nonproper time interval Δt, read off as the difference in arrival times of A' at the clocks at A and B. Since the separation of these clocks is L_0 in spaceship S and A' moves at speed v relative to this ship, we have

$$\Delta t = \frac{L_0}{v} = \frac{100 \text{ m}}{3.96 \times 10^7 \text{ m/s}} = 2.53 \times 10^{-6} \text{ s}.$$

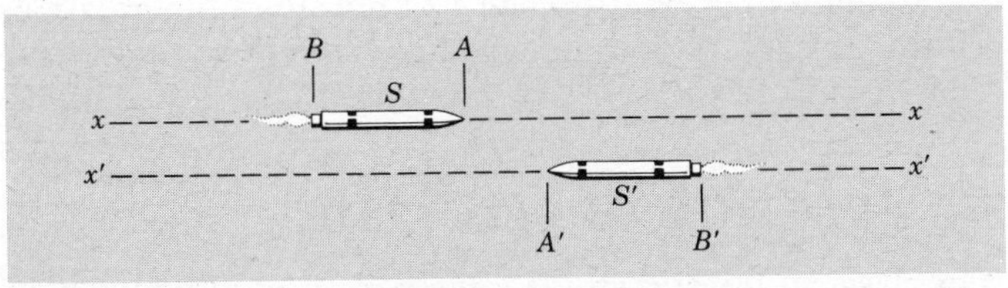

Figure 2-8. Example 8.

2-5 *THE OBSERVER IN RELATIVITY*

There are many shorthand expressions in relativity that can easily be misunderstood by the uninitiated. Thus the phrase "moving clocks run slow" means that a clock moving at a constant velocity relative to an inertial frame containing synchronized clocks will be found to run slow *when timed by those clocks.* We compare *one moving clock* with *two synchronized stationary clocks.* Those who assume that the phrase means anything else often encounter difficulties.

Similarly, we often refer to "an observer." The meaning of this term also is quite definite, but it can be misinterpreted. *An observer is really an infinite set of recording clocks distributed throughout space, at rest and synchronized with respect to one another.* The space-time coordinates of an event (x, y, z, t) are recorded by the clock at the location (x, y, z) of the event at the time (t) it occurs. Measurements thus recorded throughout space-time (we might call them local measurements) are then available to be picked up and analyzed by an experimenter. Thus, the observer can also be thought of as the experimenter who collects the measurements made in this way. Each inertial frame is imagined to have such a set of recording clocks, or such an observer. The relations between the space-time coordinates of a physical event measured by one observer (S) and the space-time coordinates of the *same* physical event measured by another observer (S') are the equations of transformation.

A misconception of the term "observer" arises from confusing "measuring" with "seeing." For example, it had been commonly assumed for some time that the relativistic length contraction would cause rapidly moving objects to appear to the eye to be shortened in the direction of motion. The location of all points of the object measured at the same time would give the "true" picture according to our use of the term "observer" in relativity. But, in the words of V. F. Weisskopf [Ref. 3]:

> When we see or photograph an object, we record light quanta emitted by the object when they arrive simultaneously at the retina or at the photographic film. This implies that these light quanta have *not* been emitted simultaneously by all points of the object. The points further away from the observer have emitted their part of the picture earlier than the closer points. Hence, if the object is in motion, the eye or the photograph gets a distorted picture of the object, since the object has been at different locations when different parts of it have emitted the light seen in the picture.

To make a comparison with the relativistic predictions, therefore, we must first allow for the time of flight of the light quanta from the different parts of the object. Without this correction, we see a distortion due to *both* the optical *and* the relativistic effects. Circumstances sometimes exist in which the object appears to have suffered no contraction at all. Under other special circumstances the Lorentz contraction can be seen unambiguously (see Refs. 4 and 5). But the term "observer" does *not* mean "viewer" in relativity, and we shall continue to use it only in the sense of "measurer" described above.

2-6 *THE RELATIVISTIC ADDITION OF VELOCITIES*

In classical physics, if we have a train moving with a velocity **v** with respect to the ground and a passenger on the train moves with a velocity **u′** with respect to the train, then the passenger's velocity relative to the ground **u** is just the vector sum of the two velocities (see Eq. 1-5); that is,

$$\mathbf{u} = \mathbf{u}' + \mathbf{v}. \qquad (2\text{-}19)$$

This is simply the classical, or Galilean, velocity addition theorem (See *Physics,* Part I, Sec. 4-6). How do the velocities add in special relativity theory?

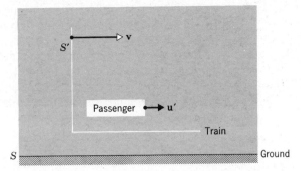

Figure 2-9. A schematic view of the system used in deriving the equations for the relativistic addition of velocities.

Consider, for the moment, the special case wherein all velocities are along the common x-x′ direction of two inertial frames S and S'. Let S be the ground frame and S' the frame of the train, whose speed relative to the ground is v (see Fig. 2-9). A passenger is walking along the aisle toward the front of the train with a speed u' relative to the train. His position on the train as time goes on can be described by $x' = u't'$. What is the speed of the passenger observed from the ground? Using the Lorentz transformation equations (Table 2-2, Eqs. 1′ and 4′), we have

$$x' = \gamma(x - vt) = u't' \quad \text{and} \quad t' = \gamma\left(t - \frac{vx}{c^2}\right).$$

Combining these yields

$$x - vt = u'\left(t - \frac{vx}{c^2}\right),$$

which can be written as

$$x = \frac{(u' + v)}{(1 + u'v/c^2)}t. \tag{2-20}$$

If we call the passenger's speed relative to ground u, then his ground location as time goes on is given by $x = ut$. Comparing this to Eq. 2-20, we obtain

$$u = \frac{u' + v}{1 + u'v/c^2}. \tag{2-21}$$

This is the *relativistic*, or Einstein *velocity addition theorem*.

Note that if $u' = 0$, Eq. 2-21 gives $u = v$, an expected result if the passenger stops walking. If $v = 0$, we find that $u = u'$, also an expected result if the train stops. If u' and v are very small compared to c, Eq. 2-21 reduces to the classical result, Eq. 2-19, $u = u' + v$, for then the second term in the denominator of Eq. 2-21 is negligible compared to one. On the other hand, if $u' = c$, it always follows that $u = c$ no matter what is the value of v. Of course, $u' = c$ means that our "passenger" is a light pulse, and we know that an assumption used to derive the transformation formulas was exactly this result, that is, that all observers measure the same speed c for light. Formally, we get, with $u' = c$,

$$u = \frac{c + v}{1 + cv/c^2} = \frac{c + v}{c(c + v)}c^2 = c.$$

Hence, any velocity (less than c) relativistically added to c gives a resultant c. In this sense, c plays the same role in relativity that an infinite velocity plays in the classical case.

The Einstein velocity addition theorem can be used to explain the observed result of the experiments designed to test the various emission theories of Chapter 1. The basic result of these experiments is that the velocity of light is independent of the velocity of the source (see Section 1-8). We have seen that this is a basic postulate of relativity, so we are not surprised that relativity yields agreement with these experiments. If, however, we merely looked at the formulas of relativity, unaware of their physical origin, we could obtain this specific result from the velocity addition theorem directly. Let the source be the S' frame. In that frame the pulse (or wave) of light has a speed c in vacuum according to the emission theories. Then, the pulse (or wave) speed measured by the S observer, for whom the source moves, is given by Eq. 2-21, and is also c. That is, $u = c$ when $u' = c$, as shown above.

It follows also from Eq. 2-21 that the addition of two velocities, each smaller than c, cannot exceed the velocity of light.

N. David Mermin has given a simple and convincing proof of Eq. 2-21 whose only nonclassical feature is the postulate of the constancy of the speed of light. The proof follows directly from this postulate and does not invoke either the Lorentz transformation equations, or the time dilation or the length contraction results [15].

EXAMPLE 9.

The Relative Speed of Two Fast Electrons. In Example 3 of Chapter 1, we found that when two electrons leave a radioactive sample in opposite directions, each having a speed $0.67c$ with respect to the sample, the speed of one electron relative to the other is $1.34c$ according to classical physics. What is the relativistic result?

We may regard one electron as the S frame, the sample as the S' frame, and the other electron as the object whose speed in the S frame we seek (see Fig. 1-3). Then

$$u' = 0.67c \qquad v = 0.67c$$

$$\text{and} \quad u = \frac{u' + v}{1 + u'v/c^2} = \frac{(0.67 + 0.67)c}{1 + (0.67)^2} = \frac{1.34}{1.45}c = 0.92c.$$

The speed of one electron relative to the other is less than c.

Does the relativistic velocity addition theorem alter the numerical result of Example 1 of Chapter 1? Explain.

EXAMPLE 10.

Relativity Explains the Fresnel Drag Coefficient. Show that the Einstein velocity addition theorem leads to the observed Fresnel drag coefficient of Eq. 1-12.

In this case, v_w is the velocity of water with respect to the apparatus and c/n is the velocity of light relative to the water. That is, in Eq. 2-21, we have

$$u' = \frac{c}{n} \quad \text{and} \quad v = v_w.$$

Then, the velocity of light relative to the apparatus is

$$u = \frac{c/n + v_w}{1 + v_w/nc}.$$

For $v_w/nc \ll 1$ (in the experiment its value was 1.8×10^{-8}), we can neglect terms of second-order in this quantity so that, using the binomial expansion, we have

$$u = \left(\frac{c}{n} + v_w\right)\left(1 + \frac{v_w}{nc}\right)^{-1}$$

$$= \left(\frac{c}{n} + v_w\right)\left(1 - \frac{v_w}{nc} + \cdots\right)$$

$$\cong \frac{c}{n} + v_w\left(1 - \frac{1}{n^2}\right).$$

This is exactly Eq. 1-12, the observed first-order effect. Notice that there is no need to assume any "drag" mechanism, or to invent theories on the interaction between matter and the "ether." The result is an inevitable consequence of the velocity addition theorem and illustrates the powerful simplicity of relativity.

It is interesting and instructive to note that there *are* speeds in excess of *c*. Although matter or energy (that is, signals) cannot have speeds greater than *c*, certain kinematical processes *can* have superlight speeds [6]. For example, the succession of points of intersection of the blades of a giant scissors, as the scissors is rapidly closed, may be generated at a speed greater than *c*. Here geometric points are involved, the motion being an illusion, whereas the material objects involved (atoms in the scissors blades, for example) always move at speeds less than *c*. Other similar examples are the succession of points on a fluorescent screen as an electron beam sweeps across the screen, or the light of a searchlight beam sweeping across the cloud cover in the sky. The electrons, or the light photons, which carry the energy, move at speeds not exceeding *c*. There is no contradiction with relativity theory in any of these situations.

It has been proposed [7] that there can exist, or be created, particles with speeds *always* greater than *c*. These hypothetical particles were named *tachyons*, from the Greek word for swift. This suggestion had the appeal of symmetry; that is, the existence of tachyons would allow us to classify particles by speed: normal particles that travel with $v < c$ always; photons and massless particles, for which $v = c$ always; and particles (tachyons) with $v > c$ always. Objections that infinite energy would be needed to create such particles, and certain causal paradoxes, can be resolved so that there are no compelling arguments against the existence of tachyons. Experimental evidence to date, however, suggests that their existence is unlikely.

Thus far, we have considered only the transformation of velocities parallel to the direction of relative motion of the two frames of reference (the x-x' direction). To signify this, we should put x subscripts on u and u' in Eq. 2-21, obtaining

$$u_x = \frac{u'_x + v}{1 + u'_x(v/c^2)}. \tag{2-22a}$$

For velocity components perpendicular to the direction of relative motion the result is more involved. Imagine that an object is observed to be at positions y_1 and y_2 in frame S at times t_1 and t_2, respectively. Its y component of velocity in S is then $u_y = (y_2 - y_1)/(t_2 - t_1)$ or $\Delta y/\Delta t$. To find its y component of velocity in frame S', we start from the Lorentz transformation equations (Table 2-3, Eqs. 2 and 4), writing

$$\Delta y = \Delta y'$$

and

$$\Delta t = \gamma\left(\Delta t' + \frac{v\,\Delta x'}{c^2}\right)$$

so that

$$u_y = \frac{\Delta y}{\Delta t} = \frac{\Delta y'}{\gamma(\Delta t' + v\,\Delta x'/c^2)}.$$

Substituting for γ and rearranging leads to

$$u_y = \frac{(\Delta y'/\Delta t')\,\sqrt{1 - v^2/c^2}}{1 + (v/c^2)(\Delta x'/\Delta t')}.$$

But

$$\frac{\Delta y'}{\Delta t'} = v'_y \quad \text{and} \quad \frac{\Delta x'}{\Delta t'} = v'_x,$$

Table 2-4
THE RELATIVISTIC VELOCITY
TRANSFORMATION EQUATIONS

$u_x' = \dfrac{u_x - v}{1 - u_x v/c^2}$	$u_x = \dfrac{u_x' + v}{1 + u_x' v/c^2}$
$u_y' = \dfrac{u_y \sqrt{1 - v^2/c^2}}{1 - u_x v/c^2}$	$u_y = \dfrac{u_y' \sqrt{1 - v^2/c^2}}{1 + u_x' v/c^2}$
$u_z' = \dfrac{u_z \sqrt{1 - v^2/c^2}}{1 - u_x v/c^2}$	$u_z = \dfrac{u_z' \sqrt{1 - v^2/c^2}}{1 + u_x' v/c^2}$

so

$$u_y = \frac{u_y' \sqrt{1 - v^2/c^2}}{1 + v\, u_x'/c^2}, \qquad (2\text{-}22b)$$

which is the relationship we seek.

In just the same way we find, for u_z,

$$u_z = \frac{u_z' \sqrt{1 - v^2/c^2}}{1 + v u_x'/c^2}. \qquad (2\text{-}22c)$$

In Table 2-4 we summarize the relativistic velocity transformation equations. The inverse relations were found by merely changing v to $-v$ and interchanging the primed and unprimed quantities. We shall have occasion to use these results, and to interpret them further, in later sections. For the moment, however, let us note certain aspects of the transverse velocity transformations. The perpendicular, or transverse, components (that is, u_y and u_z) of the velocity of an object as seen in the S frame are related both to the transverse components (that is, u_y' and u_z') and to the parallel component (that is, u_x') of the velocity of the object in the S' frame. The result is not simple because neither observer is a proper one. If we choose a frame in which $u_x' = 0$, however, then the transverse results become $u_z = u_z'\sqrt{1 - v^2/c^2}$ and $u_y = u_y'\sqrt{1 - v^2/c^2}$. But no length contraction is involved for transverse space intervals, so what is the origin of the $\sqrt{1 - v^2/c^2}$ factor? We need only point out that velocity, being a ratio of length interval to time interval, involves the time coordinate too, so time dilation is involved. Indeed, this special case of the transverse velocity transformation is a direct time-dilation effect.

2-7 *ABERRATION AND DOPPLER EFFECT IN RELATIVITY*

Up to now we have shown how relativity can account for the experimental results of various light-propagation experiments listed in Table 1-2 (for example, the Fresnel drag coefficient and the Michelson-Morley result) and at the same time how it predicts new results also confirmed by experiment (time dilation in the decay of pions or other mesons, also in Table 1-2). Here we deduce the aberration result described in Section 1-7. In doing this, we shall also come upon another new result predicted by relativity and confirmed by experiment, namely, a transverse Doppler effect.

Consider a train of plane monochromatic light waves of unit amplitude emitted from a source at the origin of the S' frame, as shown in Fig. 2-10. The rays, or

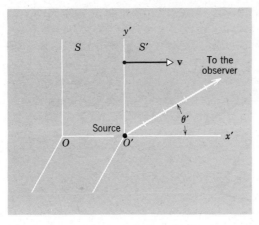

Figure 2-10. A ray, or wave normal, of plane monochromatic light waves is emitted from the origin of the S' frame. The bars signify wavefronts separated by one wavelength from adjacent wavefronts. The direction of propagation makes an angle θ' with the x' axis, the rays being parallel to the x'-y' plane. The source is at rest in the S' frame, and the observer is at rest in the S frame.

wave normals, are chosen to be in (or parallel to) the x'-y' plane, making an angle θ' with the x' axis. An expression describing the propagation would be of the form

$$\cos 2\pi \left(\frac{x' \cos \theta' + y' \sin \theta'}{\lambda'} - \nu' t' \right), \tag{2-23}$$

for this is a single periodic function, amplitude unity, representing a wave moving with velocity $\lambda' \nu' \, (= c)$ in the θ' direction. Notice, for example, that for $\theta' = 0$ it reduces to $\cos 2\pi (x'/\lambda' - \nu' t')$, and for $\theta' = 90°$ it reduces to $\cos 2\pi (y'/\lambda' - \nu' t')$, well-known expressions for propagation along the positive x' and positive y' directions, respectively, of waves of frequency ν' and wavelength λ'. Analysis of the quantities in the parentheses of these last two equations shows [see *Physics*, Part I, Section 19-3] that the wave speed is indeed $\lambda' \nu'$ which, for electromagnetic waves in free space, is equal to c.

In the S frame these wavefronts will still be planes, for the Lorentz transformation is linear so a plane transforms into a plane. Hence, in the unprimed, or S, frame, the expression describing the propagation will have the same form:

$$\cos 2\pi \left(\frac{x \cos \theta + y \sin \theta}{\lambda} - \nu t \right). \tag{2-24}$$

Here, λ and ν are the wavelength and frequency, respectively, measured by an observer fixed in the S frame, and θ is the angle a ray makes with the x axis. We know, if expressions 2-23 and 2-24 are to represent electromagnetic waves, that $\lambda \nu = c$, just as $\lambda' \nu' = c$, for c is the velocity of electromagnetic waves, the same for each observer.

Now let us apply the Lorentz transformation equations directly to expression 2-23, putting

$$x' = \frac{x - vt}{\sqrt{1 - \beta^2}}, \quad y' = y, \quad \text{and} \quad t' = \frac{t - (v/c^2)x}{\sqrt{1 - \beta^2}}.$$

We obtain

$$\cos 2\pi \left[\frac{1}{\lambda'} \frac{(x - vt)}{\sqrt{1 - \beta^2}} \cos \theta' + \frac{y \sin \theta'}{\lambda'} - \nu' \frac{[t - (v/c^2)x]}{\sqrt{1 - \beta^2}} \right]$$

or, on rearranging terms and using $\lambda' \nu' = c$,

$$\cos 2\pi \left[\frac{\cos \theta' + \beta}{\lambda' \sqrt{1 - \beta^2}} x + \frac{\sin \theta'}{\lambda'} y - \frac{(1 + \beta \cos \theta') \nu'}{\sqrt{1 - \beta^2}} t \right].$$

As expected, this has the form of a plane wave in the S frame and must be identical to expression 2-24, which represents the same thing. Hence, the coefficient of x, y, and t in each expression must be equated, giving us

$$\frac{\cos \theta}{\lambda} = \frac{\cos \theta' + \beta}{\lambda' \sqrt{1 - \beta^2}}, \tag{2-25}$$

$$\frac{\sin \theta}{\lambda} = \frac{\sin \theta'}{\lambda'}, \tag{2-26}$$

$$\nu = \frac{\nu'(1 + \beta \cos \theta')}{\sqrt{1 - \beta^2}}. \tag{2-27}$$

We also have the relation

$$\lambda \nu = \lambda' \nu' = c, \tag{2-28}$$

a condition we knew in advance.

In the procedure we have adopted here, we start with a light wave in S' for which we know λ', ν', and θ' and we wish to find what the corresponding quantities λ, ν, and θ are in the S frame. That is, we have three unknowns, but we have four equations (Eqs. 2-25 to 2-28) from which to determine the unknowns. The unknowns have been overdetermined, which means simply that the equations are not all independent. If we eliminate one equation, for instance, by dividing one by another (that is, we combine two equations), we shall obtain three independent relations. It is simplest to divide Eq. 2-26 by Eq. 2-25; this gives us

$$\tan \theta = \frac{\sin \theta' \sqrt{1 - \beta^2}}{\cos \theta' + \beta}, \tag{2-29a}$$

which is *the relativistic equation for the aberration of light.* It relates the directions of propagation, θ and θ', as seen from two different inertial frames. The inverse transformation can be written at once as

$$\tan \theta' = \frac{\sin \theta \sqrt{1 - \beta^2}}{\cos \theta - \beta}, \tag{2-29b}$$

wherein β of Eq. 2-29a becomes $-\beta$ and we interchange primed and unprimed quantities. Experiments in high-energy physics involving photon emission from high-velocity particles confirm the relativistic formula exactly.

EXAMPLE 11.

Relativity and Stellar Aberration. Show that the classical expression for the aberration effect for an overhead star (Eq. 1-11) is an excellent first approximation to the correct relativistic expression (Eqs. 2-29).

In the S frame (attached to the sun) let the one direction of propagation of light from the star be along the negative y direction. Hence $\theta = 270°$. In S' (attached to the earth), the propagation direction is θ', given by Eq. 2-29b with $\theta = 270°$. That is,

$$\tan \theta' = \frac{\sin 270° \sqrt{1 - \beta^2}}{\cos 270° - \beta} = \frac{\sqrt{1 - \beta^2}}{\beta}.$$

When v is much less than c, we have $\beta \ll 1$. Thus β^2 will be negligible compared to one. Neglecting terms of the second order, we can write

$$\tan \theta' = \frac{\sqrt{1 - \beta^2}}{\beta} \cong \frac{1}{\beta}$$

as a first approximation to the exact relativistic result.

Let us now replace θ' in this relationship by $\theta - \alpha$, where $\theta = 270°$ and α is the (very small) aberration angle displayed in Fig. 1-9 and Fig. 2-11. By trigonometry (see Problem 77), we can easily show that the relation $\tan \theta' = (1/\beta)$ is exactly equivalent to the relation $\tan \alpha = \beta$, which (see Eq. 1-11) is the prediction of classical theory. Thus the exact classical formula does indeed appear as a first-order approximation to the relativistic formula, a not unexpected result.

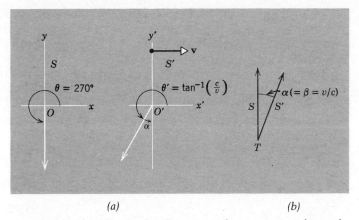

Figure 2-11. (*a*) In *S*, the direction of propagation from the source is the negative *y* direction ($\theta = 270°$). In *S'*, the same ray makes an angle θ' with the *x'* axis. (*b*) The line of sight of the telescope in *S* is vertical and in *S'* is inclined forward by an angle $\alpha\ (= \beta = v/c)$ in order to see the source.

The third of our four equations above (Eqs. 2-25 to 2-28) gives us directly the one remaining phenomenon we promised to discuss; that is, *the relativistic equation for the Doppler effect,*

$$\nu = \frac{\nu'(1 + \beta \cos \theta')}{\sqrt{1 - \beta^2}},$$
(2-27*a*)

which we can also write inversely as

$$\nu' = \frac{\nu(1 - \beta \cos \theta)}{\sqrt{1 - \beta^2}}.$$

We shall find it useful to recast this as

$$\nu = \nu' \frac{\sqrt{1 - \beta^2}}{1 - \beta \cos \theta}.$$
(2-27*b*)

Let us first consider the special case of $\theta = 0$, which corresponds, as Fig. 2-11 shows, to a wave propagating in the positive direction of the *x-x'* axis and to the source and the observer *approaching* each other. For $\theta = 0$, then, Eq. 2-27*b* becomes

$$\nu = \nu_0 \frac{\sqrt{1 - \beta^2}}{1 - \beta} = \nu_0 \sqrt{\frac{1 + \beta}{1 - \beta}} \quad \text{(approaching)}.$$
(2-30*a*)

Note that we have made a small change in notation, replacing ν', which is the frequency (a *proper* frequency) measured in a frame in which the source is at rest, by ν_0.

If the source and the observer are *separating* from each other, we can find the Doppler-shifted frequency by putting $\theta = 180°$ in Eq. 2-27*b* or, equivalently, by changing the sign of β in Eq. 2-30*a*. Either choice leads to

$$\nu = \nu_0 \frac{\sqrt{1 - \beta^2}}{1 + \beta} = \nu_0 \sqrt{\frac{1 - \beta}{1 + \beta}} \quad \text{(separating)}.$$
(2-30*b*)

For moving terrestrial light sources of macroscopic dimensions it will always be true that $v \ll c$ or, equivalently, that $\beta \ll 1$. Under these circumstances it is instructive to expand Eqs. 2-30 above in a power series in β, so that we may more easily compare them with the predictions of the classical theory of the Doppler effect and with experiment. Let us write Eq. 2-30a in the form

$$\nu = \nu_0 (1 + \beta)^{1/2} (1 - \beta)^{-1/2}$$

and expand each of the quantities in parentheses by the binomial theorem. We obtain

$$\nu = \nu_0 (1 + \tfrac{1}{2}\beta - \tfrac{1}{8}\beta^2 + \cdots)(1 + \tfrac{1}{2}\beta + \tfrac{3}{8}\beta^2 + \cdots).$$

If we multiply this out, discarding terms of higher order than β^2, we find

$$\nu = \nu_0 (1 + \beta + \tfrac{1}{2}\beta^2 + \cdots) \qquad \text{(approaching).} \tag{2-31a}$$

Operating on Eq. 2-30b in the same way leads to

$$\nu = \nu_0 (1 - \beta + \tfrac{1}{2}\beta^2 - \cdots) \qquad \text{(separating).} \tag{2-31b}$$

It is instructive to compare these relativistic Doppler effect formulas with those derived from the classical ether or emission theories of the propagation of light (see *Physics*, Part II, Sec. 42-5). In each case, if the classical formulas are expanded in a power series in β, they agree with Eq. 2-31a (if source and observer are approaching) or with Eq. 2-31b (if source and observer are separating) *as far as the first two terms are concerned.*

The classical and the relativistic formulas differ in their predictions about the third term, that is, about the coefficients of β^2 in the power expansions, and it is at this level that comparisons between experiment and theory must be made. The experiments are difficult. If $\beta \ll 1$, then β^2 will be even smaller. Agreement between experiment and theory, clearly favoring the relativistic predictions (Eqs. 2-31) over the classical ones, was first obtained in 1938 by Ives and Stillwell (*Physics*, Part II, Sec. 42-5), following a suggestion first made by Einstein in 1907. They used a beam of excited hydrogen atoms of well-defined speed and direction as the source of the radiation whose Doppler frequency shift they studied. The experiment was repeated in 1961 with higher accuracy by Mandelberg and Witten [8], again confirming the relativistic predictions.

More striking, however, is the fact that the relativistic formula predicts a *transverse Doppler effect*, an effect that is purely relativistic, for there is no transverse Doppler effect in classical physics at all. This prediction follows from Eq. 2-27b when we set $\theta = 90°$, obtaining

$$\nu = \nu_0 \sqrt{1 - \beta^2} \qquad \text{(transverse).} \tag{2-32a}$$

If our line of sight is 90° to the motion of the source, then we should observe a frequency ν that is *lower* than the proper frequency ν_0 of the source.

For easier comparison with the longitudinal Doppler effect (in the important case of $\beta \ll 1$), it is helpful to expand Eq. 2-32a as a power series in β, using the binomial theorem. Doing so yields

$$\nu = \nu_0 (1 - \beta^2)^{1/2}$$
$$= \nu_0 (1 - \tfrac{1}{2}\beta^2 + \cdots) \qquad \text{(transverse).} \tag{2-32b}$$

In comparison with Eqs. 2-31 the transverse Doppler formula contains no term in β. Recall that, in the formulas for the longitudinal Doppler effect (Eqs. 2-31), it is precisely this first-order term that we associate with the classical theory. Thus the absence of such a term in the transverse Doppler formula is totally consistent with the fact that classical theory does not predict such an effect.

Ives and Stillwell, in 1938 and 1941, were the first to confirm the existence of the transverse Doppler effect, using moving hydrogen atoms as "clocks." More recently, Walter Kundig [9] has obtained excellent quantitative data confirming the relativistic formula to within the experimental error of 1.1 percent. In Kundig's experiment a radioactive source emitting 14.4-keV gamma rays (for which the proper frequency is 2.19×10^{18} Hz) was located on the axis of the rotor of a centrifuge. At the centrifuge rim was placed a resonant absorbing foil that is critically sensitive to the source frequency when the foil is at rest; a gamma-ray detector is placed behind the foil. When the centrifuge is operating, the absorbing foil is in rapid transverse motion with respect to the source and we expect that the characteristic absorption frequency of the foil will shift to a lower value, as predicted by the transverse Doppler formula (Eq. 2-32a). By sensitive Mössbauer techniques it is possible to change the effective frequency of the radiation emitted by the source and thus to measure this transverse Doppler frequency shift, as a function of rotor speed. Figure 2-12 shows that the experimental points fall very closely indeed on the curve predicted by relativity theory.

It is instructive to note that the transverse Doppler effect has a simple time-dilation interpretation. The moving source is really a moving clock, beating out electromagnetic oscillations. We have seen that moving clocks appear to run slow. Hence, we see a given number of oscillations in a time that is longer than the proper time. Or, equivalently, we see a smaller number of oscillations in our unit time than is seen in the unit time of the proper frame. Therefore, we observe a lower frequency than the proper frequency. The transverse Doppler effect is another physical example confirming the relativistic time dilation.

In both the Doppler effect and aberration, the theory of relativity introduces an intrinsic simplification over the classical interpretation of these effects in that the two separate cases, which are different in classical theory (namely, source at rest—moving observer and observer at rest—moving source), are identical in relativity. This, too, is in accord with observation. Notice, also, that a single derivation yields at once three effects, namely, aberration, longitudinal Doppler effect, and transverse Doppler effect.

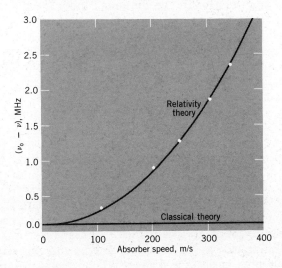

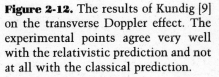

Figure 2-12. The results of Kundig [9] on the transverse Doppler effect. The experimental points agree very well with the relativistic prediction and not at all with the classical prediction.

EXAMPLE 12.

Signaling from Space. *A*, on earth, signals with a flashlight every 6 min. *B* is on a space station that is stationary with respect to the earth. *C* is on a rocket traveling from *A* to *B* with a constant velocity of 0.6*c* relative to *A*; see Fig. 2-13. (*a*) At what intervals does *B* receive signals from *A*? (*b*) At what intervals does *C* receive signals from *A*? (*c*) If *C* flashes a light every time she receives a flash from *A*, at what intervals does *B* receive *C*'s flashes?

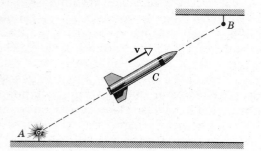

Figure 2-13. Example 12.

(*a*) There is no relative motion of frames *A* and *B*, so, in effect, they are in the same inertial reference frame. Thus *B* receives signals from *A* every 6 min.

(*b*) Here the source (*A*) and the observer (*C*) are separating from each other. We use the longitudinal Doppler effect formula, Eq. 2-30*b*, written as

$$\nu_C = \nu_A \sqrt{\frac{1 - \beta}{1 + \beta}},$$

where ν_C refers to the observer (rocket frame) and ν_A to the source (earth frame). Hence,

$$\nu_C = \nu_A \sqrt{\frac{1 - 0.6}{1 + 0.6}} = \frac{\nu_A}{2}.$$

But the period T_A (= $1/\nu_A$) equals 6 min. Therefore,

$$T_C = \frac{1}{\nu_C} = \frac{2}{\nu_A} = 2T_A = 12 \text{ min.}$$

Thus *C* receives signals from *A* at 12-min intervals.

(*c*) *C* sends signals to *B* at the same frequency ν_C that she receives them from *A*. Let ν_B be the frequency of the signals received by *B* from *C*. Because *C* (who is now the source) and *B* (the observer) are *approaching* each other, we use the Doppler formula Eq. 2-30*a*, which gives us

$$\nu_B = \nu_C \sqrt{\frac{1 + \beta}{1 - \beta}} = \left(\nu_A \sqrt{\frac{1 - \beta}{1 + \beta}} \right) \sqrt{\frac{1 + \beta}{1 - \beta}} = \nu_A.$$

Since $\nu_B = \nu_A$, it follows that $T_B = T_A$ and thus *B* receives signals from *C* at 6-min intervals. This is the same rate at which *B* receives signals directly from *A*. Explain why this is plausible.

2-8 THE COMMON SENSE OF SPECIAL RELATIVITY

We are now at a point where a retrospective view can be helpful. Special relativity theory makes still more predictions than we have discussed so far that contradict classical views. Later we shall see that, in those cases too, experimental results confirm the relativistic predictions. Indeed, in many branches of physics, whether the subject is elementary particles, nuclei, atoms, stars, or the universe itself, relativity is used in an almost commonplace way as the correct description of the real microscopic world. Furthermore, relativity is a consistent theory, as we have shown already in many ways and shall continue to show later. However, because our everyday macroscopic world is classical to a good approximation and students have not yet lived with or used relativity enough to become sufficiently familiar with it, there may remain misconceptions about the theory that are worth discussing now.

The Limiting Speed c of Signals   We have seen that, if it were possible to transmit signals with infinite speed, we could establish in an absolute way whether or not two events are simultaneous. The relativity of simultaneity depended on the existence of a finite speed of transmission of signals. Now, we probably would grant that it is unrealistic to expect that any physical action could be transmitted with infinite speed. It does indeed seem fanciful that we could initiate a signal that would travel to all parts of our universe in zero time. It is really the classical physics (which at bottom makes such an assumption) that is fictitious (science fiction) and not the relativistic physics, which postulates a limiting speed. Furthermore, when experiments are carried out, the relativity of time measurements is confirmed. Nature does indeed show that relativity is a practical theory of measurement and not a philosophically idealistic one, as is the classical theory.

We can look at this in another way. From the fact that experiment denies the absolute nature of time, we can conclude that signals cannot be transmitted with infinite speed. Hence, there must be a certain finite speed that cannot be exceeded and that we call the limiting speed. The principle of relativity shows at once that this limiting speed is the speed of light, since the result that no speed can exceed a given limit is certainly a law of physics and, according to the principle of relativity, the laws of physics are the same for all inertial observers. Therefore this given limit, the limiting speed, must be exactly the same in all inertial reference frames. We have seen, from experiment, that the speed of light has exactly this property.

Viewed in this way, the speed of electromagnetic waves in vacuum assumes a role wider than the travel rate of a particular physical entity. It becomes instead a limiting speed for the motion of anything in nature.

Absolutism and Relativity   The theory of relativity could have been called the theory of absolutism, with some justification. The fact that the observers who are in relative motion assign different numbers to length and time intervals between the pair of events, rather than finding these numbers to be absolutes, upsets the classical mind. This is so in spite of the fact that even in classical physics the measured values of the momentum or kinetic energy of a particle, for example, also are different for two observers who are in relative motion. What is troublesome, apparently, is the philosophic notion that length and time in the abstract are absolute quantities and the belief that relativity contradicts this notion. Now, without going into such a philosophic byway, it is important to note that relativity simply says that the *measured* length or time interval between a pair of events is affected by the relative *motion* of the events and measurer. Relativity is a theory of measurement, and motion affects measurement. Let us look at various aspects of this.

That relative motion should affect measurements is almost a "common-sense" idea—classical physics is full of such examples, including the aberration and Doppler effects already discussed. Furthermore, to explain such phenomena in relativity, we need not talk about the structure of matter or the idea of an ether in order to find changes in length and duration due to motion. Instead, the results follow directly *from the measurement process itself.* Indeed, we find that the phenomena are *reciprocal.* That is, just as A's clock seems to B to run slow, so does B's clock seem to run slow to A; just as A's meter stick seems to B to have contracted in the direction of motion, so likewise B's meter stick seems to A to have contracted in exactly the same way.

Moreover, we should note that in a narrower sense there *are* absolute lengths and times in relativity. The *rest length* of a rod is an absolute quantity, the same for all inertial observers: If a given rod is measured by different inertial observers by bringing the rod to rest in their respective frames, each will measure the same length. Similarly for clocks, the *proper time* (which might better have been called "local time") is an absolute quantity: The frequency of oscillation of an ammonia molecule, for instance, would be measured to be the same by different inertial observers who bring the molecule to rest in their respective frames.

The separation in space of two events (length of a rod) and the time interval between them (rate of a clock) are absolute quantities in classical (Galilean) physics, even for observers in relative motion. At first glance it may seem a step backward to learn that these quantities have surrendered their absolute character in relativity theory in that they have different values for different inertial observers. We learned in Section 2-3, however, that relativity has given us a broader perspective in this matter by replacing two separately absolute quantities by a single absolute, the *spacetime interval*, which has the same value for all observers. The result is a new understanding of the nature of space and time, an understanding that is at the same time both simpler and more profound.

Where relativity theory is clearly "more absolute" than classical physics is in the relativity principle itself: The *laws of physics* are absolute. We have seen that the Galilean transformations and classical notions contradicted the invariance of electromagnetic (and optical) laws, for example. Surely, giving up the absoluteness of the laws of physics, as classical notions of time and length demand, would leave us with an arbitrary and complex physical world. By comparison, relativity is absolute and simple.

The Reality of the Length Contraction Is the length contraction "real" or apparent? We might answer this by posing a similar question. Is the frequency, or wavelength, shift in the Doppler effect real or apparent? Certainly the proper frequency (that is, the rest frequency) of the source is measured to be the same by all observers who bring the source to rest before taking the measurement. Likewise, the proper length is invariant. When the source and observer are in relative motion, the observer definitely measures a frequency (or wavelength) shift. Likewise, the moving rod is definitely measured to be contracted. The effects are real in the same sense that the measurements are real. We do not claim that the proper frequency has changed because of our measured shift. Nor do we claim that the proper length has changed because of our measured contraction. The effects are apparent (that is, caused by the motion) in the same sense that proper quantities have not changed.

We do not speak about theories of matter to explain the contraction but, instead, we invoke the measurement process itself. For example, we do not assert, as Lorentz sought to prove, that motion produces a physical contraction through an effect on the elastic forces in the electronic or atomic constitution of matter (motion is *relative*, not absolute), but instead we remember the fish story. If a fish is swimming in water and his length is the distance between his tail and his nose, measured simultaneously, observers who disagree on whether measurements are simultaneous or not will certainly disagree on the measured length. Hence, length contraction is due to the relativity of simultaneity.

Since length measurements involve a comparison of two lengths (moving rod and measuring rod, for example), we can see that the Lorentz length contraction is really not a property of a single rod by itself but instead is a relation between two such rods in relative motion. The relation is both observable and reciprocal. Just as A's meter stick seems to B to have contracted in the direction of motion, so likewise B's meter stick seems to A to have contracted, in exactly the same way.

Rigid Bodies and Unit Length In classical physics, the notion of an ideal rigid body was often used as the basis for length (that is, space) measurements. In principle, a rigid rod of unit length is used to lay out a distance scale. Even in relativity we can imagine a standard rod defining a unit distance, this same rod being brought to rest in each observer's frame to lay out space-coordinate units. However, the concept of an ideal rigid body is untenable in relativity, for such a body would be capable of transmitting signals instantaneously; a disturbance at one end would be propagated with infinite velocity through the body, in contradiction to the relativistic principle that there is a finite upper limit to the speed of transmission of a signal.

Conceptually, then, we must give up the notion of an ideal rigid body. This causes no problems because time measurements prove to be primary and space measurements secondary. We know that this is so in relativity, for the simultaneity concept is used in the definition of length. But a similar situation also exists in classical physics. Some years have passed since distances were measured in terms of comparison with a presumed rigid measuring rod, the standard meter. This definition was replaced in 1960 by a definition of the meter in terms of the wavelength of the radiation emitted by a specified light source. Since 1983 the meter has been defined as "the length equal to the distance traveled in a time interval of 1/299,792,458 of a second by plane electromagnetic waves in a vacuum." Thus length is now measured by timing a light beam, the speed of light having a *defined* value of 299,792,458 m/s. It is interesting that this definition of the meter, arrived at after careful consideration by a representative international body, contains no mention of either a standard frequency or a standard reference frame. Those making the choice evidently took it for granted (following Maxwell) that the speed of electromagnetic radiation is independent of frequency and (following Einstein) that it is the same for all inertial observers.

Rigid-body measuring concepts have never been directly applicable in certain situations, measurements on the atomic and the astronomical scales being two limiting examples. As the unit "light-year" suggests, the timing concept that now forms the official basis of all length measurement, has served as the basis of much practical distance measurement, the precise measurement of the earth–moon distance by radar techniques being one example. It seems clear that rigid-body measuring rods—not permissible conceptually in relativity theory—are not required in practice, even for measurements in the classical realm.

It is fitting, in emphasizing the common sense of relativity, to conclude with this quotation from Bondi [10] on the presentation of relativity theory:

> At first, relativity was considered shocking, anti-establishment and highly mysterious, and all presentations intended for the population at large were meant to emphasize these shocking and mysterious aspects, which is hardly conducive to easy teaching and good understanding. They tended to emphasize the revolutionary aspects of the theory whereas, surely, it would be good teaching to emphasize the continuity with earlier thought. . . .
>
> It is first necessary to bring home to the student very clearly the Newtonian attitude. Newton's first law of dynamics leads directly to the notion of an *inertial observer*, defined as an observer who finds the law of inertia to be correct. . . . The utter equivalence of inertial observers to each other for the purpose of Newton's first law is a direct and logical consequence of this law. The equivalence with regard to the second law is not a logical necessity but a very plausible extension, and with this plausible extension we arrive at Newton's principle of relativity: *that all inertial observers are equivalent as far as dynamical experiments go.* It will be obvious that the restriction to dynamical experiments is due simply to this principle of relativity having been derived from the laws of dynamics. . . .
>
> The next step . . . is to point out how absurd it would be if dynamics were in any sense separated from the rest of physics. There is no experiment in physics that involves dynamics alone and nothing else. . . . Hence, Newton's principle of relativity is empty because it refers only to a class of experiment that does not exist— the purely dynamical experiment. The choice is therefore presented of either throwing out this principle or removing its restriction to dynamical experiments. The first

alternative does not lead us any further, and clearly disregards something of significance in our experience. The second alternative immediately gives us Einstein's principle of relativity: *that all inertial observers are equivalent.* It presents this principle, not as a logical deduction, but as a reasonable guess, a fertile guess from which observable consequences may be derived so that this particular hypothesis can be subjected to experimental testing. Thus, the principle of relativity is seen, not as a revolutionary new step, but as a natural, indeed an almost obvious, completion of Newton's work.

Questions

1. Distinguish between sound and light as to their value as synchronizing signals. Is there a lack of analogy?

2. Give an example from classical physics in which the motion of a clock affects its rate, that is, the way it runs. (The magnitude of the effect may depend on the detailed nature of the clock.)

3. Explain how the result of the Michelson-Morley experiment was put into our definition (procedure) of simultaneity (for synchronizing clocks).

4. According to Eqs. 2-4 and 2-5, each inertial observer finds the center of the spherical electromagnetic wave to be at his own origin at all times, even when the origins do not coincide. How is this result related to our procedure for synchronizing clocks?

5. What assumptions, other than the relativity principle and the principle of the constancy of the speed of light, were made in deducing the Lorentz transformation equations?

6. Two observers, one at rest in S and one at rest in S', each carry a meter stick oriented parallel to their relative motion. *Each* observer finds upon measurement that the *other* observer's meter stick is shorter than his own meter stick. Does this seem like a paradox to you? Explain. (*Hint:* Compare the following situation. Harry waves goodbye to Walter who is in the rear of a station wagon driving away from Harry. Harry says that Walter gets smaller. Walter says that Harry gets smaller. Are they measuring the same thing?)

7. Although in relativity (where motion is relative and not absolute) we find that "moving clocks run slow," this effect has nothing to do with the motion altering the way a clock works. What does it have to do with?

8. Two events occur at the same place and at the same time for one observer. Will they be simultaneous for all other observers? Will they also occur at the same place for all other observers?

9. Events A and B occur at the same point in a certain inertial reference frame, with event A preceding event B. Will A precede B in all other frames? In any other frame? Will the events occur at the same point in any other frame? Will the time interval between the events be the same in any other frame?

10. Two events are simultaneous but separated in space in one inertial reference frame. Will they be simultaneous in any other frame? Will their spatial separation be the same in any other frame?

11. We have seen that if several observers watch two events, labeled A and B, one of them may say that event A occurred first but another may claim that it was event B that did so. What would you say to a friend who asked you which event *really did* occur first?

12. Let event A be the departure of an airplane from San Francisco and event B be its arrival in New York. Is it possible to find two observers who disagree about the time order of *these* events? Explain.

13. A rod has a tiny flashbulb embedded in each end. Each of these bulbs has been arranged to flash independently, at a single unpredictable time. The flashes enable an observer to measure the coordinates of the end points of the rod in his reference frame. Consider first an observer in frame S, in which the rod is at rest and lying along the x axis. Would you label the difference between his coordinate readings as the "length of the rod"? As the "rest length of the rod"? Answer these questions for a observer S' with respect to whom the rod is moving at speed v.

14. A number of observers, in different inertial reference frames, measure the spacetime coordinates of two events. For what combination(s) of these measurements would all of these observers obtain the same numerical value?

15. How would you recognize a given pair of events as spacelike? Timelike? Lightlike?

16. Can a given pair of events appear spacelike to one inertial observer and timelike to another?

17. Explain, using the velocity addition theorem of relativity, how we can account for the result of the Michelson-Morley experiment and for the double-star observations.

18. In Example 10, what would happen if v_w were chosen equal to $-c/n$?

19. The Galilean velocity transformation equation (Eq. 2-19) is so instinctively familiar from everyday experience that it is sometimes claimed to be "obviously correct, requiring no proof." Many so-called refutations of relativity theory turn out to be based on this claim. How would you refute someone who made such a claim?

20. Compare the results obtained for length- and time-interval measurements by observers in reference frames whose relative velocity is c. In what sense, from this point of view, does c appear as a limiting velocity, to be approached (but not attained) by material bodies?

21. Is the Doppler effect simply a time-dilation effect and nothing more, or is there something else to it?

22. An observer makes simultaneous measurements of the positions of the end points of a rod that is lying at rest along the x axis of his reference frame. An observer in another reference frame who views these same measuring events will always find (can you prove it?) that the difference between his position measurements is *greater* than the rest length of the rod. How can you square this with the fact that moving rods have measured lengths that are *less than* their rest lengths?

23. Consider a spherical light wavefront spreading out from a point source. As seen by an observer at the source,

what is the *difference in velocity* of portions of the wavefront traveling in opposite directions? What is the *relative velocity* of one of these portions of the wavefront with respect to the other?

24. The sweep rate of the tail of a comet can exceed the speed of light. Explain this phenomenon and show that there is no contradiction with relativity.

25. Explain, in qualitative terms, the "headlight effect" described in Problem 82.

26. List several experimental results not predicted or explained by classical physics that are predicted or explained by the theory of relativity.

27. We have stressed the utility of relativity at high speeds. Relativity is also useful in cosmology, where great distances and large time intervals are involved. Show, from the form of the Lorentz transformation equations, why this is so.

28. In relativity the time and space coordinates are intertwined and treated on a more or less equivalent basis. Are time and space fundamentally of the same nature, or is there some essential difference between them, preserved even in relativity? (*Hint:* What significance do you attach to the minus sign in Eq. 2-16?)

29. We have stressed certain measurements of events in spacetime that are different for different inertial observers. Make a list of those things that these different observers *agree* on.

30. Some say that relativity complicates things. Give examples to the contrary, wherein relativity simplifies matters. Consider the Fresnel drag experiment as one example.

Problems

In all problems that involve two reference frames it is assumed, unless otherwise stated, that the frames are in the standard configuration of Fig. 1-1. That is, an observer in the S frame would see the S' frame moving with speed v in the direction of increasing x. The x and x' axes coincide and the y-y' and z-z' axes remain parallel. Further, it is assumed that each observer sets his or her clocks to zero at the instant the two origins pass each other.

1. The view from the other frame. In Fig. 2-1 we took the point of view of observer O in the S frame and found that events AA' and BB' happened to be simultaneous in that frame. Figure 2-14 shows how these *same two events*

appear to observer O' in the S' frame. Note, and comment on, the following features: (a) O' is midway between points A' and B'. (b) S moves to the left with speed v. (c) The first stroke occurs on the right, making marks B, B'. (d) Later, a stroke occurs on the left, making marks A, A'. (e) As seen by O', the distances AB and $A'B'$ are *necessarily* unequal. (f) In agreement with the observations of observer O (for these same events) described in connection with Fig. 2-1, the right wavefront passes O', then both wavefronts pass O, and finally the left wavefront passes O'. (g) How does the situation described in this figure differ from that described in Fig. 2-2, which *also* purports to show the point of view of the S' observer?

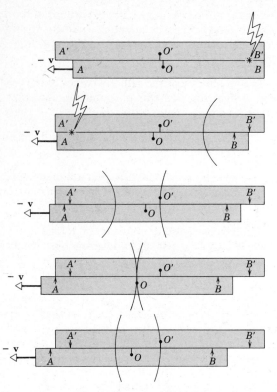

Figure 2-14. Problem 1.

2. A little algebra. Show that Eqs. 2-6 for a_{11}, a_{41}, and a_{44} are indeed solutions of the equations preceeding them.

3. Three simplifying physical arguments. Consider the two middle equations of Eqs. 2-1:

$$y' = a_{21}x + a_{22}y + a_{23}z + a_{24}t$$
$$z' = a_{31}x + a_{32}y + a_{33}z + a_{34}t.$$

(a) As part of the derivation of the Lorentz transformation equations given in Section 2-2, the S and S' frames have been so arranged that their x and x' axes coincide, as in Fig. 2-1. Show how to deduce from this fact that

$$a_{21} = a_{24} = a_{31} = a_{34} = 0.$$

(b) From the fact that the y and y' axes (and the z and z' axes) have been taken to be parallel, show that

$$a_{23} = a_{32} = 0.$$

(c) Let observer S place a rod, whose rest length is 1 m, along her y axis and let observer S' measure its length as it moves past. Then let S' place the same rod along the y' axis and let S measure it. By applying the principle of relativity to their results, show that we must have

$$a_{22} = a_{33} = 1.$$

(d) With these substitutions, to what do the expressions given above for y' and z' reduce?

4. A feeling for the Lorentz factor. The Lorentz factor γ (see Eq. 2-10) is a direct measure of, among other quantities, the relativistic length contraction and the time-dilation effect; see Eqs. 2-12a and 2.14b. What must be the relative speed parameter β of two reference frames if this factor is to be (a) 1.01? (b) 10? (c) 100? (d) 1000?

5. Making absolutely sure. In Table 2-2 the Lorentz transformation equations in the right-hand column can be derived from those in the left-hand column simply by (1) exchanging primed and unprimed quantities and (2) changing the sign of v. Verify this procedure by deriving one set of equations directly from the other by algebraic manipulation.

6. The speed of light really is the same in all frames. Equation 2-4 describes an expanding spherical wavefront of light, triggered at $t = 0$ and viewed in the S frame. Equation 2-5 describes the same expanding wavefront as viewed in the S' frame. Show that either equation can be derived from the other by direct application of the Lorentz transformation equations.

7. Two observers view the same event (I). Observer S assigns the following spacetime coordinates to an event:

$$x = 100 \text{ km} \qquad y = 10 \text{ km}$$
$$z = 55 \text{ km} \qquad t = 200 \text{ }\mu\text{s}.$$

What are the coordinates of this event in frame S', which moves in the direction of increasing x with speed $0.95c$? Check your answers by using the inverse Lorentz transformation equations to obtain the original data.

8. Two observers view the same event (II). Observer S reports that an event occurred on his x axis at $x = 3.0 \times 10^8$ m at a time $t = 2.50$ s. (a) Observer S' is moving in the direction of increasing x at a speed of $0.40c$. What coordinates would he report for the event? (b) What coordinates would he report if he were moving in the direction of *decreasing* x at this same speed?

9. A moving clock. A clock moves along the x axis at a speed of $0.60c$ and reads zero as it passes the origin. What time does it read as it passes the 180-m mark on this axis?

10. A moving rod. A rod lies parallel to the x axis of reference frame S, moving along this axis at a speed of $0.60c$. Its rest length is 1.0 m. What will be its measured length in frame S?

11. Hidden symmetry in the Lorentz equations. The spacetime symmetry of relativity is partially concealed because the space variables and the time variable are measured in different units. Let us introduce a new variable w ($= ct$) that is a linear measure of time but is expressed in length units. (*a*) Show that the first and fourth Lorentz transformation equations can be written in the forms

$$x' = \gamma(x - \beta w) \qquad x = \gamma(x' + \beta w')$$
$$w' = \gamma(w - \beta x) \qquad w = \gamma(w' + \beta x'),$$

which are certainly more symmetrical in appearance than the standard representation of Table 2-2. (*b*) The variable *w* can be expressed in "meters of time." Show that, with this understanding, 1 μs $\equiv$ 300 m. Use this representation of the Lorentz transformation equations to solve Problem 7.

12. Meters of time. In a frame *S*, two events occur along the *x* axis. They are separated in space by Δx ($= x_2 - x_1 =$ 720 m) and in the time coordinate (see Problem 11) by Δw ($= w_2 - w_1 = 1500$ m). (*a*) What is the relative speed parameter β of a second frame *S'* in which these events are found to occur at the same place? (*b*) What is the time interval between them in this second frame? Recall from Problem 11 that 1 μs is equivalent to 300 "meters of time."

13. A fast spaceship. The length of a spaceship is measured to be exactly half its rest length. (*a*) What is the speed of the spaceship relative to the observer's frame? (*b*) By what factor do the spaceship's clocks run slow, compared to clocks in the observer's frame?

14. A slow airplane. An airplane whose rest length is 40.0 m is moving at a uniform velocity with respect to the earth at a speed of 630 m/s. (*a*) By what fraction of its rest length will it appear to be shortened to an observer on earth? (*b*) How long would it take by earth clocks for the airplane's clock to fall behind by 1 μs? (Assume that only special relativity applies.)

15. Riding a fast electron. A 100-MeV electron, for which $\beta = 0.999987$, moves along the axis of an evacuated tube that has a length of 3.00 m as measured by a laboratory observer *S* with respect to whom the tube is at rest. An observer *S'* moving with the electron, however, would see this tube moving past her with speed *v* ($= \beta c$). What length would this observer measure for the tube? (*Hint:* See Example 3*b*.)

16. Watching the earth drift by. The rest radius of the earth is 6400 km and its orbital speed about the sun is 30 km/s. By how much would the earth's diameter appear to be shortened to an observer stationed so as to be able to watch the earth move past him at this speed?

17. A world that never was. Consider a universe in which the speed of light *c* is equal to 100 mi/h. A Lincoln Continental traveling at a speed *v* relative to a fixed radar speed trap overtakes a Volkswagen traveling at the speed limit of 50 mi/h. The Lincoln's speed is such that its length, as measured by the fixed observer, is the same as

that of the Volkswagen. By how much is the Lincoln exceeding the speed limit? At rest, a Lincoln is twice as long as a Volkswagen.

18. A moving, slanting, rod. A thin rod of length L', at rest in the *S'* frame, makes an angle of θ' with the *x'* axis, as in Fig. 2-15. (*a*) What is its length *L* as measured by an observer in the *S* frame, for whom the rod is moving at a speed of βc in the direction of increasing *x*? (*b*) What angle θ does this moving rod make with the *x* axis? (*c*) Evaluate these quantities for $L' = 1.00$ m, $\theta' = 30°$, and $\beta = 0.40$.

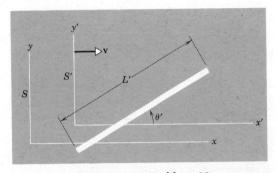

Figure 2-15. Problem 18.

19. Timing a spaceship. A spaceship of rest length 100 m drifts past a timing station at a speed of 0.80*c*. What time interval between the passage of the front and the back end of the ship will the station monitor record?

20. A journey to Vega. A space traveler takes off from earth and moves at speed 0.990*c* toward the star Vega, which is 26 ly distant. How much time will have elapsed by earth clocks (*a*) When the traveler reaches Vega? (*b*) When the earth observers receive word from him that he has arrived? (*c*) How much older will the earth observers calculate the traveler to be when he reaches Vega than he was when he started the trip?

21. To the galactic center! (*a*) Can a person, in principle, travel from earth to the galactic center (which is about 28,000 ly distant) in a normal lifetime? Explain, using either time-dilation or length-contraction arguments. (*b*) What constant velocity would be needed to make the trip in 30 y (proper time)?

22. Pretty small but quite measureable. An airline pilot synchronizes his watch with earth clocks and then takes off on a nonstop flight of 6000 mi at a steady speed of 600 mi/h. On landing, how far behind would his watch be compared to earth clocks, assuming that special relativity alone applies? (An around-the-world flight using atomic clocks confirmed this effect, as well as a separate general-relativistic effect associated with the earth's gravitational field [11]).

23. Decay in flight (I). A pion is created in the higher reaches of the earth's atmosphere when an incoming high-energy cosmic-ray particle collides with an atomic nucleus. A pion so formed descends toward earth with a speed of 0.990c. In a reference frame in which they are at rest, pions decay with a mean life of 26 ns. As measured in a frame fixed with respect to the earth, how far (on the average) will such a typical pion move through the atmosphere before it decays?

24. Decay in flight (II). The mean lifetime of muons stopped in a lead block in the laboratory is measured to be 2.2 μs. The mean lifetime of high-speed muons in a burst of cosmic rays observed from the earth is measured to be 16 μs. Find the speed of these cosmic ray muons.

25. Decay in flight (III). An unstable high-energy particle enters a detector and leaves a track 1.05 mm long before it decays. Its speed relative to the detector was 0.992c. What is its proper lifetime? That is, how long would it have lasted before decay had it been at rest with respect to the detector?

26. Decay in flight (IV). In the target area of an accelerator laboratory there is a straight evacuated tube 300 m long. A momentary burst of 1 million radioactive particles enters at one end of the tube, moving at a speed of 0.80c. Half of them arrive at the other end without having decayed. (a) How long is the tube as measured by an observer moving with the particles? (b) What is the half-life of the particles (that is, the time during which half of the particles initially present have decayed) in this same reference frame? (c) With what speed is the tube measured to more in this frame?

27. Decay in flight (V). (a) If the average (proper) lifetime of a muon is 2.2 μs, what average distance would it travel in free space before decaying, as measured in reference frames in which its velocity is 0.00c; 0.60c; 0.90c; 0.990c? (b) Compare each of these distances with the distance the muon sees itself traveling through.

28. Simultaneous—but to whom? An experimenter arranges to trigger two flashbulbs simultaneously, a blue flash located at the origin of his reference frame and a red flash at $x = 30$ km. A second observer, moving at a speed of 0.25c in the direction of increasing x, also views these flashes. (a) What time interval between them does he find? (b) Which flash does he say occurs first?

29. Simultaneity—the general case. Two events, one at position x_1, y_1, z_1 and another at a different position x_2, y_2, z_2 occur at the same time according to observer S. (a) Do these events appear to be simultaneous to observer S', who moves relative to S at speed v? (b) If not, what is the time interval that S' measures between these events? (c) How is this interval affected as $v \rightarrow 0$? As $v \rightarrow c$? As the separation between the events goes to zero?

30. A string of lights across the desert. Observer S sees a series of light flashes extending indefinitely in a straight line across a desert. By measurement he declares them all to have occurred simultaneously and finds further that adjacent flashes were uniformly separated by 3.0 km. Observer S' is moving along this line with a speed 0.50c. What are the results of *his* measurements of the space-time coordinates of these light flashes?

31. Careful measurements on a moving flatcar (I). A flatcar moves on a track at a constant speed v (see Fig. 2-16). Observers A and B are on the ends of the car and observers C and D are stationed along the track. We define event AC as the occurrence of A passing C, and the others similarly. (a) Of the four events BD, BC, AD, AC, which are useful for observers along the track who wish to measure the rate of a clock carried by A? (b) Let Δt be the time interval between these two events for the track observers. What time interval does the moving clock show? (c) Suppose that events BC and AD are simultaneous to the track observers. Are they simultaneous to the observers on the flatcar? If not, which event is earlier?

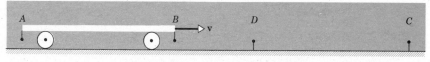

Figure 2-16. Problems 31 and 32.

32. Careful measurements on a moving flatcar (II). In Problem 31 (see Fig. 2-16), event AD turns out to be simultaneous with BC in the track frame. (a) The track observers set out to measure AB, the length of the car. They can do so either by using the events BD and AD and working through time measurements or by using events BC and AC. In either case, the car observers are not apt to regard

these results as valid. Explain why for each case. (b) Suppose that the car observers seek to measure the distance DC by making simultaneous marks on a long meter stick. Where (relative to A and B) would an observer E have to be situated such that AD is simultaneous with EC in the car frame? Explain why in terms of synchronization. Can you see why there is a length contraction?

33. S and S' time two events. Inertial frame S' moves at a speed of 0.60c with respect to frame S. Two events are recorded. In frame S, event 1 occurs at the origin at $t = 0$ and event 2 occurs on the x axis at $x = 3.0$ km and at $t = 4.0$ μs. What time of occurrence does observer S' record for these same events? Explain the difference in the time order.

34. Two flashes at different places—or are they? An observer S sees a flash of red light 1200 m from his position and a flash of blue light 720 m closer to him and on the same straight line. He measures the time interval between the occurrence of the flashes to be 5.00 μs, the red flash occurring first. (a) What is the relative velocity **v** (magnitude and direction) of a second observer S' who would record these flashes as occurring at the same place? (b) From the point of view of S', which flash occurs first? (c) What time interval between them would S' measure?

35. The limit of possibility. In Problem 34, observer S sees the two flashes in the same positions, but they now occur closer together in time. How close together in time can they be and still have it possible to find a frame S' in which they occur at the same place?

★ **36. What time is it anyway?** Observers S and S' stand at the origins of their respective frames, which are moving relative to each other with a speed 0.60c. Each has a standard clock, which, as usual, they set to zero when the two origins coincide. Observer S keeps the S' clock visually in sight. (a) What time will the S' clock record when the S clock records 5.00 μs? (b) What time will observer S actually read on the S' clock when his own clock reads 5.00 μs?

37. A long train struck curiously by lightning. Assume, in Fig. 2-1, that S' is a train having a speed of 100 mi/h and that it is 0.50 mi long (rest length). What is the elapsed time between the arrival of the two wavefronts at O'? Do this in two ways: (a) Make the discussion we had in connection with Fig. 2-1 quantitative by finding expressions for the arrival times of the two signals at O' and subtracting them. Bear in mind that, from the point of view of S', the distance between the marks at A and B in S is contracted by the Lorentz factor γ. (b) Treat the formation of the marks at A and B in frame S as two "events" and apply the Lorentz transformation equations to find their time separation in frame S'. To what prediction does your expression reduce if the train is at rest on the tracks? . . . if the speed of light suddenly becomes infinitely great?

38. A comforting thought. An observer at rest in the laboratory (S frame) sees a uranium nucleus at the origin of this frame emit an alpha particle at time $t = 0$. The alpha particle travels along the +x axis to position x_1, where, at time t_1, it is absorbed by a radon nucleus, to form an atom of radium. Show that a second observer in frame S', moving along the x-x' axis at speed v, cannot see the alpha particle absorbed by the radon nucleus before it is emitted by the uranium nucleus. Assume that all speeds involved are less than the speed of light.

39. Two events—same time difference, same separation, different sequence. Observer S notes that two colored flashes of light, separated by 2400 m, occur along the positive branch of the x axis of his reference frame. A blue flash occurs first, followed after 5.00 μs by a red flash, the latter being the most distant from the origin of his reference frame. A second observer S' obtains exactly the same numerical values for both the time difference and the absolute spatial separation between the two events but declares that the *red* flash occurs first. (a) What is the relative speed of S' with respect to S? (b) Which flash will S' find to be the more distant from the origin of *her* reference frame?

40. Can it be done? In Problem 39, assume no other change but that observer S determines that the blue flash occurs first. In particular, the spatial order of the two flashes is to remain unchanged, the red flash still being farther from the S origin. Is it still possible to find a frame S' in which the time interval between the events would remain unchanged but the order of events would be reversed?

41. The interval is invariant—prove it! Show that the (square of) the spacetime interval $(\Delta s)^2$ associated with two events (assumed to occur on the x-x' axis) is invariant under a Lorentz transformation. That is (see Eq. 2-16), show that

$$(c \, \Delta t)^2 - (\Delta x)^2 = (c \, \Delta t')^2 - (\Delta x')^2.$$

(*Hint:* Use Table 2-3.)

42. The interval is invariant—check it out! Two events occur on the x axis of reference frame S, their spacetime coordinates being:

Event	x	t
1	720 m	5.0 μs
2	1200 m	2.0 μs

(a) What is the square of the spacetime interval $(\Delta s)^2$ for these two events? (Recall that the speed of light can be written as 300 m/μs.) (b) What are the coordinates of these events in a frame S' that moves at speed 0.60c in the direction of increasing x? Calculate the square of the interval in this frame and compare it to the value calculated for frame S. (c) What are the coordinates of these events in a frame S" that moves at speed 0.95c in the direction of decreasing x? Again, calculate $(\Delta s)^2$ and compare its value with the values found in (*a*) and (*b*). Do your calculations bear out the invariance of the spacetime interval?

43. An event pair—timelike or spacelike? Two events occur on the *x* axis of reference frame *S*, their spacetime coordinates being:

Event	x	t
1	200 m	5.0 μs
2	1200 m	2.0 μs

(*a*) What is the square of the spacetime interval $(\Delta s)^2$ for these two events? (*b*) What is the *proper distance* interval $\Delta\sigma$ between them? (*c*) If two events possess a (mathematically real) proper distance interval, it should be possible to find a frame *S'* in which these events would be seen to occur simultaneously. Find this frame. (*d*) Can you calculate a (mathematically real) *proper time* interval $\Delta\tau$ for this pair of events? (*e*) Would you describe this pair of events as timelike? Spacelike? Lightlike? (Compare this problem carefully with Problem 42, noting that $(\Delta s)^2$ in that problem is positive and here it turns out to be negative.)

44. An event pair—spacelike or timelike? In Problem 42 the spacetime coordinates of two *x*-axis events are given and three reference frames from which they might be viewed are described. (*a*) Using data from the solution to that problem, calculate the *proper time* interval $\Delta\tau$ for this pair of events, from the point of view of each of the three frames. Do your calculations support the claim that the proper time interval is an invariant quantity? The proper time interval that you have calculated should be *smaller* than any of the actual time intervals in the three given frames. Is it? (*b*) If two events have a (mathematically real) proper time interval between them, it ought to be possible to find a reference frame in which these events would be seen to occur at the same place. Find this frame. (*c*) Can you calculate a (mathematically real) *proper distance* interval $\Delta\sigma$ for this pair of events? (*d*) Would you describe this pair of events as timelike? Spacelike? Lightlike?

45. Reversing an argument. In our physical derivation of the length contraction of a moving rod (Section 2.1), we assumed that the time dilation was given. In a similar manner, derive the time dilation for a moving clock, assuming that the length contraction is given.

46. Length contraction—another approach. We could define the length of a moving rod as the product of its velocity by the time interval between the instant that one end point of the rod passes a fixed marker and the instant the other end point passes the same marker. Show that this definition also leads to the length contraction result

of Eq. 2-12*b*. (*Hint:* Let the rod be at rest in frame *S'* and let the marker be fixed at one position in frame *S*.)

47. In a relativistic world, earth satellites are slow movers. To circle the earth in low orbit a satellite must have a speed of about 17,000 mi/h. Suppose that two such satellites orbit the earth in opposite directions. (*a*) What is their relative speed as they pass? Evaluate using the classical Galilean velocity transformation equation. (*b*) What fractional error was made because the (correct) relativistic transformation equation was not used?

48. Double checking. (*a*) Derive Eq. 2-22*a* the way Eq. 2-22*b* was derived. (*b*) In Table 2-4 we arrived at the inverse velocity transformation relations (right-hand column) by changing the sign of *v* and interchanging the primed and unprimed quantities in the equations in the left-hand column. Verify this procedure by deriving the equation for u'_y directly, using the same procedure that we used in deriving the equation for u_y (that is, Eq. 2-22*b*).

49. A moving particle. A particle moves along the *x'* axis of frame *S'* with a speed of 0.40*c*. Frame *S'* moves with a speed of 0.60*c* with respect to frame *S*. What is the measured speed of the particle in frame *S*?

50. S and S' watch a moving particle. Frame *S'* moves relative to frame *S* at 0.60*c* in the direction of increasing *x*. In frame *S'* a particle is measured to have a velocity of 0.40*c* in the direction of increasing *x'*. (*a*) What is the velocity of the particle with respect to frame *S*? (*b*) What would be the velocity of the particle with respect to *S* if it moved (at 0.40*c*) in the direction of *decreasing x'* in the *S'* frame? In each case, compare your answers with the predictions of the classical velocity transformation equation.

51. Two fast particles rush toward each other. One cosmic-ray particle approaches the earth along its axis with a velocity of 0.80*c* toward the North Pole and another, with a velocity 0.60*c*, toward the South Pole. What is the relative speed of approach of one particle with respect to the other? (*Hint:* It is useful to consider the earth and one of the particles as the two inertial reference frames.)

52. Interesting information but not very helpful. The meteorite watch officer on a spaceship reports that two fast micrometeorites are approaching the ship on parallel tracks, one at a speed of 0.90*c* and the other at 0.70*c*. What is the speed of either of them with respect to the other?

53. Faster than a speeding bullet! A spaceship whose rest length is 300.0 m has a speed of 0.800*c* with respect to a certain reference frame. A micrometeorite, also with a speed of 0.80*c* in this frame, passes the spaceship on an antiparallel track. How long does it take this object to pass the spaceship?

54. The expanding universe (I). Galaxy A is reported to be receding from us with a speed of 0.35c. Galaxy B, located in precisely the opposite direction, is also found to be receding from us at this same speed. What recessional speed would an observer on Galaxy A find (a) for our galaxy? (b) For Galaxy B?

55. The expanding universe (II). It is concluded from measurements of the red shift of the emitted light that quasar Q_1 is moving away from us at a speed of 0.80c. Quasar Q_2, which lies in the same direction in space but is closer to us, is moving away from us at speed 0.40c. What velocity for Q_2 would be measured by an observer on Q_1?

56. Fast, but not as fast as you might think. Starfleet spacecruisers *Lorentz* and *Minkowski* are proceeding outward from Lunar Base on a straight line with *Lorentz* leading. *Lorentz* has a speed of 0.60c with respect to *Minkowski*, which, in turn, has a speed of 0.60c with respect to Lunar Base. What is the speed of *Lorentz* with respect to Lunar Base?

57. The unreachable goal! A spaceship, at rest in a certain reference frame S, is given a speed increment of 0.50c. It is then given a further 0.50c increment in this new frame, and this process is continued until its speed with respect to its original frame S exceeds 0.999c. How many increments does it require?

58. Directions change too. A particle moves with speed u at an angle θ with respect to the x axis in frame S. Frame S' moves along this axis with speed v. What speed u' and angle θ' will the particle appear to have to an observer in S'?

59. Watching the decay of a moving nucleus. A radioactive nucleus moves with a uniform velocity of 0.050c along the x axis of a reference frame (S) fixed with respect to the laboratory. It decays by emitting an electron whose speed, measured in a reference frame (S') moving with the nucleus, is 0.800c. Consider first the cases in which the emitted electron travels (a) along the common x-x' axis and (b) along the y' axis and find, for each case, its velocity (magnitude and direction) as measured in frame S. (c) Suppose, however, that the emitted electron, viewed now from frame S, travels along the y axis of that frame with a speed of 0.800c. What is its velocity (magnitude and direction) as measured in frame S'?

60. A philosophical difficulty. Suppose that event A *causes* event B, the effect being propagated from A to B with a speed greater than c. Show, using the relativistic velocity transformation equation, that there exists an inertial frame S', which moves relative to S with a velocity less than c, in which the order of these events would be reversed. Hence, if concepts of cause and effect are to be

preserved, it is impossible to send signals with a speed greater than that of light.

★ **61. A surprising result.** In Fig. 2-17, A and B are trains on perpendicular tracks, shown radiating from station S. The velocities are in the station frame (S frame). (a) Find $\mathbf{v}_{AB}$, the velocity of train B with respect to train A. (b) Find $\mathbf{v}_{BA}$, the velocity of train A with respect to train B. (c) Comment on the fact that these two relative velocities do not point in opposite directions [12].

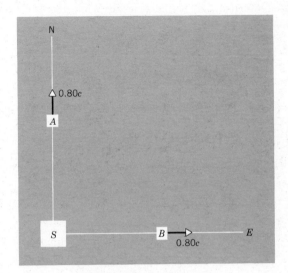

Figure 2-17. Problem 61.

62. A neat formulation. A particle has a speed u in frame S and a corresponding speed u' in frame S', where

$$u^2 = u_x^2 + u_y^2$$

and

$$u'^2 = u_x'^2 + u_y'^2.$$

(a) Verify by direct substitution and by use of the appropriate velocity transformation equations from Table 2-4 that the following relationship holds:

$$(c^2 - u^2)(c^2 + u_x'v)^2 = c^2(c^2 - u'^2)(c^2 - v^2).$$

(b) Show that this formulation contains within itself the result that if u' < c and also v < c then u must be less than c. (c) Show also that the equation contains the result that if u' = c or if v = c, then u must also be equal to c. See [13].

★ **63. How to put a long pole into a short garage.** Suppose that a pole vaulter, holding a 16-ft pole parallel to his direction of motion, runs toward an open garage that is 8 ft deep. The far end of the garage is a massive concrete barrier. (a) At what speed must the pole vaulter run if, at the instant the front end of the pole touches the barrier, the rear end is within the garage entrance, so that the entire pole is contained within the garage? (b) In the preceding we assumed a reference frame in which the garage was at rest. Consider, however, a reference frame fixed with respect to the pole. In this frame the pole has its rest length (= 16 ft), but the garage, which now rushes toward the runner, is contracted (to 4 ft!) by the same Lorentz factor that operated on the pole in (a). How can a 16-ft pole fit into a 4-ft garage? If it can't, there is a violation of the principle of relativity, a serious matter indeed. (*Hint:* No body is truly rigid. When the front of the pole hits the barrier, the rear end of the pole keeps on going, at unchanged speed, until it "gets the word" by means of a compression wave sent down the pole. The question is, is there enough time for the rear end of the pole to get inside the garage before the front end of the pole reaches it? Can you show that the answer is "yes"? See [14].)

64. A very large Doppler shift. A spaceship, moving away from the earth at a speed of $0.90c$, reports back by transmitting on a frequency (measured in the spaceship frame) of 100 MHz. To what frequency must earth receivers be tuned to receive these signals?

65. The Doppler shift in terms of wavelengths. Show that the Doppler shift formulas (Eqs. 2-30) can be written in the forms

$$\lambda = \lambda_0(1 - \beta + \tfrac{1}{2}\beta^2 - \cdots) \qquad \text{(approaching)}$$

and

$$\lambda = \lambda_0(1 + \beta + \tfrac{1}{2}\beta^2 + \cdots) \qquad \text{(separating)},$$

in which λ_0 is the proper wavelength, that is, the wavelength measured by an observer for whom the source is at rest. These formulations are especially useful for $\beta \ll 1$. Compare Eqs. 2-31.

66. A red-shifted quasar. Observations on the light from a certain quasar show a red shift of a spectral line of laboratory wavelength 500 nm to a wavelength of 1300 nm. What is the quasar's speed of recession from us, according to a Doppler-effect interpetation?

67. Doppler shifts and the rotating sun. Because of the rotation of the sun, points on its surface at its equator have a speed of 1.85 km/s with respect to its center. Consider groups of atoms on opposite edges of the sun's equator as seen from the earth, emitting light of proper wavelength 546 nm. What wavelength difference is observed on the earth for the light from these two groups of atoms? (*Hint:* Use the formulas displayed in Problem 65.)

68. A Doppler shift revealed as a color change. A spaceship is receding from the earth at a speed of $0.20c$. A light on the rear of the ship appears blue ($\lambda = 450$ nm) to passengers on the ship. What color would it appear to an observer on earth?

69. The exact and the approximate Doppler formulas compared. (a) Calculate the Doppler wavelength shifts $\lambda - \lambda_0$ expected for the sodium D_1 line ($\lambda_0 = 589.6$ nm) for source and observer approaching each other at relative speeds of $0.050c$, $0.40c$, and $0.80c$. (b) Calculate the same quantities using the formula developed in Problem 65, discarding terms of order β^2 or higher. Compare the two sets of results.

70. Quasar, quasar, burning bright. . . . In the spectrum of quasar 3C9, some of the familiar hydrogen lines appear but they are shifted so far forward toward the red that their wavelengths are observed to be three times as large as that observed in the light from hydrogen atoms at rest in the laboratory. (a) Show that the classical Doppler equation gives a velocity of recession greater than c. (b) Assuming that the relative motion of 3C9 and the earth is entirely one of recession, find the recession speed predicted by the relativistic Doppler equation.

71. The Ives-Stillwell experiment. Neutral hydrogen atoms are moving along the axis of an evacuated tube with a speed of 2.0×10^6 m/s. A spectrometer is arranged to receive light emitted by these atoms in the direction of their forward motion. This light, if emitted from resting hydrogen atoms, would have a measured (proper) wavelength of 486.133 nm. (a) Calculate the expected wavelength for light emitted from the forward-moving (approaching) atoms, using the exact relativistic formula (see Eqs. 2-30). (b) By use of a mirror this same spectrometer can also measure the wavelength of light emitted by these moving atoms in the direction opposite to their motion. What wavelength is expected under this arrangement, in which the light source and the observer are—effectively—separating? (c) Calculated the difference between the average of the two wavelengths found in (a) and (b) and the unshifted (proper) wavelength. Show, by analyzing the formulas displayed in Problem 65, that this difference measures the β^2 term in these formulas. By this technique, Ives and Stillwell (see *Physics*, Part II, Sec. 42-5) were able to distinguish between the predictions of the classical and the relativistic Doppler formulas.

72. The transverse Doppler effect. Show that the transverse Doppler shift formula (Eq. 2-32) can be written in the form

$$\lambda = \lambda_0(1 + \tfrac{1}{2}\beta^2 + \tfrac{3}{8}\beta^4 + \cdots),$$

in which λ_0 is the proper wavelength, that is, the wavelength that would be measured by an observer for whom the source is at rest. This formulation is especially useful for $\beta \ll 1$. Compare Problem 65.

73. A case of purely transverse motion. Give the Doppler wavelength shift $\lambda - \lambda_0$, if any, for the sodium D_2 line (589.00 nm) emitted from a source moving in a circle with constant speed $(= 0.10c)$ as measured by an observer fixed at the center of the circle.

74. A case of (not quite) transverse motion. Calculate the wavelength shift

$$\Delta\lambda = \lambda - \lambda_0$$

for $\lambda_0 = 589.00$ nm, $\beta = 0.10$ and (a) $\theta = 90°$ (b) $88°$ (c) $85°$. (d) Why does such a small departure from purely transverse motion generate such a large change in the measured Doppler shift? What lessons are there here for the experimenter who wishes to measure the transverse Doppler effect?

75. The Doppler effects for sound and light compared. In the case of wave propagation in a medium (sound in air, say), the Doppler shifts for the source moving through the medium and for the observer moving through the medium are different, even though the relative speeds of the source with respect to the observer may be the same. For light in free space, however, the two situations are completely equivalent; the same Doppler shift results. Show that if we take the geometric mean of the two former results, we get exactly the relativistic Doppler shift of Eq. 2-27. (See *Physics*, Secs. 20-7 and 42-5; recall that the geometric mean of two quantities, a and b, is $\sqrt{ab}$.)

76. A moving radar transmitter and a moving clock. A radar transmitter T is fixed to a reference frame S' that is moving to the right with speed v relative to reference frame S (see Fig. 2-18). A mechanical timer (essentially a clock) in frame S', having a period τ_0 (measured in S') causes transmitter T to emit radar pulses, which travel at the speed of light and are received by R, a receiver fixed in frame S. (a) What would be the period τ of the timer relative to observer A, who is fixed in frame S? (b) Show that the receiver R would observe the time interval between pulses arriving from T, not as τ or as τ_0, but as

$$\tau_R = \tau_0 \sqrt{\frac{c + v}{c - v}}.$$

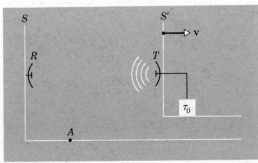

Figure 2-18. Problem 76.

(c) Explain why the observer at R measures a different period for the transmitter than does observer A, who is in the same reference frame. (*Hint:* A clock and a radar pulse are not the same.)

77. The classical aberration formula. In Example 11 the aberration of light from a star that is directly overhead is shown to be given (for $\beta << 1$) by

$$\tan \theta' = \frac{1}{\beta}.$$

The aberration angle α is shown in Fig. 2-11 to be related to θ' by

$$\alpha = 270° - \theta'.$$

Show, by combining these relations, that

$$\tan \alpha = \beta$$

results. This (see Eq. 1-11) is the prediction of classical theory for the aberration of starlight.

78. A moving rod and a moving laser beam. (a) A rod makes an angle of $30°$ with the x' axis of frame S', which is moving with speed $0.80c$ with respect to frame S. What angle does the rod make with the x axis of frame S? (b) A laser beam, generated by a laser gun fixed in frame S', also makes an angle of $30°$ with the x' axis. What angle does *it* make with the x axis, as determined by an observer in frame S? Why are these angles so different?

79. A particle and a light pulse. A particle has a speed u' in the S' frame, its track making an angle θ' with the x' axis. The particle is viewed by an observer in frame S, the two frames having a relative speed parameter β. (a) Show that the angle θ made by the track of the particle with the x axis is given by

$$\tan \theta = \frac{u' \sin \theta'}{\gamma(u' \cos \theta' + \beta c)}.$$

(b) Show that this equation reduces to the standard aberration formula (Eq. 2-29a) if the "particle" is, in fact, a light pulse, so that $u' = c$.

80. The aberration of light—a different formulation. Show that, by combining Eqs. 2-25 and 2-27 (rather than 2-25 and 2-26), the aberration formula can be written in the form

$$\cos \theta = \frac{\cos \theta' + \beta}{1 + \beta \cos \theta'}.$$

Test that this formula is equivalent to Eq. 2-29a by finding the value of θ corresponding to $\theta' = 30°$ and $\beta = 0.80$, using each formula.

81. A moving nucleus emits a gamma ray. A radioactive nucleus moves with a uniform velocity of $0.050c$ in the laboratory frame. It decays by emitting a gamma ray, which we may view as a pulse of electromagnetic radiation. What are the magnitude and direction of the velocity of this pulse as observed in the laboratory frame? Assume that the pulse is emitted (a) parallel to the direction of the motion of the nucleus, as judged by an observer on the nucleus; (b) at 45° to this direction; (c) at right angles to this direction.

82. The headlight effect. A source of light, at rest in the S' frame, emits radiation uniformly in all directions. (a) Show that the fraction of light emitted into a cone of half-angle θ' is given by

$$f = 0.50(1 - \cos \theta').$$

Calculate f for $\theta' = 30°$. (b) The source is viewed from frame S, the relative velocity of the two frames being $0.80c$. Find the value of θ (in frame S) to which this value of f corresponds, using the appropriate aberration formula. Repeat the calculation for $\beta = 0.90$ and for $\beta = 0.990$. Can you see why this aberration phenomenon is often referred to as the "headlight effect"?

83. The headlight effect—a high-speed limit. A source of light, at rest in the S' frame, emits uniformly in all directions. The source is viewed from frame S, the relative speed parameter relating the two frames being β. (a) Show that at high speeds (that is, as $\beta \to 1$), the forward-pointing cone into which the source emits half of its radiation has a half-angle $\theta_{0.5}$ given closely, in radian measure, by

$$\theta_{0.5} = \sqrt{2(1 - \beta)}.$$

(b) What value of $\theta_{0.5}$ is predicted for the gamma radiation emitted by a beam of energetic neutral pions, for which $\beta = 0.993$? (c) At what speed would a light source have to move toward an observer to have half of its radiation concentrated into a narrow forward cone of half-angle 5.0°?

84C. Using your calculator—Lorentz transformations for an event pair. Write a program for your handheld, programmable calculator to handle Lorentz transformations for an event pair. Accept as inputs: (1) the speed parameter β, (2) the Δx (or the $\Delta x'$) coordinate difference, and (3) the Δt (or the $\Delta t'$) coordinate difference. Display as outputs, in succession: (1) the Lorentz factor γ, (2) the $\Delta x'$ (or the Δx) coordinate difference, (3) the $\Delta t'$ (or the Δt) coordinate difference, (4) the spacetime interval, and (5) a signal as to whether the event pair is spacelike or timelike. (Hint: See Table 2-3. Also, if your calculator does not display letters you can manage (5) above by displaying a string of 5's for "spacelike" and a string of 7's for "timelike".)

85C. Using your calculator. Test the program that you have written in Problem 84C in the following ways. (a) Two events are simultaneous in frame S. Show, by trial, that they cannot be simultaneous in frame S' no matter what values you assign to β (except zero) or to Δx (except zero). Show also, by trial, that this event pair will be spacelike in all frames. (b) Two events occur at the same position in frame S. Show, by trial, that they cannot occur at the same position in frame S' no matter what values you assign to β (except zero) or to Δt (except zero). Show, also by trial, that this event pair will be timelike in all frames. (c) A rod is at rest in frame S'. You (in frame S) measure its length to be 5.00 m by making simultaneous measurements of its endpoints. What is its rest length if $\beta = 0.600$? Is your result consistent with the length contraction phenomenon? (d) A clock is at rest in frame S. You (also in frame S) measure a time interval of 2.50 μs between two events. What interval would S' observe between these same events if $\beta = 0.600$? Is your result consistent with the time dilation phenomenon? (e) Put $\Delta x = 5.00 \times 10^8$ m and $\Delta t = 4.00$ s. What is the magnitude and the nature of the spacetime interval associated with this event pair? Show, by experimenting with various values of β (including zero and negative values) that the interval is the same in all inertial frames. Change the character of the interval by changing Δx and by changing Δt.

86C. Using your calculator—the Doppler effect. Use your handheld, programmable calculator to write a program that will do the following: Accept as inputs (1) the proper wavelength λ_0 of the source or (2) the corresponding proper frequency ν_0, and also (3) the speed parameter β; if the source and the observer are approaching put $\beta > 0$ but if they are separating put $\beta < 0$. Display as successive outputs (1) the wavelength λ, (2) the proper wavelength λ_0, (3) the wavelength shift $(\lambda - \lambda_0)$, (4) the frequency ν, (5) the proper frequency ν_0, and (6) the frequency shift $(\nu - \nu_0)$. See Eqs. 2-30.

87C. Using your calculator. Check out the program that you have written in Problem 86C as follows: (a) A galaxy is receding from us at $0.30c$ where c is the speed of light. What is the expected shift in wavelength for the sodium spectrum line in the light from this galaxy? The laboratory wavelength of this line is 589 nm. (b) An automobile is receding from a Doppler radar speed detector at 100 mi/hr (= 44.7 m/s). What will be the frequency shift of the reflected radar beam? The proper wavelength of the radar radiation is 3.00 cm.

references

1. Robert Resnick, *Introduction to Special Relativity*, Wiley, New York, 1968.

2. David H. Frisch and James H. Smith, "Measurement of Relativistic Time Dilation Using μ-Mesons," *Am. J. Phys.*, **31,** 342 (1963). See also the related film, "Time Dilation—An Experiment with μ-Mesons," Educational Services, Inc., Watertown, Mass.

3. V. T. Weisskopf, "The Visual Appearance of Rapidly Moving Objects," *Phys. Today* (September 1960).

4. N. C. McGill, "The Apparent Shape of Rapidly Moving Objects in Special Relativity," *Contemp. Phys.* (January 1968).

5. G. D. Scott and H. J. van Driel, "Geometric Appearances at Relativistic Speeds," *Am. J. Phys.*, **38,** 971 (1970).

6. Milton A. Rothman, "Things That Go Faster Than Light," *Scientific American* (July 1960).

7. Gerald Feinberg, "Particles That Go Faster Than Light," *Scientific American* (February 1970).

8. Hirsch I. Mandelberg and Louis Witten, "Experimental Verification of the Relativistic Doppler Effect," *J. Optical Soc. Am.*, **52,** 529 (1962).

9. Walter Kundig, "Measurement of the Transverse Doppler Effect in an Accelerated System," *Phys. Rev.*, **129,** 2371 (1963).

10. H. Bondi, "The Teaching of Special Relativity," *Phys. Educ.*, **1,** 223 (1966).

11. J. C. Hafele and Richard E. Keating, "Around-the-World Atomic Clocks: Predicted Relativistic Time Gains," *Science*, **177,** 166 (1972), and J. C. Hafele and Richard E. Keating, "Around-the-World Atomic Clocks: Observed Relativistic Time Gains," *Science*, **177,** 168 (1972).

12. Suggested by Professor William Doyle of Dartmouth College.

13. See Wolfgang Rindler, *Essential Relativity* (Van Nostrand Reinhold, New York, 1969), sec. 36.

14. Wolfgang Rindler, "Length Contraction Paradox," *Am. J. Phys.*, **29,** 365 (1961).

15. N. David Mermin, "Relativistic Addition of Velocities Directly from the Constancy of the Velocity of Light," *Am. J. Phys.*, **51,** 1130 (1983).

16. Harry Woolf, Ed. *Some Strangeness in the Proportion: A Centennial Symposium to Celebrate the Achievements of Albert Einstein* (Addison-Wesley, Reading, Mass., 1980). See especially Wolfgang K. H. Panofsky, "Special Relativity in Engineering," and Edward M. Purcell, "Comments on 'Special Relativity Theory in Engineering.' "

CHAPTER 3

relativistic dynamics

From this equation it directly follows that:—If a body gives off the energy L in the form of radiation, its mass diminishes by L/c². The fact that the energy withdrawn from the body becomes energy of radiation evidently makes no difference, so that we are led to the more general conclusion that . . . The mass of a body is a measure of its energy content. . . .

Albert Einstein (1905)

3-1 MECHANICS AND RELATIVITY

In Chapter 1 we saw that experiment forced us to the conclusion that the Galilean transformations had to be replaced and the basic laws of mechanics, which were consistent with those transformations, needed to be modified. In Chapter 2 we obtained the new transformation equations, the Lorentz transformations, and examined their implications for kinematic phenomena. Now we must consider dynamic phenomena and find how to modify the laws of classical mechanics so that the new mechanics is consistent with relativity.

Basically, classical Newtonian mechanics is inconsistent with relativity because its laws are invariant under a Galilean transformation and *not* under a Lorentz transformation. This formal result is plausible, as well, from other considerations. For example, in Newtonian mechanics a force can accelerate a particle to indefinite speeds, whereas in relativity the limiting speed is *c*. We need a new law of motion that is consistent with relativity. When we obtain such a law of motion, we must also ensure that it reduces to the Newtonian form as $\beta \, (= v/c) \rightarrow 0$, since, in the domain where $\beta \ll 1$, Newton's laws are consistent with experiment. Thus, the relativistic law of motion will be a generalization of the classical one.

We shall proceed by studying collisions. The laws of conservation of momentum and energy are valid classically during such interactions. If we require that these conservation laws also be valid relativistically (that is, invariant under a Lorentz transformation) and hence that they be general laws of physics, we must modify them from the classical form in such a way that they also reduce to the classical form as $\beta \rightarrow 0$. In this way, we shall obtain the relativistic law of motion.

3-2 THE NEED TO REDEFINE MOMENTUM

The first thing we wish to show is that if we want to find a quantity such as momentum (for which there is a conservation law in classical physics) that is also subject to a conservation law in relativity, we cannot use the same expression for momentum as the classical one. We must, instead, redefine momentum in order that a law of conservation of momentum in collisions be invariant under a

93

Lorentz transformation. Ultimately this leads to a more general view of the concept of mass as well.

Let us begin by considering a head-on collision between two identical particles. We choose the simplest case, namely, one in which the particles stick together after impact—a so-called totally inelastic collision. We shall first analyze the collision from the point of view of classical mechanics (see *Physics*, Part I, Sect. 10-4).

In Fig. 3-1*a* we view the collision from a reference frame S chosen so as to present the collision as symmetrically as possible. Initially, we see that the particles, which have the same mass m, are approaching each other with the same speed u. After the collision the particles form one larger particle whose mass is $2m$ (because of the conservation of mass) and whose speed in the S frame is zero (because of the conservation of momentum). The total system momentum p, as measured in frame S, is zero both before and after the collision.

Let us now view this same collision from a frame S', moving to the left as viewed from S with a speed v that we deliberately choose to be equal to u; see Fig. 3-1*b*. With this choice for the separation speed of the two frames, one of the colliding particles is initially at rest in S' and the resulting compound particle moves to the right with speed u. The other colliding particle has a speed indicated by u' in Fig. 3-1*b*. From the Galilean velocity transformation law, however, we see at once that $u' = 2u$. Note that again the mass is conserved (at the value $2m$) and so is the total system momentum (now at the value $p' = 2mu$) in frame S'.

Now let us examine this same collision from the relativistic point of view. We shall carry out the analysis just as before with the exception that we shall use the relativistic velocity transformation law in place of the classical Galilean transformation law we used earlier. In frame S momentum is still conserved, simply because of the symmetry of the collision as viewed from this frame. In frame S', however, the speed shown as u' in Fig. 3-1*b* is no longer equal to $2u$, but is given

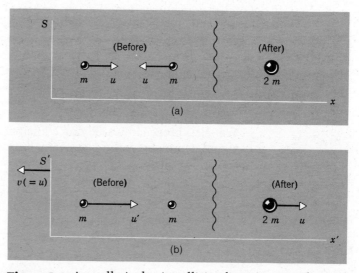

Figure 3-1. A totally inelastic collision between two identical particles is viewed from two reference frames. (*a*) Frame S, in which the center of mass of the system is stationary. (*b*) Frame S', which moves to the left with respect to frame S with speed $v(= u)$.

by (see Eq. 2-21)

$$u' = \frac{u + v}{1 + uv/c^2} = \frac{2u}{1 + u^2/c^2}.$$

After the collision the speed of the compound particle will again be u. We see now that, although mass is again conserved at the value $2m$, the total momentum is *not* conserved in frame S'. Before the collision it is $p' = mu' = 2mu/(1 + u^2/c^2)$, whereas after the collision it is simply $p' = 2mu$.

Our conclusion is that, if we compute momentum according to the classical formulas $\mathbf{p} = m\mathbf{u}$ and $\mathbf{p'} = m\mathbf{u'}$, then when momentum is conserved in a collision in one frame, it is not conserved in the other frame. This result contradicts the basic postulate of special relativity that the laws of physics are the same in all inertial reference frames. If the conservation of momentum in collisions is to be a law of physics, then the classical definition of momentum cannot be correct in general.

In the next section, we shall show that it is possible to preserve the *form* of the classical definition of the momentum of a particle, $\mathbf{p} = m\mathbf{u}$, where $\mathbf{p}$ is the momentum, m the mass, and $\mathbf{u}$ the velocity of a particle, and also to preserve the classical law of the conservation of momentum of a system of interacting particles, providing that we modify the classical concept of mass. We need to let the mass of a particle be a function of its speed u; that is, $m = m_0/\sqrt{1 - u^2/c^2}$, where m_0 is the classical mass and m is the relativistic mass of the particle. Clearly, as u/c tends to zero, m tends to m_0. The relativistic momentum then becomes $\mathbf{p} = m\mathbf{u} = m_0\mathbf{u}/\sqrt{1 - \beta^2}$ and reduces to the classical expression $\mathbf{p} = m_0\mathbf{u}$ as $\beta \to 0$. Let us now deduce these results.

3-3 RELATIVISTIC MOMENTUM

In our analysis above, we considered the mass of a particle to be independent of its motion. However, we have already learned that the measured length of a rod and the measured rate of a clock are affected by the motion of the rod or the clock relative to an observer. Would it be so surprising, then, to find that the measured mass of a particle also depends on its state of motion?

Let us then assume that the mass of a particle is *not* a constant but is a function of the speed of the particle. Then, in the totally inelastic collision we have been considering, we should use the symbol m_0 for the mass of a particle at rest, m for the mass of the same particle if its speed is u, and m' for its mass if its speed is u'. For the masses of the compound particle we can use M_0 and M. Figure 3-2 shows the collision of Fig. 3-1 with the masses and the speeds relabeled in this way.

Because we assume that (relativistic) mass is conserved in the S' frame of Fig. 3-2b, we can write

$$M = m' + m_0, \tag{3-1}$$

and, for the conservation of (relativistic) momentum,

$$Mu = m'u'. \tag{3-2}$$

Now u' is related to u by the relativistic law for the addition of velocities, or (see Eq. 2-21)

$$u' = \frac{u + v}{1 + uv/c^2}.$$

$2m_0(\delta-1)c^2$

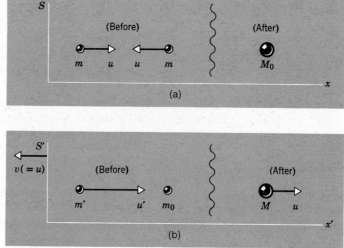

Figure 3-2. The same collision as in Fig. 3-1, with the masses and speeds relabeled with symbols appropriate for a relativistic analysis.

Recalling that $v = u$ (see Fig. 3-2), we can rewrite this as

$$u' = \frac{2u}{1 + u^2/c^2}.$$ (3-3)

The above three numbered equations involve five variables, m_0, m', M, u', and u. If we could eliminate M and u from these equations, we would be left with a relationship involving only m', m_0, and u'. If we look at the left-hand particle in Fig. 3-2*b*, we see that these are precisely the variables we need if we are to relate the relativistic mass (m') of this particle to its rest mass (m_0) and to its speed (u').

We start by eliminating M between Eqs. 3-1 and 3-2, obtaining

$$u = u' \left(\frac{m'}{m' + m_0} \right).$$

Combining this with Eq. 3-3 leads, after some rearrangement (see Problem 18), to

$$m' = \frac{m_0}{\sqrt{1 - (u')^2/c^2}},$$

which is the relationship we seek. We can extend our conclusion beyond the confines of the specific problem we have been analyzing and write as a general result, true for *any* particle of rest mass m_0 and speed u:

$$m = \frac{m_0}{\sqrt{1 - u^2/c^2}},$$ (3-4*a*)

which tells us how the relativistic mass m of a body varies with its speed u. If we represent u/c by β, we can express this relationship as

$$m = \frac{m_0}{\sqrt{1 - \beta^2}}$$ (3-4*b*)

or, equivalently, as

$$m = m_0 \gamma,$$ (3-4*c*)

in which γ is the familiar Lorentz factor. Recall that this factor enters in a similar simple and direct way into the relativistic formulas for length contraction (see Eq. 2-12b) and time dilation (see Eq. 2-14b).

Inspection of Eqs. 3-4 shows us that (1) the relativistic mass m is always greater than the rest mass m_0, and (2) as $\beta \rightarrow 0$ (or, equivalently, as $\gamma \rightarrow 1$), then $m \rightarrow m_0$, as we expect in this Newtonian low-speed limit.

Hence, if we want to preserve the *form* of the classical momentum conservation law while requiring that the law be relativistically invariant, we must define the mass of a moving body by Eqs. 3-4. That is, momentum still has the form $m\mathbf{u}$, but mass is defined as $m = m_0/\sqrt{1 - u^2/c^2}$. Note that u is the speed of the body relative to S, which we can regard as the laboratory frame, and that u has no necessary connection with changing reference frames. By accepting Eqs. 3-4 as our definition of the mass of a moving body, we implicitly assume that the mass of a body does not depend on its acceleration relative to the reference frame, although it does depend on its speed. Mass remains a scalar quantity in the sense that its value is independent of the *direction* of the velocity of the body. The rest mass m_0 is often called the *proper mass*, for it is the mass of the body measured, like proper length and proper time, in the inertial frame in which the body is at rest.

We have presented above a derivation of an expression for relativistic momentum that obviously centers around a very special case, a totally inelastic one-dimensional collision. Such a derivation enables us to make an educated guess as to what the general result may be. We have avoided rather involved general derivations that lead, in any case, to exactly the same results. When the general case is done, u becomes the absolute value of the velocity of the particle; that is, $u^2 = u_x^2 + u_y^2 + u_z^2$.

Hence, to conclude, in order to make the conservation of momentum in collisions a law that is experimentally valid in all reference frames, we must define momentum not as $m_0\mathbf{u}$, but as

$$\mathbf{p} = \frac{m_0\mathbf{u}}{\sqrt{1 - u^2/c^2}}. \tag{3-5}$$

The components of the momentum then are

$$p_x = \frac{m_0 u_x}{\sqrt{1 - u^2/c^2}}, \qquad p_y = \frac{m_0 u_y}{\sqrt{1 - u^2/c^2}}, \qquad p_z = \frac{m_0 u_z}{\sqrt{1 - u^2/c^2}}, \tag{3-6}$$

which we write out explicitly to emphasize that the magnitude u of the total velocity appears in the denominator of each component equation.

EXAMPLE 1.

Relativistic Mass and Rest Mass. For what value of u/c $(=\beta)$ will the relativistic mass of a particle exceed its rest mass by a given fraction f?

From Eq. 3-4b, we have

$$f = \frac{m - m_0}{m_0} = \frac{m}{m_0} - 1 = \frac{1}{\sqrt{1 - \beta^2}} - 1,$$

which, solved for β, is

$$\beta = \frac{\sqrt{f(2 + f)}}{1 + f}.$$

The table below shows some computed values, which hold for all particles regardless of their rest mass.

f	β
0.001 (0.1%)	0.045
0.01	0.14
0.1	0.42
1 (100%)	0.87
10	0.996
100	0.99995
1000	0.9999995

EXAMPLE 2.

Relativistic Mass is Conserved. In writing Eq. 3-1 we assumed that relativistic mass is conserved when the collision of Fig. 3-2 is viewed from the S' frame. Show that relativistic mass is also conserved in the S frame.

The conservation of mass in the S frame requires that

$$M_0 = m + m = 2m,$$

in which the meanings of the symbols will be clear from Fig. 3-2. Now M_0 is related to M, and m to m_0, by the relativistic mass relationship (Eq. 3-4a). Using that equation, we can recast the above expression as

$$M = \frac{2m_0}{1 - u^2/c^2}. \tag{3-7}$$

To demonstrate that Eq. 3-7 is true, we turn to Eqs. 3-1 and 3-2 and eliminate m' between them. The result is

$$M = \frac{m_0}{1 - u/u'}.$$

We now use Eq. 3-3 to eliminate u' from this expression, obtaining

$$M = \frac{m_0}{1 - \frac{1}{2}(1 + u^2/c^2)} = \frac{2m_0}{1 - u^2/c^2},$$

which is precisely Eq. 3-7, the relationship we sought to prove.

The implications of our new relativistic definition of mass for the relationship between mass and energy are considered in later sections.

We can summarize the relativistic definition of momentum that we have introduced in this section by writing, for the x-coordinate,

$$p_x = m_0\gamma u_x = [m_0\gamma]\, dx/dt = m\, dx/dt$$

in which we have combined the Lorentz factor γ with the rest mass m_0 to generate the relativistic mass m. In some more advanced treatments of relativity, however, the Lorentz factor is combined with the time element dt and the concept of relativistic mass is not introduced. Thus:

$$p_x = m_0\gamma u_x = m_0\,[dx/(dt/\gamma)] = m_0\, dx/d\tau,$$

in which $d\tau$ is an element of proper time; see Eq. 2-14b. This approach has the advantage that both m_0 (which in this treatment no longer requires a subscript) and $d\tau$ are invariant quantities. The choice between the two approaches reduces to a matter of taste. We believe that, in an introductory treatment, the pedagogic advantages of the first approach outweigh the formal advantages of the second (see Reference 6, Section 3-4). Note that in neither case is the relativistic definition of momentum in question.

3-4 THE RELATIVISTIC FORCE LAW AND THE DYNAMICS OF A SINGLE PARTICLE

Newton's second law must now be generalized to

$$\mathbf{F} = \frac{d}{dt}(\mathbf{p}) = \frac{d}{dt}\left(\frac{m_0\mathbf{u}}{\sqrt{1 - u^2/c^2}}\right) \tag{3-8}$$

in relativistic mechanics. When the law is written in this form, we can immediately deduce the law of the conservation of relativistic momentum from it; when $\mathbf{F}$ is zero, $\mathbf{p} = m_0\mathbf{u}/\sqrt{1 - u^2/c^2}$ must be a constant. In the absence of external forces, the momentum is conserved. Notice that this new form of the law, Eq.

3-8, is *not* equivalent to writing

$$\mathbf{F} = m\mathbf{a} = \left(\frac{m_0}{\sqrt{1 - u^2/c^2}}\right)\left(\frac{d\mathbf{u}}{dt}\right),$$

in which we simply multiply the acceleration by the relativistic mass.

We find that experiment agrees with Eq. 3-8. When, for example, we investigate the motion of high-speed charged particles, it is found that the equation correctly describing the motion is

$$q(\mathbf{E} + \mathbf{u} \times \mathbf{B}) = \frac{d}{dt}\left(\frac{m_0\mathbf{u}}{\sqrt{1 - u^2/c^2}}\right), \tag{3-9}$$

which agrees with Eq. 3-8. Here, $q(\mathbf{E} + \mathbf{u} \times \mathbf{B})$ is the Lorentz electromagnetic force, in which $\mathbf{E}$ is the electric field, $\mathbf{B}$ is the magnetic field, and $\mathbf{u}$ is the particle velocity, all measured in the same reference frame, and q and m_0 are constants that describe the electrical (charge) and inertial (rest mass) properties of the particle, respectively (see *Physics*, Part II, Sec. 33-2). Notice that the form of the Lorentz force law of classical electromagnetism remains valid relativistically, as we should expect from the discussion of Chapter 1.

Later we shall turn to the question of how forces transform from one Lorentz frame to another. For the moment, however, we confine ourselves to one reference frame (the laboratory frame) and develop other concepts in mechanics, such as work and energy, which follow from the relativistic expression for force (Eq. 3-8). We shall confine ourselves to the motion of a single particle. In succeeding sections we shall consider many-particle systems and conservation laws.

In Newtonian mechanics we define the kinetic energy, K, of a particle to be equal to the work done by an external force in increasing the speed of the particle from zero to some value u (see *Physics*, Part I, Sec. 7-5). That is,

$$K = \int_{u=0}^{u=u} \mathbf{F} \cdot d\mathbf{l},$$

where $\mathbf{F} \cdot d\mathbf{l}$ is the work done by the force $\mathbf{F}$ in displacing the particle through $d\mathbf{l}$. For simplicity, we can limit the motion to one dimension—say, x—the three-dimensional case being an easy extension. Then, classically,

$$K = \int_{u=0}^{u=u} F\,dx = \int m_0\left(\frac{du}{dt}\right)dx = \int m_0\,du\,\frac{dx}{dt} = m_0\int_0^u u\,du = \tfrac{1}{2}m_0u^2.$$

Here we write the particle mass as m_0 to emphasize that, in Newtonian mechanics, we do not regard the mass as varying with the speed, and we take the force to be $m_0a = m_0(du/dt)$.

In relativistic mechanics, it proves useful to use a corresponding definition for kinetic energy in which, however, we use the relativistic equation of motion, Eq. 3-8, rather than the Newtonian one. Then, relativistically,

$$K = \int_{u=0}^{u=u} F\,dx = \int \frac{d}{dt}(mu)\,dx = \int d(mu)\,\frac{dx}{dt}$$

$$= \int (m\,du + u\,dm)u = \int_{u=0}^{u=u} (mu\,du + u^2\,dm), \tag{3-10}$$

in which both m and u are variables. These quantities are related, furthermore, by Eq. 3-4a, $m = m_0/\sqrt{1 - u^2/c^2}$, which we can rewrite as

$$m^2c^2 - m^2u^2 = m_0^2c^2.$$

Taking differentials in this equation yields

$$2mc^2\,dm - m^2 2u\,du - u^2 2m\,dm = 0,$$

which, upon division by $2m$, can also be written as

$$mu\,du + u^2\,dm = c^2\,dm.$$

The left side of this equation is exactly the integrand of Eq. 3-10. Hence, we can write the relativistic expression for the kinetic energy of a particle as

$$K = \int_{u=0}^{u=u} c^2\,dm = c^2 \int_{m=m_0}^{m=m} dm = mc^2 - m_0 c^2$$

$$= (m - m_0)c^2. \tag{3-11a}$$

By using Eq. 3-4, we obtain equivalently

$$K = m_0 c^2 \left(\frac{1}{\sqrt{1 - u^2/c^2}} - 1 \right) \tag{3-11b}$$

or

$$K = m_0 c^2 (\gamma - 1). \tag{3-11c}$$

Also, if we take $mc^2 = E$, where E is called the *total energy* of the particle—a name whose aptness will become clear later—we can express Eq. 3-11a compactly as

$$E = m_0 c^2 + K, \tag{3-12}$$

in which $m_0 c^2$ is called the *rest energy* of the particle. The rest energy (by definition) is the energy of the particle at rest, when $u = 0$ and $K = 0$. The total energy of the particle (Eq. 3-12) is the sum of its rest energy and its kinetic energy.

The relativistic expression for K must reduce to the classical result, $\frac{1}{2}m_0 u^2$, when $u/c \ll 1$. Let us check this. From Eq. 3-11b,

$$K = m_0 c^2 \left(\frac{1}{\sqrt{1 - u^2/c^2}} - 1 \right)$$

$$= m_0 c^2 \left[\left(1 - \frac{u^2}{c^2} \right)^{-1/2} - 1 \right],$$

and the binomial theorem expansion in (u/c) gives

$$K = m_0 c^2 \left[1 + \frac{1}{2}\left(\frac{u}{c}\right)^2 + \frac{3}{8}\left(\frac{u}{c}\right)^4 + \cdots - 1 \right]$$

$$= \frac{1}{2} m_0 u^2 \left[1 + \frac{3}{4}\left(\frac{u}{c}\right)^2 + \cdots \right]. \tag{3-11d}$$

We see that, as $u \to 0$, the second term in square brackets above become negligible in comparison to the first term and the expression for K approaches the classical value, $\frac{1}{2}m_0 u^2$, thereby confirming the Newtonian limit of the relativistic result.

It is interesting to notice also that, as $u \to c$ in Eq. 3-11b, the kinetic energy K tends to infinity. That is, from Eq. 3-10, an infinite amount of work would need to be done on the particle to accelerate it up to the speed of light. Once again we find c playing the role of a limiting velocity. Note also from Eq. 3-11a, $K = (m - m_0)c^2$, that a change in the kinetic energy of a particle is related to a change in its relativistic mass.

EXAMPLE 3.

A Special Speed. What is the speed of a particle whose kinetic energy K is equal to its rest energy $m_0 c^2$?

The kinetic energy of a particle is defined from Eq. 3-11a, or

$$K = (m - m_0)c^2.$$

If we substitute $m_0 c^2$ for K in this equation we find, after cancellation and minor rearrangement, that

$$m = 2m_0,$$

in which m is the relativistic mass of the particle. Substituting for m from Eq. 3-4b gives us

$$\frac{m_0}{\sqrt{1 - \beta^2}} = 2m_0,$$

or, cancelling and rearranging,

$$2\sqrt{1 - \beta^2} = 1.$$

Solving yields $\beta = 0.866$, and thus $u = \beta c = 0.866c$ is the speed of a particle whose kinetic energy is equal to its rest energy. Because the rest mass m_0 of the particle cancelled out in our derivation, the nature of the particle does not matter. The same speed holds for all particles, be they electrons, protons, or—for that matter—baseballs.

We often seek a connection between the kinetic energy K of a rapidly moving particle and its momentum p. This can be found by eliminating u between Eq. 3-11b and Eq. 3-5. You can verify that the result is

$$(K + m_0 c^2)^2 = (pc)^2 + (m_0 c^2)^2, \tag{3-13a}$$

which, with the total energy $E = K + m_0 c^2$, can also be written as

$$E^2 = (pc)^2 + (m_0 c^2)^2. \tag{3-13b}$$

The right triangle of Fig. 3-3 is a useful mnemonic device for remembering Eq. 3-13.

The relationship between K and p (Eq. 3-13a) should reduce to the Newtonian expression $p = \sqrt{2m_0 K}$ for $u/c \ll 1$. To see that it does, let us expand Eq. 3-13a, obtaining

$$K^2 + 2Km_0 c^2 = p^2 c^2. \tag{3-13c}$$

When $u/c \ll 1$, the kinetic energy K of a moving particle will always be much less than its rest energy $m_0 c^2$. Under these circumstances, the first term on the left above (K^2) can be neglected in comparison with the second term ($2Km_0 c^2$), and the equation becomes $p = \sqrt{2m_0 K}$, as required.

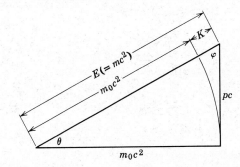

Figure 3-3. A mnemonic device, using a right triangle and the Pythagorean relation, to help in remembering the relations between total energy E, rest energy $m_0 c^2$, and momentum p; see Eq. 3-13b. Shown also is the relation $E = m_0 c^2 + K$ between total energy, rest energy, and kinetic energy. You can show that $\sin \theta = \beta$ and $\sin \phi = 1/\gamma$.

The relativistic expression, Eq. 3-13b, often written as

$$E = c\sqrt{p^2 + m_0^2 c^2},\tag{3-14}$$

is useful in high-energy physics to calculate the total energy of a particle when its momentum is given, or vice versa. By differentiating Eq. 3-14 with respect to p, we can obtain another useful relation:

$$\frac{dE}{dp} = \frac{pc}{\sqrt{m_0^2 c^2 + p^2}} = \frac{pc^2}{c\sqrt{m_0^2 c^2 + p^2}} = \frac{pc^2}{E}.$$

But with $E = mc^2$ and $p = mu$ this reduces to

$$\frac{dE}{dp} = u,\tag{3-15}$$

a result that, incidentally, is also valid in classical dynamics.

EXAMPLE 4.

Approximations at Low Speeds. Show that when $u/c < 0.1$, then (a) the ratio $K/m_0 c^2$ is less than $1/200$ ($= 0.005$) and the classical expressions for (b) the kinetic energy and (c) the momentum of a particle may be used with an error of less than 1 percent.

(a) The kinetic energy of a particle is given by Eq. 3-11b, or

$$K = m_0 c^2 \left(\frac{1}{\sqrt{1 - \beta^2}} - 1\right).$$

or, for $\beta = 0.1$,

$$\frac{K}{m_0 c^2} = \frac{1}{\sqrt{1 - (0.1)^2}} - 1 = 1.0050 - 1 = 0.0050.$$

If $\beta < 0.1$, it is easy to show that this ratio is correspondingly less than 0.0050.

(b) Now let us investigate the accuracy of the classical kinetic energy relationship ($K_c = \frac{1}{2} m_0 u^2$) for $\beta < 0.1$. Using the series expansion of Eq. 3-11d, we can write for K, the relativistic expression for the kinetic energy,

$$K = K_c(1 + \tfrac{3}{4}\beta^2 + \cdots).$$

For e, the percent error involved in our choice of formula, we then have, keeping only the first two terms in the expression for K,

$$e = \frac{K - K_c}{K} \times 100 = \frac{K_c(1 + \tfrac{3}{4}\beta^2) - K_c}{K_c(1 + \tfrac{3}{4}\beta^2)} \times 100$$

$$= \frac{300\beta^2}{4 + 3\beta^2}.$$

For $\beta = 0.1$ this yields

$$e = \frac{(300)(0.1)^2}{4 + (3)(0.1)^2} = 0.7\%,$$

which is less than 1 percent.

(c) We now turn our attention to the classical momentum formula, $p_c = m_0 u$. The relativistic formula for the momentum follows from Eq. 3-4c as

$$p = mu = \gamma m_0 u = \gamma p_c.$$

For e, the percent error found by using the classical formula, we have

$$e = \frac{p - p_c}{p} \times 100 = \frac{\gamma p_c - p_c}{\gamma p_c} \times 100 = 100 \left(\frac{\gamma - 1}{\gamma}\right).$$

For $\beta = 0.1$ we can easily show that $\gamma (= 1/\sqrt{1 - \beta^2}) = 1.0050$. Thus,

$$e = 100 \left(\frac{1.0050 - 1}{1.0050}\right) = 0.5\%,$$

which is less than 1 percent.

EXAMPLE 5.

Approximations at High Speeds. Show that when $u/c > 99/100 (= 0.99)$, then (a) the ratio $K/m_0 c^2$ is greater than 6 and (b) the relativistic relation $p' = E/c$ for a zero-rest-mass particle may be used, with an error of less than 1 percent, for the momentum of a particle whose rest mass is actually m_0.

(a) The kinetic energy of a particle is given by Eq. 3-11c, or

$$K = m_0 c^2 (\gamma - 1).$$

The Lorentz factor γ for $\beta = 0.99$ is readily found from

$$\gamma = \frac{1}{\sqrt{1 - \beta^2}} = \frac{1}{\sqrt{1 - (0.99)^2}} = 7.09.$$

Thus we have

$$K/m_0 c^2 = \gamma - 1 = 7.09 - 1 = 6.09,$$

which is greater than 6. If $\beta > 0.99$, it is easy to show that this ratio will be correspondingly greater than 6.09.

(b) The total energy E of a particle is equal to mc^2 in which m is the relativistic mass. From $p' = E/c$, we then have

$$p' = \frac{E}{c} = \frac{mc^2}{c} = mc,$$

whereas the formula for the momentum of a particle whose rest mass is *not* zero is simply (see Eq. 3-5) $p = mu$. Thus e, the percent error involved in using the approximate formula, is

$$e = \frac{p - p'}{p} \times 100 = \frac{mu - mc}{mu} \times 100 = 100 \left(\frac{\beta - 1}{\beta} \right).$$

For $\beta = 0.99$ this yields

$$e = 100 \left(\frac{0.99 - 1}{0.99} \right) = -1.01\%,$$

which is essentially 1 percent. It is easy to show that if $\beta > 0.99$, the error will be even smaller.

From this example and the preceding one we see that relativistic effects are small, though not necessarily negligible, when $u < 0.1\,c$; and that when $u > 0.99\,c$, purely relativistic effects predominate.

As a final consideration in the relativistic dynamics of a single particle, we look at the acceleration of a particle under the influence of a force. In general, the force is given by

$$\mathbf{F} = \frac{d\mathbf{p}}{dt} = \frac{d}{dt}(m\mathbf{u})$$

or

$$\mathbf{F} = m\frac{d\mathbf{u}}{dt} + \mathbf{u}\frac{dm}{dt}. \qquad (3\text{-}16)$$

We know that $m = E/c^2$, so

$$\frac{dm}{dt} = \frac{1}{c^2}\frac{dE}{dt} = \frac{1}{c^2}\frac{d}{dt}(K + m_0 c^2) = \frac{1}{c^2}\frac{dK}{dt}.$$

But

$$\frac{dK}{dt} = \frac{(\mathbf{F} \cdot d\mathbf{l})}{dt} = \mathbf{F} \cdot \frac{d\mathbf{l}}{dt} = \mathbf{F} \cdot \mathbf{u},$$

so

$$\frac{dm}{dt} = \frac{1}{c^2}\mathbf{F} \cdot \mathbf{u}.$$

We can now substitute this into Eq. 3-16 and obtain

$$\mathbf{F} = m\frac{d\mathbf{u}}{dt} + \frac{\mathbf{u}(\mathbf{F} \cdot \mathbf{u})}{c^2}.$$

The acceleration $\mathbf{a}$ is defined by $\mathbf{a} = d\mathbf{u}/dt$, so the general expression for acceleration is

$$\mathbf{a} = \frac{d\mathbf{u}}{dt} = \frac{\mathbf{F}}{m} - \frac{\mathbf{u}}{mc^2}(\mathbf{F} \cdot \mathbf{u}). \qquad (3\text{-}17)$$

What this equation tells us at once is that, in general, the acceleration $\mathbf{a}$ is *not* parallel to the force in relativity, since the last term above is in the direction of the velocity $\mathbf{u}$.

EXAMPLE 6.

Accelerating Electrons. A certain linear accelerator is accelerating electrons by letting them fall through a potential difference of 4.50 MV. Find (a) the kinetic energy, (b) the mass, and (c) the speed of the electrons as they emerge from the accelerator into the laboratory.

(a) The charge on the electron is $-e$, where $e (= 1.60 \times 10^{-19}$ C) is the electronic charge. The kinetic energy is given by

$$K = q(V_i - V_f)$$
$$= (-1.00 \text{ electronic charge})(0 - 4.50 \times 10^6 \text{ V})$$
$$= 4.50 \text{ MeV}.$$

In SI units we have, for this same quantity,

$$K = (-1.60 \times 10^{-19} \text{ C})(0 - 4.50 \times 10^6 \text{ V})$$
$$= 7.20 \times 10^{-13} \text{ J}.$$

(b) We can rearrange Eq. 3-11a $[K = (m - m_0)c^2]$ in the form

$$\frac{m}{m_0} = \frac{K}{m_0 c^2} + 1$$
$$= \frac{(7.20 \times 10^{-13} \text{ J})}{(9.11 \times 10^{-31} \text{ kg})(3.00 \times 10^8 \text{ m/s})^2} + 1$$
$$= 8.78 + 1 = 9.78.$$

Thus the relativistic mass m of such accelerated electrons is almost 10 times their rest mass. We can find m from

$$m = 9.78 m_0 = (9.78)(9.11 \times 10^{-31} \text{ kg})$$
$$= 8.91 \times 10^{-30} \text{ kg}.$$

(c) Finally, we can rearrange Eq. 3-4b $(m = m_0/\sqrt{1 - \beta^2})$ to read

$$\beta = \sqrt{1 - \left(\frac{m_0}{m}\right)^2}$$
$$\beta = \sqrt{1 - \left(\frac{1}{9.78}\right)^2}$$
$$= 0.9948.$$

Thus,

$$u = \beta c = (0.9948)(3.00 \times 10^8 \text{ m/s})$$
$$= 2.98 \times 10^8 \text{ m/s}.$$

These electrons are moving very close indeed to the speed of light. At such speeds (see Problem 5), a relatively small fractional increase in speed corresponds to a sizable fractional increase in kinetic energy.

The calculations of this example illustrate the point made in Example 5, namely, that if $\beta > 0.99$, then relativistic effects are central to the situation and, in particular, the kinetic energy K will be at least six times greater than the rest energy $m_0 c^2$. In part (c) we see that β does indeed exceed 0.99; by inspection of the calculation in (b), we see that $K/m_0 c^2$ is 8.78, which is greater than 6.

EXAMPLE 7.

A Charged Particle Moving in a Magnetic Field. (a) Show that, in a region in which there is a uniform magnetic field, a charged particle entering at right angles to the field moves in a circle whose radius is proportional to the particle's momentum.

The force on the particle is

$$\mathbf{F} = q\mathbf{u} \times \mathbf{B},$$

which is at right angles both to $\mathbf{u}$ and to $\mathbf{B}$. The scalar product $\mathbf{F} \cdot \mathbf{u}$ in Eq. 3-17 is thus zero and the acceleration, which is now entirely in the direction of the force, is given by

$$\mathbf{a} = \frac{\mathbf{F}}{m} = \frac{q}{m} \mathbf{u} \times \mathbf{B}.$$

Because the acceleration is always at right angles to the particle's velocity $\mathbf{u}$, the speed of the particle is constant and the particle moves in a circle. Let the radius of the circle be r, so that the centripetal acceleration is u^2/r. We equate this to the acceleration obtained from above, $a = quB/m$, and find

$$\frac{quB}{m} = \frac{u^2}{r}$$

or

$$r = \frac{mu}{qB} = \frac{p}{qB}. \qquad (3\text{-}18)$$

Hence, the radius is proportional to the momentum $p (= mu)$.

Notice that both the equation for the acceleration and the equation for the radius (Eq. 3-18) are identical in form to the classical results, but that the rest mass m_0 of the classical formula is replaced by the relativistic mass $m = m_0/\sqrt{1 - u^2/c^2}$.

How would the motion change if the initial velocity of the charged particle had a component parallel to the magnetic field?

(b) Compute the radius, both classically and relativistically, of a path of a 10-MeV electron moving at right angles to a uniform magnetic field of strength 2.0 T.

Classically, we have $r = m_0 u/qB$. The classical relation between kinetic energy and momentum is $p = \sqrt{2 m_0 K}$, so

$$p = \sqrt{2 m_0 K}$$
$$= \sqrt{2(9.11 \times 10^{-31} \text{ kg})(10 \text{ MeV})(1.60 \times 10^{-13} \text{ J/MeV})}$$
$$= 1.7 \times 10^{-21} \text{ kg} \cdot \text{m/s}.$$

Then

$$r = \frac{m_0 u}{qB} = \frac{p}{qB} = \frac{1.7 \times 10^{-21} \text{ kg} \cdot \text{m/s}}{(1.60 \times 10^{-19} \text{ C})(2.0 \text{ T})}$$

$$= 5.3 \times 10^{-3} \text{ m} = 5.3 \text{ mm}.$$

Relativistically, we have $r = mu/qB$. The relativistic relation between kinetic energy and momentum (Eq. 3-13a) may be written as

$$p = \frac{1}{c} \sqrt{(K + m_0 c^2)^2 - (m_0 c^2)^2}.$$

Here, the rest energy of an electron, $m_0 c^2$, equals 0.511 MeV, so that

$$p = \frac{1}{3.0 \times 10^8} \sqrt{(10 + 0.511)^2 - (0.511)^2} \frac{\text{MeV} \cdot \text{s}}{\text{m}}$$

$$\times (1.60 \times 10^{-13} \text{ J/MeV})$$

$$= 5.6 \times 10^{-21} \text{ kg} \cdot \text{m/s}.$$

Then

$$r = \frac{mu}{qB} = \frac{p}{qB} = \frac{5.6 \times 10^{-21} \text{ kg} \cdot \text{m/s}}{(1.60 \times 10^{-19} \text{ C})(2.0 \text{ T})}$$

$$= 1.8 \times 10^{-2} \text{ m} = 18 \text{ mm},$$

which is substantially larger than the classical prediction. Experiment supports the relativistic prediction conclusively.

3-5 SOME EXPERIMENTAL RESULTS

Early (1909) experiments in relativistic dynamics by Bucherer made use of Eq. 3-18. Electrons (from the β decay of radioactive particles) enter a velocity selector, which determines the speed of those that emerge, and then enter a uniform magnetic field, where the radius of their circular path can be measured. Bucherer's results are shown in Table 3-1.

The first column gives the measured speeds in terms of the fraction of the speed of light. The second column gives the ratio e/m computed from the measured quantities in Eq. 3-18 as $e/m = u/rB$. It is clear that the value of e/m varies with the speed of the electrons. The third column gives the calculated values of $e/m \sqrt{1 - \beta^2} = e/m_0$, which are seen to be constant. The results are consistent with the relativistic relation

$$r = \frac{m_0 u}{qB\sqrt{1 - \beta^2}}$$

rather than the classical relation $r = m_0 u/qB$ and can be interpreted as confirming Eq. 3-4b, $m = m_0/\sqrt{1 - \beta^2}$, for the variation of mass with speed. Many similar experiments have since been performed, greatly extending the range of u/c and always resulting in confirmation of the relativistic results (see Fig. 3-4).

You may properly ask why, in measuring a variation of e/m with speed, we attribute the variation solely to the mass rather than to the charge, for instance, or some other more complicated effect. We might have concluded, for example,

Table 3-1
BUCHERER'S RESULTS

β ($= u/c$) (Measured)	e/m ($= u/rB$) 10^{11} C/kg (Measured)	e/m_0 ($= e/m\sqrt{1 - \beta^2}$) 10^{11} C/kg (Calculated)
0.32	1.66	1.75
0.38	1.63	1.76
0.43	1.59	1.76
0.52	1.51	1.76
0.69	1.28	1.77

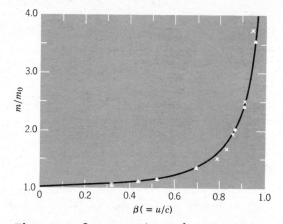

Figure 3-4. Some experimental measurements of the ratio m/m_0 for the electron, at various speed parameters $\beta(= u/c)$. The symbols stand for the work of Kaufmann ($\times$, 1901), Bucherer ($\triangle$, 1908), and Bertozzi ($\bullet$, 1964).

that $e = e_0\sqrt{1 - \beta^2}$. Actually, we have implicitly assumed above that the charge on the electron is independent of its speed. This assumption is a direct consequence of relativistic electrodynamics, wherein the charge of a particle is not changed by its motion. That is, charge is an invariant quantity in relativity. This is plausible, as a little thought shows, for otherwise the neutral character of an atom would be upset merely by the motion of the electrons in it. As a clincher, of course, we turn to experiment; we then find that experiment not only verifies relativity theory as a whole, but also confirms directly this specific result of the constancy of e (see [1] for an analysis of such an experiment). Beyond this, experiments with neutrons—neutral particles not involving electric charge—demonstrate directly the same variation of mass with velocity as for charged particles.

An experiment in relativistic dynamics was also carried out by Bertozzi [2]. In this experiment electrons are accelerated to high speed in the electric field of a linear accelerator and emerge into a vacuum chamber. Their speed can be measured by determining the time of flight in passing two targets of known separation. As we vary the voltage of the accelerator, we can plot the values of eV, the kinetic energy of the emerging electrons, versus the measured speed u. In the experiment, an independent check was made to confirm the relation $K = eV$. This is accomplished by stopping the electrons in a collector, where the kinetic energy of the absorbed electrons is converted into heat energy, which raises the temperature of the collector, and determining the energy released per electron by calorimetry. It is found that the average kinetic energy per electron before impact, measured in this way, agrees with the kinetic energy obtained from eV.

Figure 3-5 shows the results of this experiment. We see that the measured values of K agree very well with the relativistic prediction but not at all with the classical prediction. Note also that in all cases the measured value of u is less than c or, what is the same thing, the value of β is less than unity. This is further direct confirmation of the role of c as a limiting speed. We see that, as $u \rightarrow c$, small changes in u correspond to increasingly large changes in the kinetic energy K. To attain a given speed we always need more kinetic energy than is classically predicted; by extrapolation, we would need infinite energy to accelerate an electron to the speed of light.

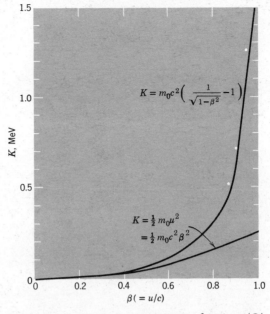

Figure 3-5. Bertozzi's experimental points (●) are seen to fit the relativistic expression for the kinetic energy of an electron at various speeds, rather than the classical expression.

Note carefully that the relativistic formula for kinetic energy is *not* $\frac{1}{2}mu^2$; this shows the danger in assuming that we can simply substitute the relativistic mass for the rest mass in generalizing a classical formula to a relativistic one. This is not so for the kinetic energy.

Although the experimental checks of relativistic dynamics that we have described are direct and to the point, perhaps the most convincing evidence that relativistic dynamics is valid lies in the design and successful operation of high-energy particle accelerators (see *Physics*, Part II Sect. 33-7). The proton synchrotron [3] at the Fermi National Accelerator Laboratory (Fermilab) at Batavia, Illinois, is a prime example; see Fig. 3-6. It is a multimillion-dollar device whose design is based squarely on the relativistic predictions that (1) the speed of light is a limiting speed for material particles and (2) the mass of a particle increases with velocity as described by Eq. 3-4. If these predictions were not correct, this huge accelerator, built around a ring-shaped magnet ~4 miles in circumference, simply would not work.

Consider a proton moving with speed u at right angles to a uniform magnetic field B. As we have seen, it will move in a circular path whose radius R is given by Eq. 3-18, or $R = mu/Be$, in which m is the relativistic mass. The angular frequency ω at which the proton circulates in this field is called the proton's *cyclotron frequency* in that field and is given by

$$\omega = \frac{u}{R} = \frac{Be}{m}. \tag{3-19}$$

The circulating protons can be accelerated by allowing them to pass repeatedly through a small region in which an alternating electric field is established, the field giving the proton an accelerating impulse or "kick" on every passage. It

Figure 3-6. The proton synchrotron at Fermilab; a view through the tunnel. (Courtesy of Fermilab)

would be simplest to provide these accelerating impulses from a fixed-frequency oscillator. However, there is the complication that ω, the cyclotron frequency of the circulating protons, is *not* constant but, as Eq. 3-19 shows, decreases with time throughout the accelerating cycle because of the relativistic increase of mass with speed. This can be compensated for, however, by adjusting the magnetic field B in Eq. 3-19 in a cyclic way so that ω does indeed remain constant and thus in resonance at all times with an oscillator of fixed frequency.

There is a problem, though. Under the scheme just outlined, the orbit radius, given by (see Eq. 3-19)

$$R = \frac{u}{\omega}, \tag{3-20}$$

would increase as the speed increases. Now, it is highly desirable to maintain the orbit radius constant during the acceleration process, because the steering magnets can then be ring-shaped, resulting in great cost savings. Inspection of Eq. 3-20 shows that R can indeed be held constant if the oscillator frequency (which must always be equal to the cyclotron frequency ω) is *also* increased during the accelerating cycle, to compensate for the increase in u.

Table 3-2 shows some characteristics of the proton synchrotron at the Fermi National Laboratory.

Table 3-2
THE PROTON SYNCHROTRON AT FERMILAB

Maximum proton energy	500 GeV
Repetition rate	5 pulses/min
Internal beam intensity	6×10^{12} protons/s
Radius of the ring	1000 m
Radius of curvature of the proton path	750 m
Steel weight	9000 tons
Copper weight (in magnet coils)	850 tons
Magnetic field at injection	40 mT
Magnetic field at 300 GeV	1.35 T
Mean power to magnets	36 MW
Oscillator range	53.08–53.10 MHz

EXAMPLE 8.

The Fermilab Accelerator. The proton synchrotron at Fermilab is accelerating protons to a kinetic energy of 500 GeV. At this energy, find (a) the ratio m/m_0, (b) the speed parameter β, (c) the magnetic field B at the orbit position, and (d) the period T of the circulating 500-GeV protons. (See Table 3-2 for other needed data.)

(a) From Eq. 3-4c we see that the ratio m/m_0 is simply the Lorentz factor γ. From Eq. 3-11c [$K = (\gamma - 1)m_0c^2$], we can write

$$\gamma = \frac{K}{m_0c^2} + 1 = \frac{500 \times 10^3 \text{ MeV}}{938 \text{ MeV}} + 1$$
$$= 532 + 1 = 533.$$

Thus the relativistic mass of a 500-GeV proton is 533 times its rest mass.

(b) Knowing γ, we can solve Eq. 2-10 for the speed parameter β, obtaining

$$\beta = \sqrt{1 - \left(\frac{1}{\gamma}\right)^2} = \sqrt{1 - \left(\frac{1}{533}\right)^2}$$
$$= 0.9999982.$$

(c) The magnetic field follows from Eq. 3-18, or

$$B = \frac{mu}{eR} = \frac{(\gamma m_0)(\beta c)}{eR} = \frac{\gamma \beta m_0 c}{eR}$$
$$= \frac{(533)(1)(1.67 \times 10^{-27} \text{ kg})(3.00 \times 10^8 \text{ m/s})}{(1.60 \times 10^{-19} \text{ C})(750 \text{ m})}$$
$$= 2.23 \text{ T}.$$

This must be the value of B at the end of the accelerating cycle for 500-GeV protons. The value of R used to calculate B must then be the actual radius of curvature of the proton path, not the radius of the magnet ring; these differ because the "ring" contains a number of straight segments (see Table 3-2).

(d) At 500 GeV, the proton speed is virtually c and the period is given by

$$T = \frac{2\pi R'}{u} = \frac{(2\pi)(1000 \text{ m})}{(3.00 \times 10^8 \text{ m/s})} = 20.9 \text{ } \mu\text{s}.$$

Note that the magnet ring used in *this* calculation is the actual effective radius of the magnetic ring; see Table 3-2.

3-6 *THE EQUIVALENCE OF MASS AND ENERGY*

In Section 3-3 we found that the laws of the conservation of momentum and of mass could be preserved by generalizing the classical definition of mass to a relativistic mass m; see Eqs. 3-4. In Section 3-4 we introduced the concept of the total energy E of a single particle, defining it equivalently as

$$E = mc^2 \tag{3-21a}$$

or

$$E = m_0c^2 + K, \tag{3-21b}$$

in which $m_0 c^2$ is the rest energy and K is the kinetic energy of the (single) particle. In this section we seek to extend the concept of total energy from single particles to systems of interacting particles.

Let us return to our examination of the totally inelastic collision of two identical particles, described in Fig. 3-2. Equation 3-21a suggests that, if relativistic mass is conserved in such a collision (and we have shown that it is), then the total energy E of the system of particles will also be conserved. Equation 3-21b suggests further that, if E is conserved for this system of particles, then any net change in the kinetic energies of the interacting particles must be balanced by an equal but opposite net change in the rest energies of those particles. Let us verify that these energy exchanges do in fact occur. We shall be led from this study to important conclusions about the nature of energy conservation in relativity and about the equivalence of mass and energy.

Conservation of relativistic mass in the S frame of Fig. 3-2 (see Example 2) requires that

$$M_0 = 2m.$$

We can substitute for the relativistic mass m from Eq. 3-4b, obtaining

$$M_0 = \frac{2m_0}{\sqrt{1 - \beta^2}} = 2m_0\gamma,$$

which shows at once that, although relativistic mass is conserved in this collision, *rest mass* is not; M_0, the rest mass after the collision, is greater than $2m_0$, the rest mass before the collision. Thus we can write, for the increase in rest mass during the collision,

$$\text{increase in rest mass} = M_0 - 2m_0 = 2m_0\gamma - 2m_0$$
$$= 2m_0(\gamma - 1). \tag{3-22a}$$

Now let us look at the kinetic energy situation. In frame S of Fig. 3-2a the individual particles have a combined kinetic energy before the collision given by Eq. 3-11c as $2m_0(\gamma - 1)c^2$. After the collision, no kinetic energy remains. In place of the "lost" kinetic energy there appears internal (thermal) energy, recognizable by the rise in temperature of the colliding particles. Thus we can write

$$\text{increase in internal energy} = \text{decrease in kinetic energy}$$
$$= 2m_0(\gamma - 1)c^2. \tag{3-22b}$$

Comparison of these last two equations allows us to write, for this collision at least,

$$\text{(decrease in kinetic energy)} = \text{(increase in rest mass)}(c^2),$$

or equivalently

$$\text{(increase in thermal energy)} = \text{(increase in rest mass)}(c^2).$$

Thus we see that the decrease in kinetic energy for this isolated system is balanced by a corresponding increase in rest energy, just as we expected. We see also that the thermal energy that appears in the system is associated with an increase in the rest mass of the system. Hence rest mass is equivalent to energy (rest-mass energy) and must be included in applying the conservation of energy principle.

All of the foregoing justifies our making a great extrapolation from the simple special case we have examined and asserting a general principle, namely, that mass and energy are two aspects of a single invariant quantity, which we can call *mass-energy*. We can find the energy equivalent of a given mass by multiplying by c^2. In the same way, we can find the mass equivalent of a given amount of energy by dividing by c^2.

The relation

$$E = mc^2 \tag{3-23}$$

expresses the fact that mass-energy can be expressed in energy units (E) or equivalently in mass units ($m = E/c^2$). In fact, it has become common practice to refer to masses in terms of electron volts, such as saying that the rest mass of an electron is 0.511 MeV, for convenience in energy calculations. Likewise, particles of zero rest mass (such as photons, see below) may be assigned an effective mass equivalent to their energy. Indeed, the mass that we associate with various forms of energy really has all the properties we have heretofore given to mass, properties such as inertia, weight, contribution to the location of the center of mass of a system, and so forth. We shall exhibit some of these properties later in the chapter (see also Ref. 4).

Equation 3–23, $E = mc^2$, is, of course, one of the famous equations of physics. It has been confirmed by numerous practical applications and theoretical consequences. Einstein, who derived the result originally in another context, made the bold hypothesis that it was universally applicable. He considered it to be the most significant consequence of his special theory of relativity.

Thus we see that the conclusions we have drawn here from our study of two-body collisions are perfectly consistent with the single-particle equations we developed in Section 3-4. Consider Eq. 3-11*a*:

$$mc^2 = m_0c^2 + K.$$

We can differentiate this result, obtaining

$$\frac{dK}{dt} = c^2\frac{dm}{dt}, \tag{3-24}$$

which states that a change in the kinetic energy of a particle causes a proportionate change in its (relativistic) mass. That is, mass and energy are equivalent, their units differing by a factor c^2.

If the kinetic energy of a body is regarded as a form of external energy, then the rest-mass energy may be regarded as the energy of the body. This internal energy consists, in part, of such things as molecular motion, which changes when heat energy is absorbed or given up by the body, or intermolecular potential energy, which changes when chemical reactions (such as dissociation or recombination) take place. Or the internal energy can take the form of atomic potential energy, which can change when an atom absorbs radiation and becomes excited or emits radiation and is deexcited, or nuclear potential energy, which can be changed by nuclear reactions. The largest contribution to the internal energy is, however, the total rest-mass energy contributed by the "fundamental" particles, which is regarded as the primary source of internal energy. This too, may change, as, for example, in electron-positron creation and annihilation. The rest mass (or proper mass) of a body, therefore, is not a constant, in general. Of course, if there are no changes in the internal energy of a body (or if we consider a closed system through which energy is not transferred), then we may regard the rest mass of the body (or of the system) as constant.

This view of the internal energy of a particle as equivalent to rest mass suggests an extension to a collection of particles. We sometimes regard an atom as a particle and assign it a rest mass, for example, although we know that the atom consists of many particles with various forms of internal energy. Likewise, we can assign a rest mass to any collection of particles in relative motion, in a frame in which the center of mass is at rest (that is, in which the resultant momentum is zero). The rest mass of the system as a whole would include the contributions of the internal energy of the system to the inertia.

Returning our attention now to collisions or interactions between bodies, we have seen that the total energy is conserved and that the conservation of total energy is equivalent to the conservation of (relativistic) mass. Although we showed this explicitly for a totally inelastic collision only, it holds regardless of the nature of the collision. To retain the conservation laws we must consider mass and energy as two different aspects of a single entity, namely, mass-energy. In the energy balance we must consider the masses of particles, and in the mass balance we must consider the energy of radiation, for example. The formula $E = mc^2$ can be regarded as giving the rate of exchange between two interchangeable currencies, E and m.

In classical physics we had two separate conservation principles: (1) the conservation of (classical) mass, as in chemical reactions, and (2) the conservation of energy. In relativity, these merge into one conservation principle, that of conservation of mass-energy. The two classical laws may be viewed as special cases that would be expected to agree with experiment only if energy transfers into or out of the system are so small compared to the system's rest mass that the corresponding fractional change in rest mass of the system is too small to be measured.

EXAMPLE 9.

Pulling Apart a Deuteron. A convenient mass unit to use when dealing with atoms or nuclei is the *atomic mass unit* (abbreviation u). It is defined so that the atomic mass of one atom of the most common carbon isotope (carbon-12) is exactly 12 such units and has the value 1 u = 1.66 × 10^{-27} kg. The rest mass of the proton (the nucleus of a hydrogen atom) is 1.00728 u, and that of the neutron (a neutral particle and a constituent of all nuclei except hydrogen) is 1.00867 u. A deuteron (the nucleus of heavy hydrogen) is known to consist of a proton and a neutron and to have a rest mass of 2.01355 u. What energy is required to break up a deuteron into its constituent particles?

The rest mass of a deuteron is *less than* the combined rest masses of a proton and a neutron by

$$\Delta m_0 = [(1.00728 + 1.00867) - 2.01355]u = 0.00240 \text{ u},$$

which is equivalent, in energy terms, to

$$\begin{aligned}\Delta m_0 c^2 &= (0.00240 \text{ u})(1.66 \times 10^{-27} \text{ kg/u})(3.00 \times 10^8 \text{ m/s})^2 \\ &= (3.59 \times 10^{-13} \text{ J})(1 \text{ MeV}/1.60 \times 10^{-13} \text{ J}) \\ &= 2.23 \text{ MeV}.\end{aligned}$$

When a proton and a neutron at rest combine to form a deuteron, this amount of energy is *given off* in the form of electromagnetic (gamma) radiation. If the deuteron is to be broken up into a proton and a neutron, this same amount of energy must be *added to* the deuteron.

Notice that

$$\frac{\Delta m_0}{M_0} = \frac{0.00240 \text{ u}}{2.01355 \text{ u}} = 0.12\%.$$

This fractional rest-mass change is characteristic of the magnitudes found in nuclear reactions.

In solving problems involving nuclear interactions, in which the energies are typically measured in MeV and the masses in atomic mass units, it is convenient to realize (see Problem 61) that c^2 can be written as

$$c^2 = 931 \text{ MeV/u}.$$

In this example, then, we could simply have multiplied the mass change (= 0.00240 u) by c^2 written in this form and obtained at once for the energy (0.00240 u)(931 MeV/u) or 2.23 MeV.

EXAMPLE 10.

Pulling Apart a Hydrogen Atom. The energy E_b required to break a hydrogen atom apart into its constituents—a proton and an electron—is called its *binding energy*. Its measured value is 13.58 eV. The rest mass M_0 of a hydrogen atom is 1.00783 u. By how much does the rest mass of this atom change when it is ionized? Is the change an increase or a decrease?

From the relation $E = mc^2$, we can write

$$\Delta m_0 = \frac{E_b}{c^2} = \frac{13.58 \text{ eV}}{931 \times 10^6 \text{ eV/u}} = 1.46 \times 10^{-8} \text{ u},$$

in which 931×10^6 eV/u, as we have seen in Example 9, is simply a convenient way of writing c^2. The change is an

increase because energy must be *added to* the atom to ionize it. We also note that

$$\frac{\Delta m_0}{M_0} = \frac{1.46 \times 10^{-8} \text{ u}}{1.00783 \text{ u}} \cong 1.5 \times 10^{-8} = 1.5 \times 10^{-6}\%.$$

Such a fractional change in rest mass is actually smaller than the experimental errors involved in measuring the rest masses themselves. Thus, in interactions involving atoms (including all chemical reactions), the changes in rest mass are too small to detect and the classical principle of the conservation of (rest) mass is practically correct. As Example 9 shows, this statement is not true for reactions involving the nuclei of atoms.

In a paper [5] entitled "Does the Inertia of a Body Depend upon its Energy Content," Einstein writes:

> If a body gives off the energy L in the form of radiation, its mass diminishes by L/c^2. The fact that the energy withdrawn from the body becomes energy of radiation evidently makes no difference, so that we are led to the more general conclusion that the mass of a body is a measure of its energy content. . . . It is not impossible that with bodies whose energy-content is variable to a high degree (for example, with radium salts) the theory may be successfully put to the test. If the theory corresponds to the facts, radiation conveys inertia between the emitting and absorbing bodies.

Experiment has abundantly confirmed Einstein's theory.

Today, we call such a pulse of radiation a photon and may regard it as a particle of zero rest mass. The relation $p = E/c$, taken from classical electromagnetism, is consistent with the result of special relativity for particles of "zero rest mass" since, from Eq. 3-14, $E = c\sqrt{p^2 + m_0^2 c^2}$, we find that $p = E/c$ when $m_0 = 0$. This is also consistent with the fact that photons travel with the speed of light because, from the relation $E = mc^2 = m_0 c^2/\sqrt{1 - u^2/c^2}$, the energy E would go to zero as $m_0 \to 0$ for $u < c$. In order to keep E finite (neither zero nor infinite) as $m_0 \to 0$, we must let $u \to c$. Strictly speaking, however, the term zero rest mass is a bit misleading, because it is impossible to find a reference frame in which photons (or anything that travels at the speed of light) are at rest (see Question 7). However, if m_0 is determined from energy and momentum measurements as $m_0 = \sqrt{(E/c^2)^2 - (p/c)^2}$, then $m_0 = 0$ when (as for a photon*) $p = E/c$. The result, that a particle of zero rest mass can have a finite energy and momentum and that such particles must move at the speed of light, is also consistent with the meaning we have given to rest mass as internal energy. For if rest mass is internal energy,

---

* For students who are unfamiliar with the relation $p = E/c$, found in electromagnetism, the argument can be run in reverse. Start with the relativistic relation $E = m_0 c^2/\sqrt{1 - u^2/c^2}$. This implies that E approaches infinity if $u = c$, unless $m_0 = 0$. Therefore photons, which by definition have $u = c$, must have $m_0 = 0$. Then, from $E = c(p^2 + m_0^2 c^2)^{1/2}$, it follows that photons must satisfy the relation $p = E/c$. That this same result is found independently in classical electromagnetism illustrates the consistency between relativity and classical electromagnetism.

existing when a body is at rest, then a "body" without mass has no internal energy. Its energy is all external, involving motion through space. Now, if such a body moved at a speed less than c in one reference frame, we could always find another reference frame in which it *is* at rest. But if it moves at a speed c in one reference frame, it will move at this same speed c in all reference frames. It is consistent with the Lorentz transformation, then, that a body of zero rest mass should move at the speed of light and be nowhere at rest.

EXAMPLE 11.

The Sun is Losing Mass. The earth receives radiant energy from the sun at the rate of 1340 W/m². At what rate is the sun losing rest mass because of its radiation? The sun's mass is now about 2.0×10^{30} kg.

If we assume that the sun radiates uniformly in all directions, we can calculate its luminosity (= dE/dt) from

$$\frac{dE}{dt} = (1340 \text{ W/m}^2)(4\pi R^2)$$
$$= (1340 \text{ W/m}^2)(4\pi)(1.50 \times 10^{11} \text{ m})^2$$
$$= 3.79 \times 10^{26} \text{ W},$$

in which R is the mean earth–sun distance.

The rate of mass loss is then

$$\frac{dm}{dt} = \frac{dE/dt}{c^2}$$
$$= \frac{3.79 \times 10^{26} \text{ J/s}}{(3.00 \times 10^8 \text{ m/s})^2}$$
$$= 4.21 \times 10^9 \text{ kg/s}.$$

At this rate, the fractional rate of loss of solar mass is

$$f = \frac{(4.21 \times 10^9 \text{ kg/s})(3.16 \times 10^7 \text{ s/y})}{2.0 \times 10^{30} \text{ kg}} = 6.7 \times 10^{-14} \text{ y}^{-1}.$$

From this it can be shown that, if the sun had maintained its present luminosity since its formation about 5×10^9 y ago, it would have lost only about 0.03 percent of its rest mass.

EXAMPLE 12.

Another Approach to $E = mc^2$. Here we present an "elementary derivation of the equivalence of mass and energy" attributable to Einstein. Consider a body B at rest in frame S (Fig. 3-7a). It emits simultaneously two pulses of radiation, each of energy $E/2$, one in the $+y$ direction and one in the $-y$ direction. The energy of B therefore decreases by an amount E in the emission process; from

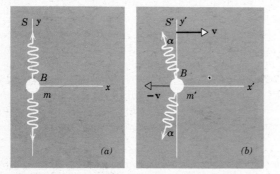

Figure 3-7. (a) Body B, at rest in frame S, emits two light pulses in opposite directions. (b) The same phenomenon as observed from frame S'. The light beams are deflected through an aberration angle α.

symmetry considerations, B must remain at rest both during and after this process.

Now consider these same events as viewed from frame S', which is moving in the $+x$ direction with speed v (assumed $\ll c$; see Fig. 3-7b). In this frame, body B is seen to move in the $-x'$ direction with speed v. The two pulses of radiation, because of the aberration effect, now make a small angle α with the vertical (y') axis. It is central to the proof to note that, because the state of motion of B was not changed by the emission process in frame S, it cannot be changed by this process in frame S'. Thus the velocity of B in frame S' must remain $-\mathbf{v}$ after the two pulses have been emitted. If it were otherwise, there would be a violation of the principle of relativity, namely, that the laws of physics must be the same in all inertial reference frames.

Let us apply the law of conservation of momentum in frame S'. We assume that $v \ll c$, so we can use the classical form of this law. Before doing so we must point out that radiation has momentum, the amount being equal to the associated energy divided by c, the speed of light. Thus, each of our light pulses has a momentum, in its direction of motion, of magnitude $E/2c$. The assigning of momentum to radiation is a result from the classical theory of electromagnetism.

Before the emission process the momentum of the system in frame S' is $-mv$, associated entirely with the body. After the emission process it is

$$-m'v - 2\left(\frac{E}{2c}\right)\sin\alpha.$$

Setting the momentum of the system before emission equal to its value after emission leads to

$$mv = m'v + \left(\frac{E}{c}\right)\sin\alpha.$$

Because $(v \ll c)$ the aberration angle α is very small, and we can replace $\sin\alpha$ by α in the above without appreciable error. Also (see Eq. 1-11), classical aberration theory predicts that (again for $v \ll c$) $\alpha \cong v/c$. Replacing $\sin\alpha$ in the above equation by v/c gives

$$m - m' = \frac{E}{c^2}.$$

Thus we see that body B, at rest in frame S, loses energy E by emitting two light pulses and remains at rest, its (rest) mass decreased by E/c^2. Generalizing from this allows us to put

$$E = (\Delta m)c^2$$

as a relation valid for any process in which energy E is emitted from or absorbed by a body.

3-7 *RELATIVITY AND ELECTROMAGNETISM*

In the earlier sections we investigated the dynamics of a single particle using the relativistic equation of motion that was found to be in agreement with experiment for the motion of high-speed charged particles. There we introduced the relativistic mass and the total energy, including the rest-mass energy. However, all the formulas we used were applicable in one reference frame, which we called the laboratory frame. Often, as when analyzing nuclear reactions, it is useful to be able to transform these relations to other inertial reference frames, such as the center-of-mass frame. In such cases one uses the equations that connect the values of the momentum, energy, mass, and force in one frame S to the corresponding values of these quantities in another frame S', which moves with uniform velocity **v** with respect to S along the common x-x' axes.

Here we simply point out that these equations of transformation can be obtained in a perfectly straightforward way from the transformation equations already derived for the velocity components (Table 2-4). Once we have the force transformations, we can then use the Lorentz force law of classical electromagnetism (Eq. 3-9) to find how the numerical value of electric and magnetic fields depends on the frame of the observer. The details (see Ref. 6) are of no special interest to us here, but some of the conclusions reached are worth discussing in order to put special relativity theory in proper perspective.

We have seen in the last two chapters how kinematics and dynamics must be generalized from their classical form to meet the requirements of special relativity. And we saw earlier the role that optical experiments played in the development of relativity theory and the new interpretation that is given to such experiments. What remains, therefore, is to investigate classical electricity and magnetism in order to discover what modifications may need to be made there because of relativistic considerations. It turns out that Maxwell's equations are invariant under a Lorentz transformation and do not need to be modified (see Sec. 4.7 of Ref. 6 for proof). This result then completes the original program of finding the transformation (the Lorentz transformation) that keeps the velocity of light constant and finding the invariant form of the laws of mechanics and electromagnetism. The (Einstein) principle of relativity appears to apply to *all* the laws of physics.

Although relativity leaves Maxwell's equations of electromagnetism unaltered, it does give us a new point of view that enhances our understanding of electromagnetism. It is shown clearly in relativity that electric fields and magnetic fields have no separate meaning; that is, **E** and **B** do not exist independently as separate quantities but are interdependent. A field that is purely electric, or purely magnetic, in one inertial frame, for example, will have both electric and magnetic components in another inertial frame. One can find, from the force transformations, just how **E** and **B** transform from one frame to another. These equations of transformation are of much practical benefit, for we can solve difficult problems by choosing a reference system in which the answer is relatively easy to find and then transforming the results back to the system we deal with in the laboratory. The techniques of relativity, therefore, are often much simpler than the classical techniques for solving electromagnetic problems.

One striking result we obtain from relativity is this. If all we knew in electromagnetism was Coulomb's law, then, by using special relativity and the invariance of charge, we could prove that magnetic fields must exist. There is no need to postulate magnetic fields separately from electric fields. The magnetic field enters relativity in a most natural way as a field that is produced by a source charge in motion and that exerts a force on a test charge that depends on its velocity relative to the observer. Magnetism is simply a new word, a short-hand designation, for the velocity-dependent part of the force. In fact, starting only with Coulomb's law and the invariance of charge, we can derive (see Ref. 7) all of electromagnetism from relativity theory—the exact opposite of the historical development of these subjects.

questions

1. In view of the fact that particle speeds are limited by the speed of light, is "acceleration" a good word to use in relativity to describe the action of a force on a particle? Can you think of a more apt name? ("Ponderation"?)

2. Can we simply substitute m for m_0 in classical equations to obtain the correct relativistic equations? Give examples.

3. Is it true that a particle that has kinetic energy must also have momentum? What if the particle has zero rest mass? Can a *system* of particles have kinetic energy but no momentum? Momentum but no kinetic energy?

4. A particle with zero rest mass (a neutrino, possibly) can transport momentum. How can this be in view of Eq. 3-5 [$p = m_0u/(1 - u^2/c^2)^{1/2}$], in which we see that the momentum is directly proportional to the rest mass?

5. Distinguish between a variable-mass problem in classical physics and the relativistic variation of mass.

6. If a particle could be accelerated to a speed greater than the speed of light, what would be some of the consequences?

7. If zero-mass particles have a speed c in one reference frame, can they be found at rest in any other frame? Can such particles have any speed other than c?

8. Does **F** equal $m\mathbf{a}$ in relativity? Does $m\mathbf{a}$ equal $d(m\mathbf{u})/dt$ in relativity?

9. What characteristic of a particle does the combination $\beta\gamma m_0c$ represent? The symbols have their usual meanings.

10. We say that a 1-keV electron is a "classical" particle, a 1-MeV electron is a "relativistic" particle, and a 1-GeV electron is an "extremely relativistic" particle. What exactly do these terms mean?

11. How many relativistic expressions can you think of in which the Lorentz factor γ enters as a simple multiplier?

12. Discuss in detail, paying careful attention to signs, how the relation $u' = 2u$, discussed in connection with Fig. 3-1b, follows from the Galilean velocity transformation law (Eq. 2-19).

13. The total energy E of the system is conserved in the collision shown in Fig. 3-2. However, the numerical value assigned to E by the observer in frame S is *not* the same as that assigned by the observer in frame S'. Is there a contradiction here?

14. Is a totally inelastic collision one in which all of the kinetic energy is lost, none remaining after the collision?

15. What determines whether a collision is elastic? Inelastic? Totally inelastic?

16. Here are some characteristics of particles or of groups of particles: rest mass, relativistic mass, total energy, kinetic energy, momentum. Which of these quantities is conserved in an elastic collision between two particles? In an inelastic collision?

17. In Section 3-5 we discussed the mode of operation of a proton synchrotron, designed to accelerate protons to very high energies. The precursor of this device was the conventional *cyclotron*, which resembled the synchrotron in broad outline except that neither the frequency of the accelerating oscillator nor the magnitude of the magnetic field changed with time. Discuss how this device might work well enough at low particle energies and point out the difficulties that would arise as the particle energy increased.

18. In a given magnetic field would a proton or an electron have the greater cyclotron frequency?

19. In the proton synchrotron at Fermilab (see Table 3-2), the radius of the magnet ring is 1000 m but the radius of the proton path is given as 750 m. How can these quantities be so different?

20. How can mass and energy be "equivalent" in view of the fact that they are totally different physical quantities, defined in different ways and measured in different units?

21. Is the rest mass of a stable composite particle (a uranium nucleus, say) greater or less than the sum of the rest masses of its constituents?

22. "The rest mass of the electron is 511 keV." What exactly does this statement mean?

23. "The relation $E = mc^2$ is essential to the operation of a power plant based on nuclear fission but has only a negligible relevance for a coal-fired plant." Is this a true statement?

24. A hydroelectric plant generates electricity because water falls under gravity through a turbine, thereby turning the shaft of a generator. According to the mass-energy concept, must the appearance of energy (the electricity) be identified with a (conventional) mass decrease somewhere? Where?

25. Exactly why is it that, pound for pound, nuclear explosions release so much more energy than do TNT explosions?

26. A hot metallic sphere cools off as it rests on the pan of a scale. If the scale were sensitive enough, would it indicate a change in rest mass?

27. A spring is kept compressed by tying its ends together tightly. It is then placed in acid and dissolves. What happens to its stored potential energy?

28. What role does potential energy play in the equivalence of mass and energy?

29. In Einstein's derivation of the $E = mc^2$ relation (see Example 12), how could he get away with using the classical (instead of the relativistic) expression for the aberration constant α, a quantity that enters so centrally into his proof?

problems

$$c = 3.00 \times 10^8 \text{ m/s} = 300 \text{ km/ms} = 0.300 \text{ m/ns}$$
$$= 0.300 \text{ mm/ps}$$
$$c^2 = 8.99 \times 10^{16} \text{ J/kg} = 932 \text{ MeV/u}$$
$$1 \text{ u} = 1.66 \times 10^{-27} \text{ kg} \qquad 1 \text{ eV} = 1.60 \times 10^{-19} \text{ J}$$
$$1 \text{ GeV} = 10^3 \text{ MeV} = 10^6 \text{ keV} = 10^9 \text{ eV} = 1.60 \times 10^{-10} \text{ J}$$
Electron rest mass $= 9.11 \times 10^{-31} \text{ kg} = 0.511 \text{ MeV}/c^2$
Proton rest mass $= 1.67 \times 10^{-27} \text{ kg} = 938 \text{ MeV}/c^2$

1. A flying brick. Consider a building brick, of volume V_0 and rest mass m_0, at rest in a certain reference frame. Let the brick be viewed by a second observer for whom it is moving with speed u. (*a*) What is the volume of the brick from the point of view of this observer? (*b*) What mass will he measure for it? (*c*) What will be its density ρ, expressed in terms of the density ρ_0 of the resting brick? For what value of u would the density measured by this observer increase by 1.0 percent over its rest value?

2. Around the world in one second. An electron is moving at a speed such that it could circumnavigate the earth at the equator in one second. (*a*) What is its speed, in terms of the speed of light? (*b*) Its kinetic energy K? (*c*) What percent error do you make if you use the classical formula to calculate K?

3. Doing work on an electron. How much work must be done to increase the speed of an electron from rest (*a*) to $0.50c$? (*b*) To $0.990c$? (*c*) To $0.9990c$?

4. True for electrons and baseballs. What is the speed of a particle (*a*) Whose kinetic energy is equal to twice its rest energy, (*b*) Whose total energy is equal to twice its rest energy?

5. A small change in speed but a large change in energy. An electron has a speed of $0.9990c$. (*a*) What is its kinetic energy? (*b*) If its speed is increased by 0.05 percent, by what percent will its kinetic energy increase?

6. The last steps are the hardest. How much work must be done to increase the speed of an electron from (*a*)

0.180c to 0.190c? (b) 0.980c to 0.990c? Note that the speed increase (= 0.010c) is the same in each case.

★ **7. The next step is always bigger.** An electron initially at rest in a certain reference frame is given a number of successive speed increases. Each increase takes it from its present speed 95 percent of the way to the speed of light. To what successive kinetic energy increases do the first six speed increases correspond?

8. A useful high-energy approximation. (a) Show that, for an extremely relativistic particle, the particle speed u differs from the speed of light c by

$$\Delta u = c - u = \left(\frac{c}{2}\right)\left(\frac{m_0 c^2}{E}\right)^2,$$

in which E is the total energy. Find this quantity for an electron whose kinetic energy is (b) 100 MeV; (c) 25 GeV. (Electrons with energies in this latter range can be generated in the Stanford Linear Accelerator.)

9. Classical physics and the speed of light. (a) What potential difference would accelerate an electron to the speed of light, according to classical physics? (b) With this potential difference, what speed would the electron actually attain? (c) What would be its mass at that speed? (d) Its kinetic energy?

10. Two ways to get it wrong. The correct relativistic expression for kinetic energy is $K = (m - m_0)c^2$. (a) Show that the greatest possible percent error committed by using the classical expression $K = \frac{1}{2}m_0 v^2$ is 100 percent. (b) Show that the greatest possible percent error committed by using $K = \frac{1}{2}mv^2$ is 50 percent.

11. Similar, but not the same. Show that the relativistic expression for K (see Eq. 3-11c) can be written as

$$K = \left(\frac{\gamma^2}{1 + \gamma}\right) m_0 u^2 \quad \text{or} \quad K = \frac{p^2}{(1 + \gamma)m_0}.$$

Note that these expressions resemble in form the classical expressions:

$$K = \left(\frac{1}{2}\right) m_0 u^2 \quad \text{and} \quad K = \frac{p^2}{2m_0}.$$

Do the relativistic expressions reduce to the classical ones in the limit of low speeds?

12. The lightest particles (?). It is now thought that the neutrino, long believed to be massless, may have a small rest mass whose energy equivalent may be (let us assume) 25 eV. (a) What is the ratio of the rest mass of the electron to that of such a neutrino? (b) By how much would the speed of a 1.0-MeV neutrino with this rest energy differ from the speed of light? (*Hint:* See Problem 8.) (c) What kinetic energy must an electron have if its speed is to be the same as that of a 1.0-MeV neutrino?

13. The fastest particles (?). Cosmic rays, originating in deep space, fall steadily upon the earth. Most are protons, and a very small number of them have energies ranging up to 10^{20} eV. For such an extremely energetic proton, calculate (a) the difference between its speed and the speed of light and (b) its relativistic mass. (c) What would be the diameter of the earth's orbit about the sun as seen by an observer moving with such a speeding proton? In the rest frame of the solar system this diameter is 3.0×10^8 km. (See "Nature's Own Particle Accelerator," M. Mitchell Waldrop, *Sky and Telescope*, September, 1981.)

14. Beta, gamma, and K (I). Find the speed parameter β and the Lorentz factor γ for an electron whose kinetic energy is (a) 1.0 keV; (b) 1.0 MeV; (c) 1.0 GeV.

15. Beta, gamma, and K (II). Find the speed parameter β and the Lorentz factor γ for a particle whose kinetic energy is 10 MeV if the particle is (a) an electron; (b) a proton; (c) an alpha particle.

16. A useful formula. Show that the speed parameter β and the Lorentz factor γ are related by

$$\beta\gamma = \sqrt{\gamma^2 - 1}.$$

Show further that this relation remains valid at both the high-speed and the low-speed limits.

★ **17. Pulsars as proton traps.** Pulsars are rapidly rotating neutron stars. They are characterized by being small (~10 km diameter), by rotating rapidly, and by having an extremely intense associated magnetic field. This field is thought to "co-rotate" with the pulsar, with the field lines and the pulsar proper behaving like a rotating rigid body. Charged particles (which may be associated with the cosmic radiation) can get trapped along these field lines and co-rotate with the pulsar. The pulsar associated with the Crab nebula has a rotation period of 33 ms. (a) What is the greatest distance from its axis at which a trapped particle can co-rotate? (b) What is the kinetic energy of a co-rotating particle (assumed to be a proton) at 90 percent of this distance?

18. Some missing algebra. Derive Eq. 3-4a ($m = m_0/\sqrt{1 - u^2/c^2}$), starting from Eqs. 3-1, 3-2, and 3-3 and supplying all the missing steps.

19. More missing algebra. Supply all of the missing algebraic steps in the proof presented in Example 2.

20. A singular value for the momentum. A particle has a momentum equal to $m_0 c$. (a) What is its speed? (b) Its mass? (c) Its kinetic energy?

21. A memory-jogging triangle. (a) Suppose that the mnemonic triangle of Fig. 3-3 represents a proton. By measurement, what is the kinetic energy of the proton? (b) What is the kinetic energy if the particle is an electron? (c)

Draw triangles to represent a 10-keV electron and a 2.5-MeV electron.

22. Three useful relationships. (a) Verify Eq. 3-13a connecting K and p by eliminating u between Eqs. 3-5 and 3-11b. (b) Derive the following useful relations among p, E, K, and m_0 for relativistic particles, starting from Eqs. 3-13a and 3-13b:

$$\text{(1)} \quad K = c\sqrt{m_0^2 c^2 + p^2} - m_0 c^2.$$

$$\text{(2)} \quad p = \frac{\sqrt{K^2 + 2m_0 c^2 K}}{c}.$$

$$\text{(3)} \quad m_0 = \frac{\sqrt{E^2 - p^2 c^2}}{c^2}.$$

(c) Show that these relationships remain valid as $u \to 0$.

23. Energy and momentum—a graphical study. Plot the total energy E against the momentum p for a particle of rest mass m_0 under three assumptions: (a) that the kinetic energy of the particle is given by the classical expression $p^2/2m_0$; (b) that the particle is a relativistic particle; and (c) that the particle is a relativistic particle but has zero rest mass. (d) In what region does curve (b) approach curve (a), and in what region does curve (b) approach curve (c)? (e) Explain briefly the physical significance of the intercepts of the curves with the axes and of their slopes (derivatives). (In classical physics, the energy of a single particle is defined only to within an arbitrary constant. Assume, in (a), that this arbitrary constant is the same as that fixed by relativity, namely, that the energy of a particle at rest, E_0, equals $m_0 c^2$.)

24. Three particles compared. Consider the following, all moving in free space: a 2.0-eV photon, a 0.40-MeV electron and a 10-MeV proton. (a) Which is moving the fastest? (b) The slowest? (c) Which has the greatest momentum? (d) The least? (*Note:* A photon is a light-particle of zero rest mass.)

25. Conditions at zero rest mass. (a) Show that a particle that travels at the speed of light must have zero rest mass. (b) Would such a particle—such as light—travel with speed c in all inertial reference frames? (c) Show that for a particle of zero rest mass, $u = c$, $\gamma = 1$, $K = E$, and $p = K/c$.

26. A practical unit for momentum. High-energy particle physicists commonly use the GeV $(= 10^9 \text{ eV} = 1.60 \times 10^{-10} \text{ J})$ as a practical laboratory unit in which to measure the kinetic energy of a particle. (a) Show that a similarly practical unit for the momentum of a particle is the GeV/c and that 1 GeV/c $= 5.33 \times 10^{-19}$ kg·m/s. What is the momentum of a proton whose kinetic energy K is (b) 5.0 GeV? (c) 500 GeV?

27. Momentum and energy for high-speed particles. In particle physics the momenta of energetic particles are usually reported in units such as GeV/c. One reason for this is that, as $u \to c$, the relationship $p = E/c$ (which is strictly true only for particles with zero rest mass) becomes more and more correct. In the extreme relativistic realm, then, one number gives both the energy and the momentum of the particle. Verify these considerations by filling in the following table, which refers to a proton.

$\beta \ (=u/c)$		0.80	0.90	0.99	0.999	0.9999
E,	GeV					
p,	GeV/c					

28. Two fast particles compared. A particle has a speed of 0.990c in a laboratory reference frame. What are its kinetic energy, its total energy, and its momentum if the particle is (a) a proton or (b) an electron?

29. Finding the rest mass. (a) If the kinetic energy K and the momentum p of a particle can be measured, it should be possible to find its rest mass m_0 and thus identify the particle. Show that

$$m_0 = \frac{(pc)^2 - K^2}{2Kc^2}.$$

(b) Show that this expression reduces to an expected result as $u/c \to 0$, in which u is the speed of the particle. (c) Find the rest mass of a particle whose kinetic energy is 55.0 MeV and whose momentum is 121 MeV/c; express your answer in terms of the rest mass of the electron.

30. Decay in flight. The average lifetime of muons at rest is 2.20 μs. A laboratory measurement on the decay in flight of the muons in a beam emerging from a particle accelerator yields an average lifetime of 6.90 μs. (a) What is the speed of these muons in the laboratory? (b) The relativistic mass (in terms of m_e, the rest mass of the electron)? (c) The kinetic energy? (d) The momentum? The rest mass of a muon is 207 times greater than that of an electron.

31. Action in the upper atmosphere. In a high-energy collision of a primary cosmic-ray particle near the top of the earth's atmosphere, 120 km above sea level, a pion is created with a total energy E of 1.35×10^5 MeV, traveling vertically downward. In its proper frame this pion decays 35 ns after its creation. At what altitude above sea level does the decay occur? The rest energy of a pion is 139.6 MeV.

★ **32. A center-of-mass reference frame.** Proton A, whose speed is 0.990c in laboratory reference frame S, is approaching proton B, which is at rest in that frame. The center of mass of these two protons (a point) moves in the laboratory with a certain constant velocity. Consider a second reference frame S' moving in the laboratory with this same velocity. Calculate the kinetic energy K, the total energy E, and the momentum p of each proton in each reference frame, recording your answers in the table below.

Proton	Frame	K	E	p
A	S			
B	S			
A	S'			
B	S'			

33. Solving the equation of motion. Equation 3-17 shows Newton's second law of motion extended to relativistic form. (a) For the special case in which the force **F** acting on the particle and its velocity **u** point in the same direction, show that this equation reduces to

$$\left(1 - \frac{u^2}{c^2}\right)^{-3/2} du = \frac{F}{m_0} dt,$$

in which we assume that F does not depend on u.
(b) Show that the solution to this differential equation is

$$u = \frac{(F/m_0)t}{\sqrt{1 + (F/m_0 c)^2 t^2}}.$$

(c) Show that this solution can also be written as

$$t = \frac{(m_0/F)u}{\sqrt{1 - u^2/c^2}},$$

in which we have required that $u = 0$ when $t = 0$.
(d) Do these last two equations reduce to expected results for $u \to 0$? For $t \to \infty$?

34. The force law in relativity—two special cases. In general (see Eq. 3-17), the acceleration **a** of a particle is not parallel to the force **F** acting on it. (a) Show that there are two simple special cases in which the acceleration *is* parallel to the force, namely, when **F** is parallel to the velocity **u** and when **F** is perpendicular to **u**. (b) Give a physical example of each such special case. (c) In view of the fact that we can always resolve a force into two such components (one parallel to **u** and one perpendicular to it), why is **F** not *always* parallel to **a**?

★ **35. Accelerated motion in a straight line.** In Example 6, assume that the accelerating potential (= 4.50 MV) is applied (with a uniform gradient) between the ends of an accelerating tube 3.00 m long. In that example the speed

of the electron upon emerging from the accelerator was shown to be $0.9948c$. What was the duration of the acceleration? (*Hint:* Use the result of Problem 33.)

★ **36. A familiar collision analyzed.** In the collision of Fig. 3-2, assume that $m_0 = 1.000$ u and $u = 0.200c$. (a) Calculate M_0 and u'. (b) What is the change in kinetic energy during the collision from the point of view of frame S? (c) From the point of view of frame S'? (d) What is the change in rest energy during the collision?

37. A simple inelastic collision. Two identical objects, each of rest mass m_0, moving with equal but opposite velocities of $0.60c$ in the laboratory reference frame, collide and stick together. The resulting particle has a rest mass M_0. Express M_0 in terms of m_0.

38. Another simple inelastic collision. A body of rest mass m_0, traveling initially at a speed of $0.60c$, makes a completely inelastic collision with an identical body that is initially at rest. (a) What is the rest mass of the resulting single body? (b) What is its speed?

39. The previous collision generalized. Let the initial speed of the particle of rest mass m_0 in Problem 38 be generalized to βc, all other conditions of that problem remaining unchanged. (a) Show that the rest mass M_0 of the resulting single particle is given by

$$M_0 = \sqrt{2(\gamma + 1)}m_0.$$

(b) Show that the speed U of that particle is given by

$$U = \sqrt{\frac{\gamma - 1}{\gamma + 1}}\, c.$$

(c) Verify that, for $\beta = 0.60$, these formulas reproduce the numerical answers given in Problem 38.

40. Not the simplest reference frame! Two high-speed particles, each having rest mass m_0, approach each other on course for a head-on collision. One has a speed of $0.80c$ and the other a speed of $0.60c$. Assuming that the resulting collision is *completely inelastic*, answer the following questions in terms of m_0 and c. (a) What is the momentum after the collision? (b) What value does Newtonian mechanics predict for this quantity? (c) What is the total energy after the collision? (d) What is the total rest mass after the collision? (e) What is the total kinetic energy after the collision?

41. The completely inelastic collision. Verify that total energy E is conserved, in each reference frame, for the completely inelastic collision of Fig. 3-2.

42. An elastic collision. Consider the following head-on elastic collision. Particle 1 has rest mass $2m_0$ and particle 2 has rest mass m_0. Before the collision, particle 1 moves toward particle 2, which is initially at rest, with speed u (= $0.600c$). After the collision each particle moves in the forward direction with speeds of u_1 and u_2, respectively.

(a) Apply the laws of conservation of total energy (or, equivalently, of relativistic mass) and of relativistic momentum to this collision and solve the resulting equations to find u_1 and u_2. (b) Calculate the initial and the final kinetic energy and show that kinetic energy is indeed conserved in this elastic collision. (*Hint:* The relationship $\beta\gamma = \sqrt{\gamma^2 - 1}$ may be found useful.)

★ **43. The great neutron shoot-out.** Sue and Jim are two experimenters at rest with respect to one another at different points in space. They "fire" neutrons at each other, each neutron leaving its "gun" with a relative speed of 0.60c. Jim makes five observations about what is going on: (a) "My separation from Sue is 10 km." (b) "The speed of 'my' neutrons is 0.60c." (c) "Two of our neutrons have collided; relativistic momentum and kinetic energy are conserved." (d) "After this collision, one of the neutrons was scattered through an angle of 30°." (e) "I am firing neutrons at the rate of 10,000 s^{-1}." For each of these observations, state the corresponding observation that would be reported by a third person (Sam, say), who is in a frame S chosen so that Sue's neutrons are at rest in it.

44. The relativistic rocket. A rocket with initial mass M_i (fuel + payload) is accelerated from rest to a final speed v that is comparable to the speed of light. It can be shown that the mass of the rocket M_p (payload alone) when the speed v has been achieved is given by

$$\frac{M_i}{M_p} = \left(\frac{1 + \beta}{1 - \beta}\right)^{c/2u},$$

in which u is the speed of the exhaust relative to the rocket and $\beta = v/c$. Hence, to keep the ratio M_i/M_p small, large exhaust speeds are needed. (a) Using nuclear fusion to generate the exhaust, exhaust speeds of $c/7$ are theoretically possible. Suppose that, for interstellar travel, a final cruising speed v of 0.99c is desired. Calculate M_i/M_p. (b) For a round trip to a star—Sirius, say—the rocket must be accelerated from rest near the earth to cruising speed v (= 0.99c), brought to a stop at Sirius, accelerated from rest for the return trip when the visit is over, and finally brought to rest at home base. Calculate M_i/M_p for this round trip, where M_p is now the mass of the rocket when the round trip is over. (c) The greatest possible exhaust speed ($u = c$) occurs if the exhaust is electromagnetic radiation, generated perhaps by matter–antimatter annihilation. Recalculate the ratios in (a) and (b) for such a rocket. (d) In view of all of the above, what do you think of the possibility of interstellar travel?

45. Fast particles curve less than slow ones. What is the radius of curvature of an electron moving at right angles to a uniform magnetic field of 0.10 T at a speed of (a) 0.9c? (b) 0.99c? (c) 0.999c? (d) 0.9999c?

46. Identifying particles. Ionization measurements show that a particular nuclear particle carries a double charge (=2e) and is moving with a speed of 0.710c. Its measured radius of curvature in a magnetic field of 1.00 T is 6.28 m. Find the rest mass of the particle and identify it. (*Hint:* Light nuclear particles are made up of neutrons [which carry no charge] and protons [charge = +e], in roughly equal numbers. Take the rest mass of either of these particles to be 1.00 u.)

47. A curving relativistic electron. A 2.50-MeV electron moves at right angles to a magnetic field in a path whose radius of curvature is 3.0 cm. (a) What is the magnetic field B? (b) By what factor does the relativistic mass of the electron exceed its rest mass?

48. A curving cosmic-ray proton. A 10-GeV proton in the cosmic radiation approaches the earth in the plane of its geomagnetic equator, in a region over which the earth's average magnetic field is 5.5×10^{-5} T. What is the radius of its curved path in that region?

49. Cosmic rays in the sun's magnetic field. A cosmic-ray proton of kinetic energy 10 GeV may experience an effective magnetic field due to the sun of $\sim 2 \times 10^{-10}$ T. What will be the radius of curvature of the path of such a proton? Compare this radius to the radius of the earth's orbit about the sun (1.5×10^{11} m).

50. Cosmic-ray protons in the galactic magnetic field. Cosmic ray particles (presumably protons) with energies as large as 10^{20} eV have been detected as they enter our atmosphere. What is the radius of curvature of such an extremely energetic proton as it moves in the extremely weak magnetic field (10^{-9}–10^{-10} T) that permeates our galaxy? Compare your answer with the radius of (the luminous portion of) the galaxy, which is about 80,000 ly. Assume for simplicity that the galactic magnetic field is uniform and that the proton moves at right angles to it.

51. The ultimate terrestrial accelerator (I). Imagine a proton synchrotron extending around the earth at its equator, with a maximum magnetic field of 5.0 T. (a) What would be the energy of a proton circulating in such a path? (b) What would be the ratio of its relativistic mass to its rest mass?

52. The ultimate terrestrial accelerator (II). The linear electron accelerator at Stanford is about two miles long and has an acceleration gradient of about 15 MeV/m. Imagine a similar accelerator built into a tunnel through the earth, along a diameter. (a) What electron energy would be achieved by such a device, assuming that the same acceleration gradient could be maintained? (b) What would be the ratio of the relativistic mass to the rest mass for such electrons?

53. The binding energy of carbon-12. The nucleus ^{12}C consists of six protons (^{1}H) and six neutrons (n) held in close association by strong nuclear forces. The rest masses are

$$^{12}C \qquad 12.000000 \text{ u},$$
$$^{1}H \qquad 1.007825 \text{ u},$$
$$n \qquad 1.008665 \text{ u}.$$

How much energy would be required to separate a ^{12}C nucleus into its constituent neutrons and protons? This energy is called the *binding energy* of the ^{12}C nucleus. (*Note*: The masses given are really those of neutral atoms, but the extranuclear electrons have relatively negligible binding energy and are of equal number both before and after the breakup of the ^{12}C nucleus.)

54. A neutron-capture process. A helium-3 nucleus (nuclear mass = 3.01493 u) captures a slow neutron (mass = 1.00867 u) to form a nucleus of helium-4 (nuclear mass = 4.00151 u), according to the scheme

$$^{3}He + n \rightarrow {}^{4}He + \gamma.$$

The symbol γ represents an emitted gamma ray. What is its energy?

55. Spontaneous decay. A body of mass m at rest breaks up spontaneously into two parts with rest masses m_1 and m_2 and respective speeds u_1 and u_2. Show that $m > m_1 + m_2$.

★ **56. The decay of a pion.** A charged pion at rest in the laboratory decays according to the scheme

$$\pi \rightarrow \mu + \nu,$$

in which μ represents a muon (rest energy = 105.7 MeV) and ν a neutrino (rest energy taken as zero). If the measured kinetic energy of the muon is 4.1 MeV, what rest energy may be calculated for the pion?

★ **57. Fixed-target collisions.** (*a*) A proton accelerated in a proton synchrotron to a kinetic energy K strikes a second (target) proton at rest in the laboratory. The collision is entirely inelastic in that the rest energy of the two protons, plus all of the kinetic energy consistent with the law of conservation of momentum, is available to generate new particles and to endow them with kinetic energy. Show that the energy available for this purpose is given by

$$\mathcal{E} = 2m_0 c^2 \sqrt{1 + \left(\frac{K}{2m_0 c^2}\right)}.$$

(*b*) How much energy is made available when 100-GeV protons are used in this fashion? (*c*) What proton energy would be required to make 100 GeV available? (*Note*: Compare Problem 58.)

58. Center-of-mass collisions. (*a*) In modern experimental high-energy physics, energetic particles are made to circulate in opposite directions in so-called storage rings

and permitted to collide head-on. In this arrangement each particle has the same kinetic energy K in the laboratory. The collisions may be viewed as totally inelastic, in that the rest energy of the two colliding protons, plus all available kinetic energy, can be used to generate new particles and to endow them with kinetic energy. Show that the available energy in this arrangement can be written in the form (compare Problem 57)

$$\mathcal{E} = 2m_0 c^2 \left(1 + \frac{K}{m_0 c^2}\right).$$

(*b*) How much energy is made available when 100-GeV protons are used in this fashion? (*c*) What proton energy would be required to make 100 GeV available? (*Note*: Compare your answers with those in Problem 57, which describes another—less energy-effective—bombarding arrangement.)

59. Nuclear recoil during gamma emission. The nucleus of a carbon atom initially at rest in the laboratory is in a so-called excited, or unstable, state. It goes to its stable ground state by emitting a pulse of radiation (gamma ray) whose energy is 4.43 MeV and simultaneously recoiling. Thus

$$^{12}C^{\star} \rightarrow {}^{12}C + \gamma,$$

in which γ represents the gamma ray and the asterisk signifies an excited state. The atom in its final state has a rest mass of 12.0 u. (*a*) What is the momentum of the recoiling carbon atom, as measured in the laboratory? (*b*) What is the kinetic energy of the recoiling atom? (*Hint*: The speed of the relatively massive recoiling carbon atom will be so small that the classical expressions for its kinetic energy and momentum can be used without appreciable error.)

60. Atomic recoil during photon emission. An excited atom, with excitation energy E and rest mass m_0, is initially at rest. It releases this energy by emitting a pulse of radiation (photon) of energy E' and simultaneously recoiling. Because of the kinetic energy given to the recoiling atom, the energy E' of the photon will be (slightly) less than E. Show that, to a close approximation,

$$E' = E\left(1 - \frac{E}{2m_0 c^2}\right).$$

(*Hint*: The speed of the relatively massive recoiling atom will be low enough that the classical expressions for the kinetic energy and the momentum of the atom may be used without appreciable error. The answer given involves an approximation based on the binomial expansion.)

61. Deriving a conversion factor. (*a*) Prove that 1 u = 931 MeV/c^2. (*b*) Find the energy equivalent to the rest mass of an electron and (*c*) of a proton.

62. A real fast ball! A baseball has a mass of 140 g. (*a*) What energy must be supplied to it if its relativistic mass is to exceed its rest mass by 0.1 percent? (*b*) For how long a time would the full output of a 1000-MW power plant be required to supply this energy?

63. A low-speed space probe. *Voyager 2*, an 810-kg planetary probe, flew by Saturn in 1981, achieving speeds in excess of 54,000 mi/h. At this speed, what mass increment is associated with the kinetic energy of the probe?

64. A good student project (?). A metric ton of water (= 1000 kg) is heated from the freezing point to the boiling point. By how much does its mass increase?

65. Lots of energy in a penny. (*a*) What is the equivalent in energy units of the mass of a penny (= 3.1 g)? (*b*) How long would a 1000-MW power plant have to run to generate this amount of energy?

66. Water is heavier than ice (but not by much). Find the fractional increase in mass when an ice cube melts. The energy required is 3.35×10^5 J/kg.

67. About the weight of 30 gal of gasoline. The United States consumed about 2.2×10^{12} kWh of electrical energy in 1979. How much matter would have to vanish to account for the generation of this energy? Does it make any difference to your answer if this energy is generated in oil-burning, nuclear, or hydroelectric plants?

68. Drop a tablet in your tank! A 5-grain aspirin tablet has a mass of 320 mg. For how many miles would the energy equivalent of this mass, in the form of gasoline, power an automobile? Assume 30 mi/gal and a heat of combustion of 1.3×10^8 J/gal for the gasoline.

69. A ton of sunlight. The sun radiates energy at the rate of 4.0×10^{26} W. How many "tons of sunlight" (mass equivalent) does the earth intercept each day?

70. An enormous source of energy. Quasars are thought to be the nuclei of active galaxies in the early stages of their formation. A typical quasar radiates energy at the rate of 10^{41} W. At what rate is the mass of this quasar being reduced to supply this energy? Express your answer in solar mass units per year, where one solar mass unit, (smu = 2×10^{30} kg), is the mass of our sun.

71. Nuclear and TNT explosions compared. (*a*) How much energy is released in the explosion of a fission bomb containing 3.0 kg of fissionable material? Assume that 0.10 percent of the rest mass is converted to released energy. (*b*) What mass of TNT would have to explode to provide the same energy release? Assume that each mole of TNT liberates 3.4 MJ of energy on exploding. The molecular weight of TNT is 0.227 kg/mol. (*c*) For the same mass of explosive, how much more effective are nuclear explosions than TNT explosions? That is, compare the fractions of the rest mass that are converted to energy in each case.

72C. Using your calculator (I). Store in your hand-held programmable calculator the rest energies of the electron, the muon, and the proton (see Appendix), and also make provision to enter the rest energy of any other particle that you may choose. Write a program that will allow you to select one of these particles and that will accept as input data any *one* of the following five properties of that particle: its kinetic energy, its total energy, its momentum, its Lorentz factor γ, or its speed parameter β. The output is to be the remaining four quantities. (*Note:* Your calculator may overflow if you try to calculate β for particles—20-GeV electrons, say—whose speeds are very close to the speed of light. See Problem 74C.)

73C. Checking it out (I). Check out the program you wrote in Problem 72 C by answering these questions. (*a*) What is the momentum of an electron whose kinetic energy is 2.00 eV? 2.00 keV? 2.00 MeV? 2.00 GeV? (*b*) For what kinetic energies will an electron and a proton each have a momentum of 5.00 MeV/*c*? (*c*) An electron, a muon, and a proton each have a kinetic energy of 10.0 MeV. What are their speed parameters? (*d*) An electron, a muon, and a proton each have a relativistic mass that is three times their rest mass. What are their kinetic energies?

74C. Using your calculator (II). For particles whose speed parameters are sufficiently close to unity, the program called for in Problem 72C will yield a meaningless result for β because of calculator overflow. In such cases it is useful to calculate $1 - \beta$, using an approximate formula appropriate to the extreme relativistic case. To study the transition from the relativistic to the extreme relativistic case, write a program for your hand-held calculator that will accept as an input the kinetic energy of an electron and will display as successive outputs: (1) the speed parameter β, calculated exactly; (2) the quantity $1 - \beta$, also calculated exactly; (3) the quantity $1 - \beta$, calculated using an approximate formula appropriate to the extreme relativistic case; and (4) the percent difference between these last two quantities. (*Hint:* Store the rest energy of the electron, which is 0.511003 MeV, in your calculator. For the extreme relativistic formula, use $(1 - \beta) = (1/2)(m_0 c^2/K)^2$, in which K is the kinetic energy and $m_0 c^2$ is the rest energy.)

75C. Checking it out (II). (*a*) Run the program that you have written in Problem 74 for electron kinetic energies extending from a few keV to several hundred GeV and get a feeling for the transition from the relativistic to the extremely relativistic case. (*b*) At what electron energy do the exact and the approximate formulas for $1 - \beta$ differ by 10 percent? By 1.0 percent? (*c*) At about what electron kinetic energy do the predictions of the exact formula for β break down totally because of calculator overflow?

references

1. R. Kollath and D. Menzel, "Measurement of the Charge on Moving Electrons," *Z. Phys.* **134,** 530 (1953).

2. W. Bertozzi, "Speed and Kinetic Energy of Relativistic Electrons," *Am. J. Phys.,* **32,** 551 (1964).

3. R. R. Wilson, "The Batavia Accelerator," *Sci. Am.* (February 1974); R. R. Wilson, "The Tevatron," *Phys. Today* (October 1977).

4. R. T. Weidner, "On Weighing Photons," *Am. J. Phys.,* **35,** 443 (1967).

5. See *The Principle of Relativity* (Dover, New York, 1953), p. 29. This book is a collection of English translations of original papers by Einstein, Lorentz, and others.

6. Robert Resnick, *Introduction to Special Relativity* (Wiley, New York, 1968).

7. Edward M. Purcell, *Electricity and Magnetism* (McGraw-Hill, New York, 1965).

Introduction to Chapters 4 to 7

At a meeting of the German Physical Society in 1900, Max Planck read his paper "On the Theory of the Energy Distribution Law of the Normal Spectrum." This paper, which at first attracted little attention, was the start of a revolution in physics; its date of presentation has come to be called the "birthday of quantum theory." It was not until a quarter of a century later that the modern theory of quantum mechanics, the basis of our present understanding, was developed. Many paths converged on this understanding, each showing another aspect of the breakdown of classical physics. In the next four chapters we examine the major milestones along the way. The phenomena we discuss are all important in their own right and span all the classical disciplines, such as mechanics, thermodynamics, statistical mechanics, and electromagnetism (including light). Their contradiction of classical laws and their resolution on the basis of quantum ideas show us the need for a sweeping new theory and give us a deeper conceptual understanding of the theory that eventually emerged.

CHAPTER 4

the quantization of energy

After 1900 Planck strove for many years to bridge, if not to close, the gap between the older and quantum physics. The effort failed, but it had value in that it provided the most convincing proof that the two could not be joined.

Max von Laue

4-1 INTRODUCTION

In this and succeeding chapters we describe the development of the quantum theory of physics. Like relativity theory, quantum theory is a generalization of classical physics that includes the classical laws as special cases. Just as relativity theory extends the range of physical laws to the region of high speeds, so quantum theory extends this range to the region of small dimensions. Just as the speed of light c, a universal fundamental constant, characterizes relativity theory, so the Planck constant h, a second universal fundamental constant, characterizes quantum theory. Quantum theory arose from the study of the radiation emitted by heated objects. These studies led, as we shall see, to the concept that the internal energy of a physical system such as an atom or a molecule cannot have values chosen from a continuous range but can have only particular values, chosen from a discrete set. Energy—as we say—is *quantized*.

4-2 THERMAL RADIATION

At the turn of the century, one of the great unsolved problems in physics was the nature of the radiation, called *thermal radiation*, given off by a body because of its temperature. All bodies not only emit such radiation but also absorb such radiation from other bodies in their surroundings. If a body is at first hotter than its surroundings, it will cool off because its rate of emitting energy will exceed its rate of absorbing energy. Normally it will come to thermal equilibrium with its surroundings, a condition in which its rates of emission and absorption of energy are equal.

At ordinary temperatures we see most bodies not by their own emitted light but rather by the light that they reflect. At high enough temperatures, however, bodies become self-luminous. We can see them glow in the dark, hot coals and incandescent lamp filaments being familiar examples. But even at lamp-filament temperatures, well over 90 percent of the emitted thermal radiation is invisible to us, being in the infrared part of the electromagnetic spectrum.

The spectrum of the thermal radiation from a hot solid body is continuous, its details depending strongly on the temperature. If we were to steadily raise the temperature of such a body, we would notice two things: (1) The higher the temperature, the more thermal radiation is emitted—at first the body appears dim, then it glows brightly; and (2) the higher the temperature, the shorter is the wavelength of that part of the spectrum radiating most intensely—the predominant color of the hot body shifts from dull red through bright yellow-orange to bluish "white heat." Since the characteristics of its spectrum depend on the temperature, we can estimate the temperature of a hot body, such as a glowing steel ingot or a star, from the radiation it emits. The eye sees chiefly the color corresponding to the most intense emission in the visible range.

The radiation emitted by a hot body depends not only on the temperature but also on the material of which the body is made, its shape, and the nature of its surface. For example, at 2000 K a polished flat tungsten surface emits radiation at a rate of 23.5 W/cm²; for molybdenum, however, the corresponding rate is 19.2 W/cm². In each case the rate increases somewhat if the surface is roughened. Other differences appear if we measure the distribution in wavelength of the emitted radiation. Such details make it hard to understand thermal radiation in terms of simpler physical ideas; it reminds us of the complications that come up in trying to understand the properties of real gases in terms of a simple atomic model. The "gas problem" was managed by introducing the notion of an ideal gas. In much the same spirit, the "radiation problem" can be made manageable by introducing an "ideal radiator" for which the spectrum of the emitted thermal radiation depends *only* on the temperature of the radiator and not on the material, the nature of the surface, or other factors.

We can make such an ideal radiator, in fact, by forming a cavity within a body, the walls of the cavity being held at a uniform temperature. We must pierce a small hole through the wall so that a sample of the radiation inside the cavity can escape into the laboratory to be examined. It turns out that such thermal radiation, called *cavity radiation*, has a very simple spectrum whose nature is indeed determined only by the temperature of the walls and not in any way by the material of the cavity, its shape, or its size. Cavity radiation (radiation in a box) helps us to understand the nature of thermal radiation, just as the ideal gas (matter in a box) helped us to understand matter in its gaseous form.

Figure 4-1 shows a cavity radiator made of a thin-walled cylindrical tungsten tube about 1 mm in diameter and heated to incandescence by passing a current through it. A small hole has been drilled in its wall. It is clear from the figure that the radiation emerging from this hole is much more intense than that from the

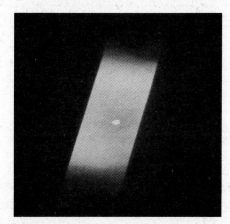

Figure 4-1. A thin-walled incandescent tungsten tube with a small hole drilled in its wall, photographed by its emitted light. The radiation emerging from the hole is cavity radiation.

outer wall of the cavity, even though the temperatures of the outer and inner walls are more or less equal.

Cavity radiation is also called *black-body radiation,* because it is the radiation that would be emitted by an ideal black body, that is, by a body that absorbed all the radiation that fell on it. If we look at an actual cavity by reflected light rather than by its own emitted light, the hole in the cavity wall appears very black indeed—much blacker, in fact, than charcoal or other objects normally considered to be quite black. In Example 1 we show that a "perfect" absorber (that is, a black body or a cavity aperture) is also a "perfect" emitter (that is, it radiates more intensely than any actual body at the same temperature).

There are three interrelated properties of cavity radiation—all well verified by experiment—that any proposed theory of cavity radiation must explain:

1. *The Stefan-Boltzmann law.* The total radiated output per unit area of the cavity aperture, called its *radiancy* $R(T)$, is given by

$$R(T) = \sigma T^4, \tag{4-1}$$

in which $\sigma (= 5.670 \times 10^{-8} \text{ W/m}^2 \cdot \text{K}^4)$ is a universal constant called the *Stefan-Boltzmann constant.*

2. The *spectral radiancy* function $\mathscr{R}(\lambda, T)$ tells us how the intensity of the cavity radiation varies with wavelength for any given temperature. It is defined so that the product $\mathscr{R}(\lambda) \, d\lambda$ gives the radiated power per unit area for the cavity radiation that lies in the wavelength band between λ and $\lambda + d\lambda$. You can find the radiancy $R(T)$ for any temperature by adding up (that is, by integrating) the spectral radiancy over the entire spectrum at that temperature. That is,

$$R(T) = \int_0^\infty \mathscr{R}(\lambda) \, d\lambda \qquad (T = \text{a constant}). \tag{4-2}$$

Figure 4-2 shows the spectral radiancy for cavity radiation at four selected temperatures. Examination of Eq. 4-2 shows that we can interpret the radiancy $R(T)$

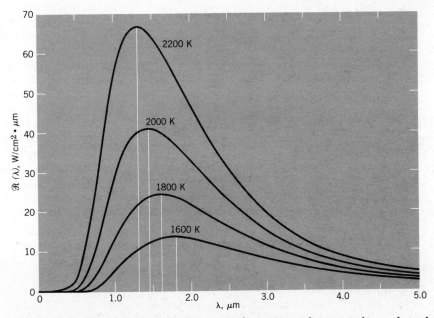

Figure 4-2. The spectral radiancy curves for cavity radiation at four selected temperatures. Note that as the temperature increases, the wavelength of maximum spectral radiancy shifts to lower values.

as the area under the appropriate curve of Fig. 4-2. We see from that figure that, as the temperature increases, so does this area—and thus the radiancy—as Eq. 4-1 predicts.

3. *The Wien Displacement law.* We can see from the spectral radiancy curves of Fig. 4-2 that λ_m, the wavelength at which the spectral radiancy is a maximum, decreases as the temperature increases. Wilhelm Wien was able to show from classical theory that, in fact, the product $\lambda_m T (= w)$ is a universal constant; its experimental value proves to be

$$w = \lambda_m T = 2898 \ \mu\text{m} \cdot \text{K}. \tag{4-3}$$

Equation 4-3 is often called the *Wien displacement law* and *w* the *Wien constant*.

EXAMPLE 1.

A "Perfect" Absorber is a "Perfect" Emitter. Figure 4-3 shows a cross section through a cavity radiator, its walls maintained at a temperature *T*. A ray emerging in an arbitrary direction is traced backward to its origins on the cavity walls. The emerging ray is confined to a selected narrow band of wavelengths by means of a band-pass filter *F*.

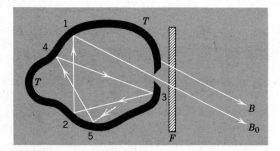

Figure 4-3. Example 1. A cross section through a cavity radiator whose walls are held at temperature *T*. Filter *F* selects an arbitrary narrow wavelength interval for study.

The *brightness* of a surface is a measure of its spectral radiancy. Let B_0 represent the brightness of the hole, viewed in the direction shown, and let *B* be the brightness of the outside wall of the cavity in that same direction. If radiation in the selected wavelength band falls onto the inner cavity wall, suppose that a fraction *a* is absorbed and a fraction $r(= 1 - a)$ is reflected. We see that

the brightness B_0 of the hole is made up of *B*, the brightness of the wall near position 1 in Fig. 4-3; of *rB*, the brightness of the wall at position 2, once reflected; of $r^2 B$, the brightness of the wall at position 3, twice reflected, and so on. Thus

$$B_0 = B + rB + r^2 B + r^3 B + \cdots$$
$$= B(1 + r + r^2 + r^3 + \cdots).$$

But the quantity in parentheses above can be written as $1/(1 - r)$, as you can show simply by long-hand division, so that

$$B = (1 - r)B_0 = aB_0.$$

Because all known materials have $a < 1$, it follows that *B*, the brightness of the radiation from the outer wall of the cavity, will always be less than B_0, the brightness of the hole, as in Fig. 4-1. This will be true at all wavelengths, so the spectral radiancy curve for any given material will always lie below the curve for a cavity radiator at the same temperature. A cavity aperture, being a "perfect" absorber $(a = 1)$, is also a "perfect" emitter.

You can notice this effect by looking at a bed of glowing charcoal briquettes, which do not appear uniformly bright. The "hot spots" correspond to pockets between the briquettes, which would appear dark if you looked at the (unlit) briquettes by daylight. The same principle is used to increase the efficiency of a lamp filament simply by coiling it tightly. It then approximates a cavity radiator more than does a straight wire and shines more brightly for the same power input.

EXAMPLE 2.

Measuring the Temperatures of Stars. The "surfaces" of stars are not sharply defined boundaries like the surface of the earth. Most of the radiation emerging from the star is in good thermal equilibrium with the hot gases that make

up the star's outer layers. Without too much error, then, we can treat starlight as cavity radiation. The (continuous) spectra of three stars reveal the following values for λ_m, the wavelength at which the spectral radiancy is a maximum:

Star	λ_m	Appearance
Sirius	240 nm	blue-white
The Sun	500 nm	yellow
Betelgeuse	850 nm	red

(*a*) What are the surface temperatures of these stars? From Eq. 4-3 we find, for Sirius,

$$T = \frac{w}{\lambda_m}$$

$$= \left(\frac{2898\ \mu m \cdot K}{240\ nm}\right)\left(\frac{1000\ nm}{1\ \mu m}\right) = 12{,}000\ K.$$

The temperatures for the Sun and for Betelgeuse work out in the same way to be 5800 K and 3400 K, respectively. At 5800 K the Sun's surface is near the temperature at which most of its radiation lies within the visible region of the spectrum. This suggests that over ages of human evolution our eyes have adapted to the Sun to become most sensitive to those wavelengths that it radiates most intensely.

(*b*) What are the radiances of these three stars?

For Sirius we have, from the Stefan-Boltzmann law (Eq. 4-1),

$$R = \sigma T^4 = (5.67 \times 10^{-8}\ W/m^2 \cdot K^4)(12{,}000\ K)^4$$
$$= 1.2 \times 10^9\ W/m^2 = 120\ kW/cm^2.$$

The radiancies for the Sun and for Betelgeuse work out to be 6.4 kW/cm² and 0.767 kW/cm², respectively.

(*c*) The radius of the Sun is 7.0×10^8 m and that of Betelgeuse is over 500 times larger, or 4.0×10^{11} m. What is the total radiated power output [that is, the *luminosity* (*L*)] of these stars?

We find the luminosity of a star by multiplying its radiancy by its surface area. Thus, for the Sun,

$$L = R(4\pi r^2) = (6.4\ kW/cm^2)(4\pi)(700 \times 10^8\ cm)^2$$
$$= 3.9 \times 10^{26}\ W.$$

For Betelgeuse the luminosity works out to be 1.5×10^{31} W, about 38,000 times larger. The enormous size of Betelgeuse, which is classified as a "red giant," much more than makes up for the relatively low radiancy associated with its low surface temperature.

Interestingly enough, the colors of stars are not strikingly apparent to the average observer because the retinal cones, which are responsible for color vision, do not function well in dim lights. If this were not so, the night sky would be spangled with color. As it is, although there are many yellow stars, none seem as visually yellow as our sun.

EXAMPLE 3.

Fire or Ice? Assume that the earth is in thermal equilibrium and radiates energy into space at the same rate at which it receives it from the sun. (*a*) At what orbit radius around the sun would the oceans freeze? (*b*) At what orbit radius would they boil? By what percent do these radii differ from the earth's actual mean orbit radius of 1.50×10^{11} m? Assume the earth to behave like a black body. The sun's luminosity (rate of energy output) is 3.9×10^{26} W.

If *L* is the sun's luminosity and *a* is the earth–sun distance, then the rate at which the earth intercepts the sun's energy is $L(\pi R^2/4\pi a^2)$, where *R* is the earth's radius. If the earth's mean temperature is *T*, the rate at which it radiates energy into space is given by (see Eq. 4-1) $(4\pi R^2)(\sigma T^4)$. Setting these two rates equal to each other gives us

$$L\left(\frac{\pi R^2}{4\pi a^2}\right) = (4\pi R^2)(\sigma T^4)$$

or

$$a = \frac{1}{4T^2}\sqrt{\frac{L}{\pi\sigma}}.$$

Note that the radius of the earth cancels out, so our conclusions hold for a planet of any size.

(*a*) For the oceans to freeze, we put *T* = 273 K. We have, then, for the orbit radius,

$$a = \frac{1}{4(273\ K)^2}\sqrt{\frac{3.9 \times 10^{26}\ W}{(\pi)(5.67 \times 10^{-8}\ W/m^2 \cdot K^4)}}$$
$$= 1.57 \times 10^{11}\ m,$$

which exceeds the earth's actual orbit radius by ~5 percent.

(*b*) For the oceans to boil, we put *T* = 373 K and obtain $a = 0.84 \times 10^{11}$ m. This is less than the earth's actual orbit radius by 44 percent. We are closer to freezing than to boiling, as perhaps the experience of the Ice Ages attests.

4-3 THE THEORY OF CAVITY RADIATION

In experimental work the spectral radiancy $\mathcal{R}(\lambda, T)$ of the cavity radiation is the natural quantity of interest. In developing a theory of such radiation, however, it seems more straightforward to consider the radiation while it is still inside the cavity and—for reasons that will become clear later—to work with the frequency of the radiation rather than with its wavelength. The natural quantity of interest then becomes the *spectral energy density* $\rho(\nu, T)$. It is defined so that, at a given temperature, the product $\rho(\nu)\, d\nu$ is the energy per unit volume of the cavity whose frequencies lie in the interval ν to $\nu + d\nu$.

In Supplementary Topic D we show that the spectral energy density can be calculated from the spectral radiancy (and from the relation $c = \lambda \nu$) by

$$\rho(\nu, T) = \left(\frac{2}{\nu}\right)^2 \mathscr{R}(\lambda, T). \tag{4-4a}$$

Conversely, the spectral radiancy can be calculated from the spectral energy density (and, again, from the relation $c = \lambda \nu$) by

$$\mathscr{R}(\lambda, T) = \left(\frac{c}{2\lambda}\right)^2 \rho(\nu, T). \tag{4-4b}$$

Using these two equations we can easily transform any function $\rho(\nu, T)$ derived from theory into the appropriate function $\mathscr{R}(\lambda, T)$ for comparison with experiment, and conversely.

In September 1900 there were two theoretical predictions of the spectral energy density, both based on classical physics. The first, due to Wilhelm Wien, involved a conjecture that the spectral energy density function can be related to the Maxwell speed distribution function for the molecules of an ideal gas. It is given by

$$\rho(\nu, T) = a\nu^3 e^{-b\nu/T} \qquad \text{(Wien)}, \tag{4-5}$$

in which a and b are arbitrary constants.[*]

The second prediction, due originally to Lord Rayleigh but later derived independently by Einstein and modified slightly by James Jeans, is developed rigorously from its classical base and, unlike Eq. 4-5, involves no conjectural assumptions. Their result, which contains no arbitrary constants, is

$$\rho(\nu, T) = \left(\frac{8\pi kT}{c^3}\right) \nu^2 \qquad \text{(Rayleigh-Einstein-Jeans)}, \tag{4-6}$$

in which $k\,(= 1.38 \times 10^{-23} \text{ J/K})$ is the Boltzmann constant. Note that this expression does not pass through a maximum but increases without limit as $\nu \to \infty$.

Early in October 1900, Max Planck, who had been working on the cavity radiation problem for some time and who was in active contact with the principal experimental workers in the field, made a truly inspired interpolation between these two classical predictions and put forward a third prediction, namely,

$$\rho(\nu, T) = a\nu^3 \frac{1}{e^{b\nu/T} - 1} \qquad \text{(Planck)}. \tag{4-7}$$

At this stage, a and b appear as arbitrary constants and the equation itself must be viewed as empirical, having no theoretical base.

For a given temperature, which of these three formulas for the spectral energy density agrees best with experiment? Here are the laboratory findings:

1. Wien's formula agrees well at high frequencies (short wavelengths), but fails otherwise.

---

[*] Wien's frequency distribution function (Eq. 4-5) must not be confused with Wien's displacement law (Eq. 4-3).

2. The Rayleigh-Jeans formula (as it is usually called) agrees well at very low frequencies (very long wavelengths), but fails otherwise.
3. Planck's formula agrees extremely well over the entire range of the experimental variables.

Figure 4-4 shows how very well Planck's formula (the solid curve) fits the experimental data (circles) for radiation from a cavity at 1595 K. The solid curve is Planck's theoretical prediction for the spectral radiancy, found by substituting his Eq. 4-7 into Eq. 4-4a. Planck's formula is also consistent with the Wien displacement law of Eq. 4-3 (see Problem 20) and with the Stefan-Boltzmann fourth power law of Eq. 4-1 (see Problem 22).

Figure 4-5 shows plots of the spectral energy density as given by Eqs. 4-5, 4-6, and 4-7, for an arbitrarily selected temperature of 2000 K. If we take the Planck curve as representing the experimental data, we see that the Wien curve does indeed fit well at high frequencies and the Rayleigh-Jeans curve at very low frequencies, just as claimed above. The dramatic failure of the Rayleigh-Jeans formula at high frequencies, in view of the straightforward derivation of this formula from well-established classical laws, is a serious matter. This formula predicts that as $\nu \to \infty$ (or, equivalently, as $\lambda \to 0$), the spectral energy density should approach an infinitely large value. In fact, it approaches zero. This failure of the Rayleigh-Jeans formula has been described as an "ultraviolet catastrophe."

Planck's formula (Eq. 4-7) is very important, but it is empirical; it should be possible to derive it from simpler and broader general principles. Planck worked hard at this task, and in December 1900, about two months after he had first advanced his formula, he was able to do so. In the process he recast his formula slightly, presenting the two arbitrary constants it contains in a different form. In his new notation, Planck's formula becomes (compare Eq. 4-7)

$$\rho(\nu, T) = \left(\frac{8\pi h\nu^3}{c^3}\right) \frac{1}{e^{h\nu/kT} - 1} \tag{4-8a}$$

With the aid of Eq. 4-4a we can also display this as a spectral radiancy, or

$$\mathcal{R}(\lambda, T) = \left(\frac{2\pi hc^2}{\lambda^5}\right) \frac{1}{e^{hc/\lambda kT} - 1} \tag{4-8b}$$

The two adjustable constants a and b in Eq. 4-7 are here replaced by quantities involving two constants of direct physical significance, the Boltzmann constant k and a totally new constant h, now called the *Planck constant*. Planck obtained

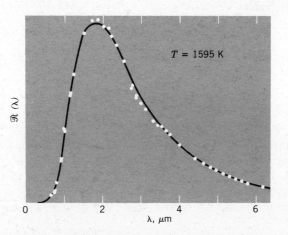

Figure 4-4. Planck's theoretical spectral radiancy prediction (solid line) compared with the experimental results (circles) of Coblentz (1916) for a cavity radiator at 1595 K.

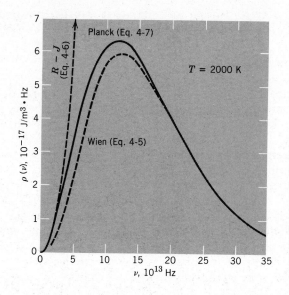

Figure 4-5. Three theoretical predictions for the spectral energy density of cavity radiation at 2000 K. The arbitrary constants in the Planck formula (see Eq. 4-7) were chosen to give the best fit to the experimental data. These same constants (see Eq. 4-5) were used in plotting the Wien curve. The Rayleigh-Jeans curve (R-J; Eq. 4-6) requires no adjustment of constants.

the best fit of Eq. 4-8b to the experimental data by choosing (in modern units) $k = 1.34 \times 10^{-23}$ J/K and $h = 6.55 \times 10^{-34}$ J·s. These quantities are within 3 and 1 percent, respectively, of their modern accepted values.

EXAMPLE 4.

Checking the Limiting Cases. Show that the Planck spectral energy density formula (a) reduces to the Wien formula in its high-frequency limit and (b) reduces to the Rayleigh-Jeans formula in its low-frequency limit.

(a) At frequencies such that $h\nu \gg kT$, the exponential quantity in the denominator of Eq. 4-8a becomes much greater than unity, so we can neglect the term "1" in that denominator. This leads at once to

$$\rho(\nu, T) \cong \left(\frac{8\pi h}{c^3}\right) \nu^3 e^{-h\nu/kT} \quad \text{(Wien)}.$$

Comparison with Eq. 4-5 shows that this is the Wien formula, the constants a and b having the values

$$a = \frac{8\pi h}{c^3} \quad \text{and} \quad b = \frac{h}{k}.$$

(b) In the other limit we consider frequencies such that $h\nu \ll kT$. Here the quantity $h\nu/kT(= x)$ is very much less than unity, and we can use the expansion

$$e^x = 1 + x + \frac{x^2}{2} + \frac{x^3}{3 \cdot 2} + \cdots \cong 1 + x,$$

ignoring terms beyond the first power in x. Substituting into Eq. 4-8a leads to

$$\rho(\nu, T) \cong \left(\frac{8\pi h\nu^3}{c^3}\right) \frac{1}{(1 + h\nu/kT) - 1}$$

$$= \left(\frac{8\pi kT}{c^3}\right) \nu^2 \quad \text{(Rayleigh-Jeans)},$$

which, as comparison with Eq. 4-6 shows, is the Rayleigh-Jeans formula.

4-4 THE RAYLEIGH-JEANS RADIATION LAW

Before we analyze Planck's radiation formula, let us look in broad outline at the derivation of the Rayleigh-Jeans radiation law (Eq. 4-6). We shall then be able to see more clearly where classical laws fail when applied to the cavity radiation problem and under what circumstances they need to be replaced by the more general laws of quantum theory.

Consider a cavity whose walls are held at a temperature T. The atoms that make up these walls radiate energy into the cavity volume and absorb energy from it. At thermal equilibrium these two rates are equal, and within the cavity we find an equilibrium distribution of trapped electromagnetic radiation, covering a wide range of frequencies. We can look at this trapped radiation as an assembly of standing waves, each with a node at the cavity walls and each with a different frequency or, correspondingly, a different wavelength. There is an analogy here to standing sound waves in a resonant acoustical cavity such as an organ pipe. In each case—radiation and sound—only certain discrete wavelengths (or frequencies) can meet the boundary conditions at the walls and can exist in the cavity as a standing wave.

For a typical cavity the number of standing electromagnetic waves in the cavity volume is very large indeed. The frequencies of the individual standing waves, though discrete, lie very close together, and there will be many such waves in any specified range of frequencies, however narrow. We can represent by $n(\nu)\,d\nu$ the number of waves per unit volume whose frequencies lie in the interval ν to $\nu + d\nu$. Lord Rayleigh showed that $n(\nu)$ is given by

$$n(\nu) = \left(\frac{8\pi}{c^3}\right)\nu^2, \tag{4-9}$$

in which c is the speed of light. If we just knew the average energy associated with each of these discrete standing waves, we could easily calculate the spectral energy density.

Rayleigh assumed that the classical equipartition of energy theorem (see *Physics*, Part I, Sec. 23-8) applied to the cavity radiation. We can state this theorem as:

For a system in thermal equilibrium, all independent modes of sharing energy do so equally, the average energy per degree of freedom being $\frac{1}{2}kT$, where k is the Boltzmann constant.

Let us illustrate this theorem by an example from the kinetic theory of gases. It can readily be shown in that context that the average kinetic energy of an atom of a monatomic gas is $\frac{3}{2}kT$. Such an atom has three degrees of freedom, associated with the three independent components of its translational motion. This is totally consistent with the assignment of $\frac{1}{2}kT$ to each of these translational modes.

There are two degrees of freedom associated with each standing electromagnetic wave in the cavity volume, associated with the two independent polarization directions that a transverse wave can have. Thus each standing wave, regardless of its frequency, has the same average energy, given by

$$\bar{\mathscr{E}} = 2 \times \tfrac{1}{2}kT = kT. \tag{4-10}$$

To get the spectral energy density we simply multiply the number of standing waves per unit volume by the average energy of each wave, or

$$\rho(\nu, T) = \left(\frac{8\pi\nu^2}{c^3}\right)(kT) = \left(\frac{8\pi kT}{c^3}\right)\nu^2, \tag{4-6}$$

which is the Rayleigh-Jeans law. We stress that this law is a straightforward prediction from well-established classical principles. We saw in Fig. 4-4, however, that, except at very low frequencies, Eq. 4-6 simply does not agree with experiment. It is not a matter of a small discrepancy; something is seriously wrong.

4-5 THE QUANTIZATION OF ENERGY

We turn now to Planck's radiation law (Eq. 4-8a) and to the significance of the quantity h that appears in it. The assumptions that Planck made in deriving this formula and the consequences of those assumptions were not immediately clear to Planck's contemporaries or, for that matter (as Planck confirmed later) to Planck himself. Planck's work on this difficult problem stimulated Einstein to carry out his own investigation. In what follows we describe the situation as it appeared in 1906 or shortly thereafter, some six years after Planck advanced his theory. It seems to be true that the concept of energy quantization was not well understood at any earlier date [1].

Planck represented the atoms that made up the cavity walls by harmonically oscillating electrons, which could both emit and absorb electromagnetic radiation. He assumed that these oscillators had a broad distribution of natural resonant frequencies, corresponding to the distribution in frequency of the radiation in the cavity. In fact, it is possible to assign one oscillator in the cavity wall to each standing wave in the cavity volume. Thus counting the oscillators is the same as counting the standing waves, a problem already solved by Rayleigh (Eq. 4-9).

Classically, the energy that any such oscillator can have is a smoothly continuous variable. We certainly assume this for large-scale oscillators such as mass–spring systems or pendulums. It turns out, however, that, in order to derive the Planck equation one must assume that the electronic oscillators in the cavity walls behave in quite a different way: *The oscillators may no longer have any energy but only energies chosen from a discrete set, defined by*

$$\mathscr{E} = nh\nu \qquad n = 0, 1, 2, \ldots, \tag{4-11}$$

where $h (= 6.63 \times 10^{-34}$ J·s) is a new constant of the greatest importance and n is a positive integer.

We say that the energy of an elementary oscillator is *quantized* and that n is a *quantum number*. Equation 4-11 tells us that the oscillator energy levels are evenly spaced, the interval between adjacent levels being $h\nu$. As Fig. 4-6 shows, the higher the frequency of the oscillator, the more widely separated are its allowed energy levels. Figure 4-7b shows that at low enough frequencies, the allowed levels become so close together that they approximate the classical continuous distribution of Fig. 4-7a. Recall that it is precisely at such low frequencies that the classical Rayleigh-Jeans formula agrees with experiment.

Equation 4-11 tells us the energy that an oscillator of a given frequency *may* have. What we need to know, however, is the energy $\bar{\mathscr{E}}$, averaged over time, that the oscillator actually *does* have for any given temperature of the cavity walls. Classically, as we have seen, this average oscillator energy is simply kT, the same for all frequencies. With energy quantization in the picture, however, Planck showed that the average energy depends on the frequency and is given by

$$\bar{\mathscr{E}} = \frac{h\nu}{e^{h\nu/kT} - 1}. \tag{4-12}$$

To find the spectral energy density we simply multiply this quantity by $n(\nu)$, the number of standing waves (and thus of their associated oscillators) per unit volume and per unit frequency interval; see Eq. 4-9. Doing so yields

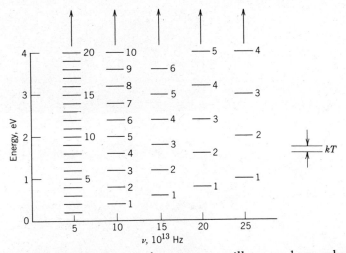

Figure 4-6. The energy of a quantum oscillator can have only discrete values, governed by Eq. 4-11. Energy levels for oscillators with five different frequencies are shown, labeled with their quantum numbers. On the right is shown the energy interval kT, the time-averaged oscillator energy according to classical physics, evaluated for 2000 K.

$$\rho(\nu,\, T) = \left(\frac{8\pi\nu^2}{c^3}\right)\left(\frac{h\nu}{e^{h\nu/kT} - 1}\right)$$

$$= \left(\frac{8\pi h\nu^3}{c^3}\right)\frac{1}{e^{h\nu/kT} - 1}, \qquad [4\text{-}8a]$$

the Planck formula for the spectral energy density.

One can understand the results of the energy quantization assumptions physically in this way. The electronic oscillators in the cavity walls are pictured classically as radiating their energy smoothly as their motion gradually subsides. In the quantum view, however, an oscillator ejects its radiation in spurts. Thus the energy of an oscillator does not decrease continuously but in jumps. The

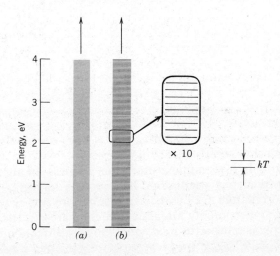

Figure 4-7. (a) Showing the continuous energy distribution available to a classical oscillator. (b) The allowed energy levels of a low-frequency quantum oscillator ($\nu \cong 0.5 \times 10^{13}$ Hz; compare the frequency scale of Fig. 4-6). A magnified section is displayed. The interval kT for 2000 K is again shown; see Fig. 4-6.

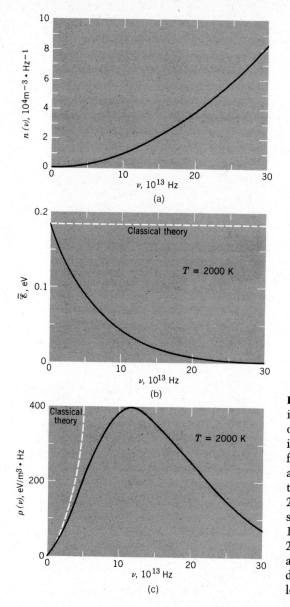

Figure 4-8. (*a*) The number of standing waves (and thus of their associated oscillators) per unit volume of the cavity and per unit frequency interval as a function of frequency (Eq. 4-9). (*b*) The average energy per oscillator as a function of frequency for a temperature of 2000 K (Eq. 4-12). The dashed line shows the classical prediction (Eq. 4-10). (*c*) The spectral energy density at 2000 K formed by multiplying the above two curves (Eq. 4-8*a*). Again, the dashed curve shows the classical Rayleigh-Jeans prediction (Eq. 4-6).

allowed energy values of an oscillator must then be discrete and, as it exchanges energy with the cavity radiation, the oscillator emits or absorbs energy only in discrete amounts. Since the discrete energies that an oscillator can emit or absorb are directly proportional to its frequency, the oscillators of low frequency can emit or absorb energy in small packets, whereas the oscillators of high frequency emit or absorb energy only in large packets. Now imagine that the cavity wall is at a relatively low temperature. Then there is enough thermal energy to excite the low-frequency oscillators, but not the high-frequency ones. The high-frequency oscillators need to receive much more energy to begin radiating and fewer of them proportionally are activated compared to the low-frequency oscillators (the energy is *not* partitioned equally over all frequencies). Hence the walls radiate principally in the long-wavelength region and hardly at all in the ultraviolet.

As the temperature of the wall is raised, there is sufficient thermal energy to activate a larger number of high-frequency oscillators and the resulting radiation shifts its character toward higher frequencies, that is, toward the ultraviolet. Hence the quantum assumptions quite naturally lead to the experimental observations discussed earlier and avoid the ultraviolet catastrophe of the classical analysis.

We can summarize the situation in still another way. Figure 4-8a shows that as the frequency increases, the number of available oscillators also increases, in proportion to the square of the frequency (see Eq. 4-9). It is the fact that, classically, *all* of these oscillators have the *same* average energy (= kT) that causes the dramatic failure of the Rayleigh-Jeans law at high frequencies. The problem, then, is to avoid this ultraviolet catastrophe by preventing most of these high-frequency oscillators from being activated. As Fig. 4-8b shows, quantizing the oscillators according to the scheme of Eq. 4-11 does precisely that. As the frequency increases, so does the level spacing, and the oscillators spend shorter and shorter periods of time in activated states so that their average energy decreases rapidly. We find the spectral energy density curve of Fig. 4-8c by multiplying the curves of Figs. 4-8a and 4-8b, and we see that the reduction in average energy as the frequency increases more than compensates for the fact that more oscillators become available.

Planck's hypothetical harmonic oscillators were the first physical systems for which it was realized that energy quantization—a notion that has no place in classical physics—is a fundamental characteristic. We now know that *all* physical systems, be they elementary particles confined to a limited region of space, nuclei, atoms, or molecules, exist with quantized energies. The quantized levels are rarely uniformly spaced in energy, but they all involve the Planck constant h.

EXAMPLE 5.

A Macroscopic Oscillator. A mass m of 0.30 kg, suspended from a spring whose force constant k is 3.0 N/m, is oscillating with an amplitude A of 0.10 m. As time goes on the oscillations damp out because of friction, and the total energy of the system gradually decreases.

(*a*) Is the energy decrease continuous or discrete? In other words, does energy quantization apply to a macroscopic harmonic oscillator?

The frequency of oscillation of the mass–spring system is given by (see Physics, Part I, Sec. 15-3)

$$\nu = \frac{1}{2\pi} \sqrt{\frac{k}{m}} = \frac{1}{2\pi} \sqrt{\frac{3.0\ \text{N/m}}{0.30\ \text{kg}}} = 0.50\ \text{Hz}.$$

The total mechanical energy of the system initially is

$$E = \tfrac{1}{2}kA^2 = \tfrac{1}{2}(3.0\ \text{N/m})(0.10\ \text{m})^2 = 1.5 \times 10^{-2}\ \text{J}.$$

Now let us assume that the energy of the system is quantized so that the energy decreases by discrete jumps of magnitude $\Delta E = h\nu$. The magnitude of these jumps is then

$$\Delta E = h\nu = (6.63 \times 10^{-34}\ \text{J}\cdot\text{s})(0.50\ \text{s}^{-1})$$
$$= 3.3 \times 10^{-34}\ \text{J}.$$

Thus

$$\frac{\Delta E}{E} = \frac{3.3 \times 10^{-34}\ \text{J}}{1.5 \times 10^{-2}\ \text{J}} \cong 2 \times 10^{-32}.$$

Energy measurements of such precision simply cannot be made.

(*b*) What is the quantum number n for the initial energy state of this macroscopic oscillator?

From the quantization relation Eq. 4-11, we have

$$n = \frac{E}{h\nu} = \frac{1.5 \times 10^{-2}\ \text{J}}{3.3 \times 10^{-34}\ \text{J}} \cong 5 \times 10^{31},$$

an enormous number!

This example shows that the quantization of energy is not apparent for ordinary large-scale oscillators. The smallness of h makes the graininess in the energy too fine to be distinguished from an energy continuum. Indeed, h might as well be zero for classical systems and, in fact, one way to reduce quantum formulas to their classical limits is to let $h \to 0$ in those formulas. The fact that Planck's quantization postulate gives the correct cavity radiation formula, however, suggests that for oscillators of atomic dimensions ΔE and E can be of comparable order, so that the quantized nature of energy reveals itself.

EXAMPLE 6.

Energy Levels and Oscillators. In Fig. 4-5 we see (Planck curve) that the maximum spectral energy density for a cavity at 2000 K occurs at a frequency of 1.2×10^{14} Hz. At this temperature and this frequency, what are the values of (a) kT? (b) $h\nu$? (c) $h\nu/kT$? (d) The average oscillator energy according to classical (Rayleigh-Jeans) theory? (e) The average oscillator energy according to quantum (Planck) theory? (f) The spectral energy density according to classical (Rayleigh-Jeans) theory? (g) The spectral energy density according to quantum (Planck) theory?

(a) $kT = (8.62 \times 10^{-5} \text{ eV/K})(2000 \text{ K})$
$= 0.172 \text{ eV}.$

It can be shown that kT is the translational kinetic energy associated with those molecules in a gas that are moving, at any given temperature, with their most probable speeds.

(b) $h\nu = (4.14 \times 10^{-15} \text{ eV s})(1.2 \times 10^{14} \text{ Hz})$
$= 0.497 \text{ eV}.$

This quantity (see Fig. 4-6) is the spacing between adjacent oscillator levels at the frequency in question.

(c) $\dfrac{h\nu}{kT} = \dfrac{0.497 \text{ eV}}{0.172 \text{ eV}} = 2.89.$

We see that the levels are spaced farther apart than kT, our measure of the thermal agitation energy. The energy fluctuations will thus often not be large enough to activate the lowest ($n = 1$) oscillator level.

(d) The average oscillator energy in classical theory, in which there are no levels and all oscillator energies are accessible, is simply $kT (= 0.172 \text{ eV})$; see Eq. 4-10.

(e) In quantum theory (see Eq. 4-12), the average oscillator energy is given by

$$\bar{\mathscr{E}} = \frac{h\nu}{e^{h\nu/kT} - 1} = \frac{0.497 \text{ eV}}{e^{2.89} - 1} = 0.0292 \text{ eV}.$$

This is less than the classical value given in (d) by the factor $0.172/0.0292 = 5.9$. Compare the ratio in (c).

(f) From Eq. 4-6, we see that the spectral energy density according to classical theory is

$$\rho(\nu, T) = \frac{8\pi kT\nu^2}{c^3}$$

$$= \frac{(8\pi)(1.38 \times 10^{-23} \text{ J/K})(2000 \text{ K})(1.2 \times 10^{14} \text{ Hz})^2}{(3.00 \times 10^8 \text{ m/s})^3}$$

$$= 37.0 \times 10^{-17} \text{ J/m}^3 \cdot \text{Hz}.$$

(g) From Eq. 4-8a we see that the spectral energy density according to quantum theory is

$$\rho(\nu, T) = \frac{8\pi h\nu^3}{c^3} \frac{1}{e^{h\nu/kT} - 1}$$

$$= \frac{(8\pi)(6.63 \times 10^{-34} \text{ J} \cdot \text{s})(1.2 \times 10^{14} \text{ Hz})^3}{(3.00 \times 10^8 \text{ m/s})^3} \frac{1}{e^{2.89} - 1}$$

$$= 6.41 \times 10^{-17} \text{ J/m}^3 \cdot \text{Hz}.$$

You can check this value from Fig. 4-5. The spectral energy density predicted by quantum theory (which agrees with experiment) is thus about 5.8 times lower than that predicted by classical theory (which does not agree with experiment). This difference comes about *entirely* because the oscillator energies are assumed to be quantized. This has the effect of reducing the classically predicted average oscillator energy [see part (e)] and thus avoiding the ultraviolet catastrophe that would otherwise occur.

4-6 ENERGY QUANTIZATION AND THE HEAT CAPACITIES OF SOLIDS

Energy quantization was very slow to be accepted by the physics community, not an unusual fate for a radically new idea. It is perhaps not hard to understand why. The physical systems whose energies were first quantized were the oscillators assumed by Planck and others to form the active elements of the walls of a cavity radiator. In fact, there are no such oscillators. It is the atoms that make up the walls that are in thermal equilibrium with the cavity radiation, and they are much more complex than the simple harmonic oscillators that Planck assumed. It was later shown by Einstein (see Supplementary Topic E) that Planck's radiation formula (Eq. 4-8a) can, in fact, be derived without making any detailed assumptions whatever about the nature of the constituents of the cavity walls, except that their energies are quantized! Thus it cannot be said that energy quantization first appeared in connection with a comfortably familiar physical system. Furthermore, energy quantization first appeared only in a single context, that of the cavity radiation. Although this was a very important problem, one could never be sure that the energy quantization idea was not in some way an artifact, unique to this problem. Indeed, many workers thought so.

Energy quantization began to be accepted, however, after Einstein showed in 1907 that the same ideas that worked so well for cavity radiation could be applied to solve still another problem that could not be solved by classical physics—the anomalous heat capacities of solids. Here, as we shall see, the systems whose energies are quantized are not hypothetical oscillators but real atoms in familiar solids.

If heat Q is added to a solid specimen of mass m and if a temperature rise ΔT results, the specific heat capacity is defined as $Q/m\,\Delta T$. As part of the definition we must also specify the conditions under which the heat is added because the same amount of heat added to the same specimen under different experimental conditions will produce different temperature changes.

From the theoretical point of view, the specific heat capacity under conditions of constant volume is the simplest situation. At constant volume, all of the added heat appears as internal energy associated with the oscillations of the atoms about their lattice sites. The dimensions of the lattice as a whole do not change, so no internal energy transfers are involved on this account. The specific heat capacity at constant volume is thus defined as

$$c_v = \frac{(Q/m)}{\Delta T} \quad \text{(specimen at constant volume)}. \quad (4\text{-}13)$$

Unfortunately, it is almost impossible to measure c_v directly for solids, the specific heat capacity at constant pressure, c_p, being the usual measured quantity. The difference between them, however, is small, especially at low temperatures, and can be calculated from thermodynamics. Table 4-1 lists some values of c_v obtained in this way for some elemental solids at or near room temperature. The

**Table 4-1
THE MOLAR HEAT
CAPACITIES OF SOME
SELECTED SOLIDS**[a]

Solid	c_v (J/mol·K)
Aluminum	23
Beryllium	11
Bismuth	25
Boron	13
Cadmium	25
Carbon (diamond)	6
Copper	24
Gold	25
Lead	25
Platinum	25
Silver	24
Tungsten	24

[a] All measurements were made at room temperature; three "anomalous" values have been deliberately offset for emphasis.

unit of mass that appears in Eq. 4-13 is the mole, so comparisons can be made from element to element on the basis of the same number of atoms. When this is done, the measured quantity is usually called the *molar heat capacity*.

Table 4-1 shows a regularity first pointed out by Dulong and Petit in 1819 and known as the Dulong and Petit rule: With a few exceptions (beryllium, boron, and carbon), the values all seem to be essentially the same, averaging about 24–25 J/mol·K. Values that differ substantially from this were called "anomalous" in those early years.

Figure 4-9 clarifies the situation. It is a plot of c_v as a function of temperature for lead, aluminum, and beryllium. We see that all three elements approach the same limiting value as high temperatures. That beryllium appears "anomalous" in Table 4-1 simply reflects the fact that, for this element, room temperature (which is where the measurements of Table 4-1 were taken) does not happen to be a very high temperature. Indeed, if the individual temperature scales for the three elements plotted in Fig. 4-9 are adjusted by a multiplying factor characteristic of the element, it is possible to make all three curves coincide. Thus we see that there is a single unique molar heat capacity curve for solids and that the difference between beryllium on the one hand and lead and aluminum on the other is one of degree and not of kind. The existence of a single curve, the same for all solids, suggests that a theoretical understanding in terms of fundamental concepts can be found.

The Prediction of Classical Theory   Let us first see what classical physics predicts for the heat capacities of solids. The atoms in a solid are arranged in a three-dimensional lattice by virtue of the forces that act between them. Each atom, bound to its lattice site by these elastic forces, oscillates about that site with an amplitude that depends on the temperature, increasing as the temperature in-

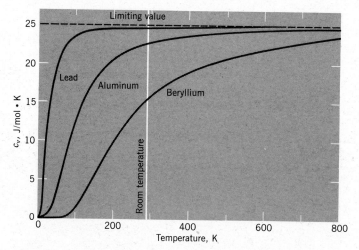

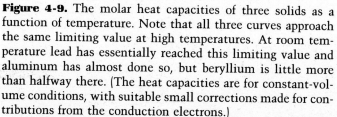

Figure 4-9. The molar heat capacities of three solids as a function of temperature. Note that all three curves approach the same limiting value at high temperatures. At room temperature lead has essentially reached this limiting value and aluminum has almost done so, but beryllium is little more than halfway there. (The heat capacities are for constant-volume conditions, with suitable small corrections made for contributions from the conduction electrons.)

creases. Each such oscillating atom behaves like a tiny mass–spring oscillator with six independent degrees of freedom, corresponding to the fact that its motion involves both potential and kinetic energy and is associated with three independent directional axes.

According to the classical equipartition of energy principle, we associate an energy of $\frac{1}{2}kT$, where k is the Boltzmann constant, with each such degree of freedom. Thus each atom in the solid has an associated energy of $(6)(\frac{1}{2}kT)$ or $3kT$. The energy per mole of a solid is then

$$u = (3kT)(N_A) = 3RT, \tag{4-14}$$

in which $N_A(= 6.02 \times 10^{23} \text{ mol}^{-1})$ is the Avogadro constant and $R(= 8.31$ J/mol·K) is the universal gas constant.

If the specimen is held at constant volume—which we assume—then all the added heat energy goes toward increasing the energy of the atomic oscillators. Thus we can replace Q/m in Eq. 4-13 (the heat added per mole) by Δu, the change in the internal energy per mole. Doing so yields $c_v = \Delta u/\Delta T$, which becomes

$$c_v = \frac{du}{dT} \tag{4-15}$$

in the differential limit. Substituting from Eq. 4-14 yields finally, as the prediction of classical physics for the molar heat capacity of a solid,

$$c_v = \frac{d}{dT}(3RT) = 3R. \tag{4-16}$$

Thus we see that classical theory predicts the molar heat capacity to be a constant, the same for all substances and independent of temperature. Substituting the numerical value of R yields $c_v = 24.9$ J/mol·K. This agrees very well indeed with the limiting value of c_v at high temperatures, as a glance at Table 4-1 and Fig. 4-9 shows. There is no suggestion, however, of the variation at lower temperatures that is the outstanding feature of Fig. 4-9.

The Prediction of Quantum Theory Einstein [2] was the first to see that energy quantization could account for the variation of the molar heat capacity of solids with temperature. He started by making an important simplifying assumption, namely, that every atom in the solid oscillates at a single frequency characteristic of the element and that, furthermore, the atoms oscillate independently, their motions not being coupled to each other. He assumed that the energies of these oscillators were quantized and that the average energy per atom at any given temperature T is not $3kT$, as classical theory predicts, but

$$\bar{\mathcal{E}} = \frac{3h\nu}{e^{h\nu/kT} - 1}. \tag{4-17}$$

Aside from the numerical factor, which appears because we are dealing with a three-dimensional oscillator rather than a one-dimensional one, this is precisely the same assumption made for the one-dimensional oscillators of the cavity radiation problem; see Eq. 4-12. It can be shown that Eq. 4-17 reduces to the classical assumption of $3kT$ as $T \to \infty$.

The internal energy per mole is found by multiplying the above quantity by the Avogadro constant, or

$$u(T) = \frac{3N_A h\nu}{e^{h\nu/kT} - 1}. \tag{4-18}$$

Again, in the limit of high temperatures this expression reduces precisely to Eq. 4-14, the classical result. Finally, we can find the molar heat capacity by differentiating the above relationship with respect to T, as Eq. 4-15 instructs us to do. Doing so leads to

$$c_v(T) = \left(\frac{3N_A h^2 \nu^2}{kT^2}\right) \frac{e^{h\nu/kT}}{(e^{h\nu/kT} - 1)^2} \qquad (4\text{-}19a)$$

as the quantum expression for the molar heat capacity. We can rewrite this expression in a simpler form by introducing a temperature T_E, characteristic of each element, from

$$T_E = \frac{h\nu}{k}$$

in which ν is the characteristic atomic vibrational frequency of the atom in question. With this substitution, and recalling that $N_A k = R$, Eq. 4-19a becomes

$$c_v(T) = 3R \frac{x^2 e^x}{(e^x - 1)^2} \qquad (4\text{-}19b)$$

in which $x = T_E/T$. This expression reduces, as it must, to $3R$ for temperatures high enough that $T \gg T_E$.

Figure 4-10 shows a plot of Eq. 4-19b, along with some experimental data for aluminum. Recall that the quantity T_E in that equation, often called the *Einstein temperature* of the solid, is the only adjustable parameter. It was given the value of 290 K, chosen so as to make theory and experiment agree at an arbitrarily selected temperature of 100 K. We see that the theory agrees well with experiment at other temperatures except in the lower range, where there is a small but definite disagreement. Nevertheless it can fairly be said that Einstein, by applying the energy quantization ideas first introduced into physics by the cavity radiation problem, succeeded in accounting for the main features of the variation of the molar heat capacities of solids with temperature.

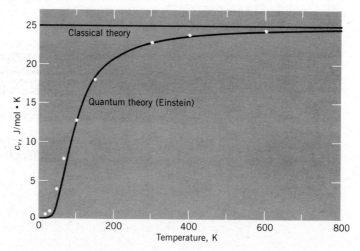

Figure 4-10. A plot of Eq. 4-19b in which the Einstein temperature T_E has been assigned the value of 290 K. The circles are experimental points for aluminum. Note the overall agreement except at low temperatures. The prediction of the classical theory is also shown.

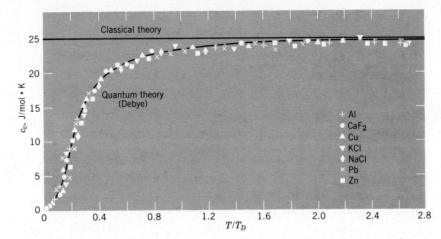

Figure 4-11. Showing the excellent agreement of the Debye quantum theory of heat capacity with experiment, for a number of solid materials. The horizontal axis displays a dimensionless temperature parameter T/T_D, in which T_D, the Debye temperature, has a different value for each substance.

It is possible to understand physically how energy quantization enters into the heat capacity problem. At high temperatures $(kT \gg h\nu)$, the atomic oscillators are excited by their thermal environment to energy levels with high quantum numbers. Adding another single increment (or quantum) of energy produces an energy change that is small compared to the energy the oscillator already has. We are then near the classical situation, in which the energy varies continuously, and we are not surprised to learn that Einstein's formula reduces to $c_v = 3R$. At low temperatures, however $(kT \ll h\nu)$, the thermal agitation of the environment is often not great enough to excite the atomic oscillators up to even their lowest allowed energy level. Quantum conditions prevail. The ability of the solid to absorb energy decreases, which is reflected as a drop in the heat capacity.

In 1912 the Dutch physicist Peter Debye refined and extended Einstein's theory of heat capacities by taking into account the interactions between adjacent atoms, thus treating the solid as an assembly of coupled oscillators. See Supplementary Topic F. The disagreement between theory and experiment at low temperatures was removed, and it is now possible to say that the quantum theory of the molar heat capacities of solids is very well established. Figure 4-11 shows the excellent agreement of Debye's theory with experiment for a number of solids. The horizontal temperature scale has been replaced by the dimensionless ratio T/T_D, where T_D is the so-called *Debye temperature* of the solid. Like the Einstein temperature we discussed above, it too is an adjustable parameter, having a different value for each material.

EXAMPLE 7.

Atoms as Harmonic Oscillators. In the plot of Eq. 4-19b that appears in Fig. 4-10, the Einstein temperature T_E for aluminum was taken to be 290 K. (a) What is the characteristic oscillation frequency of the aluminum atoms about their equilibrium positions in the crystalline lattice? (b) What is the energy interval between adjacent levels for these quantized atomic oscillators? (c) What is the effective spring constant for the oscillatory motion?

(a) From the text we see that the characteristic oscillator frequency is given by

$$\nu = \frac{T_E k}{h} = \frac{(290 \text{ K})(1.38 \times 10^{-23} \text{ J/K})}{6.63 \times 10^{-34} \text{ J} \cdot \text{s}}$$

$$= 6.0 \times 10^{12} \text{ Hz}.$$

The wavelength corresponding to this frequency $(= c/\nu = 50 \; \mu\text{m})$ is in the infrared range of the electromagnetic spectrum.

(*b*) The energy interval between adjacent levels is given by

$$E = h\nu = (4.14 \times 10^{-15} \text{ eV} \cdot \text{s})(6.0 \times 10^{12} \text{ Hz})$$
$$= 0.025 \text{ eV} = 25 \text{ meV}.$$

(*c*) The spring constant κ (see PHYSICS, Part I, Section 15-3) is defined from the relation

$$\nu = \frac{1}{2\pi} \sqrt{\frac{\kappa}{m}},$$

in which m is the mass of an aluminum atom. We can find m by dividing the atomic weight of aluminum (= 27 g/mol or 0.027 kg/mol) by the Avogadro constant, which gives the number of atoms per mole. Thus

$$m = \frac{A}{N_A} = \frac{0.027 \text{ kg/mol}}{6.02 \times 10^{23} \text{ mol}^{-1}} = 4.5 \times 10^{-26} \text{ kg}.$$

Solving for κ above and substituting yields then

$$\kappa = 4\pi^2 \nu^2 m$$
$$= (4\pi^2)(6.0 \times 10^{12} \text{ Hz})^2(4.5 \times 10^{-26} \text{ kg}) = 64 \; N/m.$$

Interestingly, this value is only 50–60 times smaller than that of a typical office postage scale.

Let us apply the atomic oscillator concept to the heat capacity curves for beryllium, aluminum, and lead; see Fig. 4-9. The atomic weights of these three elements are 9.0, 27, and 207 g/mol respectively. From the analysis above we expect that light atoms will oscillate at relatively high frequencies which means that their oscillator level spacings $(= h\nu)$ will also be relatively high. This in turn means that such elements will achieve their limiting heat capacity of $3R$ only at relatively high temperatures, just as Fig. 4-9 shows. Table 4-1 confirms that all of the elements shown there with "anomalous" heat capacities are light elements.

EXAMPLE 8.

Heating Up a Metal Block. A 1.00-kg block of beryllium (see Fig. 4-9) is heated from room temperature ($T = 300$ K) to 700 K. According to the predictions of the Einstein theory of heat capacity, how much thermal energy is required? The Einstein temperature for beryllium may be taken as 690 K; the atomic weight of beryllium is 9.01 g/mol.

The required energy is simply the difference between the thermal energy of the beryllium lattice at the two temperatures. From Eq. 4-18 we have, for the energy per atom,

$$u(T) = \frac{3h\nu}{e^{h\nu/kT} - 1}.$$

The number of atoms in the specimen is $N_A(M/A)$, where N_A is the Avogadro number, M is the specimen mass, and A is the atomic weight. Also, we can replace the quantity $h\nu$ by kT_E, where T_E is the Einstein temperature of beryllium. Thus, for the energy at any temperature T for this specimen, we have

$$U(T) = \left(\frac{3kT_E}{e^{T_E/T} - 1}\right)\left(\frac{N_A M}{A}\right)$$

$$= \left[\frac{(3)(1.38 \times 10^{-23} \text{ J/K})(690 \text{ K})}{e^{690/T} - 1}\right]$$

$$\times \left[\frac{(6.02 \times 10^{23} \text{ mol}^{-1})(1.00 \text{ kg})}{0.00901 \text{ kg/mol}}\right]$$

$$= \frac{1.91 \times 10^6 \text{ J}}{e^{690/T} - 1}.$$

Note that as $T \to 0$, this relation shows us that $U(T) \to 0$. In dealing in this section with the energy of the atomic oscillators we have assumed, arbitrarily, that the energy is zero at the absolute zero of temperature. If we had done otherwise it would not matter in the present case, because we intend to subtract two energies and the arbitrary additive constant would cancel out. For the two temperatures in questions the above relation yields

$$U(700 \text{ K}) = \frac{1.91 \times 10^6 \text{ J}}{e^{690/700} - 1} = 1.137 \times 10^6 \text{ J}$$

and

$$U(300 \text{ K}) = \frac{1.91 \times 10^6 \text{ J}}{e^{690/300} - 1} = 0.213 \times 10^6 \text{ J},$$

and for the difference,

$$\Delta U = U(700 \text{ K}) - U(300 \text{ K})$$
$$= (1.137 - 0.213) \times 10^6 \text{ J} = 924 \text{ kJ},$$

which is the result we seek.

4-7 *THE FRANCK-HERTZ EXPERIMENT*

In the preceding section we have presented evidence that atoms, oscillating about fixed lattice points in a solid, do so with quantized energies. The allowed energy states for such atomic oscillators are evenly spaced, the energy difference between them being given by $h\nu$, where h is the Planck constant and ν is the oscillator frequency. We have said nothing yet about the structure or the internal energy of the oscillating atom itself and have, in fact, treated it for our purposes so far as a simple mass point.

In this section we expand our horizons to consider the atom itself, perhaps in relative isolation in a low-density gas or vapor. An atom is made up of a relatively massive, positively charged nucleus surrounded by a number of (negatively charged) electrons, the atom as a whole being electrically neutral. The internal energy of an atom depends on the precise arrangement of its electrons about its central nucleus. It will turn out that the energies of the various possible atomic structural arrangements are not distributed in a continuous fashion but are *quantized*, only values belonging to a discrete set being possible. An atom is a much more complex structure than a simple harmonically oscillating mass point, so we should not be surprised to learn that the allowed internal energy values permitted for atoms are *not* evenly spaced in energy, as they are for the oscillator.

Suppose that two atoms collide, as in a gas, each of them being originally in its lowest energy (or ground) state. It could happen that some of their translational kinetic energy is used to increase the internal energy of one or both of the atoms. Should this happen, we call the collision *inelastic*, for translational kinetic energy is not conserved in the collision. If the internal energy of an atom cannot have any of a continuous set of values above its ground state, however, but only certain discrete ones, then an inelastic collision could take place only if enough kinetic energy were available to raise one of the colliding atoms from its ground state to the next higher allowed internal energy state. Otherwise, the collision would be *elastic*, none of the kinetic energy being absorbed internally. If an atom is raised from its ground state to some higher one, however, the atom can thereafter give up this absorbed energy by emitting radiation whose energy equals the difference in energy between these two states.

Evidence that the internal energy of an atom is quantized already existed in experiments with gas collisions, though it was not generally recognized at the time. In collisions, the available kinetic energy is of the order of kT, which has the value of ~0.04 eV at room temperature. This is very much smaller than the energy difference between the ground state and the first excited states for typical atoms. This energy difference is a few electron volts, and effective gas temperatures of the order of 10^4 K are required to reach it. Nevertheless, inelastic collisions *were* observed under such conditions, but their significance was not recognized. In retrospect we may say that early evidence that the internal energy of an atom is quantized existed in experiments with gas collisions and that the seeds of energy quantization lay in the kinetic theory of gases.

In 1914, however, James Franck and Gustav Hertz [3] obtained clear and direct evidence that the internal energy states of atoms are quantized. Their experiments showed that electrons, moving through a gas, lost energy only in discrete amounts in collisions with the gas atoms. The Franck-Hertz apparatus, shown schematically in Fig. 4-12, consisted of a glass tube, filled with mercury vapor, containing a heated filament F, a cylindrical metal plate P, and an accelerating grid G near the plate. The filament temperature is set so that thermally excited

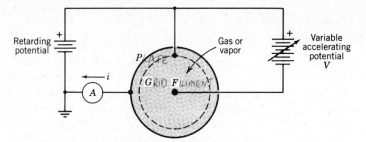

Figure 4-12. A schematic diagram of the apparatus used by Franck and Hertz. The apparatus has cylindrical symmetry, the view shown being a cross section at right angles to the cylinder axis. *F* is a heated filament that lies along this axis. *G* is a metallic mesh grid and plate *P* is a solid metal cylinder. The whole is enclosed in a cylindrical glass tube (not shown) filled with mercury vapor at low pressure. Ammeter *A* measures the current of electrons that leave the filament and succeed in reaching the plate.

electrons are emitted from its surface with but little kinetic energy; the grid–filament potential difference *V* (the accelerating potential) is set to accelerate these electrons toward the plate. A small retarding potential is put between the plate and the grid so that electrons that pass through the openings in the grid with very small kinetic energy will not reach the plate. Then the experimenter measures the variation in the electron-beam current through the plate, by means of an ammeter in the plate circuit, as the accelerating potential *V* is increased.

The results obtained with the tube containing mercury vapor are shown in Fig. 4-13. For low values of the accelerating potential *V* the behavior is similar to

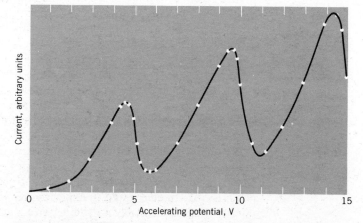

Figure 4-13. The current, measured in the Franck-Hertz experiment using the apparatus shown in Fig. 4-12, as a function of the accelerating potential *V*. The break at 4.9 V corresponds to electrons exciting the mercury atoms just in front of the grid. The breaks at 9.8 V and at 14.7 V correspond, respectively, to electrons exciting the first energy state of the mercury atoms on two or three (respectively) separate occasions on their trip from filament to grid.

what one expects in an evacuated tube; that is, the current increases as *V* is increased. However, when *V* reaches 4.9 V, the current drops sharply. The interpretation is that electrons lose most of their kinetic energy in inelastic collisions with the mercury atoms in the tube once they reach the kinetic energy of 4.9 eV just in front of the grid; they then have little energy left and the retarding potential prevents them from reaching the plate—hence the drop in plate current. The mercury atoms absorb the energy and are said to be in an excited state. As the accelerating potential increases above 4.9 V, the electrons have enough kinetic energy after the excitation process to overcome the retarding potential and to reach the plate, so the current rises again. When the accelerating potential is made large enough that electrons have enough kinetic energy to make *two* separate inelastic collisions with mercury atoms in their trip from the filament toward the grid, the current falls again. Hence, as they increased *V*, Franck and Hertz observed a series of drops in current spaced at about equal intervals at 4.9 eV.

All this suggests that the first excited energy state of mercury is about 4.9 eV above its lowest energy state and that mercury atoms are not able to absorb energy from the incident electrons unless they have at least this much energy. The mercury atoms that are excited will eventually give up this energy, by emit-

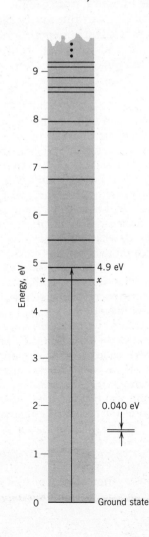

Figure 4-14. An energy-level diagram for mercury, showing the quantized states up to about 9 eV above the ground state. The vertical arrow represents a transition of a mercury atom from its ground state ($E = 0$) to its first excited state ($E = 4.9$ eV). The spacing between the two parallel lines on the right represents, to scale, the mean translational kinetic energy ($= 0.040$ eV) of a mercury (or other) atom at room temperature. (The level marked *xx* lies lower than the 4.9-eV level and might seem to have a claim to be the "first excited state" of the mercury atom. However, direct transitions to this state from the ground state are not observed, and their absence is quite consistent with the detailed predictions of quantum mechanics.)

ting radiation as they make a transition from the first excited state back to the lowest (ground) state. In later experiments Hertz arranged to observe the spectrum emitted by mercury vapor in the tube. He found that when the kinetic energy of the incident electrons was less than 4.9 eV, no spectral lines were emitted at all. However, as the 4.9 eV energy was reached, a single spectral line of mercury appeared in the spectrum. Even when the electron kinetic energy was somewhat greater than 4.9 eV, only this one line, the 254-nm line in the ultraviolet region of the spectrum, was seen. This is exactly the behavior expected if the internal energy of the mercury atom is quantized.

With grid arrangements different than the one shown in Fig. 4-12, the electrons can be made to gain energy quickly enough that they might not collide with a mercury atom until their energy is able to excite still higher states than the first. Indeed, Hertz found with such an arrangement that other lines appeared in the spectrum just as the accelerating voltage was made large enough to excite the state believed to give rise to that line in transitions to a lower state. Therefore, the Franck-Hertz experiments give striking evidence for the quantization of the internal energy of atoms. In fact, by measuring the breaks in the current-versus-voltage curve, the energy differences between the allowed quantum states can be measured directly in a Franck-Hertz experiment. It is found (see Fig. 4-14) that the allowed energy states of the mercury atom are not spaced equally, as was the case for Planck's harmonic oscillators in the cavity radiation problem. We shall see in Chapter 7 that as the Franck-Hertz experiments were being done, Niels Bohr was independently advancing his ideas of atomic structure to construct a theory of the atom that predicted quantization of energy.

EXAMPLE 9.

An Electron Excites a Mercury Atom (Barely!). An electron (mass m) collides head-on with a mercury atom (mass M) that is initially at rest in a Franck-Hertz tube. Let $\mathcal{E}(= 4.9$ eV) be the energy difference between the ground state of the mercury atom and its first excited state. (a) Show that, to excite the mercury atom during the collision, the initial kinetic energy of the electron must have *at least* the value

$$K_{min} = \mathcal{E}\left(1 + \frac{m}{M}\right).$$

(b) Evaluate K_{min} for an electron striking a mercury atom.

(a) Let an electron of speed v fall on the resting mercury atom as in Fig. 4-15a. After the collision, the mer-

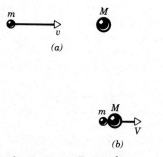

(a)

(b)

Figure 4-15. Example 9.

cury atom will recoil with speed V. What will be the corresponding motion of the electron? Because we are investigating conditions at the threshold of excitation, we must arrange matters so that as little of the initial kinetic energy as possible remains as kinetic energy of the moving particles after the collision. In the laboratory frame there will have to be *some* kinetic energy after the collision, because momentum must be conserved. The question is, what is the minimum possible amount consistent with momentum conservation?

It is helpful to consider the kinetic energies of the two particles after the collision as being made up of (1) the kinetic energy associated with the movement of the center of mass of the two particles and (2) the kinetic energy associated with the motions of the two particles with respect to each other in the center-of-mass reference frame. This second part of the kinetic energy must be set equal to zero if we seek the excitation threshold condition. This means simply that, as viewed in the laboratory reference frame, the two particles must move with the same speed V after the collision, as shown in Fig. 4-15b.

The law of conservation of linear momentum gives us

$$mv = MV + mV.$$

Conservation of total energy yields, assuming that excitation takes place and that energy $\mathcal{E}$ is stored internally in the mercury atom,

$$K_{\min} = \mathscr{E} + \tfrac{1}{2}MV^2 + \tfrac{1}{2}mV^2.$$

Eliminating V between these two equations yields

$$K_{\min} = \mathscr{E} + \frac{1}{2}(M + m)\left(\frac{mv}{M + m}\right)^2$$

$$= \mathscr{E} + K_{\min}\left(\frac{m}{M + m}\right),$$

which can be rearranged to give

$$K_{min} = \mathscr{E}\left(1 + \frac{m}{M}\right),$$

the sought-for result.

(b) We find the mass M of a mercury atom by dividing the atomic weight A of mercury by the Avogadro constant N_A. We note in passing that mercury has several stable isotopes, so this procedure generates a weighted-average mass. However, as we shall see, this consideration is not important in the Franck-Hertz experiment as ordinarily carried out. The mass ratio m/M then becomes

$$\frac{m}{M} = \frac{mN_A}{A}$$

$$= \frac{(9.11 \times 10^{-31}\ \text{kg})(6.02 \times 10^{23}\ \text{mol}^{-1})}{(200\ \text{g/mol})(10^{-3}\ \text{kg/g})} \cong 3 \times 10^{-6},$$

and we have

$$K_{min} = \mathscr{E}\,(1 + 3 \times 10^{-6}) \cong \mathscr{E}.$$

It is clear that, for electrons colliding with mercury atoms, we can take the mass of the mercury atom to be infinitely great with little error as far as the excitation threshhold is concerned.

4-8 EPILOGUE

From Planck's analysis of cavity radiation, from Einstein's analysis of the heat capacities of solids, and from the Franck-Hertz experiments, we are faced with a conclusion that contradicts both classical theory and familiar experience with macroscopic objects, namely:

> In some systems energy cannot be changed in a continuous way or by an arbitrary amount but can take on only certain discrete values. When such a system exchanges energy with its environment, it does so in discrete amounts and in a discontinuous way. In three words:
>
> Energy is quantized.

Let us close this chapter with some remarks about the quantization of energy concept:

1. *Energy quantization was slow to be accepted.* The more a new concept challenges entrenched ideas, the longer it takes for it to be accepted by the scientific community. The energy quantization concept was no exception. Indeed, Planck himself was at first unsure whether his introduction of the constant h was only a mathematical device or a matter of deep physical significance. Much later, in a letter to the American physicist R. W. Wood, Planck called his limited postulate "an act of desperation." "I knew," he wrote, "that the problem [of the equilibrium of matter and radiation] is of fundamental significance for physics; I knew the formula that reproduces the energy distribution in the normal spectrum; a theoretical interpretation *had* to be found, at any cost, no matter how high." For more than a decade Planck tried to fit the quantum idea into classical theory. With each attempt he appeared to retreat from his original boldness but, always, he generated new ideas and techniques that quantum theory later adopted. What appears to have finally convinced him of the correctness and deep significance of his quantum hypothesis [4] was its support of the definiteness of the statistical concept of entropy and the third law of thermodynamics.

2. *The origins of energy quantization are firmly rooted in experiment.* We have seen that it was the painstakingly acquired experimental results of cavity radia-

tion that provided the initial testing ground for Planck's quantum hypothesis. The energy quantization concept received a big boost when Einstein showed that the experimental results of the anomalous heat capacities of solids could be understood by the same hypothesis. The Franck-Hertz experiments provided still further evidence for the quantization of energy, in this case the internal energy of the atoms themselves. And, as we shall see, the subsequent experimental evidence for energy quantization became overwhelming, penetrating all areas of physics.

3. *Energy quantization led to a totally new view of the physical world.* The introduction of the constant h into physics by Planck turned out to be a step of revolutionary significance, marking the beginning of the development of all that we now lump under the label of "modern physics." It is interesting to note that during his period of doubt, Planck was editor of the *Annalen der Physik* and, in 1905, received Einstein's first relativity paper. He stoutly defended Einstein's work in relativity and thereafter became one of young Einstein's patrons in scientific circles. However, he resisted for some time the very ideas on the quantum theory of radiation advanced by Einstein that subsequently confirmed and extended Planck's own work. Einstein, whose deep insight into electromagnetism and statistical mechanics was perhaps unequaled by anyone at the time, saw as a result of Planck's work the need for a sweeping change in classical statistics and electromagnetism. He advanced predictions and interpretations of many physical phenomena that were later strikingly confirmed by experiment. In Chapter 5 we turn to one of these phenomena and follow another road on the way to quantum theory.

questions

1. By simply looking at the sky at night, can you pick out stars that are hotter than the sun? Cooler than the sun? What do you look for? Is the star's brightness a clue?

2. Betelgeuse, the prominent red star in the constellation Orion, has a surface temperature that is much lower than that of the sun, yet it radiates energy into space at a much higher rate than the sun does. How can that be?

3. Why is the radiation emerging from a hole in a uniformly heated cavity called "black-body" radiation? Explain this term. Can a given surface be an effective black body (in the infrared, say) and not look black to the eye?

4. You put a platinum crucible into a laboratory furnace whose temperature, which is in the incandescent range, is still rising. As you look through the peephole in the furnace door, you can easily see the cool crucible and the details of the furnace interior. However, when the furnace comes to thermal equilibrium, with its walls and the crucible at the same temperature, the crucible is no longer visible and details of the furnace interior have vanished. All that you can see is a uniform red glow. Explain.

5. Why do sunspots, when viewed through a dense filter, appear black? Even though they are substantially cooler than their surroundings, their temperatures ($\sim$4000 K) are still much higher than that of an incandescent lamp filament.

6. Less than 10 percent of the energy supplied to a 100-W lamp appears in the form of visible light. What happens to the rest of it? What could be done to increase this percentage? Why hasn't it already been done?

7. The relation $R = \sigma T^4$ (Eq. 4-1) is exact for black bodies and holds for all temperatures. Why don't we use this relation as the basis of a practical secondary definition of temperature at, for instance, the boiling point of water?

8. Give an example of a good absorber that is (necessarily) a good emitter and a poor absorber than is (necessarily) a poor emitter. How would you classify a good reflector? A poor reflector?

9. The words "PHYSICS IS GOOD FOR YOU" are painted with black glaze on a white ceramic tile. What would this tile look like if you heated it to incandescence in a dark room? What would the tile look like if you put it in a furnace, heated uniformly to a temperature in the incandescent range, and viewed it through a peephole in the furnace door?

10. A piece of iron glows with a bright red color at 1100 K. At this same temperature, however, a piece of quartz does not glow at all. Explain. (*Hint:* See Example 1 and note that quartz is transparent to visible light.)

11. Compare the definitions and the dimensions of (*a*) the radiancy $R(T)$, (*b*) the spectral radiancy $\mathcal{R}(\lambda, T)$, (*c*) the spectral radiancy $\mathcal{R}(\nu, T)$, (*d*) the spectral energy density $\rho(\nu, T)$, and (*e*) the spectral energy density $\rho(\lambda, T)$.

12. (*a*) Compare the radiation trapped in a cavity with the molecules of an ideal gas confined to a box. The quantity of interest in the first case is the spectral energy density $\rho(\nu, T)$, which tells how the energy of the radiation is distributed by frequency at any given temperature. For the ideal gas we are interested in how the molecules are distributed by speed at any given temperature. Define carefully a function $n(v, T)$ that would give us this information about the gas molecules. (*b*) Consider now the radiation escaping from a hole in the cavity and also the molecules escaping from a hole in the box. What quantity for the beam of molecules would give us information similar to that provided by the spectral radiancy $\mathcal{R}(\lambda, T)$ for the beam of radiation?

13. What is the ultraviolet catastrophe, and just how does the assumption of the quantization of energy succeed in avoiding it? Explain it in your own words.

14. (*a*) At a given temperature the classical Rayleigh-Jeans radiation law agrees with experiment if the frequency is low enough; see Fig. 4-5. Can you explain this in terms of energy quantization of the oscillators in the cavity walls? (*b*) At a given frequency, would you expect the Rayleigh-Jeans formula to agree with experiment at high or at low temperatures? Explain this answer also in terms of energy quantization.

15. For you to be able to detect energy quantization by watching, say, a swinging pendulum, what order of magnitude would the Planck constant have to be? (*Hint:* See Example 5.)

16. Show that the Planck constant has the dimensions of angular momentum. Does that mean that angular momentum is necessarily quantized?

17. Are there quantized quantities in classical physics? If so, give examples. Is energy quantized in classical physics?

18. Electric charge is quantized. In what ways does the quantization of charge differ from the quantization of energy?

19. Consider these three statements: (*a*) Planck's radiation law reduces to the classical Rayleigh-Jeans radiation law in the limit of low frequencies. (*b*) Quantum theory

reduces to classical theory in the limit of large quantum numbers. (*c*) Quantum laws reduce to their classical equivalents in the limit as $h \to 0$. Show that the last two statements are consistent with the first.

20. You are given access to a cavity radiator whose walls are held at a temperature of, say, 1800 K, and you are asked to use it to determine a value of the Planck constant h. What measurements would you make, and how would you extract a value of h from your data?

21. Spectral radiancy curves for cavity radiators at different temperatures never intersect; see Fig. 4-2. Suppose, however, that they did. Can you show that this would violate the second law of thermodynamics?

22. Suppose that your skin temperature is about 300 K. In what region of the electromagnetic spectrum do you emit thermal radiation most intensely?

23. Absolute measurements of the spectral radiancy $\mathcal{R}(\lambda)$ at a given temperature T are relatively difficult to carry out. However, relative measurements $\mathcal{R}(\lambda)/\mathcal{R}(\lambda_0)$ are comparatively easy. The relative measurement is simply the dimensionless ratio of the spectral radiancy at wavelength λ to that at an arbitrary reference wavelength λ_0. How could you use such relative measurements to find the temperature of a source of cavity radiation?

24. Explain in your own words just why assuming that the energy of the atomic oscillators in a solid is quantized leads to a reduction in the heat capacity at low temperatures.

25. Why is the heat capacity at constant pressure for a solid greater than its heat capacity at constant volume? Why are theories of heat capacity (classical or quantum) most simply developed in terms of the heat capacity at constant volume? Why is this quantity so difficult to measure in the laboratory?

26. It is generally true that the higher the melting point of a material, the higher its Einstein temperature (or its Debye temperature). Can you explain this rough correlation, perhaps in terms of the relative spring constants of the atomic oscillators?

27. You are given access to a well-equipped laboratory set up for the purpose of measuring the heat capacities of solids. What steps would you follow if you were asked to determine an experimental value of the Planck constant h in that laboratory?

28. The classical law of equipartition of energy (see *Physics*, Part I, Sec. 23-8) leads to the Rayleigh-Jeans radiation law when applied to cavity radiation and to the Dulong-Petit law when applied to the heat capacities of solids. In both cases there is serious disagreement with experiment.

Can you relate these two failures of the equipartion law and explain why energy quantization leads, in each case, to theories that *do* agree with experiment.

29. "For the cavity radiation problem and the heat capacity of solids problem, the disagreements between experiment and classical theory in certain ranges of the variables are not small but are total, and beyond all dispute." Can you identify, in each case, the specific disagreements to which this statement refers?

30. In the cavity radiation problem and in the heat capacity of solids problem, an understanding of the experimental facts followed when it was assumed that the energies of certain oscillators were quantized. Compare and contrast these oscillators in each case.

31. A Franck-Hertz tube is filled with hydrogen gas, in the form of molecular hydrogen (H_2), for which the dissociation energy is 4.7 eV. The energy of the lowest excited state of atomic hydrogen is 13.6 eV. Will a line spectrum be observed?

32. Explain the relation between the rate at which electrons gain energy in the Franck-Hertz experiment and

which upper energy state of the mercury atom can be excited. (*Hint:* Consider the mean free time between collisions.)

33. Discuss the remarkable fact that discreteness in energy was first found by analyzing a *continuous* spectrum (cavity radiation) rather than a discrete, or line, spectrum, such as that from excited mercury atoms.

34. Consider (*a*) an aluminum atom oscillating around a lattice site in a solid and (*b*) a mercury atom in a Franck-Hertz apparatus. In each case we assumed that the energy of the atom was quantized. Specify carefully just what you mean by "the energy of the atom" in each of these cases.

35. In the Franck-Hertz experiment (see Fig. 4-12), (*a*) what is the purpose of the retarding potential between the grid G and the plate P? (*b*) Why is the grid located so close to the plate? (*c*) What factors determine the operating temperature of the filament F? (*d*) Why (see Fig. 4-13) do the amplitudes of successive peaks increase as the accelerating potential increases? (*e*) Exactly what happens inside the tube to cause the plate current to drop at critical values of the accelerating potential?

problems

$$h = 6.63 \times 10^{-34} \text{ J} \cdot \text{s} = 4.14 \times 10^{-15} \text{ eV} \cdot \text{s}$$
$$k = 1.38 \times 10^{-23} \text{ J/K} = 8.62 \times 10^{-5} \text{ eV/K}$$
$$h/k = 4.80 \times 10^{-11} \text{ K} \cdot \text{s}$$
$$w = \lambda_m T = 2898 \text{ } \mu\text{m} \cdot \text{K} \qquad \nu_m/T = 5.878 \times 10^{10} \text{ Hz/K}$$
$$\sigma = 5.67 \times 10^{-8} \text{ W/m}^2 \cdot \text{K}^4 \qquad N_A = 6.02 \times 10^{23} \text{ mol}^{-1}$$

1. How hot is the sun? The sun has a luminosity (total radiated power) of 3.9×10^{26} W. What is its effective surface temperature, assuming that the sun radiates like a black body? The solar radius is 7.0×10^8 m.

2. Energy leaks through a hole. A cavity whose walls are held at 2000 K has a small hole, 1.00 mm in diameter, drilled in its wall. At what rate does energy escape through this hole from the cavity interior?

3. A tale of three spheres. A tungsten sphere, 2.30 cm in diameter, is heated to 2200 K. At this temperature, tungsten radiates only 30 percent of the energy radiated by a black body of the same size and temperature. (*a*) Calculate the temperature of a perfectly black sphere of the same size that radiates at the same rate as the tungsten sphere. (*b*) Calculate the diameter of a perfectly black sphere at the same temperature as the tungsten sphere that radiates at the same rate.

4. About sunshine. Averaged over the earth's surface, the rate per unit area at which solar radiation falls on the earth (measured above the atmosphere) is 355 W/m^2. (*a*)

Consider the earth as a black body radiating energy into space at this rate. What surface temperature would the earth have under these circumstances? (*b*) How does the average rate given above differ from the *solar constant* (= 1340 W/m^2), which is the rate per unit area at which solar energy falls on the earth at normal incidence, measured above the atmosphere?

5. Moving the earth! Let us assume that the earth, whose mean surface temperature may be taken to be 280 K, is in thermal equilibrium and radiates energy into space at the same rate that it receives energy from the sun. If the earth were removed to an orbit of twice its present radius, what would be its expected mean surface temperature?

★ **6. It's a matter of contrast.** Sunspots are storms in the solar atmosphere; as they are cooler than the sun's surface, they are almost black in appearance. Suppose that, during a period of maximum solar activity, the solar constant (see Problem 4) is measured to fall to 98.6 percent of its usual value. If the sun's surface is at 5800 K, what is the temperature of the sunspots if they cover 2.0 percent of the sun's surface area? Is it reasonable that they should appear black, even at a temperature that is much higher than that of a tungsten lamp filament?

7. The sun is wasting away! (*a*) Use the Stefan-Boltzmann law find the rate at which the sun loses rest mass by virtue of its radiation. The sun's surface temperature and radius are 5800 K and 7.0×10^8 m, respectively. (*b*) The sun's present rest mass is 2.0×10^{30} kg, and its present age may be taken as 4.7×10^9 y. At the rate just calculated, how long would it take for the sun to lose 0.001 percent of its rest mass by radiation? Express your answer both in years and as a percentage of the sun's present age.

8. Some temperatures and some wavelengths. Calculate the wavelength of maximum spectral radiancy and identify the region of the electromagnetic spectrum to which it belongs for each of the following: (*a*) The 3.0-K cosmic background radiation, a remnant of the primordial fireball. (*b*) Your body, assuming a skin temperature of 20°C. (*c*) A tungsten lamp filament at 1800 K. (*d*) The sun, at an assumed surface temperature of 5800 K. (*e*) An exploding thermonuclear device, at an assumed fireball temperature of 10^7 K. (*f*) The universe immediately after the Big Bang, at an assumed temperature of 10^{38} K. Assume black-body conditions throughout.

9. A cold cavity. Low-temperature physicists would not consider a temperature of 2.0 mK (= 0.0020 K) to be particularly low. At what wavelength would a cavity whose walls were at this temperature radiate most copiously? To what region of the electromagnetic spectrum would this radiation belong? What are some of the practical difficulties of operating a cavity radiator at such a low temperature?

10. Birth of a planetary system? In 1983 the Infrared Astronomical Satellite (IRAS) detected a cloud of solid particles surrounding the star Vega, radiating maximally at a wavelength of 32 μm. What is the temperature of this cloud of particles?

11. How hot is Rigel? The star Rigel in the constellation Orion appears bluish-white, corresponding to an apparent wavelength of about 400 nm. Estimate its surface temperature.

12. Warming a cold object. A black body has its maximum spectral radiancy at a wavelength of 25 μm, in the infrared region of the spectrum. The temperature of the body is now increased so that the radiancy $R(T)$ of the body is doubled. (*a*) What is this new temperature? (*b*) At what wavelength will the spectral radiancy now have its maximum value?

13. Red light from a furnace. A furnace, whose inner walls are at 1500 K, has a peephole, 2.0 cm in diameter, in its door. The hole is covered with a filter that passes light in the wavelength range 630–640 nm. Find the power transmitted by the filter in this range of wavelengths. (*Hint:* Assume that the spectral radiancy is roughly constant over this narrow wavelength range.)

14. A plasma radiates. A plasma in a gas discharge tube is observed to have a spectral radiancy of 5.7×10^4 W/cm$^2 \cdot \mu$m at a wavelength of 350 nm. What is its effective temperature, calculated on the assumption that the plasma radiates like a black body?

15. The sun and Sirius compared. The sun and the bright star Sirius have their maximum spectral radiancies at 500 nm and 240 nm, respectively. (*a*) What are the maximum values of their spectral radiancies? (*b*) What is the ratio of their spectral radiancies (Sirius/Sun) at 500 nm? (*c*) At 240 nm? Assume black-body conditions.

16. Checking the curve. Consider radiation emerging from a cavity whose walls are maintained at 2000 K. (*a*) At what wavelength is the spectral radiancy a maximum? (*b*) What is the value of this maximum spectral radiancy? (*c*) By trial-and-error methods, find two wavelengths at which the spectral radiancy has half its maximum value. Verify your calculations from Fig. 4-2.

★ **17. It makes a big difference.** (*a*) A black body has a spectral radiancy at 400 nm that is 3.50 times its spectral radiancy at 200 nm. What is its temperature? (*b*) What would be its temperature if its spectral radiancy at 200 nm were 3.50 times its spectral radiancy at 400 nm?

★ **18. About visible light.** The center of the visible range of the electromagnetic spectrum may be taken to be 550 nm. (*a*) What is the temperature of a cavity radiator whose spectral radiancy is a maximum at this wavelength? Can you think of a reason that this temperature turns out to be a little less than the sun's surface temperature, which is 5800 K? (*b*) What is this maximum spectral radiancy? (*c*) At what temperature would the spectral radiancy at this wavelength have half this value? (*d*) Twice this value?

19. Holding the wavelength constant. (*a*) Show that, for a given wavelength λ, the fractional change in the spectral radiancy of a black body with temperature is given by

$$\frac{1}{\mathcal{R}}\left(\frac{\partial \mathcal{R}}{\partial T}\right)_{\lambda} = \frac{1}{T}\left(\frac{xe^x}{e^x - 1}\right),$$

in which $x = hc/\lambda kT$. (*b*) Evaluate this quantity for $\lambda = 550$ nm (the center of the visible spectrum) and $T = 2000$ K.

★ **20. Finding the Wien constant.** (*a*) Starting from Eq. 4-8*b*, Planck's equation for the spectral radiancy $\mathcal{R}(\lambda, T)$, show that the Wien constant $w(= \lambda_m T$; see Eq. 4-3) can be written as

$$\lambda_m T = \frac{hc}{4.965\,k}.$$

(*b*) Substitute numerical values for the constants and evaluate. Compare your result with the value displayed in Eq. 4-3. (*Hint:* In finding the maximum value of the spectral radiancy for a given temperature, an equation will be encountered whose numerical solution is the quantity 4.965 that appears above. Use a hand calculator and trial-and-error methods.)

21. Finding the maximum spectral radiancy. (*a*) Show that, at a given temperature, the maximum value of the spectral radiancy $\mathcal{R}(\lambda, T)$ is given by

$$\mathcal{R}_m(T) = \left(\frac{133.2k^5}{h^4c^3}\right) T^5 = AT^5,$$

in which

$$A = 1.281 \times 10^{-15} \text{ W/cm}^2 \cdot \mu\text{m} \cdot \text{K}^5.$$

(*b*) Evaluate for $T = 2000$ K and verify your answer by comparison with Fig. 4-2. (*Hint:* Use the result of Problem 20.)

22. Radiancy and spectral radiancy. Verify that the radiancy at any given temperature is the area under the spectral radiancy curve for that temperature; see Eq. 4-2. In the process, show that the Stefan-Boltzmann constant is related to other physical constants by

$$\sigma = \frac{2\pi^5k^4}{15h^3c^2}.$$

Evaluate and compare with the accepted value of 5.67×10^{-8} W/m$^2 \cdot$ K^4. (*Hint:* Make a change in variables, letting $\lambda = hc/xkT$, in which x replaces λ as the variable. A definite integral will be encountered that has the value

$$\int_0^\infty \frac{x^3\,dx}{e^x - 1} = \frac{\pi^4}{15}.)$$

23. A universal spectral radiancy curve (*I*). Wien succeeding in showing, entirely on the basis of thermodynamic arguments, that the spectral radiancy law for cavity radiation must be of the form

$$\mathcal{R}(\lambda, T) = T^5F(\lambda T),$$

in which $F(\lambda T)$ is a function that cannot be further specified on the basis of classical physics alone. (*a*) Show that Planck's radiation law (Eq. 4-8*b*) can indeed be written in this form. (*b*) Show how, by plotting $\mathcal{R}/T^5$ against λT, the entire family of spectral radiancy curves can be reduced to a single universal curve. Make a plot of this curve, using the following experimental data for $T = 2000$ K:

λ	$\mathcal{R}$	λ	$\mathcal{R}$
0.50 μm	0.7 W/cm$^2 \cdot \mu$m	2.00 μm	33.0 W/cm$^2 \cdot \mu$m
0.75	10.8	2.50	22.8
1.00	28.1	3.00	15.4
1.25	38.9	4.00	7.2
1.50	41.1	5.00	3.7

24. A universal spectral radiancy curve (*II*). Show that the area under the universal spectral radiancy curve described in Problem 23 is simply σ, the Stefan-Boltzmann constant (Eq. 4-1).

25. A universal spectral radiancy curve (*III*). Use the universal spectral radiancy curve that you drew in Problem 23 to generate a specific spectral radiancy curve for 3000 K.

26. Wien and Rayleigh-Jeans on the spectral radiancy. (*a*) Start from Planck's expression for the spectral radiancy (Eq. 4-8*b*) and show that Wien's formula for that quantity is

$$\mathcal{R}(\lambda, T) = \frac{2\pi hc^2}{\lambda^5} e^{-hc/\lambda kT} \quad \text{(Wien)}.$$

(*b*) Show likewise that the Rayleigh-Jeans formula for the spectral radiancy is

$$\mathcal{R}(\lambda, T) = \frac{2\pi ckT}{\lambda^4} \quad \text{(Rayleigh-Jeans)}.$$

27. Testing the approximate formulas. (*a*) For a cavity radiator at 3660 K (the melting point of tungsten), find the wavelength at which the spectral radiancy is a maximum. Calculate the spectral radiancy at this wavelength, using (*b*) the Planck formula (Eq. 4-8*b*), (*c*) the Wien formula (see Problem 26), and (*d*) the Rayleigh-Jeans formula (see Problem 26). Verify (see Fig. 4-5) that the Planck value lies between the Wien and the Rayleigh-Jeans values.

28. Wien's formula can be pretty accurate. (*a*) Derive an expression for the Wien constant following the procedures of Problem 20 but using the (approximate) Wien formula for the spectral radiancy (see Problem 26) rather than the (exact) Planck formula. (*b*) By what percent does your result differ from the exact result given in Problem 20?

29. Sunshine. The visible region of the electromagnetic spectrum may be taken to extend from 400 nm to 700 nm. For a black body at 5800 K—the sun's approximate surface temperature—what fraction of the radiated output lies in this range? (*Hint:* For simplicity of calculation, use Wien's approximate spectral radiancy formula, displayed in Problem 26, rather than the exact Planck law.)

30. Dividing up the energy. What fraction of the energy radiated by a cavity radiator has wavelengths that lie below λ_m? Above λ_m? Note that your answer does not depend on the temperature of the cavity. (*Hint:* For simplicity of calculation, use Wien's approximate spectral radiancy formula, displayed in Problem 26, rather than the exact Planck law.)

31. Changing the variable. The spectral radiancy of a black body is usually expressed as a function of the wavelength, as in Eq. 4-8b and Fig. 4-2. Show that the spectral radiancy may also be written in terms of the frequency, as

$$\mathcal{R}(\nu, T) = \frac{2\pi h \nu^3}{c^2} \frac{1}{e^{h\nu/kT} - 1}.$$

(*Hint:* See Supplementary Topic D.)

32. Checking out two formulas. A black body is maintained at a temperature of 5000 K. (*a*) Calculate the spectral radiancy $\mathcal{R}(\lambda)$ for a wavelength of 600 nm, using Eq. 4-8b. (*b*) Calculate the spectral radiancy $\mathcal{R}(\nu)$ for the same conditions, using the formula displayed in Problem 31. (*c*) Calculate the energy per unit area radiated in a wavelength band 5.00 nm wide centered at $\lambda = 600$ nm. Do so using each of the results obtained in (*a*) and (*b*) and show that you get the same numerical answer. (*Hint:* To what frequency band does this wavelength band correspond? See Supplementary Topic D.)

★ **33. Measuring the temperature of a hot plasma.** Many measuring instruments (and also the human eye) respond linearly not to $\mathcal{R}(\nu)$ itself but to $\log \mathcal{R}(\nu)$. A series of measurements of the spectral radiancy of a hot plasma in a thermonuclear fusion device was taken and $\log \mathcal{R}(\nu)$ was plotted against $\log \nu$; it was found that a straight line resulted, its equation being, empirically,

$$\log \mathcal{R}(\nu) = a \log \nu + b$$

in which

$$a = 2.00 \quad \text{and} \quad b = -32.715.$$

It is assumed that $\mathcal{R}(\nu)$ is measured in W/m²·Hz. (*a*) Show that the form of this empirical result follows if the measurements lie in the Rayleigh-Jeans region of the spectrum, that is, if the temperature and the frequency range covered by the measurements are such that $kT \gg h\nu$. (*b*) Find the temperature of the source. (*Hint:* The Planck expression for $\mathcal{R}(\nu, T)$ is displayed in Problem 31; the Rayleigh-Jeans formulation can easily be derived from it.)

34. Wien's displacement law—a second look. Planck's formula for the spectral radiancy expressed as a function of frequency is displayed in Problem 31. Show that, for a given temperature, this function has a maximum value at a frequency ν_m given by

$$\frac{\nu_m}{T} = 2.821 \, k/h = 5.872 \times 10^{10} \text{ Hz/K}.$$

This is another way of expressing Wien's displacement law. (*Hint:* The equation encountered in the process of finding the maximum value of $\mathcal{R}(\nu)$ can readily be solved numerically by trial and error, using a hand calculator.)

35. A displacement law puzzle. In Problem 34 an expression for ν_m, the frequency at which $\mathcal{R}(\nu)$ has its maximum value at a given temperature, is displayed. Show that this relationship can also be written as

$$\lambda_m T = 5104 \ \mu\text{m} \cdot \text{K}.$$

Why does this not agree with the value shown in Eq. 4-3, namely,

$$\lambda_m T = 2898 \ \mu\text{m} \cdot \text{K}?$$

36. Changing the variable. The spectral energy density of a cavity radiator is usually expressed as a function of the frequency, as in Eq. 4-8a and Fig. 4-5. Show that the spectral energy density may also be written in terms of the wavelength, as

$$\rho(\lambda) = \left(\frac{8\pi hc}{\lambda^5}\right) \frac{1}{e^{hc/\lambda kT} - 1}.$$

(*Hint:* See Supplementary Topic D.)

37. Checking out two formulas. The walls of a cavity radiator are maintained at a temperature of 2000 K. (*a*) Calculate the spectral energy density $\rho(\nu)$ within the cavity for a frequency of 1.5×10^{14} Hz, using Eq. 4-8a. (*b*) Calculate the spectral energy density $\rho(\lambda)$ for the same conditions, using the formula displayed in Problem 36. (*c*) Calculate the energy per unit volume contained in a narrow frequency band 0.020×10^{14} Hz wide centered at $\nu = 1.5 \times 10^{14}$ Hz. Do so using each of the results obtained in (*a*) and (*b*) and show that you get the same numerical answer. (*Hint:* To what wavelength band does this frequency band correspond? See Supplementary Topic D.)

38. How hot is the cavity? A cavity whose volume is 13.5 cm³ has a stored energy per unit wavelength interval of 2.7×10^{-12} J/nm at a wavelength of 1500 nm. What is the temperature of the cavity walls? (*Hint:* Use the result of Problem 36.)

39. The total energy in a cavity (I). (*a*) Show by direct integration of Eq. 4-8a that the total energy per unit volume in a cavity whose walls are maintained at a temperature T is given by

$$u(T) = \left(\frac{8\pi^5 k^4}{15 c^3 h^3}\right) T^4$$
$$= aT^4,$$

in which

$$a = 7.52 \times 10^{-16} \text{ J/m}^3 \cdot \text{K}^4.$$

(*b*) Evaluate for $T = 2000$ K. (*Hint:* See the hint in Problem 22.)

40. The total energy in a cavity (*II*). It is shown in Supplementary Topic D that, at a given temperature T,

$$\rho(\lambda) = \left(\frac{4}{c}\right) \Re(\lambda).$$

Show by integrating this relationship that the total energy per unit volume in a cavity whose walls are maintained at a temperature T is given by

$$u(T) = \left(\frac{4\sigma}{c}\right) T^4.$$

Show that this expression is identical to that derived in Problem 39. (*Hint:* See Problem 22.)

★ **41. The cosmic background radiation.** The universe is filled with a uniform flux of electromagnetic radiation at microwave frequencies, a remnant from its early history. The spectrum of this radiation fits that of a black body at $T = 2.9$ K. To verify this assumption, it is planned to make careful measurements of $\Re(\lambda)$ from a wavelength below λ_m where its value is $0.2\Re(\lambda_m)$ to a wavelength above λ_m where its value is again $0.2\Re(\lambda_m)$. Over what wavelength range must the measurements be made? (*Hint:* See Problem 20; also, make a change in variables, replacing $hc/\lambda kT$ by x; solve the resulting equation by trial and error, using a hand calculator.)

42. Two elements compared. The specific heat capacities at constant pressure and at room temperature for boron and for gold are 0.136 cal/g · F° and 0.0171 cal/g · F°, respectively. For each, calculate the molar heat capacity (in J/mol·K) and state whether the substance obeys the Dulong and Petit rule.

43. The internal energy. Show that the internal energy per mole of a solid can be written, according to Einstein's theory of heat capacities, as

$$u(T) = 3RT\left(\frac{x}{e^x - 1}\right),$$

in which $x = T_E/T$, where T_E is the Einstein temperature.

44. A sample calculation. The Einstein temperature of aluminum may be taken as 290 K. According to Einstein's theory of heat capacity, what are (a) its oscillator frequency; (b) its lattice-vibration energy per mole at 150 K; (c) its molar heat capacity, under constant-volume conditions, at 150 K?

45. Checking the high-temperature limits. (a) Verify that Einstein's expression for $u(T)$, the internal energy per mole of a solid associated with the lattice vibrations (Eq. 4-18), approaches $3RT$ as $T \rightarrow \infty$. (b) Verify also that Einstein's expression for $c_v(T)$, the heat capacity at constant volume (Eq. 4-19*b*), approaches $3R$ as $T \rightarrow \infty$.

46. Checking the low-temperature limits. (a) Verify that as $T \rightarrow 0$, Einstein's expression for $u(T)$, the energy per mole of a solid associated with the lattice vibrations (Eq. 4-18), varies as

$$u(T) = 3RT_E e^{-T_E/T}.$$

(b) Verify also that as $T \rightarrow 0$, Einstein's expression for $c_v(T)$, the molar heat capacity at constant volume (Eq. 4-19*b*), varies as

$$c_v(T) = 3R \left(\frac{T_E}{T}\right)^2 e^{-T_E/T}.$$

(*Note:* Einstein's theory, because of its deliberately chosen simplifying assumption of a single oscillator frequency, is known not to agree with experiment at low temperatures; the Debye theory, which makes broader and more detailed quantum assumptions, does yield good agreement over the full range of temperatures. See Supplementary Topic F.)

47. Conditions at the Einstein temperature. In terms of Einstein's theory of heat capacity, (a) what is the molar heat capacity at constant volume of a solid at its Einstein temperature? Express your answer as a percentage of its classical value of $3R$. (b) What is the molar internal energy at the Einstein temperature? Express your answer as a percentage of its classical value of $3RT$.

48. Checking the halfway point. In terms of Einstein's theory of heat capacity, (a) at what temperature will the energy per mole of a solid achieve one-half its classical value of $3RT$? (b) At what temperature will the heat capacity at constant volume of a substance achieve one-half its classical value of $3R$? Express each answer in terms of the Einstein temperature of the solid. (c) Evaluate for aluminum, for which $T_E = 290$ K.

49. Frequencies and spring constants. The Einstein temperatures of lead, aluminum, and beryllium may be taken as 68 K, 290 K, and 690 K, respectively. For each of these elements, find (a) the resonant frequency ν of its atomic oscillators, (b) the spacing ΔE between adjacent oscillator levels, and (c) the effective spring constant κ. Interpret your results in terms of the heat-capacity curves displayed in Fig. 4-9. Note that the values for aluminum are worked out in Example 7.

★ **50. The oscillation amplitude.** What is the amplitude of oscillation of aluminum atoms (strictly, ions) oscillating about their lattice sites at room temperature ($T = 300$ K)? Express your answer in terms of absolute length units and also as a percentage of the interatomic nearest-neighbor distance for aluminum, which is 0.29 nm. (*Hint:* See Example 7; also *Physics*, Part I, Sec. 15-4.)

51. Heating a block of aluminum. A 12.0-g block of aluminum is heated from 80 K up to 180 K, under presumed constant-volume conditions. How much thermal energy is required according to (a) the classical theory of heat capacity and (b) Einstein's quantum theory of heat capacity? The Einstein temperature for aluminum may be taken to be 290 K.

52. Thermal equilibrium. 25.0 g of aluminum at 80 K are mixed thoroughly with 12 g of aluminum at 200 K in an insulated container. What is the final temperature of the mixture? Assume that Einstein's theory of heat capacities is valid and that, at these relatively low temperatures, the differences between the heat capacity at constant volume and that at constant pressure may be neglected. Assume further that there are no energy exchanges between the two aluminum specimens and the container. The Einstein temperature of aluminum may be taken to be 290 K.

53. Finding c_v (I). It can be shown from purely thermodynamic arguments [5] that the difference in heat capacities at constant pressure and at constant volume for solids is given by

$$c_p - c_v = \frac{T\beta^2}{\rho\kappa}.$$

Here β is the coefficient of thermal expansion of the substance and ρ is its density. The quantity κ is the compressibility, defined as $(1/V)(dV/dp)$, in which V is the volume and p is the external pressure. For copper at 800 K, $\beta = 6.07 \times 10^{-5}$ K^{-1}, $\rho = 1.38 \times 10^5$ mol/m^3, and $\kappa = 9.22 \times 10^{-12}$ m^2/N. The measured value of c_p is 28.0 J/mol·K. What is the heat capacity at constant volume at this temperature? (*Note:* The value found will exceed $3R$ somewhat, largely because of the small contribution of the conduction electrons to the heat capacity. The theories of Einstein and of Debye consider only the energy associated with the vibrations of the crystal lattice and do not take the conduction electrons into account.)

54. Finding c_v (II). At 0°C, nickel has the following properties: atomic weight = 58.7 g/mol, density = 8.90 g/cm^3, $c_p = 0.456$ J/g·K, coefficient of thermal expansion = 4.00×10^{-5} K^{-1}, and compressibility = 5.68×10^{-12} m^2/N. What is the heat capacity at constant volume for nickel at this temperature? (*Hint:* See Problem 53.)

55. The free electrons can also absorb energy. We have seen that the classical theory of heat capacities assigns six degrees of freedom to each oscillating atom in the lattice of a solid. If the solid is a conductor, however, there are also conduction electrons; in copper, for example, there is one such particle per copper atom. Classically, these electrons, which are not tied to the lattice sites but are free to move throughout the sample, can also absorb energy. If each such conduction electron is assigned three (translational) degrees of freedom, what is now the classical prediction for the molar heat capacity at constant volume? (*Note:* The failure of this prediction to agree with experiment was just as great a failing of the classical theory as was the departure from the Dulong and Petit law.)

56. A range of thermal energies. (a) What is the average translational kinetic energy of a gas molecule at room temperature (300 K)? (b) Assume that the first excited state of an atom is a few electron volts above its lowest (ground) state. What is the order of magnitude of the temperature needed if an appreciable number of such atoms are to be excited by colliding with each other at thermal energies?

57. Calculating the Planck constant. Assume in the Franck-Hertz experiment that the electromagnetic energy emitted by a mercury atom when it gives up the energy absorbed from a 4.9-eV electron equals $h\nu$, where ν is the frequency corresponding to the 253.6-nm mercury resonance line. Calculate the value of the Planck constant h that results from this assumption and compare it with the value arrived at from cavity radiation studies.

58. Exciting the sodium yellow line. The lowest excited state of a free sodium atom lies 2.1 eV above the ground state. (When a sodium atom moves from this state back to its ground state, it emits the familiar yellow sodium line.) At what temperature is the average translational kinetic energy of sodium atoms, present as sodium vapor, equal to this excitation energy?

59. Exciting the hydrogen spectrum. In a Franck-Hertz type of experiment, atomic hydrogen is bombarded with electrons and excitation potentials are found at 10.21 V and 12.10 V. (a) Explain the observation that three different lines of the hydrogen spectrum (and three lines only) accompany these excitations. (*Hint:* Draw an energy-level diagram.) (b) Now assume that the energy differences between states of the hydrogen atom can be written as $h\nu$, where ν is the frequency of the radiation emitted when the atom moves from the upper of these states to the lower. Find the wavelengths of the three observed spectrum lines.

60. Excitation by collision. (a) An atom, at rest and in its ground state, is struck by another atom of the same kind that is also in its ground state but that has kinetic energy K. Let $\mathscr{E}$ be the excitation energy of the atom, that is, the energy difference between its ground state and its first excited state. Show that if $K < 2\mathscr{E}$ the collision will be elastic; that is, excitation will *not* occur. (b) Show that the relationship derived in Example 9 includes this situation as a special case.

61. Energy loss in an elastic collision. What fraction of its initial kinetic energy does an electron lose in a head-on *elastic* collision with a resting mercury atom in a Franck-Hertz experiment? If the collision is not "head-on," will the fractional energy loss be greater or less than this?

62. Stopping the electron. An electron undergoes a head-on *inelastic* collision with a resting mercury atom, raising the latter to its first excited state (excitation energy $\mathscr{E}$). Suppose that, after the collision, the electron is found to be exactly at rest. (*a*) Show that, for this to happen, the initial kinetic energy K_0 must be

$$K_0 = \frac{\mathscr{E}}{1 - m/M}.$$

(*b*) In Example 9 the collision was similar except that there we required that the electron have the *minimum* kinetic energy (K_{min}) it needed to excite the resting mercury atom. What is the fractional difference between K_0 and K_{min}? Which is larger? Given the specifications of Example 9, is this what you expect?

63. The mean free path. Estimate the mean free path of electrons in mercury vapor in a Franck-Hertz tube. The pressure in the tube is 1.0 atm $(= 1.0 \times 10^5 \, \text{N/m}^2)$ and $T = 300$ K. The diameter of a mercury atom is 0.31 nm; treat the electrons as points. The electrons move much faster than the atoms, which may be considered to be at rest. Express your answer in absolute length units and also as a ratio to the diameter of the mercury atom. (*Hint:* See *Physics*, Part I, Sec. 24-1.)

64C. Using your calculator (*I*). Write a program for your hand-held, programmable calculator in which the input data are the temperature T (in K) and *either* the wavelength λ (in μm, say) *or* the frequency ν (in units of 10^{13} Hz, say) and the outputs are the spectral radiancy $\mathscr{R}$ and the spectral energy density ρ, each expressed in convenient units. Check your program by verifying the Planck curves shown in Figs. 4-2 and 4-5. (*Hint:* See Eqs. 4-8 and Supplementary Topic D. For the Planck constant h, the Boltzmann constant k, and the speed of light c, use values rounded to four significant figures; see Appendix 1.)

65C. Using your calculator (*II*). Write a program for your hand-held, programmable calculator in which the input datum is the temperature and the successive outputs are (*a*) the radiancy R, (*b*) the wavelength λ_m at which the spectral radiancy is a maximum, (*c*) the energy density $u(T)$ (see Problems 39 and 40), and (*d*) the frequency ν_m at which the spectral energy density is a maximum (see Problem 34).

66C. Using your calculator (*III*). Write a program for your hand-held, programmable calculator in which the input data are the temperature T (in K) and the Einstein temperature T_E (in K) and the outputs are the molar energy $u(T)$ (see Eq. 4-18 and Problem 43) and the molar heat capacity $c_v(T)$ (see Eq. 4-19*b*). Check your program by verifying the curves of Fig. 4-9 for lead, aluminum, and beryllium; the Einstein temperatures for these elements are 68 K, 290 K, and 690 K, respectively. (*Note:* For the gas constant R use the value 8.314 J/mol·K. Recall that the curves of Fig. 4-9 represent experimental data and do not agree exactly with the predictions of Einstein's theory at low temperatures.)

references

1. See the following for fascinating insights into the early history of the cavity radiation problem and of the concept of the quantization of energy:

(*a*) Hans Kangro, *Early History of Planck's Radiation Law* (Taylor & Francis, London, 1976).

(*b*) Thomas S. Kuhn, *Black-Body Theory and the Quantum Discontinuity, 1894–1912* (Clarendon Press, Oxford, 1978).

(*c*) Martin J. Klein, "No Firm Foundation: Einstein and the Early Quantum Theory," in Harry Woolf, Ed., *Some Strangeness in the Proportion: A Centennial Symposium to Celebrate the Achievements of Albert Einstein* (Addison-Wesley, Reading Mass., 1980).

(*d*) Abraham Pais, *Subtle Is the Lord . . . The Science and Life of Albert Einstein* (Clarendon Press, Oxford, 1982), part VI, "The Quantum Theory."

2. For a discussion of the importance of Einstein's work on the heat capacities of solids for the development of the quantum theory see:

(*a*) Martin J. Klein, "Einstein and the Development of Quantum Physics," in A. P. French, Ed., *Einstein—A Centenary Volume* (Harvard University Press, Cambridge, Mass., 1979).

(*b*) Abraham Pais, "Einstein and Specific Heats," in Pais, *Subtle is the Lord . . .*, Ref. 1*d*, Chap. 20.

3. The 1914 article by Franck and Hertz is translated and appears, with explanatory commentary, in: Henry A. Boorse and Loyd Motz, *The World of the Atom* (Basic Books, New York, 1966): see article entitled "The Quantum Theory Is Tested," vol. 1, p. 766.

4. Martin J. Klein, "Thermodynamics and Quanta in Planck's Work," *Physics Today*, **19,** 23 (1966).

5. For an excellent general account of the heat capacities of solids from an introductory point of view, see Mark W. Zemansky, *Heat and Thermodynamics*, 4th ed. (McGraw-Hill, New York, 1957).

CHAPTER 5

the particle nature of radiation

5-1 INTRODUCTION

In Chapter 4 we saw how the concept of energy quantization arose from the study of cavity radiation, that is, of radiation interacting with matter under conditions of thermal equilibrium. In 1905 Einstein first proposed [1,2] that not only is the energy of matter—by which we mean the atomic oscillators in the cavity walls—quantized but so also is the energy of the radiation that is trapped in the cavity, in equilibrium with the cavity walls. Planck and others had assumed that the radiation within the cavity was wavelike and thus continuous in nature.

In this chapter we shall examine the experimental evidence that supports Einstein's hypothesis of *light quanta* or, as we now call them, *photons*. Again we shall look at radiation interacting with matter, although in experimental situations quite different from those of cavity radiation. We shall consider cases in which photons are absorbed or deflected by matter (the photoelectric effect, the Compton effect, and pair production) and also cases in which photons are generated by matter (*bremsstrahlung*, synchrotron radiation, and pair annihilation). In all of these studies we shall find strong support for the view that light is particle-like, and we shall also see the Planck constant h showing up in many entirely new experimental situations.

5-2 THE PHOTOELECTRIC EFFECT

In 1886 and 1887, Heinrich Hertz performed the experiments that first confirmed the existence of electromagnetic waves and Maxwell's electromagnetic theory of light propagation. It is one of those fascinating and paradoxical facts in the history of science that in the course of his experiments Hertz noted the effect that Einstein later used to contradict other aspects of the classical electromagnetic

161

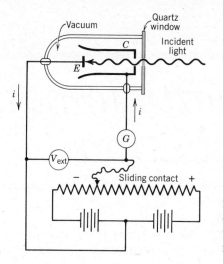

Figure 5-1. An apparatus used to study the photo-electric effect. The potential difference V_{ext} can be varied continuously through both positive and negative values.

theory. Hertz discovered that an electric discharge between two electrodes occurs more readily when ultraviolet light falls on one of the electrodes. It was shown soon after that the ultraviolet light facilitates the discharge by causing electrons to be emitted from the cathode surface. The ejection of electrons from a surface by the action of light is called the *photoelectric effect*.

Figure 5-1 shows an apparatus used to study the photoelectric effect. A glass envelope encloses the apparatus in an evacuated space. Monochromatic light, incident through a quartz window, falls on the metal plate E (the *emitter*) and liberates electrons, called *photoelectrons*. The electrons can be detected as a current if they are attracted to the metal cup C (the *collector*) by means of a potential difference V applied between E and C. The galvanometer G serves to measure this photoelectric current.

The actual potential difference that acts between emitter E and collector C is not identical with the external potential difference V_{ext} supplied by the battery. In the circuit of Fig. 5-1 in which the photoelectrons circulate, there is also a second emf—a hidden battery, if you will—associated with the fact that the emitter and the collector are (usually) made of different materials [3]. If suitable precautions are taken, this *collector–emitter contact potential difference* V_{ce} remains constant throughout the experiment. The potential difference V that acts on the emitted electrons, then, is the algebraic sum of these two quantities, or

$$V = V_{ext} + V_{ce}. \tag{5-1}$$

V_{ce} can be measured directly if the external potential difference V_{ext} is temporarily set equal to zero. In all that follows we shall assume that the contact potential difference has been suitably taken into account and we shall express all our results in terms of V as defined by Eq. 5-1.

Figure 5-2 (curve a) is a plot of the photoelectric current, in an apparatus like that of Fig. 5-1, as a function of the potential difference V. If V is made large enough, the photoelectric current reaches a certain limiting (saturation) value at which all photoelectrons ejected from E are collected by C.

If V is reduced to zero and gradually reversed in sign, the photoelectric current does not drop to zero, which suggests that the electrons are emitted from E with kinetic energy. Some will reach collector C in spite of the fact that the electric field opposes their motion. However, if this reversed potential difference is made

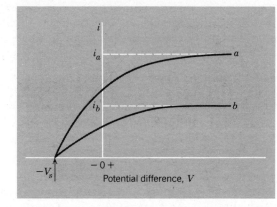

Figure 5-2. Graphs of the current i as a function of the potential difference V (see Eq. 5-1) from data taken with the apparatus of Fig. 5-1. In curve b the incident light intensity has been reduced to one-half that of curve a. The stopping potential V_s is independent of the light intensity, but the saturation currents i_b and i_a are directly proportional to it.

large enough, a value V_s (the *stopping potential*) is reached at which the photoelectric current does drop to zero. This potential difference V_s, multiplied by the electron charge, measures the kinetic energy K_{max} of the *fastest* ejected photoelectron. That is,

$$K_{max} = eV_s. \tag{5-2}$$

K_{max} turns out experimentally to be independent of the intensity of the light, as is shown by curve b in Fig. 5-2, in which the light intensity has been reduced to one-half the value used in obtaining curve a.

Suppose now that we vary the frequency, or wavelength, of the incident light. Figure 5-3 shows the stopping potential V_s as a function of the frequency of the light incident on a clean sodium surface. Note that there is a definite cutoff frequency ν_0, below which no photoelectric effect occurs. These data were reported in 1916 by R. A. Millikan from his laboratory at the University of Chicago. His monumentally painstaking work on the photoelectric effect, together with his earlier measurement of the charge of the electron, earned him the Nobel Prize in 1923, the first native-born American physicist to be so honored. Because the photoelectric effect for visible or near-visible light is largely a surface phenomenon, it is necessary to avoid oxide films, grease, or other surface contaminants. Millikan devised a technique to cut shavings from the metal surface under vacuum conditions, a "machine shop in vacuo" as he called it.

Figures 5-2 and 5-3 display the essential experimental facts about the photoelectric effect. In trying to understand these facts in terms of the classical wave

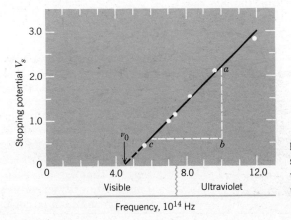

Figure 5-3. A plot of Millikan's measurements of the stopping potential at various frequencies for a sodium emitter. The cutoff frequency ν_0 is 4.39×10^{14} Hz.

theory of light three insurmountable problems arise. The difficulties are not matters of detail or interpretation. The classical theory simply fails, and that it does so is beyond any reasonable dispute. The three problems are the intensity problem, the frequency problem, and the time delay problem.

The Intensity Problem   Wave theory requires that the oscillating electric vector **E** of the light wave increases in amplitude as the intensity of the light beam is increased. Since the force applied to the electron is *e***E**, this suggests that the *kinetic energy* of the photoelectrons should also increase as the light beam is made more intense. However, Fig. 5-2 shows that $K_{max}(= eV_s)$ *is independent of the light intensity;* this has been tested over a range of intensities of 10^7.

The Frequency Problem   According to the wave theory, the photoelectric effect should occur for any frequency of the light, provided only that the light is intense enough to provide the energy needed to eject the photoelectrons. However, Fig. 5-3 shows that there exists, for each surface, a characteristic cutoff frequency ν_0. *For frequencies less than ν_0, the photoelectric effect does not occur, no matter how intense the illumination.*

The Time Delay Problem   If the energy acquired by a photoelectron is absorbed directly from the wave incident on the metal plate, the "effective target area" for an electron in the metal is limited and probably not much more than that of a circle of the order of an atomic diameter. In the classical theory the light energy is uniformly distributed over the wave front. Thus, if the light is feeble enough, there should be a measurable time lag, which we shall estimate in Example 1, between the impinging of the light on the surface and the ejection of the photoelectron. During this interval the electron should be absorbing energy from the beam until it had accumulated enough to escape. However, *no detectable time lag has ever been measured.* This disagreement is particularly striking when the photoelectric substance is a gas; under these circumstances collective absorption mechanisms can be ruled out and the energy of the emitted photoelectron must certainly be soaked out of the light beam by a single atom or molecule.

EXAMPLE 1.

Classical Wave Theory and the Time Delay Problem. A foil of potassium is placed 3.0 m from a light source whose power is 1.0 W. Assume that an ejected photoelectron may have collected its energy from a circular area of the foil whose radius is one atomic radius ($r \approx 5.0 \times 10^{-11}$ m). The energy required to remove an electron through the potassium surface is about 1.8 eV; how long would it take for such a "target" to absorb this much energy from such a light source? Assume the light energy to be spread uniformly over the wavefront.

The target area is $\pi(5.0 \times 10^{-11}$ m$)^2$; the area of a 3.0-m sphere centered on the light source is $4\pi(3.0$ m$)^2$. Thus if the light source radiates uniformly in all directions—that is, if the light energy is uniformly distributed over spherical wavefronts spreading out from the source, in agreement with classical theory—the rate P at which energy falls on the target is given by

$$P = (1.0 \text{ W})\left(\frac{25\pi \times 10^{-22} \text{ m}^2}{36\pi \text{ m}^2}\right) = 6.9 \times 10^{-23} \text{ J/s}.$$

Assuming that all this power is absorbed, we may calculate the time required for the electron to acquire enough energy to escape; we find that

$$t = \left(\frac{1.8 \text{ eV}}{6.9 \times 10^{-23} \text{ J/s}}\right)\left(\frac{1.6 \times 10^{-19} \text{ J}}{1 \text{ eV}}\right) \cong 1.2 \text{ h}.$$

Of course, we could modify the above picture to reduce the calculated time by assuming a much larger effective target area. The most favorable assumption, that energy is transferred by a resonance process from light wave to electron, leads to a target area of λ^2, where λ is the wavelength of the light. But we would still obtain a finite time lag that is within our ability to measure experimentally. (For ultraviolet light of $\lambda = 10$ nm, for example, $t \cong 1$ s). However, no time lag has been detected under any circumstances, the early experiments setting an upper limit of 10^{-9} s on any such possible delay!

5-3 EINSTEIN'S QUANTUM THEORY OF THE PHOTOELECTRIC EFFECT

In 1905, many years before Millikan's experiments were performed, Einstein called into question the classical theory of light, proposed a new theory, and cited the photoelectric effect as one application that could test which theory was correct. Planck had, at first, restricted his concept of energy quantization to the emission or absorption mechanism of a material oscillator. He believed that light energy, once emitted, was distributed in space like a wave. Einstein proposed instead that the radiant energy itself existed in concentrated bundles, which later came to be called photons. The energy E of a single photon is given by

$$E = h\nu, \tag{5-3}$$

where ν is the frequency of the radiation and h is the Planck constant. In effect, Einstein proposed a granular structure to radiation itself, radiant energy being distributed in space in a discontinuous way. Millikan, whose brilliant experiments over many years later verified Einstein's ideas in every detail, "contrary to my own expectations," spoke of Einstein's "bold, not to say reckless, hypothesis."

Applying the photon concept to the photoelectric effect, Einstein proposed that the entire energy of a photon is transferred to a single electron in a metal. When the electron is emitted from the surface of the metal, then, its kinetic energy will be

$$K = h\nu - w, \tag{5-4}$$

where $h\nu$ is the energy of the absorbed incident photon and w is the work required to remove the electron from the metal. This work is needed to overcome the attractive fields of the atoms in the surface and losses of kinetic energy caused by internal collisions of the electron. Some electrons are bound more tightly than others; some lose energy in collisions on the way out. In the case of loosest binding and no internal losses the photoelectron will emerge with the maximum kinetic energy, K_{max}. Hence,

$$K_{max} = h\nu - w_e \quad \text{(Einstein's photoelectric equation),} \tag{5-5}$$

where w_e, a characteristic energy of the emitter called the *work function*, is the minimum energy needed by an electron to pass through the metal surface.

Consider now how Einstein's photon hypothesis meets the three objections raised against the wave theory interpretation of the photoelectric effect. As for the intensity problem, there is complete agreement of the photon theory with experiment. Doubling the light intensity merely doubles the number of photons and thus doubles the photoelectric current; it does *not* change the energy distribution of the individual electrons, the energy $(= h\nu)$ of the individual photons, or the nature of the individual photoelectric process described by Eq. 5-4.

The frequency problem is resolved at once from Eq. 5-5. If K_{max} equals zero, we have

$$h\nu_0 = w_e, \tag{5-6}$$

which asserts that a photon of frequency ν_0 has just enough energy to eject the photoelectrons and none extra to appear as kinetic energy. If the frequency is reduced below ν_0, the individual photons, regardless of how many of them there are (that is, no matter how intense the illumination), will not have enough energy individually to eject photoelectrons.

The time lag problem is resolved by the photon theory because the required energy is supplied in concentrated bundles. It is *not* spread uniformly over a large area, as we assumed in Example 1, which is based on the assumption that the classical wave theory is true. If there is any illumination at all incident on the cathode, then there will be at least one photon that hits it; this photon will be immediately absorbed, by *some* atom, leading to the immediate emission of a photoelectron.

Let us rewrite Einstein's photoelectric equation (Eq. 5-5) by substituting eV_s for K_{max} (see Eq. 5-2). This yields

$$V_s = \left(\frac{h}{e}\right) \nu - \left(\frac{w_e}{e}\right).$$

Thus Einstein's theory predicts a linear relationship between the stopping potential V_s and the frequency ν, in complete agreement with experiment (see Fig. 5-3). The slope of the experimental curve in this figure should be h/e, or

$$\frac{h}{e} = \frac{ab}{bc} = \frac{2.20 \text{ V} - 0.65 \text{ V}}{10.0 \times 10^{14} \text{ s}^{-1} - 6.0 \times 10^{14} \text{ s}^{-1}} = 3.9 \times 10^{-15} \text{ V} \cdot \text{s}.$$

We can find h by multiplying this ratio by the electronic charge e. Thus,

$$h = (3.9 \times 10^{-15} \text{ V} \cdot \text{s})(1.6 \times 10^{-19} \text{ C}) = 6.2 \times 10^{-34} \text{ J} \cdot \text{s}.$$

From a more careful analysis of these and other data, including data taken with lithium surfaces, Millikan found the value $h = 6.57 \times 10^{-34}$ J·s, with an accuracy of about 0.5 percent. This early measurement was in good agreement with the value of h derived from Planck's radiation formula.

This numerical agreement in two determinations of h, using completely different phenomena and theories, is striking. The presently accepted value of h, derived from an analysis of a wide variety of experiments involving not only h but other fundamental constants as well, is

$$h = 6.626176 \times 10^{-34} \text{ J} \cdot \text{s},$$

with an uncertainty of about 5 parts per million.

To quote Millikan:

> The photoelectric effect . . . furnishes a proof which is quite independent of the facts of black-body radiation of the correctness of the fundamental assumption of the quantum theory, namely, the assumption of a discontinuous explosive emission of the energy absorbed by the electronic constituents of atoms from . . . waves. It materializes, so to speak, the quantity h discovered by Planck through the study of black body radiation and gives us a confidence inspired by no other type of phenomenon that the primary physical conception underlying Planck's work corresponds to reality.

EXAMPLE 2.

The Work Function of Sodium. Find the work function of sodium from Fig. 5-3.

The intersection of the straight line in Fig. 5-3 with the horizontal axis is the cutoff frequency, $\nu_0 = 4.39 \times 10^{14}$ s^{-1}. Substituting this into Eq. 5-6 gives us

$$w_e = h\nu_0 = (6.63 \times 10^{-34} \text{ J} \cdot \text{s})(4.39 \times 10^{14} \text{ s}^{-1})$$

$$= 2.92 \times 10^{-19} \text{ J} \left(\frac{1 \text{ eV}}{1.60 \times 10^{-19} \text{ J}}\right)$$

$$= 1.82 \text{ eV}.$$

Note that the work function w_e also can be obtained directly from Fig. 5-3 as the magnitude of the negative intercept of the extended straight line with the vertical axis. For most conducting metals, the value of the work function is of the order of a few electron volts.

EXAMPLE 3.

A Rain of Photons. At what rate per unit area do photons strike the metal plate in Example 1? Assume that the light is monochromatic, of wavelength 589 nm (yellow sodium light).

The rate per unit area that energy falls on a metal plate 3.0 m from the light source (see Example 1) is

$$r = \frac{1.0\ \text{W}}{36\pi\ \text{m}^2} = 8.8 \times 10^{-3}\ \text{J/m}^2 \cdot \text{s} \left(\frac{1\ \text{eV}}{1.6 \times 10^{-19}\ \text{J}}\right)$$

$$= 5.5 \times 10^{16}\ \text{eV/m}^2 \cdot \text{s}.$$

Each photon has an energy of

$$E = h\nu = \frac{hc}{\lambda} = \frac{(6.63 \times 10^{-34}\ \text{J} \cdot \text{s})(3.00 \times 10^8\ \text{m/s})}{(5.89 \times 10^{-7}\ \text{m})}$$

$$= 3.4 \times 10^{-19}\ \text{J} \left(\frac{1\ \text{eV}}{1.6 \times 10^{-19}\ \text{J}}\right)$$

$$= 2.1\ \text{eV}.$$

Thus the rate R at which photons strike the plate is

$$R = (5.5 \times 10^{16}\ \text{eV/m}^2 \cdot \text{s}) \left(\frac{1\ \text{photon}}{2.1\ \text{eV}}\right)$$

$$= 2.6 \times 10^{16}\ \text{photons/m}^2 \cdot \text{s}.$$

The photoelectric effect will occur in this case because the photon energy (2.1 eV) is greater than the work function for the surface (1.82 eV; see Example 2). Note that if the wavelength is sufficiently increased (that is, if ν is sufficiently decreased), the photoelectric effect will not occur, no matter how large the rate R might be.

This example suggests that the intensity of light I can be regarded as the product of N, the number of photons per unit area per unit time, and $h\nu$, the energy of a single photon. We see that even at the relatively low intensity here ($\sim 10^{-2}\ \text{W/m}^2$), the number N is extremely large ($\sim 10^{16}\ \text{photons/m}^2 \cdot \text{s}$), so that the energy of any one photon is very small. This accounts for the extreme fineness of the granularity of radiation and suggests why ordinarily it is difficult to detect at all. It is analogous to detecting the atomic structure of bulk matter, which for most purposes can be regarded as continuous, the discreteness being revealed only under special circumstances.

EXAMPLE 4.

Star Light . . . Star (Not So) Bright . . . The faintest star that an experienced observer can see under good conditions with the naked eye is about magnitude 7. For our sun to shine as dimly as magnitude 7, it would have to be removed to a distance of about 60 light-years, far beyond our nearest stellar neighbors. At what rate would photons from such a faint and distant "sun" enter the pupil of our eye? (The sun radiates energy at the rate of 3.9×10^{26} W, about 30 percent of it in the visible range. Assume an effective wavelength of 550 nm and take the radius of the fully dark-adapted pupil to be 4 mm.)

The energy of a photon corresponding to a wavelength of 550 nm can be found, as in Example 3, to be 3.6×10^{-19} J. Thus, under our assumptions, the rate at which the sun emits photons in the visible range is

$$\frac{(3.9 \times 10^{26}\ \text{W})(0.3)}{3.6 \times 10^{-19}\ \text{J/photon}} = 3.3 \times 10^{44}\ \text{photons/s}.$$

To find the rate R at which photons actually enter the eye, we must multiply this large number by the ratio of the area of the pupil to the area of a sphere whose radius is 60 ly. Thus

$$R = (3.3 \times 10^{44}\ \text{photons/s}) \frac{\pi(4 \times 10^{-3}\ \text{m})^2}{(4\pi)(60\ \text{ly} \times 9.5 \times 10^{15}\ \text{m/ly})^2}$$

$$\cong 4000\ \text{photons/s}.$$

Only about 1 percent of the photons that enter the pupil are actually absorbed by the optical sensors in the retina, the rest being absorbed in other parts of the eye. Thus, at the limit of visibility, our rough calculation shows that photons need to be delivered to the optical sensors at the rate of only about 40 per second. Experiments under carefully controlled laboratory conditions have shown that the eye can detect the presence of a faint, briefly displayed luminous spot with a probability of 60 percent if only 5 to 14 photons (total) reach the retina. The eye is a remarkably sensitive photon detector and can almost—but not quite—detect the quantum nature of light!

We saw in Chapter 4 that quantization of energy, implicit in Planck's derivation (in 1900) of his black-body radiation law, was not widely accepted—or even fully understood—at that time. It was not until after 1907, the year in which Einstein developed his quantum theory of the heat capacities of solids, that quantization of energy came to be recognized as the cornerstone of a new and revolutionary quantum physics with Planck's h as its central constant.

The light quantum hypothesis, embodied in the relation $E = h\nu$, was even slower to gain acceptance. We can point to two reasons: (1) Though advanced by Einstein in 1905, this hypothesis had to wait a decade before receiving its first thorough-going experimental test, at the hands of R. A. Millikan. (2) The idea that light has a particle aspect to its nature flew in the face of the wave theory of light, which was firmly supported by all kinds of interference, diffraction, and polarization experiments and was buttressed on the theoretical side by the phenomenal successes of Maxwell's equations of electromagnetism. Einstein's light quantum hypothesis did not, in fact, gain general acceptance until after 1923, the year in which the Compton effect (see the next section) was discovered and explained in quantum terms. For all this time Einstein stood essentially alone in his belief in light quanta. He even became convinced that his light quantum hypothesis was implicit in—and essential to—any proper derivation of the Planck radiation law and, in 1917, proceeded to make that clear in the masterly derivation of that law outlined in Supplementary Topic E.

Two quotations serve to point up the general early rejection of the light quantum hypothesis by the leading physicists of the time. In 1913, eight years after he had advanced his light quantum hypothesis, Einstein was recommended for membership in the Prussian Academy of Sciences by Planck and others. In a signed affidavit, recommending and praising Einstein, they wrote:

> In sum, one can say that there is hardly one among the great problems in which modern physics is so rich to which Einstein has not made a remarkable contribution. That he may have sometimes missed the target in his speculations, as, for example, in his hypothesis of light quanta, cannot be really held too much against him, for it is not possible to introduce really new ideas even in the most exact sciences without sometimes taking a risk.

It was in 1921, incidentally, that Einstein received the Nobel Prize for one of the consequences of taking this risk, his discovery of the law of the photoelectric effect.

We have seen the care and skill with which Millikan verified Einstein's photoelectric equation. In summarizing his experimental results in 1916, he wrote:

> Einstein's photoelectric equation has been subject to very searching tests and it appears in every case to predict exactly the observed results.

His comments in the same paper on Einstein's light quantum hypothesis, however, are that:

> . . . the semicorpuscular theory by which Einstein arrived at his equation [Eq. 5-5] seems at present to be wholly untenable.

Today the light quantum or photon hypothesis is universally accepted and is recognized as valid throughout the entire electromagnetic spectrum; see Fig. 5-4. The microwave radiation in a microwave oven, for example, can be viewed as an assembly of photons. At $\lambda = 10$ cm, a typical wavelength, the photon energy can be computed as above to be 1.2×10^{-5} eV. This quantum energy is much too low to eject photoelectrons from metal surfaces. Photons in the visible and ultraviolet regions of the spectrum, as we have seen, can eject such photoelectrons; they are

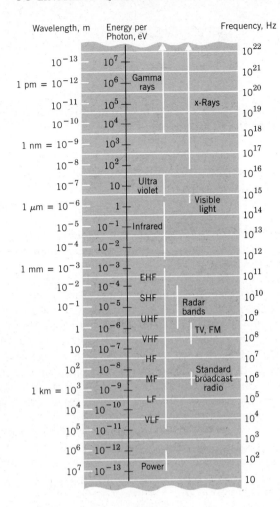

Wavelength, m	Energy per Photon, eV		Frequency, Hz

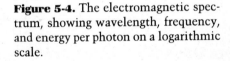

Figure 5-4. The electromagnetic spectrum, showing wavelength, frequency, and energy per photon on a logarithmic scale.

the so-called *conduction electrons*, which are bound to the metal by energies of only a few electron volts.

Photons in the x-ray or gamma-ray region of the spectrum can eject electrons from deep within the structure of the atoms on which they fall. The innermost electrons of uranium, for example, are bound with an energy of 116 keV, so the incident photons must have at least this much energy to eject them. Photons with still higher energies can even knock neutrons and protons out of the atomic nucleus in what are called *photonuclear* reactions. One of these particles may be bound into the nucleus by an energy in the range 5 to 15 MeV.

Notice that the photons are absorbed in the photoelectric process. This requires the electrons to be bound to atoms, or solids, for a truly free electron cannot absorb a photon and conserve both energy and momentum in the process (see Problem 12). We must have a bound electron, therefore, in which case the binding forces serve to transmit momentum to the atom or solid. Because of the large mass of atom, or solid, compared to the electron, the system can absorb a large amount of momentum without, however, acquiring a significant amount of energy. Our photoelectric energy equation remains valid, the effect being possible only because there is a heavy recoiling particle in addition to an ejected electron. The photoelectric effect is one important way in which photons, of energy up to and including x-ray energies, are absorbed by matter. At higher energies other photon absorption processes, soon to be discussed, become more important.

5-4 THE COMPTON EFFECT

The corpuscular (particlelike) nature of radiation received dramatic confirmation in 1923 from the experiments [4,5] of A. H. Compton, carried out at Washington University in St. Louis. Compton caused a beam of x-rays of sharply defined wavelength λ to fall on a graphite target, as shown in Fig. 5-5. For selected angles of scattering, he measured the intensity of the scattered x-rays as a function of their wavelength. Figure 5-6 shows his experimental results. We see that, although the incident beam consists essentially of a single wavelength λ, the scattered x-rays have intensity peaks at *two* wavelengths; one of them is the same as the incident wavelength, the other, λ', being larger by an amount $\Delta\lambda$. This so-called *Compton shift*, $\Delta\lambda = \lambda' - \lambda$, varies with the angle at which the scattered x-rays are observed.

The presence of a scattered wavelength λ' cannot be understood if the incident x-radiation is regarded as a classical electromagnetic wave. In the classical model the oscillating electric field vector in the incident wave of frequency ν acts on the free electrons in the scattering block and sets them oscillating at that same frequency. These oscillating electrons, like charges surging back and forth in a small radio transmitting antenna, radiate electromagnetic waves that again have this same frequency ν. Hence, in the wave picture the scattered wave should have the same frequency ν and the same wavelength λ as the incident wave.

Compton interpreted his experimental results by postulating that the incoming x-ray beam was not a wave of frequency ν but a collection of photons, each of energy $E = h\nu$, and that these photons collided with free electrons in the scattering block like a collision between billiard balls.* In this view, the "recoil" photons emerging from the target make up the scattered radiation. Since the incident photon transfers some of its energy to the electron with which it collides, the scattered photon must have a lower energy E'; it must therefore have a lower frequency $\nu' (= E'/h)$, which implies a larger wavelength $\lambda' (= c/\nu')$. This point of view accounts qualitatively for the wavelength shift, $\Delta\lambda = \lambda' - \lambda$. Notice that in the interaction the x-rays are regarded as particles, not as waves, and that, as distinguished from their behavior in the photoelectric process, the x-ray photons are scattered rather than absorbed. Let us now analyze a single photon–electron collision quantitatively.

---

* The Dutch physicist Peter Debye, perhaps influenced by a preliminary report of Compton's experimental results, published a virtually identical but totally independent analysis at about the same time.

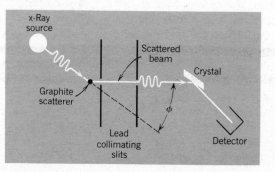

Figure 5-5. Compton's experimental arrangement. Monochromatic x-rays of wavelength λ fall on a graphite scatterer. The distribution of intensity with wavelength is measured for x-rays scattered at any selected angle ϕ. The scattered wavelengths are measured by observing Bragg reflections from a crystal. Their intensities are measured by a detector, such as an ionization chamber.

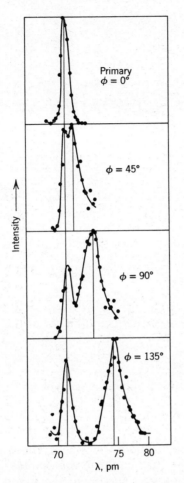

Intensity ⟶

Primary
$\phi = 0°$

$\phi = 45°$

$\phi = 90°$

$\phi = 135°$

70   75   80
λ, pm

Figure 5-6. Compton's experimental results. The vertical line on the left corresponds to λ, those on the right to λ'. Results are shown for four different scattering angles ϕ. Note the Compton wavelength shifts.

Figure 5-7 shows a collision between a photon and an electron, the electron assumed to be initially at rest and essentially free, that is, not bound to the atoms of the scatterer. Let us apply the laws of conservation of mass-energy and of linear momentum to this collision. We use relativistic expressions because the recoil speed v of the electrons may be near the speed of light.

Conservation of mass-energy yields

$$h\nu = h\nu' + (m - m_0)c^2,$$

in which $h\nu$ is the energy of the incident photon, $h\nu'$ is the energy of the scattered photon, and $(m - m_0)c^2$ is the kinetic energy acquired by the recoiling electron, initially at rest. With $m = m_0/\sqrt{1 - \beta^2}$, $\nu = c/\lambda$, and $\nu' = c/\lambda'$, this expression becomes

$$\frac{hc}{\lambda} = \frac{hc}{\lambda'} + m_0c^2 \left(\frac{1}{\sqrt{1 - v^2/c^2}} - 1 \right). \tag{5-7}$$

Now let us apply the law of conservation of linear momentum. The momentum of a photon is given by $p = E/c$, an expression derivable from relativity theory (see Example 5, Chapter 3) or from electromagnetic theory (see *Physics*. Part II, Sec. 42-2). Using $E = h\nu$, we obtain $p = h\nu/c = h/\lambda$. There are two components of momentum in the plane of the scattered photon and electron (see

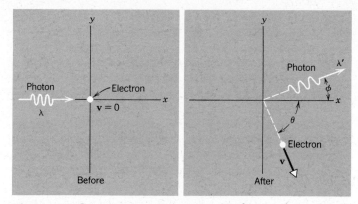

Figure 5-7. Compton's interpretation. A photon of wavelength λ falls on a free electron at rest. The photon is scattered at an angle φ with increased wavelength λ', while the electron moves off at angle θ with speed v.

Fig. 5-7); the conservation of the x component of linear momentum gives us

$$\frac{h}{\lambda} = \frac{h}{\lambda'} \cos \phi + \frac{m_0 v}{\sqrt{1 - v^2/c^2}} \cos \theta, \tag{5-8}$$

and the conservation of the y component gives us

$$0 = \frac{h}{\lambda'} \sin \phi - \frac{m_0 v}{\sqrt{1 - v^2/c^2}} \sin \theta. \tag{5-9}$$

Our immediate goal is to find $\Delta\lambda = \lambda' - \lambda$, the wavelength shift of the scattered photons, so that we may compare that expression with the experimental results of Fig. 5-6. Compton's experiment did not involve observations of the recoil electron in the scattering block. Of the five variables (λ, λ', v, ϕ, and θ) that appear in the three equations (5-7, 5-8, 5-9), we may eliminate two by combining the equations. We choose to eliminate v and θ, which deal only with the electron, thereby reducing the three equations to a single relation among the variables λ, λ', and ϕ. The result (see Problem 38) is

$$\Delta\lambda(= \lambda' - \lambda) = \frac{h}{m_0 c}(1 - \cos \phi) \quad \text{(the Compton shift).} \tag{5-10}$$

Notice that $\Delta\lambda$, *the Compton shift*, depends only on the scattering angle ϕ and *not* on the initial wavelength λ. The *change* in wavelength is independent of the incident wavelength. The constant $h/m_0 c$, called the *Compton wavelength* of the electron, has the value 2.43×10^{-12} m or 2.43 pm. Equation 5-10 predicts the experimentally observed Compton shifts of Fig. 5-6 within the experimental limits of error. The shift was found to be independent of the material of the scatterer and of the incident wavelength, as required by Compton's interpretation, and to be proportional to $(1 - \cos \phi)$. Note that $\Delta\lambda$, in the equation, varies from zero (for $\phi = 0$, corresponding to a "grazing" collision in Fig. 5-7, the incident photon being scarcely deflected) to $2h/m_0 c = 4.86$ pm (for $\phi = 180°$, corresponding to a "head-on" collision, the incident photon being reversed in direction). Subsequent experiments (by Compton and by other investigators) detected the recoil electron in the process, showed that it appeared simultaneously with the scattered x-ray, and confirmed quantitatively the predicted electron energy and direction of scattering.

It remains to explain the presence of the peak in Fig. 5-6 for which the photon wavelength does *not* change on scattering. We refer to this as the *unmodified* line, the Compton-shifted line being called the *modified* one. In our equations, we assumed that the electron with which the photon collides is free. Even when the electron is initially bound, this assumption can be justified as approximately correct if the kinetic energy acquired by the electron is much larger than its binding energy. However, if the electron is strongly bound to an atom or if the incident photon energy is small, there is a chance that the electron will not be ejected from the atom, in which case the collision can be regarded as taking place between the photon and the whole atom. The ionic core, to which the electron is bound in the scattering target, recoils as a whole during such a collision. In that case the mass M_0 of the atom is the effective mass and it must be substituted in Eq. 5-10 for the electron mass m_0. Since $M_0 \gg m_0$ ($M_0 \cong 22{,}000\,m_0$ for carbon, for instance), the Compton shift for collisions with tightly bound electrons is seen to be immeasurably small ($\sim 10^{-4}$ pm for the carbon atom), so that the scattered photon is observed to be unmodified in wavelength. Some photons then are scattered from electrons that are freed by the collision; these photons are modified in wavelength. Other photons are scattered from electrons that remain bound during the collision; these photons are not modified in wavelength.

The breadth of the peaks in Fig. 5-6 also can be simply accounted for. It is caused by the motion of the electrons, which were assumed to be initially at rest in our analysis. If we include in our calculation the components of the velocities of the atomic electrons in (or opposite to) the direction of the incident radiation, the observed line broadening can be deduced. Today, interestingly enough, the process is reversed in that measurement of the width and shape of the scattered Compton x-ray line are used to deduce the distribution in momentum of the electrons in the target. The Compton effect, in short, is used as a solid-state probe [6].

EXAMPLE 5.

A Compton Collision Analyzed. A collimated beam of x-rays of wavelength 120 pm falls on a target and the scattered radiation is observed at an angle ϕ of 90° to the incident beam. Calculate (a) the Compton wavelength shift $\Delta\lambda$, (b) the wavelength of the scattered radiation, (c) the energy E of the incident photon, (d) the energy E' of the scattered photon, (e) the kinetic energy K of the scattered electron, (f) the percentage of the energy of the incident photon lost in the collision, and (g) the angle θ that the scattered electron makes with the incident beam.

(a) The Compton wavelength shift is given from Eq. 5-10 as

$$\Delta\lambda = (h/m_0 c)(1 - \cos\phi)$$
$$= (2.43 \text{ pm})(1 - \cos 90°) = 2.43 \text{ pm},$$

in which 2.43 pm is also the Compton wavelength of the electron.

(b) The wavelength of the scattered radiation follows from

$$\lambda' = \lambda + \Delta\lambda$$
$$= 120 \text{ pm} + 2.43 \text{ pm}$$
$$= 122 \text{ pm}.$$

(c, d) The energies E of the incident photon and E' of the scattered photon are calculated exactly as in Example 3, the results being $E = 10.3$ keV and $E' = 10.1$ keV.

(e) The kinetic energy K imparted to the scattered electron is simply the difference between E and E', quantities whose values we have just calculated. However, because E and E' are so nearly equal, we do not get a very accurate value for K by simple subtraction. It is best to calculate K directly, from the conservation of energy principle. Thus

$$K = E - E' = \frac{hc}{\lambda} - \frac{hc}{\lambda'}$$

$$= \frac{hc(\lambda' - \lambda)}{\lambda\lambda'} = \frac{hc\,\Delta\lambda}{\lambda\lambda'}$$

$$= \frac{(4.14 \times 10^{-15} \text{ eV}\cdot\text{s})(3.00 \times 10^8 \text{ m/s})(2.43 \times 10^{-12} \text{ m})}{(120 \times 10^{-12} \text{ m})(122 \times 10^{-12} \text{ m})}$$

$$= 206 \text{ eV}.$$

(*f*) The percentage of the energy lost by the incident photon during the scattering process is given by

$$p = 100 \frac{E - E'}{E} = \frac{100\,K}{E}$$

$$= \frac{(100)(206\ \text{eV})}{1.03 \times 10^4\ \text{eV}} = 2.0\%.$$

(*g*) We can calculate the angle θ from Eq. 5-9, which is a statement of the conservation of linear momentum in the *y* direction. Solving for $\sin\theta$ leads to

$$\sin\theta = \frac{h\sin\phi\sqrt{1 - v^2/c^2}}{\lambda' m_0 v}.$$

We see that we need to know v, the electron velocity, before we can proceed. The kinetic energy of the electron $(= 206\ \text{eV})$ is so small that we can safely calculate v using the classical formula, $K = (1/2)m_0 v^2$. Thus

$$v = \sqrt{\frac{2K}{m_0}} = \sqrt{\frac{2(206\ \text{eV})(1.60 \times 10^{-19}\ \text{J/eV})}{9.11 \times 10^{-31}\ \text{kg}}}$$

$$= 8.51 \times 10^6\ \text{m/s}.$$

Substituting into the above equation for $\sin\theta$ yields

$\sin\theta =$

$$\frac{(6.63 \times 10^{-34}\ \text{J·s})(\sin 90°)\sqrt{1 - (8.51 \times 10^6/3.00 \times 10^8)^2}}{(122 \times 10^{-12}\ \text{m})(9.11 \times 10^{-31}\ \text{kg})(8.51 \times 10^6\ \text{m/s})}$$

$$= 0.701.$$

Note that $(v/c)^2 \ll 1$, so the quantity under the square root sign in this formula can be safely replaced by unity. This is totally equivalent to ignoring relativistic considerations in deriving Eq. 5-9, a procedure justified in this case (but not in other cases) because the kinetic energy of the electron is so small. We have then, for θ,

$$\theta = \sin^{-1} 0.701 = 44.5°.$$

EXAMPLE 6.

Head-on Compton Collisions at High Energies. A photon has a head-on Compton collision with a free electron, being scattered backwards through an angle ϕ of 180°. Find the kinetic energy K of the scattered electron if the incident photon energy is 1.50 MeV.

If the photon is scattered backwards, the electron must move forwards. This follows from Eq. 5-9, in which $\phi = 180°$ requires that $\theta = 0$. Such forward-moving electrons are sometimes called "knock-on" electrons.

In Example 5(*e*) we showed that the kinetic energy K is given by

$$K = \frac{hc\,\Delta\lambda}{\lambda\lambda'} = \frac{hc\,\Delta\lambda}{\lambda(\lambda + \Delta\lambda)}.$$

For $\phi = 180°$, the Compton shift (see Eq. 5-10) $\Delta\lambda$ is equal to $2h/m_0c$. Let us substitute this quantity and also the relation $\lambda = hc/E$ into the equation given above for K. This yields, after a little algebraic reduction,

$$K = \frac{2E^2}{2E + m_0 c^2}.$$

The rest energy of the electron $(m_0 c^2)$ is equal to 0.511 MeV. Substituting this and the value of E into the above equation yields

$$K = \frac{2(1.50\ \text{MeV})^2}{2 \times 1.50\ \text{MeV} + 0.511\ \text{MeV}} = 1.28\ \text{MeV}.$$

Table 5-1
HEAD-ON COMPTON COLLISIONS[a,b]

Radiation Source	Incident Photon Energy, E	Incident Wavelength, λ	Scattered Wavelength, λ'	Fractional Energy Loss, K/E
Cosmic fireball	0.0012 eV	1 mm	1.000000005 mm	0.00000048
Sun	2.5 eV	500 nm	500.0049 nm	0.0000097
X-Rays (dental)	10 keV	124 pm	129 pm	0.038
X-Rays (industrial)	100 keV	12.4 pm	17.3 pm	0.28
Nuclear gamma rays	1.5 MeV	0.85 pm	5.7 pm	0.85
Synchrotron x-rays	200 MeV	6.2 fm	4860 fm	0.9987
SLAC x-rays	15 GeV	0.083 fm	4850 fm	0.999983

[a] Compton events are far from equally probable over this entire range of energies; see Fig. 5-10.

[b] The last two lines represent the maximum energies of x-rays that may be generated when electrons accelerated in a typical electron synchrotron or in the Stanford Linear Accelerator (SLAC) are allowed to strike a target.

The incident photon, then, gives up 1.28 MeV, or 85 percent of its initial energy of 1.50 MeV, to the knocked-on electron.

Table 5-1 shows the results of similar calculations for head-on Compton collisions over a range of incident photon energies. Note that as the incident photon energy increases, the fraction of its energy imparted to the knock-on electron increases rapidly. For photons with energies above 200 MeV or so, virtually all of the incident photon energy is so transferred. Analysis of the final algebraic expression for K above shows that if $E \gg m_0 c^2$, then $K \to E$, just as Table 5-1 shows.

It is also interesting to inspect the values of the scattered wavelength λ' in Table 5-1. This quantity is found,

in each case, by adding the same (constant) Compton shift (= 4.85 pm) to the incident wavelength λ. When λ is large, as for visible light, we see that λ and λ' are virtually equal and the scattered electron picks up very little energy indeed. When λ is small, on the other hand, as for x-rays generated by fast electrons emerging from the Stanford Linear Accelerator, the scattered wavelength λ' (= 4850 fm = 4.85 pm) is essentially equal to the Compton wavelength shift itself. At these high energies a head-on Compton collision between an energetic photon and a resting electron truly resembles a head-on collision between two billiard balls, in which both energy and momentum are totally transfered from the ball that was initially moving to the ball that was initially at rest.

We saw earlier in the cavity radiation discussion, and also in the photoelectric effect discussion, that Planck's constant h is a measure of the granularity or discreteness in energy. Classical physics corresponds to $h = 0$, since in this case all energy spectra would be continuous. Notice that here in the Compton effect, Planck's constant h again plays a central role. If h were equal to zero there would be no Compton effect, for then $\Delta\lambda = 0$ and classical theory would be valid. The quantity h is the central constant of quantum physics.

The fact that h is not zero means that classical physics is not valid in general; but the fact that h is very small often makes quantum effects difficult to detect. For example, the quantity $h/m_0 c$ in the Compton formula has the value 2.43 pm when the scatterer is a free electron. But if m_0 is the mass of an atom, not to mention bulk matter, $h/m_0 c$ is already at least 2000 times smaller and virtually undetectable. Hence, as $m_0 \to \infty$ the quantum scattering result merges with the classical result, the scattered radiation then having the same frequency as the incident radiation. It is in the atomic and subatomic domain, where m_0 is small, that the classical results fail.

Table 5-1 illustrates the same point. If the incident radiation is in the visible, microwave, or radio part of the electromagnetic spectrum, then λ would be very large compared to $\Delta\lambda$, the Compton shift. The scattered radiation would be measured to have the same wavelength and frequency as the incident radiation within experimental limits. It is no accident that the Compton effect was discovered in the x-ray region, for the quantum nature of radiation reveals itself to us at short wavelengths. As $\lambda \to \infty$ the quantum results merge with the classical. It is in the short-wavelength region, where λ is small, that the classical results fail. Recall the "ultraviolet catastrophe" of classical physics, wherein classical predictions of black-body radiation agreed with experiment at long wavelengths but diverged radically a short wavelengths. From the point of view of the energy quantum $h\nu$, this is best interpreted as due to the smallness of h. For at long wavelengths, the frequency ν is small and the granularity in energy, $h\nu$, is so small as to be virtually indistinguishable from a continuum. But at short wavelengths, where ν is large, $h\nu$ is no longer too small to be detected and quantum effects abound.*

---

* Gamow [7] has written a delightful fantasy about a world in which the fundamental constants c (the speed of light), G (the gravitational constant), and h (the Planck constant) have values large enough so that their effects are directly apparent in daily life.

The discovery of the Compton effect and its explanation in quantum terms rapidly led the way to the general acceptance of the light quantum hypothesis, almost two decades after it had first been advanced by Einstein. This came about because Compton's explanation of his effect introduced a new feature, namely, the notion that light quanta have momentum as well as energy. The photoelectric effect, it will be recalled, dealt only with the energy. Einstein, however, became convinced as early as 1917, six years before Compton's experiments, that light quanta must possess momentum as well as energy, both being equally important characteristics of a particle. In his paper of that year on the quantum theory of radiation, in which, among other matters, he derived Planck's radiation law (see Supplementary Topic E), Einstein wrote:

> If a beam of radiation has the effect that a molecule on which it falls absorbs . . . an amount of energy $h\nu$ in the form of radiation, . . . then the momentum $h\nu/c$ is always transferred to the molecule. . . .

Six years later Compton himself wrote, in his paper on "A Quantum Theory of the Scattering of X-rays by Light Elements":

> The present theory depends essentially upon the assumption that each electron which is effective in the scattering scatters a complete quantum. It involves also the hypothesis that the quanta of radiation are received from definite directions and are scattered in definite directions. The experimental support of the theory indicates very convincingly that a radiation quantum carries with it directed momentum as well as energy.

The need for a quantum, or particle, interpretation of processes dealing with the interaction between radiation and matter seemed clear, but at the same time a wave theory of radiation seemed necessary to understand interference and diffraction phenomena. Clearly the idea that radiation is neither purely a wave phenomenon nor merely a stream of particles has to be considered seriously. But whatever radiation is, we have seen that it behaves wavelike under some circumstances and particlelike under other circumstances. Indeed, the paradoxical situation is revealed most forcefully in Compton's very experiment where (1) a crystal spectrometer is used to measure x-ray wavelengths, the measurements being interpreted by a wave theory of diffraction, and (2) the scattering affects the wavelength in a way that can be understood only by treating the x-rays as particles. It is in the very expressions $E = h\nu$ and $p = h/\lambda$ that the waves attributes (ν and λ) and the particle attributes (E and p) are combined.

5-5 PAIR PRODUCTION

In addition to the photoelectric and Compton effects, there is another process whereby photons lose their energy in interactions with matter, namely, the process of *pair production*. Pair production is also an excellent example of the conversion of radiant energy into rest mass energy as well as into kinetic energy. In this process a high-energy photon loses all of its energy in an encounter with a nucleus (see Fig. 5-8), creating an electron and a positron (the pair) and endowing them with kinetic energies K_- and K_+. (A positron, discovered first by Anderson in the cosmic radiation, has the same mass and magnitude of charge as an elec-

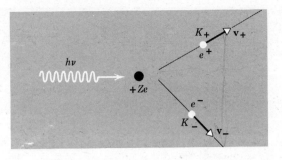

Figure 5-8. Showing the pair-production process.

tron, but its charge is positive rather than negative.) The energy taken by the recoil of the massive nucleus is negligible, so that the relativistic energy balance can be written

$$h\nu = (m_0 c^2 + K_-) + (m_0 c^2 + K_+) \qquad \text{(pair production)}. \qquad (5\text{-}11)$$

Here $2m_0 c^2$ is the (rest) energy needed to create the positron–electron pair, the positron and electron rest masses being equal. Although K_- and K_+ are approximately equal, the positron has a somewhat larger kinetic energy than the electron. This is because of the Coulomb interaction of the pair with the (positively charged) nucleus, which leads to an acceleration of the positron away from the nucleus and a deceleration of the electron.

In analyzing this process here we ignore the details of the interaction itself, considering only the situation before and after the interaction. Our guiding principles are the conservation of energy, conservation of momentum, and conservation of charge. From these (see Problem 47) it follows that a photon cannot simply disappear in empty space, creating a pair as it vanishes, but that the presence of a massive particle is necessary to conserve both energy and momentum in the process. Charge is automatically conserved, the photon having no charge and the created pair of particles having no net charge. From Eq. 5-11 we see that the minimum (threshold) energy needed by a photon to create a pair is $2m_0 c^2$ or 1.022 MeV; this corresponds to a wavelength of 1.22 pm. If the wavelength is shorter than this, corresponding to an energy greater than the threshold value, the photon endows the pair with kinetic energy as well as rest energy. The phenomenon is a high-energy one, the photons being in the very short x-ray or the gamma-ray region of the electromagnetic spectrum (these regions overlap; see Fig. 5-4). Other particle pairs, such as proton and antiproton (a particle with the same mass and magnitude of charge as the proton, but with negative rather than positive charge), can be produced as well if the initiating photon has sufficient energy. Because the electron and positron have the smallest rest mass of known particles the threshold energy of their production is the smallest. Experiment verifies the quantum picture of the pair-production process. There is no satisfactory explanation whatever of this phenomenon in classical theory.

Electron–positron pairs are produced in nature by energetic photons in the cosmic radiation and in the laboratory by photons generated when accelerated particle beams fall on a target. Figure 5-9a shows, at site I, a positron–electron pair produced in a bubble chamber filled with liquid hydrogen. A beam of charged particles, generated in the 5.7-GeV proton synchrotron at the Lawrence Radiation Laboratory of the University of California at Berkeley, enters the bubble chamber from the left. The beam contains some gamma rays, which, being uncharged, leave no tracks in the chamber. At site I one of these incoming gamma-ray photons, interacting with the nucleus of one of the hydrogen atoms in the cham-

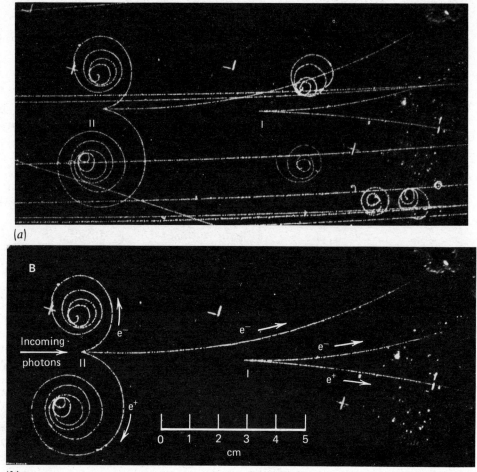

(a)

(b)

Figure 5-9. (a) A bubble-chamber photo showing a pair-production event at site I and a triplet event at site II. A beam of charged particles and gamma rays enters from the left. A uniform 1.17-T magnetic field points out of the plane of the figure. (b) The same as (a) but with extraneous tracks and markers removed by hand. Courtesy Lawrence Berkeley Radiation Laboratory, University of California; see Ref. 8.

ber, produces a pair. The chamber is immersed in a uniform magnetic field **B**, which causes the tracks to curve; the direction of **B** is perpendicularly out of the plane of the figure.

Figure 5-9b shows the essential features of Fig. 5-9a; verify that the upward-curving track in the figure is the (negatively charged) electron member of the pair, the downward-curving track being the (positively charged) positron. It is clear also that, just as we stated earlier, the positron member of the pair, having the larger radius of curvature, has the larger momentum and thus the larger kinetic energy.

At site II in Fig. 5-9 we see a three-pronged event. Here an incoming energetic photon, again leaving no track, interacts this time with the atomic electron (not with the atomic nucleus as at site I) of a hydrogen atom. The photon produces a positron–electron pair and, at the same time, "knocks on" the light atomic electron, imparting, in fact, a large fraction of the available energy to it. We know this because the members of the positron–electron pair at site II, being curled up by the magnetic field into fairly tight spirals, must have considerably less energy than the ejected atomic electron, which is curved only slightly.

EXAMPLE 7.

Pair Production—An Analysis. What is the energy of the incoming gamma-ray photon that precipitated the pair production event at site I in Fig. 5-9? The magnitude of the magnetic field in the bubble chamber is 1.17 T; the radius of curvature of the electron track is 0.16 m; and that of the positron track is 0.18 m.

The momentum of the electron is given (see *Physics*, Part II, Sec. 33-6) by

$$p = eBr$$
$$= (1.60 \times 10^{-19} \text{ C})(1.17 \text{ T})(0.16 \text{ m})$$
$$= 3.0 \times 10^{-20} \text{ kg} \cdot \text{m/s}.$$

The quantity pc then has the value

$$pc = \frac{(3.0 \times 10^{-20} \text{ kg} \cdot \text{m/s})(3.00 \times 10^8 \text{ m/s})}{(1.60 \times 10^{-13} \text{ J/MeV})}$$
$$= 56 \text{ MeV},$$

and the total relativistic energy E_- of the electron ($= K_- + m_0 c^2$) is given (see Eq. 3-13*b*) by

$$E_- = \sqrt{(pc)^2 + (m_0 c^2)^2}$$
$$= \sqrt{(56 \text{ MeV})^2 + (0.51 \text{ MeV})^2}$$
$$\cong 56 \text{ MeV}.$$

In the same way, the total relativistic energy E_+ of the positron may be calculated to be 63 MeV. From Eq. 5-11, then, the gamma-ray photon energy is

$$h\nu = (m_0 c^2 + K_-) + (m_0 c^2 + K_+)$$
$$= E_- + E_+$$
$$= 56 \text{ MeV} + 63 \text{ MeV} \cong 120 \text{ MeV}.$$

A gamma ray of this energy could easily be produced in the particle beam emerging from the accelerator by the decay in flight of moving neutral pions (see Problem 24).

So far we have looked at three ways in which photons can be absorbed when they interact with matter—the photoelectric effect, the Compton effect, and pair production. There are other possibilities. Photons falling on bulk matter, for example, can be transformed directly into thermal energy, a phenomenon familiar to sunbathers. The three processes that we have chosen to examine, however, dominate the absorption process for photon energies above the visible range.

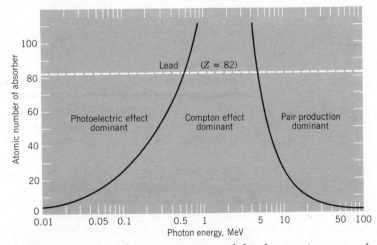

Figure 5-10. The relative importance of the three major types of photon interaction for various photon energies and for various absorbers. For points along the left branch, the photoelectric and the Compton effects occur with equal probability. For points along the right branch, the Compton effect and pair production do so. A horizontal line for lead ($Z = 82$) is shown; see Ref. 9.

We have said nothing yet about the relative probability that, for a given absorbing material and a given incident photon energy, each of the three absorption processes described above will actually occur. Quantum physics, in addition to providing our present insight into the individual nature of these processes (see Eqs. 5-5, 5-10, and 5-11), also allows us to calculate their probabilities of occurrence, the calculations being amply confirmed by experiment. Figure 5-10, for example, gives important qualitative information about the relative importance of the three processes at various photon energies and for various absorbers. We see that, for any absorber, the Compton effect always dominates the middle range of energies, with the photoelectric effect dominating at low energies and pair production at high energies. A horizontal line for lead ($Z = 82$) is shown by way of example; from it we see that the Compton effect begins to dominate over the photoelectric effect when the incident photon energy reaches about 500 keV and that pair production takes over at about 4.8 MeV. As we have noted earlier, pair production cannot occur at all for photon energies below $2m_0c^2 (= 1.02$ MeV).

5-6 *PHOTONS GENERATED BY ACCELERATING CHARGES*

In the rest of this chapter we turn our attention to ways in which photons can be produced. Photon production by the sun and by heated lamp filaments immediately come to mind. We restrict ourselves here, however, to photons generated explicitly by accelerating charges and whose wavelengths are by and large smaller than—or whose quantum energies are correspondingly greater than—those of visible light. We deal first with x-rays, as commonly generated in an x-ray tube.

X-Rays.   X-Rays, so named by their discoverer Wilhelm Roentgen because their nature was then (1895) unknown, are radiations in the electromagnetic spectrum whose wavelengths lie in the approximate range of 5 to 1000 pm. They show the typical transverse wave behavior of polarization, interference, and diffraction that is found in light and all other electromagnetic radiation. Figure 5-11*a* shows an x-ray tube, in which a beam of electrons, accelerated by a potential difference of several kilovolts, falls on a solid target from which the x-rays radiate. According to classical physics, the deceleration of the electrons when they are brought to rest in the target results in the emission of a continuous spectrum of electromagnetic radiation.

Figure 5-11*b* shows, for four different values of the incident electron energy, how the x-rays emerging from a tungsten target are distributed in wavelength. (In addition to the continuous x-ray spectrum, x-ray lines characteristic of the target material are emitted. We discuss these later.) The most notable feature of these curves is that, for a given electron energy, there exists a well-defined minimum wavelength λ_{min}; for 40.0-keV electrons, for instance, λ_{min} is 31.0 pm. Although the overall shape of the continuous x-ray distribution spectrum depends on the choice of target material as well as on the electron accelerating potential V, the value of λ_{min} depends only on V, being the same for all target materials. Classical electromagnetic theory cannot account for this fact, there being no reason why waves whose length is less than a certain critical value should not emerge from the target.

A ready explanation appears, however, if we regard the x-rays as photons. Figure 5-12 shows the elementary process that, on the photon view, is responsible for the continuous x-ray spectrum of Fig. 5-11*b*. An electron of initial kinetic energy K is decelerated during an encounter with a heavy target nucleus, the energy it loses appearing in the form of radiation as an x-ray photon. The electron

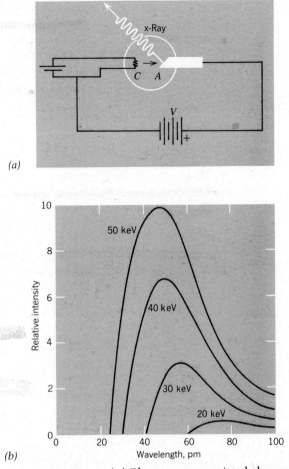

(a)

(b)

Figure 5-11. (*a*) Electrons are emitted therm-ionically from the heated cathode *C* and are accelerated toward the anode target *A* by the applied potential difference *V*. (*b*) Plots of intensity versus wavelength for the continuous x-ray spectrum emitted from a tungsten target for four different values of the energy of the incident electrons.

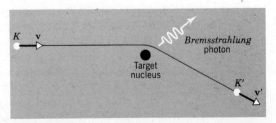

Figure 5-12. The *bremsstrahlung* process responsible for the production of x-rays in the continuous spectrum.

interacts with the charged nucleus via the Coulomb field, transferring momentum to the nucleus. The accompanying deceleration of the electron leads to photon emission. The target nucleus is so massive that the energy it acquires during the collision can safely be neglected. If K' is the kinetic energy of the electron after the encounter, then the energy of the photon is

$$h\nu = K - K',$$

and the photon wavelength follows from

$$\frac{hc}{\lambda} = K - K'. \tag{5-12}$$

Electrons in the incident beam can lose different amounts of energy in such encounters, and typically a single electron will be brought to rest only after many such encounters. The x-rays thus produced make up the continuous spectrum of Fig. 5-11b and are discrete photons whose wavelengths vary from λ_{min} to $\lambda \to \infty$, corresponding to the different energy losses in the individual encounters. The shortest-wavelength photon would be emitted when an electron loses *all* its kinetic energy in *one* deceleration process; here $K' = 0$ so that $K = hc/\lambda_{min}$. Since K equals eV, the energy acquired by the electron in being accelerated through the potential difference V applied to the x-ray tube, we have

$$eV = \frac{hc}{\lambda_{min}}$$

or

$$\lambda_{min} = \frac{hc}{eV}. \tag{5-13}$$

Thus the minimum wavelength cutoff represents the complete conversion of the electron's kinetic energy to x-ray radiation. Equation 5-13 shows clearly that if $h \to 0$, then $\lambda_{min} \to 0$, which is the prediction of classical theory. This shows that the very existence of a minimum wavelength is a quantum phenomenon.

The continuous x-ray radiation of Fig. 5-11b is often called *bremsstrahlung*, from the German *brems* (= braking, that is, decelerating) + *strahlung* (= radiation). The *bremsstrahlung* process occurs not only in x-ray tubes but wherever fast electrons collide with matter, as in cosmic rays, in the van Allen radiation belts that surround the earth, and in the stopping of electrons emerging from accelerators. Many objects in our own galaxy, as well as in other galaxies, emit *bremsstrahlung* x-rays in abundance, a fact that has become especially clear since the launching of the first orbiting x-ray satellite in 1970.

Synchrotron Radiation   A particle does not need to change its speed—as by coming to rest in an x-ray tube—in order to accelerate. Electrons moving at constant speed in a circular path are accelerating centripetally, and photons are generated under these circumstances as well. Such radiation is called *synchrotron radiation* [10,11], because it was first observed in 1947 in a synchrotron designed to accelerate electrons to relativistic speeds. Accelerators are now built specifically to serve as sources of synchrotron radiation for a wide variety of experiments in diverse fields. The National Synchrotron Light Source at the Brookhaven National Laboratory is one such facility; it employs circulating electrons at energies as high as 2.5 GeV.

Figure 5-13 shows the intensity pattern of the synchrotron radiation emerging from such an accelerator. The photons "squirt forward," tangent to the orbit of the circulating electrons, in a narrow cone whose half-angle, expressed in radian

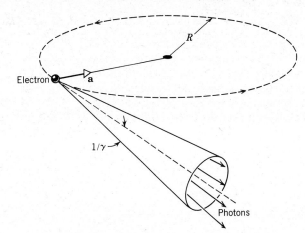

Figure 5-13. Electrons, circulating at relativistic speeds in an orbit of radius R, are accelerated radially inward and emit synchrotron radiation in a sharply defined forward-pointing cone of half-angle $1/\gamma$.

measure, is equal to $1/\gamma$, where γ is the Lorentz factor (see Eq. 2-10) of the circulating electrons. For a machine accelerating electrons to 1.0 GeV, for example, the Lorentz factor γ can be shown to be ~ 1960. The half-angle of the synchrotron radiation beam is thus 1/1960 radians or ~ 1.8 minutes of arc. The emerging photon beam, in part by virtue of its sharpness, can be exceedingly intense, very much more so than the *bremsstrahlung* radiation generated in x-ray tubes such as that of Fig. 5-11*a*. Synchrotron sources can cover a range of photon energies and wavelengths extending from the infrared through the conventional x-ray region. Much of the radiation that reaches us from distant galaxies in the radiofrequency region turns out to be synchrotron radiation, generated by charged particles spiraling in the very weak but enormously extended magnetic fields that are present in these galaxies.

EXAMPLE 8.

Evaluating the Planck Constant (Again!). The minimum x-ray wavelength produced by 40.0-keV electrons is 31.1 pm. Derive a value for the Planck constant h from these data.

From Eq. 5-13 we have

$$h = \frac{eV\lambda_{min}}{c}$$

$$= \frac{(1.60 \times 10^{-19}\ \text{C})(4.00 \times 10^4\ \text{V})(3.11 \times 10^{-11}\ \text{m})}{(3.00 \times 10^8\ \text{m/s})}$$

$$= 6.64 \times 10^{-34}\ \text{J} \cdot \text{s}.$$

This agrees well with the value deduced from Planck's radiation law, from the photoelectric effect, and from the Compton effect.

For many years the minimum-wavelength method was the most precise procedure for measuring h/e, the ratio of two important basic constants. More recently, however, an entirely new method (based on the Josephson effect) has been introduced that permits a measurement of this ratio of such accuracy that the limiting-wavelength method is no longer used. This is a good illustration of the ubiquity of the Planck constant h; it simply turns up everywhere in experiments carried out at the atomic or electronic level. The ratio h/e is combined with many other measured combinations of physical constants, the assembly of data being analyzed by elaborate statistical methods to find the "best" value for the various physical constants. The best values change (but usually only within the *a priori* estimates of accuracy) and become increasingly precise as new experimental data and higher-precision methods are used.

EXAMPLE 9.

An Energetic Photon! Electrons can be accelerated to 18 GeV in the 2-mile-long linear accelerator at the Stanford Linear Accelerator Center (SLAC). If these electrons are brought to rest in a target, what are (*a*) the wavelength and (*b*) the frequency associated with the maximum-energy photon that can be generated by the *bremsstrahlung* process?

(*a*) From Eq. 5-13, we have

$$\lambda_{min} = \frac{hc}{eV}$$

$$= \frac{(6.63 \times 10^{-34} \text{ J} \cdot \text{s})(3.00 \times 10^{8} \text{ m/s})}{(1.60 \times 10^{-19} \text{ C})(18 \times 10^{9} \text{ V})}$$

$$= 6.9 \times 10^{-17} \text{ m} = 0.069 \text{ fm}.$$

For comparison, the effective radius of the proton is about 0.8 fm, more than 10 times greater. Such energetic photons can thus be used as probes to explore the structure of particles such as the proton or the neutron.

(*b*) The frequency is given simply by

$$\nu = \frac{c}{\lambda} = \frac{3.00 \times 10^{8} \text{ m/s}}{6.9 \times 10^{-17} \text{ m}} = 4.3 \times 10^{24} \text{ Hz}.$$

5-7 PHOTON PRODUCTION BY PAIR ANNIHILATION

Closely related to pair production is the inverse process called *pair annihilation*. An electron and a positron, which are essentially at rest near one another, unite and are annihilated. Matter disappears, and in its place we get radiant energy. Since the initial momentum is zero and momentum must be conserved in the process, we cannot have only one photon created, because a single photon cannot have zero momentum. The most probable process is the creation of two photons moving with equal momenta in opposite directions. Less probable, but possible, is the creation of three photons.

In the two-photon process (see Fig. 5-14), conservation of momentum requires that the photons move in opposite directions and that each has a momentum whose magnitude is $h\nu/c$. The energies of the two photons (= $h\nu$) are thus equal, as we expect from the symmetry of the situation. Conservation of mass-energy then requires that

$$m_0 c^2 + m_0 c^2 = h\nu + h\nu.$$

Hence, $h\nu = m_0 c^2 = 0.511$ MeV, corresponding to a photon wavelength of 2.43 pm. If the electron–positron pair was not initially at rest, as we have tacitly assumed, then in general the two photons would not move in opposite directions and their energies and wavelengths would not be equal and would have values that differ from those given above.

Positrons are emitted spontaneously by many radioactive nuclei and may also be generated in pair-production processes. Whatever its source, upon passing through matter a positron loses energy in successive collisions until it combines with an electron to form a bound system called *positronium* [12]. The positronium "atom" is short-lived, decaying into photons within about 100 ps of its

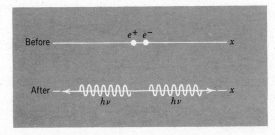

Figure 5-14. The annihilation of an electron-positron pair produces two photons.

formation. The electron and the positron presumably move about their common center of mass in a kind of "death dance" before mutual annihilation.

Analysis of the annihilation radiation emerging from a solid permits us to measure the distribution in momentum of the electrons in the solid. Recall (Section 5-4) that the Compton effect can also be used in a similar way. Both effects, then, are the basis of useful solid-state probes, the emerging photons serving as messengers to convey, by their directions and their wavelengths, information about their points of origin within the solid.

Annihilation radiation is also used as a diagnostic tool in medicine in a recent development called positron emission tomography (PET). The patient is injected with a solution containing a radioactive isotope that emits positrons. By studying the annihilation radiation that emerges from the patient's body it is possible, using elaborate detection, computer analysis, and display techniques, to follow, pictorially, the course of certain biochemical reactions in the body.

EXAMPLE 10.

Annihilation in Flight. Figure 5-15 shows the annihilation in flight of an electron–positron pair moving at speed $v(= 0.10c)$ along the x axis of a reference frame. What wavelengths and photon energies do observers stationed at points 1 and 2 measure for the annihilation photons?

The pair has initial energy $2mc^2$, rather than merely the rest energy $2m_0c^2$, so conservation of energy in the annihilation process gives us

$$2mc^2 = p_1 c + p_2 c.$$

Also, the initial momentum of the pair is now $2mv$, rather than zero as for the case in which the pair is at rest. Conservation of momentum now gives us

$$2mv = p_1 - p_2.$$

Let us combine these two expressions. We multiply the second by c and add it to the first, obtaining

$$p_1 = m(c + v) = \frac{m_0(c + v)}{\sqrt{1 - (v/c)^2}} = m_0 c \sqrt{\frac{1 + \beta}{1 - \beta}}.$$

But $p = h/\lambda$, so that

$$\lambda_1 = \frac{h}{p_1} = \frac{h}{m_0 c} \sqrt{\frac{1 - \beta}{1 + \beta}} = \lambda_0 \sqrt{\frac{1 - \beta}{1 + \beta}},$$

in which $\beta(= v/c)$ is the familiar speed parameter and λ_0 is the wavelength that would be measured in the rest frame of the moving pair. We can verify this last statement by putting $\beta = 0$ in the above expression for λ_1. In a similar way, by subtracting the second equation from the first, we find that

$$\lambda_2 = \lambda_0 \sqrt{\frac{1 + \beta}{1 - \beta}}.$$

The energy of photon 1 is given by

$$E_1 = \frac{hc}{\lambda_1} = \frac{hc}{\lambda_0} \sqrt{\frac{1 + \beta}{1 - \beta}} = E_0 \sqrt{\frac{1 + \beta}{1 - \beta}},$$

in which E_0 is the energy in the rest frame of the pair, as we may again verify by putting $\beta = 0$ in the expression for E_1. Similarly, we find for E_2,

$$E_2 = E_0 \sqrt{\frac{1 - \beta}{1 + \beta}}.$$

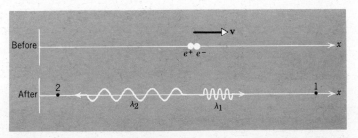

Figure 5-15. The annihilation of an electron-positron pair in flight. The photons no longer have the same wavelengths or energies.

Table 5-2
ANNIHILATION IN FLIGHT

	Rest Frame	Moving Frame
λ_1	2.43 pm	2.19 pm
λ_2	2.43 pm	2.68 pm
E_1	511 keV	565 keV
E_2	511 keV	462 keV

Table 5-2 shows the numerical values of all of these derived quantities for the case of $\beta = 0.10$. We see that the photons do *not* have the same wavelengths but are Doppler-shifted from the wavelength λ_0 they have in the rest frame of the source (the positronium atom). An observer at point 1 in Fig. 5-15 will see the source moving toward him. He will receive photon 1, whose wavelength is *smaller* than the rest wavelength and whose energy is correspondingly *greater* than the rest energy. An observer at point 2 will see the source moving away from her and will receive photon 2, whose wavelength is *larger* than the rest wavelength and whose photon energy is correspondingly *smaller*. Indeed, this example constitutes a derivation of the longitudinal Doppler effect, alternative to the one presented in Section 2-7.

In this chapter we have presented evidence for the particle nature of radiation. This evidence contradicts the classical picture of radiation as exhibiting only a wave nature. We have found that, as distinguished from its wavelike nature when it propagates, radiation is particlelike in its interaction with matter. We have avoided the details of the interaction between matter and radiation, although a detailed theory (quantum electrodynamics) does now exist for all the processes we have discussed (pair production, Compton scattering, photoelectric effect, and so on). Such a theory explains the energy dependence of the probabilities of the various processes, their dependence on the atomic number of target materials, the angular distribution of scattered radiation, and so forth. Instead, we have looked at the situation before and after the interaction, using the conservation principles of momentum and mass-energy to extract the principal features of the processes; see Table 5-3 for a summary. The breakdown of classical physics and the need for a new theory have emerged clearly. In the next chapter we continue to look at phenomena that help suggest how such a theory might be formulated.

questions

1. In his photoelectric experiments, Millikan went to great lengths to make sure that the spectral lines emitted by the light source that he used were "clean," that is, were well defined in wavelength. In particular, he made sure that they had no weak wavelength components at slightly shorter wavelengths. He did not worry so much about weak wavelength components at slightly larger wavelengths. Why this differential concern?

2. In Fig. 5-2, why doesn't the photoelectric current rise vertically to its saturation value when the applied potential difference slightly exceeds the stopping potential?

3. Sodium, like other alkali metals, is notoriously chemically active and very subject to surface contamination. Why did Millikan choose such materials for his photoelectric experiments instead of, say, aluminum or copper?

4. Why is it that even if the incident radiation is monochromatic the photoelectrons are emitted with a spread of energies?

5. Why are photoelectric measurements sensitive to the nature of the surface of the emitter?

6. Explain in your own words why the existence of a cutoff frequency cannot be explained in terms of a wave theory of light.

7. We often use the device of letting $h \to 0$ to obtain a classical equation from a corresponding quantum equation. Can we do this in the case of Einstein's photoelectric equation (Eq. 5-5)? Discuss.

8. In Einstein's photoelectric equation (Eq. 5-5), should we use the classical or the relativistic expression for K, the kinetic energy of the ejected electron?

9. In the photoelectric experiments the photoelectric current is proportional to the light intensity. Show that this is consistent with both the photon and the wave picture of light.

Table 5-3
THE PRODUCTION, SCATTERING, AND ABSORPTION OF PHOTONS—A SUMMARY

	Photoelectric Effect	Compton Effect	Pair Production	x-Ray Production	Pair Annihilation
Process described	Photon absorbed and electron released from target.	Photon scattered and electron recoils in target.	Photon absorbed and electron–positron pair created.	Electron deflected in target and photon created.	Electron and positron combine and pair of photons created.
Principal relation	$K_{max} = h\nu - w_e$.	$\Delta\lambda = \dfrac{h}{m_0 c}(1 - \cos\phi)$.	$h\nu = 2m_0 c^2 + (K_+ + K_-)$.	$h\nu = K - K'$.	$(m_0 c^2 + K_+) + (m_0 c^2 + K_-) = h\nu_1 + h\nu_2$.
Presence of other particles	Electron must be bound to other particle—as an atom.	Electron can be free.	Photon absorbed in encounter with heavy target nucleus.	Electron deflected in encounter with heavy target nucleus.	Free electron and free positron, but two or more photons created.
Effective energy and wavelength region	$h\nu > w_e$; energy a few electron volts and wavelength chiefly in ultraviolet.	Any energy, in principle. Most effective near 0.1 MeV and in long x-ray region.	Minimum $h\nu$ is $2m_0 c^2 = 1.02$ MeV. Gamma-ray region or short x-ray region.	$\lambda_{min} = hc/eV$ to $\lambda = \infty$.	Particles can be at rest. Minimum energy of photons produced 0.511 MeV and maximum wavelength 2.43 pm.
Relative probability	Less likely the higher the energy and frequency are beyond $h\nu_0 = w_e$. Goes to zero essentially at 1 MeV.	Less likely the higher the energy and frequency, but more likely than photoeffect beyond about 0.5 MeV.	Probability increases as energy increases beyond minimum. More effective than Compton effect beyond about 5 MeV.	Most effective near 1.5 to 2 times λ_{min}.	Two-photon process more probable than three-photon process.

10. Do the results of Millikan's photoelectric experiments (1915) invalidate the results of Young's interference experiments (1802)? Does the interpretation of the results of one experiment invalidate the interpretation of the results of the other?

11. A light source is a distance R from a photoelectric emitter. Would you expect the photoelectric current to vary as $1/R^2$ on the basis of the wave theory of light? On the basis of the photon theory? Assume that the source emits uniformly in all directions and that the dimensions of both the source and the emitter are small in comparison to R.

12. What is the direction of a Compton-scattered electron with maximum kinetic energy? What is the direction of the corresponding Compton-scattered photon?

13. In Compton scattering, why would you expect the Compton wavelength shift $\Delta\lambda$ to be independent, not only of the incident wavelength, but also of the material of which the target is composed?

14. Light from distant stars is Compton-scattered many times by free electrons in outer space before reaching us. Show that this shifts the light toward the red. How can this shift be distinguished from the Doppler "red shift" due to the motion of receding stars?

15. Is it energetically possible for the Compton effect to occur over the full range of energies listed in Table 5-1? If so, is the effect equally probable over this full range? Over what energy range is the effect more likely to occur than other competing effects in, say, a copper absorber (see Fig. 5-10)?

16. Explain the breadth of the Compton-scattered lines in Fig. 5-6. That is, why are the peaks representing λ and λ' not sharply defined in wavelength?

17. Why can't we observe a Compton effect with visible light?

18. Can Compton-scattered radiation ever be of shorter wavelength than the incident radiation? Explain.

19. In both the photoelectric effect and the Compton effect there is an incident photon and an ejected electron. What is the difference between these two effects?

20. Why is the Compton effect more supportive of the photon theory of light than is the photoelectric effect?

21. In the Compton effect, why is it more useful to deal with the Compton wavelength shift than with the Compton frequency shift of the scattered photons?

22. Does a television tube emit x-rays? Explain.

23. What effect(s) does decreasing the voltage applied to an x-ray tube have on the spectrum of the emitted x-rays?

24. Why is the existence of a definite minimum wavelength in the continuous spectrum emitted by an x-ray tube difficult to understand in terms of the wave theory of light? How does the photon theory explain this phenomenon?

25. What is synchrotron radiation? How does it differ from *bremsstrahlung*? In what ways is it similar?

26. What determines the frequency of the photons emitted as synchrotron radiation? Is there a minimum wavelength cutoff, as there is for x-rays?

27. In Fig. 5-9, why do the electron and the positron tracks shown at site II spiral steadily inward? Should they not move in a circular orbit of constant radius?

28. Do you think that an electron (negative charge) could annihilate with a proton (positive charge), changing their combined rest energy into the energy of gamma-ray photons? Does any law of physics that you know about prevent it? How can you be sure that it does not happen?

29. If the energy of one annihilation photon is measured as equal to m_0c^2, does the other annihilation photon necessarily also have this value? Discuss.

30. Suppose that you have a source of 200-keV x-rays and you wish to demonstrate (a) the photoelectric effect, (b) the Compton effect, and (c) pair production? What element would you choose in each case as an absorber? Would you expect any difficulty in providing your demonstrations? (See Fig. 5-10.)

31. Suppose that an electron–positron pair was in uniform motion with respect to a laboratory observer. Is it possible for the pair to annihilate with the production of only a single photon? (*Hint:* Use the principle of relativity.)

32. Could electron–positron annihilation occur with the production of only one photon if a nearby nucleus was available for recoil momentum?

33. Explain how pair annihilation with the creation of *three* photons is possible. Is it possible in principle to create even more than three?

34. A single photon of very high energy is created at the top of our atmosphere by incoming cosmic radiation. Describe how, by successive pair-production and annihilation processes (among other events), it can build up a massive cosmic-ray "shower."

35. We have seen that the threshold energy for pair production is $2m_0c^2$ (= 1.02 MeV). How can this be, if the photon energy depends—as it does—on the motion of the observer? Surely pair production either occurs or it does not. Explain.

36. Distinguish between the Planck relation $\mathscr{E} = nh\nu$ (Eq. 4-11) and the Einstein relation $E = h\nu$ (Eq. 5-3).

37. Describe several ways to measure the Planck constant.

38. A photon that reaches us from a distant galaxy has a wavelength that is shifted toward the red (longer waves) because the galaxy is receding from us as the universe expands. A longer wavelength means a smaller photon energy. What happened to this "missing" energy? [*Hint:* See discussion in *The Physics Teacher* (December 1983), p. 616.]

39. Two observers in relative motion each measure the energy of photons emitted by identical atoms in a monochromatic light source. Because of differences in the Dop-

pler shift, they find different values for the photon energy. How can this be? They are measuring the same photons!

40. It is claimed that, on the basis of the wave theory of light, it should not be possible for a person to see faint starlight. Explain.

41. Express the energy E of a photon in terms of its momentum p and its relativistic mass m.

42. Can photons be created and/or destroyed? Give examples to support your answer. Do you think that the number of photons in the universe is constant?

43. (*a*) Newton's light corpuscles were assumed to behave according to the laws of Newtonian mechanics. Is the photon concept a return to this idea? (*b*) The ether was invented as a medium in which light waves are propagated. Does the photon concept eliminate the need for an ether?

problems

$$h = 6.63 \times 10^{-34} \text{ J} \cdot \text{s} = 4.14 \times 10^{-15} \text{ eV} \cdot \text{s}$$
$$c = 3.00 \times 10^8 \text{ m/s} \qquad m_0 = 9.11 \times 10^{-31} \text{ kg}$$
$$m_0 c^2 = 0.511 \text{ MeV} \qquad e = 1.60 \times 10^{-19} \text{ C}$$
$$h/m_0 c = 2.43 \times 10^{-12} \text{ m} = 2.43 \text{ pm}$$
$$\text{Area of a sphere} = 4\pi R^2 \qquad 1 \text{ eV} = 1.60 \times 10^{-19} \text{ J}$$

1. Photoelectrons from sodium. (*a*) The energy needed to remove an electron from metallic sodium is 2.28 eV. Does sodium show a photoelectric effect for red light, with $\lambda = 680$ nm? (*b*) What is the cutoff wavelength for photoelectric emission from sodium and to what color does this wavelength correspond?

2. Photoelectrons from aluminum. Light of wavelength 200 nm falls on an aluminum surface. In metallic aluminum 4.2 eV are required to remove an electron. What is the kinetic energy of (*a*) the fastest and (*b*) the slowest emitted photoelectrons? (*c*) What is the stopping potential for this wavelength? (*d*) What is the cutoff wavelength for aluminum? (*e*) If the intensity of the incident light is 2.0 W/m², what is the average rate per unit area at which photons strike the aluminum surface?

3. Photoelectrons from lithium. The work function for a clean lithium surface is 2.4 eV. Make a plot of the stopping potential V_s versus the frequency of the incident light for such a surface. Indicate important features of the curve, such as slope and intercepts. (*Hint:* compare Fig. 5-3.)

4. Photoelectrons from an unknown surface. The stopping potential for photoelectrons emitted from a surface illuminated by light of wavelength 491 nm is 0.71 V.

When the incident wavelength is changed to a new value the stopping potential is found to be 1.43 V. (*a*) What is this new wavelength? (*b*) What is the work function for the surface? (*c*) Can you identify the material, given that it is an elemental substance?

5. Calculating the Planck constant, and more. In a photoelectric experiment in which a sodium surface is used, one finds a stopping potential of 1.85 V for a wavelength of 300 nm and a stopping potential of 0.82 V for a wavelength of 400 nm. From these data find (*a*) a value for the Planck constant, (*b*) the work function for sodium and (*c*) the cutoff wavelength for sodium.

6. Designing a photocell. You wish to pick a substance for a photocell that will be operable with visible light. Which of the following will do (work function in parentheses): tungsten (4.6 eV), aluminum (4.2 eV), tantalum (4.1 eV), barium (2.5 eV), lithium (2.4 eV)? Take the range of visible light to be 400–700 nm.

7. The photographic process. Consider monochromatic light falling on a photographic film. The incident photons will be recorded if they have enough energy to dissociate a AgBr molecule in the film. The minimum energy required to do this is about 0.6 eV. Find the cutoff wavelength, greater than which the light will not be recorded. In what region of the spectrum does this wavelength fall?

8. The photoelectric effect for tightly-bound electrons. X-rays with a wavelength of 0.071 nm eject photoelectrons from a gold foil, the electrons originating from deep within the gold atoms. The ejected electrons move in circular paths of radius r in a region of uniform magnetic field **B**. Experiment shows that $rB = 1.88 \times 10^{-4}$ T·m. Find (a) the maximum kinetic energy of the photoelectrons and (b) the work done in removing the electrons from the gold atoms that make up the foil.

9. A nuclear photo-effect.   Consider the photonuclear reaction

$$^{197}\text{Au} + \gamma \rightarrow {}^{196}\text{Au} + n$$

in which γ represents an incident gamma-ray photon and n represents a neutron. What must be the minimum energy of the incident photon for the reaction to "go"? The rest masses of ^{197}Au, ^{196}Au and n are 196.9665 u, 195.9666 u, and 1.00867 u respectively. (*Hint:* The kinetic energy of the two particles after the interaction is small and may be neglected.)

10. The contact potential difference.   (a) Show that the contact potential difference between the emitter and the collector of a photoelectric tube (see Fig. 5-1) is given by

$$eV_{ce} = w_c - w_e$$

in which w_c and w_e are the work functions of the collector and the emitter, respectively. (b) Suppose that, in Fig. 5-2, the photoelectric current were plotted against V_{ext} instead of against V (see Eq. 5-1) and the stopping potential V_s for each frequency determined from that plot. A figure like Fig. 5-3 is then constructed and a work function determined as in Example 1. Show that a work function so derived would be characteristic of the collector rather than of the emitter.

11. Is relativity needed?   The relativistic expression for kinetic energy should be used for the electron in the photoelectric effect when $v/c > 0.1$, if errors greater than about 1% are to be avoided. (a) For photoelectrons ejected from an aluminum surface, what is the smallest wavelength of an incident photon for which the classical expression may be used? (b) In what region of the electromagnetic spectrum does this wavelength fall? (The work function of aluminum is 4.2 eV.)

12. A photon and a free electron.   (a) Show that a free electron cannot absorb a photon and conserve both energy and momentum in the process. Hence, the photoelectric process requires a bound electron. (*Hint:* Assume energy conservation and show that momentum is not then conserved.) (b) In the Compton effect, however, the electron *can* be free. Explain.

13. Two important wavelengths.   (a) A spectral emission line, important in radioastronomy, has a wavelength of 21 cm. What is its corresponding photon energy? (b) At one time the meter was defined as 1,650,763.73 wavelengths of the orange light emitted by a light source containing krypton-86 atoms. What is the corresponding photon energy of this radiation?

14. Two photon sources.   An ultraviolet lightbulb, emitting at 400 nm, and an infrared lightbulb, emitting at 700 nm, each are rated at 400 W. (a) Which bulb radiates photons at the greater rate? (b) How many more photons does it generate per second than does the other bulb?

15. Photons are falling on my head . . .   Solar radiation is falling on the earth at a rate of 1340 W/m², on a surface normal to the incoming rays. At what rate per unit area do solar photons fall on such a surface? Assume an average solar photon wavelength of 550 nm.

16. Comparing two photons.   Photon A has twice the energy of photon B. How do their momenta compare? . . . their wavelengths? . . . their frequencies? . . . their speeds? Which, if any, of these quantities depend on the state of motion of the observer? Assume that the photons are traveling in free space.

17. Comparing two beams of light.   In the photon picture of radiation show that if two parallel beams of light of different wavelengths are to have the same intensity, then the rates per unit area at which photons pass through any cross-section of the beam are in the same ratio as the wavelengths.

18. A sensitive detector!   Under ideal conditions the normal human eye will record a visual sensation at 550 nm if incident photons are absorbed at a rate as low as 100 s^{-1}. To what power level does this correspond?

19. A special photon.   What are (a) the frequency, (b) the wavelength, and (c) the momentum of a photon whose energy equals the rest energy of an electron?

20. A simple equality.   Show that the energy of a photon, expressed in units of $m_0 c^2$, is the same thing as the momentum of the photon expressed in units of $m_0 c$. Here m_0 is the electron rest mass and c is the speed of light.

21. Photons from a sodium lamp.   A 100-W sodium vapor lamp radiates uniformly in all directions. (a) At what distance from the lamp will the average density of photons be 10 cm^{-3}? (b) What is the average density of photons 2.0 m from the lamp? Assume the light to be monochromatic, with a wavelength of 590 nm.

★ **22. A box full of photons.**   A cavity radiator is maintained at a wall temperature of 2000 K. How many photons per unit volume does it contain? (*Hint:* To simplify calculations, use Wien's expression for the energy density. Although Wien's expression is only an approximation to the correct Planck expression, the approximation is a good one; see Example 4, Chapter 4.)

23. Focusing laser light.   The emerging beam from a 1.0-W argon laser ($\lambda = 515$ nm) has a diameter d of 3.0 mm. (a) At what rate per unit area do photons pass through any cross-section of the incident laser beam? (b) The beam is focused by a lens system (assumed ideal) whose effective focal length f is 2.5 mm. The focused beam forms a circular diffraction pattern whose central disk has a radius R given by $1.22 f \lambda/d$. [See *Physics*, Part II, Sec. 46-5.] It can be shown that 84% of the incident power strikes this central disk, the rest falling in the fainter, concentric diffraction rings that surround the central disk. At what rate per unit area do photons fall on the central disk of the diffraction pattern?

24. Decay in flight of a pion. A neutral pion decays into two gamma rays, thus:

$$\pi^0 \rightarrow \gamma_1 + \gamma_2.$$

Suppose that the pion has a kinetic energy of 150 MeV when it decays and that one of the decay photons, by chance, moves forward along the track of the pion. What are the energies of the two photons? The rest energy of the neutral pion is 135.0 MeV.

25. What a difference a frame makes! A very energetic cosmic ray photon of energy E_1 (= 2.5×10^{14} eV) and a photon of modest energy E_2 (= 0.0010 eV) are approaching each other head-on. The reference frame in which these energies are measured is the earth. Consider now a frame S moving with speed v along the line of the photons, v being chosen so that, as measured from S, the energies of the photons are equal. (*a*) Toward which source must frame S be moving? (*b*) By how much does v differ from the speed of light? (*c*) What is the common photon energy, as measured in S?

26. Analyzing a Compton collision. Photons of wavelength 2.40 pm are incident on a target containing free electrons. (*a*) Find the wavelength of a photon that is scattered at 30° from the incident direction and also the kinetic energy imparted to the recoil electron. (*b*) Do the same if the scattering angle is 120°. (*Hint*: See Example 5.)

27. The fractional photon energy loss. (*a*) Show that $\Delta E/E$, the fractional loss of energy of a photon during a Compton collision, is given by $(h\nu'/m_0 c^2)(1 - \cos \phi)$. (*b*) Plot $\Delta E/E$ versus ϕ and interpret the curve physically.

28. Losing energy in a Compton collision (I). What fractional increase in wavelength leads to a 75% loss of photon energy in a Compton collision with a free electron?

29. Losing energy in a Compton collision (II). Through what angle must a 200-keV photon be scattered by a free electron so that it loses 10% of its energy?

30. No back-scattered Compton electrons. Prove, by qualitative arguments about momentum conservation, that the electron in a Compton collision cannot be scattered through an angle θ greater than 90°, no matter what the scattering angle ϕ of the photon may be. Prove this claim also by formal analysis of Eq. 5-8.

★ **31. A relationship between two scattering angles.** Prove that the scattering angle θ of the electron in a Compton scattering event is related to the scattering angle ϕ of the incident photon by

$$\cot(\phi/2) = (1 + \rho) \tan \theta,$$

in which $\rho = h\nu/m_0 c^2$. Show by inspection of this equation that the electron cannot be scattered backwards, that is, that θ cannot exceed 90°. (*Hint*: Start from the conservation relations, Eqs. 5-7 to 5-9, or from equations derived from them, such as the Compton shift formula, Eq. 5-10.)

32. The kinetic energy of the Compton electron. Derive a relation between the kinetic energy K of the recoil electron and the energy E (= $h\nu$) of the incident photon in the Compton effect, starting from the conservation relations Eqs. 5-7, 5-8 and 5-9. One form of the relation is

$$K = E \frac{2\rho \sin^2 \phi/2}{1 + 2\rho \sin^2 \phi/2}$$

in which

$$\rho = h\nu/m_0 c^2.$$

(*Hint*: See Example 5; note that $1 - \cos \phi = 2 \sin^2 \phi/2$.)

33. Compton recoil electrons (I). What is the maximum possible kinetic energy that can be imparted to a Compton recoil electron for a given energy of the incident photon? (*Hint*: See Problem 32.)

34. Compton recoil electrons (II). What is the maximum kinetic energy that can be imparted to a Compton recoil electron from a copper foil struck by a monochromatic photon beam in which the incident photons each have a momentum of 0.880 MeV/c? (*Hint*: See Problem 32.)

35. The momentum of the Compton recoil electron. An x-ray photon traveling in the +x-direction has an initial energy of 100 keV and falls on a free electron at rest. The photon is scattered at right angles into the +y-direction. Find the momentum components of the recoiling electron.

36. An interesting equality. Show that the Compton wavelength of a particle is equal to the wavelength of a photon whose energy is equal to the rest energy of the particle.

37. Compton collisions with protons. Find the maximum wavelength shift for a Compton collision between a photon and a free proton.

★ **38. Filling in the algebra.** Eliminate v and θ from Eqs. 5-7, 5-8 and 5-9 and derive Eq. 5-10, the expression for the Compton wavelength shift. (*Hint*: Simplify the notation by replacing $1\sqrt{1 - v^2/c^2}$ by γ; divide Eq. 5-7 by $m_0 c^2$

and Eqs. 5-8 and 5-9 by m_0c; use the relation $\sin^2 x + \cos^2 x = 1$.)

39. The Compton frequency shift. We have seen that the Compton wavelength shift is given by Eq. 5-10. Show that the Compton frequency shift can be written in the form

$$\Delta \nu = \nu - \nu' = \nu \, \frac{\rho(1 - \cos \phi)}{1 + \rho(1 - \cos \phi)},$$

in which

$$\rho = \frac{h\nu}{m_0 c^2}.$$

How is this expression related to the kinetic energy of the electron?

40. The short wavelength cutoff. (a) Show that the short wavelength cutoff in the continuous x-ray spectrum is given by $\lambda_{\min} = 1.24 \text{ nm}/V$ in which V is the accelerating potential in kilovolts. (b) What is the cutoff wavelength if a potential difference of 186 kV is applied to an x-ray tube?

41. X-ray production. What is the minimum potential difference that must be applied to an x-ray tube if (a) the x-ray photons are to have a wavelength of 1.00 Å? (b) If they are to have a wavelength equal to the Compton wavelength of the electron? (c) If they are to be capable of pair production?

★ **42. Jolting to rest.** A 20-keV electron is brought to rest by undergoing two successive bremsstrahlung events, thus transferring its kinetic energy into the energy of two bremsstrahlung photons. The wavelength of the second photon is 0.130 nm longer than the wavelength of the first photon. (a) What is the energy of the electron after its first deceleration? (b) What are the wavelengths and the energies of the two photons?

43. The best of three. Gamma rays fall on a copper absorber ($Z = 29$). (a) At what photon energy will the photoelectric effect and the Compton effect occur at the same rate? (b) At what photon energy will the Compton effect and pair production occur at the same rate? (c) Regardless of the nature of the absorber, for what range of photon energies will the Compton effect always be the most likely process? (*Hint*: see Fig. 5-10.)

44. Pair production—a special case. A particular pair is produced such that the positron is at rest and the electron, which moves in the direction of the incident photon, has a kinetic energy of 1.00 MeV. (a) What is the energy of the incident photon? (Neglect the energy transferred to the heavy nucleus in whose vicinity the pair is produced.) (b) What percentage of the momentum of the incident photon is transferred to this nucleus? (c) Why is it that the

energy transferred to this nucleus can be neglected but the momentum transferred can not?

45. Pair production with "knock-on". Figure 5-9 (site II) shows a case of pair production by an incident gamma ray in the vicinity of an electron, which is "knocked-on" to form the central track. Show that the minimum gamma ray energy required for this process to occur is $4m_0c^2$. (*Hint*: At the threshold, all three electrons will move forward with the same momentum. Can you justify this assumption?)

46. Pair production at the threshold energy. Assume that an electron-positron pair is produced by a photon whose energy is just the threshold energy $(= 2m_0c^2)$ for this process. (a) Calculate the momentum transferred to the heavy nucleus in whose vicinity the pair is created. (b) Assume the nucleus to be that of a lead atom (mass = 207 u) and compute the kinetic energy of the recoiling nucleus. Are we justified in neglecting this energy compared to the threshold energy assumed above?

★ **47. Pair production and the conservation laws.** An energetic photon creates an electron-positron pair. Show directly that, without the presence of a third body (a heavy nucleus, say) to take up some of the momentum, energy and momentum cannot both be conserved. Thus pair production cannot occur in a vacuum. (*Hint*: Set the initial and final energies equal and show that this leads to the conclusion that momentum cannot be conserved.)

48. Proton-antiproton pair production. An antiproton is a particle that has the same mass as the proton but a charge of the opposite sign. It can annihilate with a proton, in the same way that a positron and an electron can annihilate with each other. What is the threshold energy for a photon to produce a proton-antiproton pair in the vicinity of a heavy nucleus?

49. Annihilation on the run. An electron-positron pair at rest annihilate, creating two photons. (a) At what speed must an observer move along the common axis of the photons in order that the wavelength of one photon be twice that of the other? (b) At this speed, how do the energies of the two photons compare? (c) How do the momenta compare?

50. More about Example 10 (I). (a) Show that, for the special case of annihilation in flight discussed in Example 10, the difference in energy between the two photons is given by

$$E_1 - E_2 = \frac{2\beta^2}{\sqrt{1 - \beta^2}} E_0$$

in which $E_0 = m_0 c^2$.

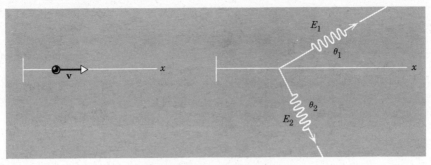

Figure 5-16. Problem 52.

(b) Show that the sum of the two energies is given by

$$E_1 + E_2 = \frac{2}{\sqrt{1 - \beta^2}} E_0.$$

(c) Do these expressions yield reasonable results for $\beta = 0$? As $\beta \to 1$? (d) For what value of β is the difference in energy between the two photons just equal to the rest energy of the particle? Does this value depend on the nature of the annihilating particles?

51. More about Example 10 (II). (a) In the pair annihilation event of Example 10, what must be the speed parameter β of an electron-positron pair if the energy of one of the annihilation photons is to be just twice that of the other? (b) What are these energies?

★ **52. Pair annihilation generalized.** Figure 5-16a shows an electron-positron pair moving along an x-axis with speed $v(= \beta c)$; Fig. 5-16b shows the two annihilation photons. (a) Show that the energies E_1 and E_2 and the angle θ_2 are given by

$$E_1 = E_0 \frac{\sqrt{1 - \beta^2}}{1 - \beta \cos \theta_1},$$

$$E_2 = E_0 \frac{1 - 2\beta \cos \theta_1 + \beta^2}{\sqrt{1 - \beta^2}\,(1 - \beta \cos \theta_1)},$$

and

$$\sin \theta_2 = \sin \theta_1 \frac{1 - \beta^2}{1 - 2\beta \cos \theta_1 + \beta^2},$$

in which $E_0 = m_0 c^2 = 511$ keV.
(b) Evaluate these three quantities for $\beta = 0.20$ and $\theta_1 = 30.0°$. (c) Show that the three expressions above reduce to reasonable results in these special cases: $\beta = 0$ and $\theta_1 = 0$; $\beta = 0$ and $0 < \theta_1 < 180°$; $0 < \beta < 1$ and $\theta_1 = 0$.

★ **53. A pair production puzzle.** In frame S of Fig. 5-17 a 1.02-MeV photon has just enough energy to create a pair in the vicinity of a resting heavy nucleus. Frame S' is an inertial frame moving at speed 0.60c with respect to S. To an observer in S', the incident photon will be Doppler-shifted; see Eqs. 2-30. (a) Calculate, from the Doppler shift, the energy of the photon as measured by the observer in S'. (b) This energy will turn out to be below the threshold energy for pair production. But pair production either occurs or it does not. The fact of its occurrence cannot depend on which frame you choose to analyze the event. Explain this apparent paradox. (Hint: What role does the heavy nucleus play?)

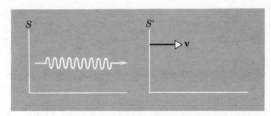

Figure 5-17. Problem 53.

54C. All about the photon. Write a program for your handheld programmable calculator that will accept as an input *either* the wavelength λ (in m) *or* the frequency ν (in Hz) *or* the energy E (in eV) and will deliver as successive outputs (a) the wavelength λ (in m), (b) the frequency ν (in Hz), (c) the photon energy E (in J), (d) the photon energy E (in eV), (e) the photon momentum p (in SI units), and (f) the relativistic mass m of the photon in units of the electron rest mass m_0.

55C. Checking it out (I). Test the program that you have written in Problem 54C by calculating (a) the energy and the corresponding wavelength of a photon whose frequency is 25 kHz; 80 MHz; 3 GHz. (b) The relativistic mass and the corresponding wavelength of a photon whose energy is 15 GeV; 2 MeV; 3 eV. (c) The energy and the corresponding frequency of a photon whose wavelength is 50 fm; 50 nm; 10 mm. (d) Locate the nine photons referred to above on the electromagnetic spectrum displayed in Fig. 5-4. (e) Verify that the wavelength, frequency, and energy scales in this figure are properly drawn.

56C. All about the Compton effect. Write a program for your handheld, programmable calculator that will accept as inputs the scattering angle ϕ shown in Fig. 5-7 and also *either* the incident wavelength λ (in m) *or* the incident photon energy E (in eV). The program should display as successive outputs (a) the wavelength λ of the incident photon, (b) the wavelength λ' of the scattered photon, (c) the wavelength shift $\Delta\lambda$, (d) the energy E of the incident photon, (e) the energy E' of the scattered photon, (f) the kinetic energy K of the scattered electron, (g) the percent energy loss of the photon to the scattered electron ($= 100K/E$), (h) the scattering angle ϕ in Fig. 5-7, and (i) the scattering angle θ in that figure. Use relativistic formulas when dealing with the scattered electron.

57C. Checking it out (II). (a) Using the program that you wrote in Problem 56C, compute all the quantities listed in that program for a relatively low-energy photon ($E = 5$ keV). Assume a scattering angle ϕ of 45°. (b) Repeat for a relatively high-energy photon ($E = 25$ MeV) and compare your findings, item by item. Assume the same value for ϕ. (c) Does your program yield expected results when the scattering angle ϕ is zero? . . . 180°? Explain in physical terms. (Assume an incident photon energy of, say, 1.0 MeV.) (d) Investigate how the scattering angle θ of the electron varies with the scattering angle ϕ of the photon for a fixed incident photon energy (1.0 MeV, say). (e) Assume that ϕ is fixed at 45° and investigate how θ varies with the energy of the incident photon.

58C. All about pair production. Consider the generalized case of pair annihilation in flight, depicted in Fig. 5-16 and Problem 52. (a) Write a program for your handheld programmable calculator that will accept any two of the five quantities: $\beta(= v/c)$, E_1, E_2, θ_1, and θ_2 and will deliver the remaining three quantities as outputs. (b) Use this program to verify parts (b) and (c) of Problem 52. (c) Use your program to fill in the blanks in the table below:

Table 5.4
PROBLEM 58C.

β	E_1	E_2	θ_1	θ_2
0.30	_____	_____	45°	_____
0.30	_____	_____	100°	_____
0.30	300 keV	_____	_____	_____
0.30	600 keV	_____	_____	_____
_____	300 keV	_____	45°	_____
_____	600 keV	_____	100°	_____

references

1. Albert Einstein, "On a Heuristic Viewpoint Concerning the Production and Transformation of Light," *Ann. Physik* **17,** 132 (1905). There is an English translation in Morris H. Shamos, Ed., *Great Experiments in Physics* (Holt-Dryden, Hinsdale, Ill., 1959).

2. Abraham Pais, *Subtle is the Lord—The Science and Life of Albert Einstein* (Oxford University Press, New York, 1982). Chapters 19 and 21 deal with Einstein's light quantum hypothesis.

3. J. Rudnick and D. S. Tannhauser, "Concerning a Widespread Error in the Description of the Photoelectric Effect," *Am. J. Phys.* **44,** 796 (1976).

4. Roger H. Stuewer, *The Compton Effect—Turning Point in Physics* (Science History Publications, New York, 1975). An excellent historical review.

5. A. H. Compton, "The Scattering of X-Rays as Particles," *Amer. J. Phys.* **29,** 817 (1961).

6. Brian G. Williams, "Compton Scattering and Heisenberg's Microscope Revisited," *Am. J. Phys.* **52,** 425 (1984).

7. George Gamow, *Mr. Tompkins in Wonderland* (Cambridge University Press, Cambridge, 1965).

8. "Introduction to the Detection of Nuclear Particles in a Bubble Chamber," prepared at the Lawrence Radiation Laboratory, The University of California at Berkeley, Ealing Press (1964). This booklet includes a number of stereo photos so that Fig. 5-10 (among other bubble chamber events) can be viewed in three dimensions. A viewer is also included.

9. Dwight E. Gray, Coordinating Ed., *American Institute of Physics Handbook* (McGraw-Hill, New York, 1972), sec. 8e, "Gamma Rays," by Robley D. Evans.

10. Ednor M. Rowe and John H. Weaver, "The Uses of Synchrotron Radiation," *Sci. Am.* (June 1977).

11. *Physics Today,* June 1983, is devoted entirely to a review of the current status of synchrotron radiation research.

12. H. C. Corben and S. DeBenedetti, "The Ultimate Atom," *Sci. Am.* (December 1954).

CHAPTER 6

the wave nature of matter and the uncertainty principle

We thus find that in order to describe the properties of Matter, as well as those of Light, we must employ waves and corpuscles simultaneously. We can no longer imagine the electron as being just a minute corpuscle of electricity: we must associate a wave with it. And this wave is not just a fiction: its length can be measured and its interferences calculated in advance.

Louis de Broglie (1929)

. . . we have to remember that what we observe is not nature in itself but nature exposed to our method of questioning. Our scientific work in physics consists in asking questions about nature in the language that we possess and trying to get an answer from experiment by the means that are at our disposal.

Werner Heisenberg (1930)

6-1 MATTER WAVES

Maurice de Broglie was a French experimental physicist who, from the outset, had supported Compton's view of the particle nature of radiation. His experiments and discussions impressed his brother Louis so much with the philosophic problems of physics at the time that Louis changed his career from history to physics. In his doctoral thesis, presented in 1924 to the Faculty of Science at the University of Paris, Louis de Broglie proposed the existence of matter waves [1]. The thoroughness and originality of his thesis were recognized at once, but because of the apparent lack of experimental evidence de Broglie's ideas were not considered to have any physical reality. It was Albert Einstein, whose attention was drawn to de Broglie's ideas by Paul Langevin, who recognized their importance and validity and in turn called them to the attention of other physicists. Five years later, his ideas by then having been dramatically confirmed by experiment, de Broglie received the Nobel Prize [2] ". . . for his discovery of the wave nature of electrons." Although he was the first physicist to receive a Nobel Prize for his doctoral dissertation, it should be pointed out that, at age 32 and with a number of substantial research publications to his credit, he was already a mature physicist when he defended his doctoral thesis.

de Broglie's hypothesis was that the apparently dual behavior of radiation as wave and as particle applied equally well to matter. Just as a quantum of radiation has a wave associated with it that governs its motion, so a particle—a quantum of matter—will have a corresponding matter wave that governs its motion. Since the observable universe is composed entirely of matter and radiation, de Broglie's suggestions are consistent with a statement of the symmetry in nature. Indeed, de Broglie proposed that the wave aspects of matter were related quantitatively to the particle aspects in exactly the same way that we found for radiation, namely, $E = h\nu$ and $p = h/\lambda$.

That is, for matter *and* for radiation alike, the total (relativistic) energy E of the entity is related to the frequency ν of the wave associated with its motion by the equation

$$E = h\nu, \qquad (6\text{-}1a)$$

and the momentum p of the entity is related to the wavelength λ of the associated wave by the equation

$$p = \frac{h}{\lambda}. \qquad (6\text{-}1b)$$

Here the particle concepts, energy E and momentum p, are connected to the wave concepts, frequency ν and wavelength λ. Radiation corresponds to particles of zero rest mass (moving at speed c), whereas matter corresponds to particles of finite rest mass (moving at speeds less than c), but the same general transformation properties apply to both entities. We saw in Chapter 5 that radiation, which earlier was regarded as wavelike, exhibited particlelike properties as well. de Broglie suggested that matter, which had been regarded as particlelike, exhibits wavelike properties also. Equation 6-1b, in the following form, is called the de Broglie relation:

$$\lambda = \frac{h}{p} = \frac{h}{mv} \quad \text{(de Broglie relation)}. \qquad (6\text{-}2)$$

The quantity m in Eq. 6-2 is the relativistic mass of the particle whose de Broglie wavelength is sought. In the low-speed classical limit, however, the rest mass m_0 can be substituted; see Example 2.

EXAMPLE 1.

The Wavelengths of Two Quite Different Particles. (*a*) What is the de Broglie wavelength of a baseball moving at 30 m/s? Its mass m is 150 g.

From Eq. 6-2,

$$\lambda = \frac{h}{p} = \frac{h}{mv} = \frac{6.63 \times 10^{-34}\,\text{J}\cdot\text{s}}{(0.15\,\text{kg})(30\,\text{m/s})}$$
$$= 1.5 \times 10^{-34}\,\text{m}.$$

(*b*) What is the de Broglie wavelength of an electron whose kinetic energy is 100 eV?

For such a slow electron we can safely use classical rather than relativistic mechanics, just as we did for the baseball. Thus we have, from Eq. 6-2,

$$\lambda = \frac{h}{p} = \frac{h}{\sqrt{2mK}}$$
$$= \frac{6.63 \times 10^{-34}\,\text{J}\cdot\text{s}}{\sqrt{(2)(9.11 \times 10^{-31}\,\text{kg})(100\,\text{eV})(1.60 \times 10^{-19}\,\text{J/eV})}}$$
$$= 1.2 \times 10^{-10}\,\text{m} = 1.2\,\text{Å}.$$

We see that the wavelength of the baseball in this example is smaller than that of the electron by $\sim 8 \times 10^{23}$, a very large factor.

The wave nature of light propagation is not revealed by experiments in geometrical optics, for the obstacles or apertures used there are very large compared to the wavelength of light. If *a* represents a characteristic dimension of such an aperture or obstacle (for example, the width of a slit or the diameter of a lens), we say that $\lambda \ll a$ defines the region of geometrical optics; see *Physics*, Part II, Sect. 44-1. Geometrical optics is characterized by ray propagation, which is similar to the trajectory motion of particles. When the dimension *a* becomes comparable to λ (that is, $\lambda \cong a$) or smaller than λ (that is, $\lambda > a$), then we are in the region of physical optics. In this case interference and diffraction effects are easily observed and wavelengths can be measured readily. To observe the wavelike aspects of the motion of matter, therefore, we need suitably small apertures or obstacles. One of the smallest sizes available at the time of de Broglie was that of an atom, or the spacing between adjacent planes of atoms in a solid, where $a \cong 10^{-10}$ m = 1 Å. Clearly then, for the baseball in Example 1 we cannot expect to measure the de Broglie wavelength (here $\lambda/a \cong 10^{-24}$); indeed we cannot detect *any* evidence of wavelike motion. For material objects of very much smaller mass, however, such as the electron in Example 1, we might detect an associated wave because the smaller momentum corresponds to a longer wavelength.

EXAMPLE 2.

de Broglie Wavelengths at Low Speeds. What is the maximum kinetic energy of an electron such that an error of 1.0 percent or less is made in calculating its de Broglie wavelength by using the classical expression for momentum in Eq. 6-2?

The relativistic expression for the momentum of a particle whose kinetic energy is K is given by Eq. 3-13c, which we may write as

$$p = \sqrt{2m_0 K + (K/c)^2}.$$

We can find the classical expression for the momentum by the device of letting $c \to \infty$; doing so above yields $p = \sqrt{2m_0 K}$, the expected classical result. The classical (approximate) and the relativistic (correct) expressions for the de Broglie wavelength (see Eq. 6-2) are then given by

$$\lambda_c = \frac{h}{\sqrt{2m_0 K}}$$

and

$$\lambda_r = \frac{h}{\sqrt{2m_0 K + (K/c)^2}},$$

respectively.

If f is the fractional error involved, we have

$$f = \frac{\lambda_c - \lambda_r}{\lambda_r} = \frac{\lambda_c}{\lambda_r} - 1$$

$$= \sqrt{\frac{2m_0 K + (K/c)^2}{2m_0 K}} - 1$$

$$= \left(1 + \frac{K}{2m_0 c^2}\right)^{1/2} - 1.$$

Let us now expand the quantity on the right by the binomial theorem (see Appendix 5), keeping only the first two terms. If $K \ll 2m_0 c^2$, which we assume, we can do so without appreciable error. The result is

$$f \cong \left(1 + \frac{K}{4m_0 c^2}\right) - 1 = \frac{K}{4m_0 c^2}$$

or

$$K = 4f(m_0 c^2).$$

Thus, if $f = 0.010$, corresponding to a 1 percent error, then

$$K = 4f(m_0 c^2) = (4 \times 0.010)(511 \text{ keV}) = 20 \text{ keV}.$$

For all electron energies less than this, the error made in using the classical momentum formula to calculate the de Broglie wavelength will be less than 1.0 percent. For $K = 100$ keV, the error will be 4.9 percent. We see also that the error made in using the classical momentum formula to calculate the electron de Broglie wavelength in Example 1*b* is truly negligible, being about 0.005 percent.

EXAMPLE 3.

The Frequency and Speed of Matter Waves. What are (a) the frequency and (b) the speed of the matter wave associated with an electron whose kinetic energy is 100 eV?

(a) The de Broglie frequency can be found from Eq. 6-1a, in which the energy E must be taken to be the total relativistic energy, including both the rest energy and the kinetic energy. Thus

$$\nu = \frac{E}{h} = \frac{m_0 c^2 + K}{h}$$

$$= \frac{(5.11 \times 10^5 \text{ eV} + 100 \text{ eV})(1.60 \times 10^{-19} \text{ J/eV})}{6.63 \times 10^{-34} \text{ J} \cdot \text{s}}$$

$$= 1.2 \times 10^{20} \text{ Hz}.$$

Note that the kinetic energy of the electron in this case (= 100 eV) is so much less than its rest energy (= 511 keV) that the de Broglie frequency is virtually independent of K. This is not at all true for the de Broglie wavelength, as the calculation of Example 1b shows.

(b) Recall that in Example 1b we showed, using Eq. 6-2, that the de Broglie wavelength for a 100-eV electron is 1.2×10^{-10} m. We can then find the speed of the matter wave from

$$v = \lambda\nu = (1.2 \times 10^{-10} \text{ m})(1.2 \times 10^{20} \text{ Hz})$$

$$= 1.4 \times 10^{10} \text{ m/s}.$$

Thus we calculate a speed that is almost 50 times greater than the speed of light! What went wrong?

The answer is that nothing went wrong and there is no violation of the theory of relativity. The speed that we have calculated is a *phase speed*. If we imagine the matter wave describing the electron as a sine wave, we must take it—viewed as an instantaneous "snapshot"—to be infinitely long in both directions. Alternatively, if we station ourselves at a point and watch the wave go by, it does so for an infinitely long period, for both past and future times.

Such a wave is featureless, in that it has no beginning and no end and all of its maxima and minima are identical. Thus no information or signal can be transmitted by the wave and there is then no reason that the wave cannot travel with a speed exceeding that of light. In Supplementary Topic G we shall explore this matter further and show that there exists also a *group speed* that can be assigned to the electron. This group speed is less than the speed of light and, in fact, coincides with the speed of the electron.

In the rest of this chapter we shall be concerned only with the wavelength and not with the frequency or the speed of the matter waves.

6-2 TESTING THE de BROGLIE HYPOTHESIS

In 1926 Walter Elsasser pointed out that the wave nature of matter might be tested in the same way that the wave nature of x-rays was first tested, namely, by causing a beam of electrons of appropriate energy to fall on a crystalline solid. The atoms of the crystal serve as a three-dimensional array of diffracting centers for the electron "wave." We should look for strong diffracted peaks, then, in characteristic directions, just as for x-ray diffraction (see *Physics*, Part II, Sec. 47-5). This idea was confirmed independently by C. J. Davisson and L. H. Germer, working at the Bell Telephone Laboratories in the United States [3,4] and by G. P.

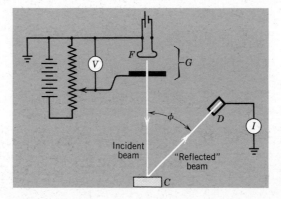

Figure 6-1. The apparatus of Davisson and Germer. Electrons from filament F are accelerated by a variable potential difference V. After "reflection" from crystal C, they are collected by detector D and read as a current I.

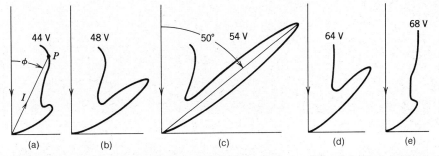

Figure 6-2. The results of five runs with the apparatus of Fig. 6-1. The accelerating potentials for each run are shown. A sharp diffraction peak appears at $\phi = 50°$ for $V = 54$ V.

Thomson, working at the University of Aberdeen in Scotland [5]. In 1937 Davisson and Thomson shared the Nobel Prize ". . . for their experimental discovery of the diffraction of electrons by crystals."

Figure 6-1 shows schematically the apparatus of Davisson and Germer. Electrons from heated filament F are accelerated through a potential difference V and emerge from the "electron gun" G with a kinetic energy eV. This collimated electron beam falls at right angles on a single crystal of nickel at C. Electrons reflected from the crystal at any selected angle ϕ enter detector D and are read as a current I. In a typical experiment the accelerating potential V is set to a selected value and the detector current I is measured as the angle ϕ is varied from ~0 to 90°.

Figure 6-2 shows the results of five runs with the apparatus of Fig. 6-1, for five different values of the accelerating potential V. The measurements are plotted in polar coordinates, each point such as P in Fig. 6-2a representing a particular value of ϕ (referred to the vertical axis) and a particular value of the detector current I (plotted along a radial line from the origin).

We see that there is a prominent intensity maximum for $V = 54$ V and for $\phi = 50°$. If either of these quantities is changed appreciably—in either direction—the intensity maximum decreases markedly. Furthermore, if the nickel target C is polycrystalline, that is, if it is composed of many small, randomly oriented crystals rather than a single large crystal, no maximum at all appears in the reflected beam.* It seems clear that the strong intensity maximum in Fig. 6-2c is caused by the constructive interference of electron wavelets scattered from the atoms of crystal C in Fig. 6-1, regularly arranged to form a periodic three-dimensional lattice.

x-Rays also show sharply defined diffracted beams, formed by Bragg reflection from a family of parallel atomic planes. The treatment of Bragg reflection for electron waves, however, is complicated by the fact that, for the relatively low electron energies considered here, such waves bend or refract as they pass through the crystal surface at other than normal incidence. In other words, the index of refraction of typical crystals for such electron waves is significantly different from unity. For x-rays of similar wavelengths, on the other hand, the

---

* An explosion of a liquid-air bottle opened the vacuum system to the air and oxidized the target. In heating the target afterwards to get it clean, Davisson and Germer recrystallized the polycrystalline target into a few large crystals which accidently created the conditions that, according to K. K. Darrow, "blew open the gate to the discovery of electron waves."

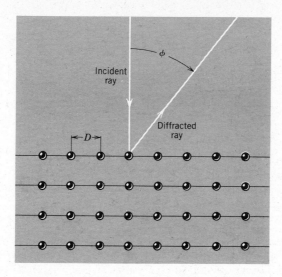

Figure 6-3. A crystal surface acts like a diffraction grating, with grating spacing D. An incident and a diffracted ray are shown. (See Supplementary Topic H for a more complete analysis of the conditions under which a diffracted ray will appear.)

index of refraction is essentially unity, so the diffracted x-ray beams pass through the crystal surface with no change in direction.

It turns out (see Supplementary Topic H) that the correct direction for the emerging diffracted electron beams is given by regarding the crystal surface as a two-dimensional diffraction grating, the parallel rows of atoms corresponding to the rulings of the grating. The grating spacing D is as indicated in Fig. 6-3. For such a grating, strong diffracted beams can be shown to occur (see *Physics*, Part II, Sec. 47-3) when the condition

$$m\lambda = D \sin \phi \qquad m = 1, 2, 3, \ldots \tag{6-3}$$

is satisfied. For the particular surface of the nickel crystal used by Davisson and Germer to provide the data shown in Fig. 6-2, it was known that $D = 2.15 \times 10^{-10}$ m = 2.15 Å. If we assume that the integer $m = 1$, which corresponds to a so-called first-order diffraction peak, Eq. 6-3 then leads to

$$\lambda = D \sin \phi$$
$$= (2.15 \text{ Å}) \sin 50° = 1.65 \text{ Å}.$$

The expected de Broglie wavelength of a 54-eV electron, calculated from Eq. 6-2 as in Example 1*b*, is 1.65 Å, in exact agreement with this measured result. This is quantitative confirmation of de Broglie's equation for λ in terms of p. The choice of $m = 1$ above is justified by the fact that if $m = 2$ (or more), then other reflection peaks for different angles ϕ would have appeared, but none were observed. The breadth of the observed peak in Fig. 6-2*c* is easily understood, also, for low-energy electrons cannot penetrate deeply into the crystal, so that only a small number of atomic planes contribute to the diffracted wave. Hence, the diffraction maximum is not sharp. Indeed, all the experimental results were in excellent qualitative and quantitative agreement with the de Broglie prediction, and provided convincing evidence that matter moves in agreement with the laws of wave motion.

In 1927, George P. Thomson showed that electron beams are diffracted in passing through thin films and independently confirmed the de Broglie relation $\lambda = h/mv$ in detail. Whereas the Davisson-Germer experiment is like Laue's in x-ray diffraction (reflection of specific wavelengths in a continuous spectrum from the regular array of atomic planes in a large single crystal), Thomson's experiment is similar to the Debye-Hull-Scherrer method of powder diffraction of x-rays (transmission of a fixed wavelength through an aggregate of very small crystals oriented at random). Thomson used higher-energy electrons, which are much more penetrating, so that many hundred atomic planes contribute to the diffracted wave. The resulting diffraction pattern consists of sharp lines. In Fig. 6-4 we show, for comparison, an x-ray diffraction pattern and an electron diffraction pattern from a polycrystalline specimen.

Figure 6-5 shows other compelling evidence that electrons behave like waves in certain experimental situations. Figure 6-5a is the familiar diffraction pattern formed on a screen when a straightedge is interposed between the screen and a narrow, linear source of visible light. Figure 6-5b shows a similar diffraction pattern for an electron beam. Is it possible to compare Figs. 6-4b and c or Figs. 6-5a and b and to doubt that electrons have a wave aspect?

It is of interest that J. J. Thomson, who in 1897 discovered the electron (which he characterized as a particle with a definite charge-to-mass ratio) and was awarded the Nobel Prize in 1906, was the father of G. P. Thomson, who in 1927 experimentally discovered electron diffraction and was awarded the Nobel Prize (with Davisson) in 1937. Max Jammer [6] says of this, "one may feel inclined to say that Thomson, the father, was awarded the Nobel Prize for having shown that the electron is a particle, and Thomson, the son, for having shown that the electron is a wave."

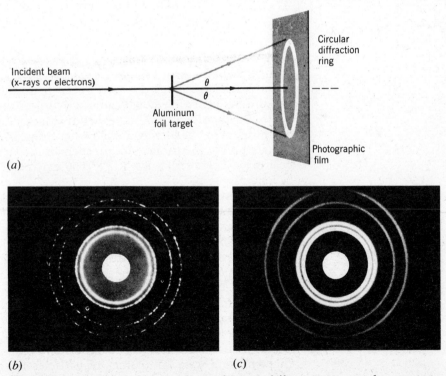

(a)

(b)   (c)

Figure 6-4. (a) An arrangement for producing a diffraction pattern characteristic of a powdered or polycrystalline aluminum target. (b) Pattern for an incident x-ray beam. (c) Pattern for an incident electron beam of the same wavelength. Courtesy Educational Development Center, Newton, Massachusetts.

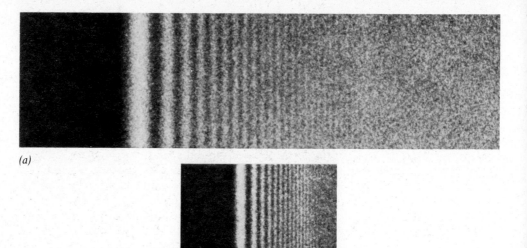

(a)

(b)

Figure 6-5. The diffraction pattern formed when an incident wave from a narrow slit falls on a straight-edge. In (*a*) the wave is visible light. (From Joseph Valasek, *Introduction to Theoretical and Experimental Optics*, Wiley, New York, 1949). In (*b*) the wave is an electron matter wave. (From *Handbuch der Physik*, vol. 32, Springer-Verlag, Berlin, 1957, p 551, Fig. 110).

Not only electrons, but *all material objects*, charged or uncharged, show wave-like characteristics in their motion under the conditions of physical optics. For example, Estermann, Stern, and Frisch performed quantitative experiments on the diffraction of molecular beams of hydrogen and atomic beams of helium from a lithium fluoride crystal; and Fermi, Marshall, and Zinn showed interference and diffraction phenomena for slow neutrons. In Fig. 6-6 we show a Laue diffraction pattern for neutrons. Indeed, even an interferometer operating with neutron beams has been constructed [7]. The existence of matter waves is established beyond question.

It is instructive to note that we had to go to long de Broglie wavelengths to find experimental evidence for the wave nature of matter. That is, we used particles of low mass and speed to bring $\lambda (= h/mv)$ into the range of measurable diffraction. Ordinary (macroscopic) matter has such short corresponding de Broglie wavelengths, because its momentum is high, that its wave aspects are practically undetectable. The particle aspects are dominant. Similarly, we had to go to very short wavelengths to find experimental evidence for the particle nature of radiation. It is in the x-ray and gamma-ray region where the corpuscular aspects of radiation stand out experimentally. In the long-wavelength region, classical wave theory is completely adequate to explain the observations. All this suggests that just as a quantum theory of radiation emerges from the experiments of earlier chapters, so a wave mechanics of particles is emerging from experiments discussed here.

Once again we see the central role played by the Planck constant h. If h were zero then, in $\lambda = h/mv$, we would obtain $\lambda = 0$. Matter would always have a wavelength smaller than *any* characteristic dimension and diffraction could

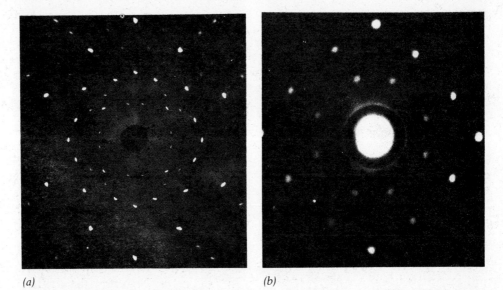

(a)   *(b)*

Figure 6-6. (a) A Laue pattern, showing x-rays diffracted by a crystal of sodium chloride. (Courtesy of W. Arrington and J. L. Katz, X-ray Laboratory, Rensselaer Polytechnic Institute.) (b) A Laue pattern, showing neutrons from a nuclear reactor diffracted by a crystal of sodium chloride. The diffracted neutron beams fell on a photographic plate that was covered with a thin indium foil. The incident neutrons induced radioactivity in the indium, thus activating the plate and producing the spots. (After Shull, Marney, and Wollan.)

never be observed; this is the classical situation. Furthermore, we see that it is the smallness of h that obscures the existence of matter waves experimentally in the macroscopic world, for we must have very small momenta to obtain measurable wavelengths. For ordinary macroscopic matter, m is large and λ is so small as to be beyond the range of experimental measurement, so classical mechanics is supreme. But in the microscopic world, de Broglie wavelengths are comparable to characteristic dimensions of systems of interest (such as atoms), so the wave properties of matter in motion are experimentally observable.

EXAMPLE 4.

The de Broglie Wavelength of a Moving Helium Atom. In the experiments with helium atoms referred to earlier, a beam of atoms of nearly uniform speed of 1640 m/s was obtained by allowing helium gas to escape through a small hole in its enclosing vessel into an evacuated chamber and then through narrow slits in parallel rotating circular disks of small separation (a mechanical velocity selector.) In addition to a regularly reflected beam, a strongly diffracted beam of helium atoms was observed to emerge from the lithium fluoride crystal surface on which the atoms were incident. The diffracted beam was detected with a highly sensitive pressure gauge. The usual crystal diffraction analysis indicated a wavelength of 0.600 Å. What was the predicted de Broglie wavelength?

The mass of a helium atom is

$$m = \frac{M}{N_A} = \frac{4.00 \times 10^{-3} \text{ kg/mol}}{6.02 \times 10^{23} \text{ atoms/mol}} = 6.65 \times 10^{-27} \text{ kg.}$$

According to the de Broglie equation, the wavelength then is

$$\lambda = \frac{h}{mv} = \frac{6.63 \times 10^{-34} \text{ J} \cdot \text{s}}{(6.65 \times 10^{-27} \text{ kg})(1640 \text{ m/s})}$$
$$= 0.609 \times 10^{-10} \text{ m} = 0.609 \text{ Å.}$$

This result, 1.5 percent greater than the value measured by crystal diffraction, was well within the limits of error of the experiment. Such experiments are very difficult to perform in any case, but especially since the intensities obtainable in atomic beams are quite low.

EXAMPLE 5. _____

Neutron Diffraction. The neutrons emerging from a graphite collimating tube that pierces the shield wall of a nuclear reactor may be shown [8] to have a distribution in wavelength that has a maximum value at a wavelength given closely by

$$\lambda = \frac{h}{\sqrt{5mkT}},$$

in which $m(= 1.675 \times 10^{-27}$ kg) is the mass of the neutron, k is the Boltzmann constant, and $T(= 300$ K) is the temperature at which the neutrons are in thermal equilibrium.

(a) Find this wavelength.

Substituting in the above equation yields

$$\lambda = \frac{6.63 \times 10^{-34} \text{ J} \cdot \text{s}}{\sqrt{5(1.675 \times 10^{-27} \text{ kg})(1.38 \times 10^{-23} \text{ J/K})(300 \text{ K})}}$$
$$= 1.13 \text{ Å}.$$

(b) The emerging neutron beam is allowed to fall on a crystal of calcite, cut so that atomic planes separated by 3.03 Å are parallel to its surface. For what angle θ with the crystal surface will the incident neutron beam form a first-order reflected beam whose wavelength is that found in (a)?

Reflected beams occur when Bragg's law, namely (see *Physics*, Part II, Sec. 47-6),

$$n\lambda = 2d \sin \theta \qquad n = 1, 2\,3, \dots,$$

is satisfied. Here d is the interplanar spacing and n is the order number. Thus

$$\theta = \sin^{-1} \frac{n\lambda}{2d} = \sin^{-1} \frac{(1)(1.13 \text{ Å})}{(2)(3.03 \text{ Å})}$$
$$= \sin^{-1} 0.187 = 10.8°.$$

(c) For what wavelength would a first-order beam be reflected—at this same angle—from a set of planes whose interplanar spacing is just half that given in (b), namely, 1.52 Å? Simple inspection of the Bragg relationship above shows that, if n and θ remain unchanged and if d is reduced by a factor of 2, then the wavelength must be reduced by this same factor. Thus the required wavelength is $\frac{1}{2}(1.13$ Å) or 0.57 Å. If neutrons of this wavelength are present in the incident beam, the reflection will occur.

Neutron diffraction, using neutrons generated in nuclear reactors, is now an important method for studying crystal structure, proving useful in situations in which x-rays cannot supply the desired information. Here are two examples:

1. x-Rays are scattered by the electrons in a crystal (rather than by the atomic nuclei) and thus give us information about the way electrons are distributed throughout the crystal lattice. It proves difficult to pin down the locations of hydrogen atoms in a lattice by this method because the hydrogen atom has only a single electron. Neutrons, however, *are* scattered by the atomic nuclei, including the hydrogen nucleus, and can thus provide this information.

2. In many cases the ions that form a crystal lattice have an inherent residual magnetic moment, associated with the inherent magnetism of their atomic electrons. x-Ray photons, having no inherent magnetic properties, cannot distinguish magnetic ions from nonmagnetic ions. Neutrons, however, have an inherent magnetic moment and can make these distinctions, thus providing information about a wide variety of magnetic substances that would be very difficult to study by any other method.

6-3 *THE ELECTRON MICROSCOPE*

In geometrical optics we assume that the wave nature of light need not be taken into account and that light forming a plane parallel wave can be represented by a single ray. The direction of such a ray can be changed if the light is permitted to pass from one transparent medium (air, for example) into another (such as glass). The angle through which the ray is bent is determined by the angle the incident ray makes with the normal to the surface separating the two media and by the relative index of refraction of the two media. An optical microscope, including its compound objective lens, can be designed on these principles alone. The wave nature of light enters only in a secondary way, namely, in that the index of refraction of a given type of optical glass relative to air varies with the wavelength. By forming the objective lens of two or more components, made of different varieties of optical glass, however, it can be arranged that an object, once brought to a focus, remains in focus regardless of the color (wavelength) of the light used to illuminate it.

The purpose of an optical microscope is to magnify, and it is natural to ask what factors determine the maximum useful magnification of a given instrument. More meaningfully, we ask: "How close together can two equally luminous point objects be and still have them resolved as two distinct objects?" We call this minimum distance the *resolution* of the instrument. Calculating the expected resolution of a microscope lies entirely beyond the scope of geometrical optics, because the wave nature of light enters in a direct and fundamental way. Point objects form point diffracting centers, and if two of them are too close together their diffraction patterns will overlap to the extent that they appear as the diffraction pattern of a single point object. The resolution Δx of a microscope is given by [9]

$$\Delta x = \frac{\lambda}{\sin \theta'} \tag{6-4}$$

in which θ is the angular aperture of the microscope, that is (see Fig. 6-10a), one-half the angle subtended by the objective lens at a focused axial point on the object stage. To improve the resolution of a microscope, then, we must illuminate the object with light whose wavelength is as small as possible, blue light giving better resolution than red light, for example. Equation 6-4 suggests that x-rays, having a wavelength smaller than the visible by a factor of 1000 or so, could give enormously better resolution. The problem is to build such an instrument, a task not yet usefully achieved, the principal drawback being that the index of refraction of matter for x-rays is very close to unity, making appropriate lens systems impractical. However, we have seen that electrons have matter waves whose wavelengths are in the x-ray range. Thus our interest turns to the possibility of an electron microscope.

The principal design features of an electron microscope follow along the same lines as those of the optical microscope, the trajectories of the electrons corresponding to the rays of geometrical optics. In an evacuated, field-free space the electron trajectory is a straight line. The trajectory can be bent as desired, however, by allowing the electron to enter a region in which a suitably arranged magnetic field is present. This corresponds, in geometrical optics, to allowing a light ray to enter a transparent medium such as a lens. Magnetic lenses can be designed that can focus electron trajectories in the same way that glass lenses focus light rays. The wave nature of electrons does not enter in any way into the design of such lenses. Figure 6-7 compares the principal design features of an optical and an electron microscope.

Just as in the optical case, however, the wave nature of electrons *does* enter— and in a fundamental way—in determining the resolution of the electron microscope; it is given precisely by Eq. 6-4 above, in which λ is now the de Broglie wavelength of the electrons. For the optical case, wavelengths in the ultraviolet—say, equal to 300 nm—represent a practical limit. For the electron microscope, however, electrons accelerated to, say, 100 keV are routinely used; the corresponding wavelength, which is calculated in Example 1b, proves to be $\sim$4 pm ($= 4 \times 10^{-3}$ nm), a reduction by a factor of $\sim$8 $\times$ 10^4. It is in this potentially higher resolution—not all of it realizable in practice for various reasons—that lies the fundamental advantage of the electron microscope over its optical counterpart. In practice the resolution of a good optical microscope is about 2000–3000 Å, and for the best electron microscopes it is about 2–3 Å, an improvement by a factor of about 1000.

Another advantage of the electron microscope is that, for the same wavelength, the electron kinetic energy is much smaller than the energy of the corre-

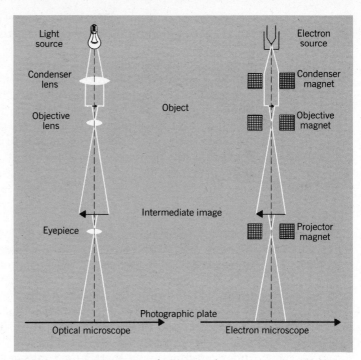

Figure 6-7. Comparison of an optical microscope and an electron microscope. The eyepiece of the optical microscope is adjusted to project a real image on a photographic plate, rather than providing a virtual image for direct visual viewing.

sponding photon (see Example 6), so that electrons will scatter elastically off specimens whose structure would be altered by photons of the same wavelength. Furthermore, one can vary the wavelength conveniently by changing the accelerating potential V. Figure 6-8a shows an electron microscope image of a thin slice of a crystal of niobium oxide, a material whose basic unit lattice cell is particularly large, its shortest edge in the plane of the figure being about 25 Å. The resolution of the microscope (3–4 Å) is good enough so that the gross structure of the crystal is readily apparent [10].

The *scanning electron microscope* [11] is a development of great interest, permitting the imaging of biological specimens in three dimensions, without the normal requirement that the specimen be in an evacuated space. The device operates much like a television camera operating in conjunction with a remote receiver. The electron beam is scanned across the specimen in a pattern like that of the background raster visible on a TV screen when no signal is being received. The scanning beam ejects secondary electrons from various points of the specimen as it passes over them. These electrons are collected by a suitable detector whose output is a time-varying potential $V(t)$ that is a measure of the efficiency of each point of the specimen as an emitter of secondary electrons. Meanwhile, a secondary electron beam located in a unit outside the microscope scans a video screen, in rigid synchronism with the primary scanning beam inside the microscope. The output potential $V(t)$ is used to control the intensity of this secondary beam and in this way an image of the specimen is "painted" on the video screen. Figure 6-8b shows one of these remarkable images.

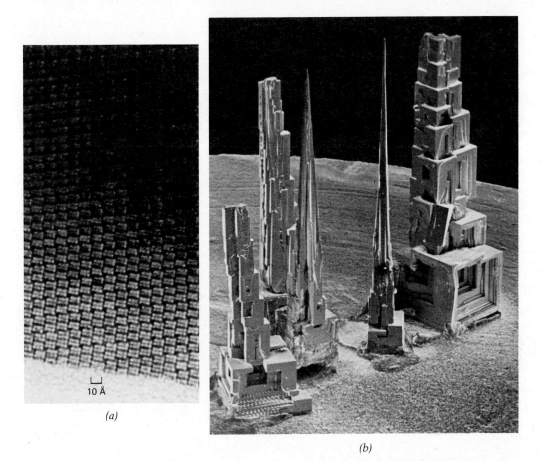

Figure 6-8. (a) An image of a thin niobium oxide crystal, formed with an electron microscope with a resolution of 3–4 Å. The scale shows a 10-Å interval. (Photo by Sumio Iijima, Department of Physics, Arizona State University.) (b) A photograph of some crystals of lead–tin telluride, taken with a scanning electron microscope. (Specimen prepared by R. W. Bicknell, photo by N. S. Griffin, Plessy Research, Allen Clark Research Centre.)

Electron wavelengths can be very much smaller than the wavelengths of x-rays produced in a conventional x-ray tube. The Stanford linear electron accelerator, for example, provides electrons at wavelengths near 10^{-16} m. The accelerator can be regarded as a microscope in the sense that it is used to study the size and structure of nuclei and their constituent neutrons and protons. Since the diameter of a single neutron or proton is about 10^{-15} m, the resolution is good enough to explore the inner structure of these particles and to shows that they are not themselves "fundamental" in the sense of being structureless.

EXAMPLE 6.

A 100-kV Electron Microscope. A typical electron microscope has an accelerating potential of 100 kV. What is the de Broglie wavelength of the electrons in its beam calculated using (a) the (approximate) classical formula and (b) the (correct) relativistic momentum formula? (c) What is the energy of a photon that has the same wavelength as that calculated in (b) above?

(a) For ease of calculation we modify the formula displayed in Example 2 by multiplying both numerator and denominator by c. We thus obtain

$$\lambda_c = \frac{h}{\sqrt{2m_0K}} = \frac{hc}{\sqrt{2(m_0c^2)(K)}}$$

$$= \frac{1240 \text{ eV·nm}}{\sqrt{2(0.511 \times 10^6 \text{ eV})(1.00 \times 10^5 \text{ eV})}}$$

$$= 3.88 \times 10^{-3} \text{ nm} = 3.88 \text{ pm}.$$

Note in the above that we have expressed the much-used quantity hc by 1240 eV·nm, a convenient formulation; see the list of constants at the beginning of the problem set.

(*b*) Again using the formula from Example 2 and modifying it as above, we have

$$\lambda_r = \frac{hc}{\sqrt{2m_0cK^2 + K^2}}$$

$$= \frac{1240 \text{ eV·nm}}{\sqrt{2(0.511 \times 10^6 \text{ eV})(1.00 \times 10^5 \text{ eV}) + (1.00 \times 10^5 \text{ eV})^2}}$$

$$= 3.70 \times 10^{-3} \text{ nm} = 3.70 \text{ pm},$$

which is the correct relativistic result. We see that it differs from the classical result by 4.9%, in full agreement with the value found using the error formula developed in Example 2. Recall that the electron microscope used to provide Fig. 6-8*a* was said to have a resolution of 3–4 Å (= 300–400 pm). Thus the resolution for this instrument is about 80–100 wavelengths. This would be a poor performance for an optical microscope, whose resolution may be roughly equal to the wavelength. In spite of this inherent disadvantage however, the resolution of the electron microscope, in absolute terms, is still about three orders of magnitude better (that is, smaller) than that of the optical microscope.

(*c*) The photon that has the same wavelength (= 3.70 pm) would be in the gamma-ray region of the spectrum. Its energy follows from

$$E = h\nu = \frac{hc}{\lambda}$$

$$= \frac{1240 \text{ eV·nm}}{3.70 \times 10^{-3} \text{ nm}} = 335 \text{ keV}.$$

Thus, if we design a gamma-ray microscope to have the same wavelength that the electrons in our electron microscope have, we must use photons that are more than three times as energetic as the electrons.

6-4 THE WAVE–PARTICLE DUALITY

The Principle of Complementarity. In our studies in classical physics we have seen that energy is transported either by waves or by particles. Water waves carry energy over the water surface and bullets transfer energy from gun to target, for example. From such experiences we built a wave model for certain macroscopic phenomena and a particle model for other macroscopic phenomena and quite naturally extended these models into visually less accessible regions. Thus we explained sound propagation in terms of a wave model and pressures of gases in terms of a particle model (kinetic theory). Our successes conditioned us to expect that entities either are particles or are waves. Indeed, these successes extended into the early twentieth century with applications of Maxwell's wave theory to radiation and the discovery of elementary particles of matter, such as the neutron and positron.

Hence, we were quite unprepared to find that to understand radiation we need to invoke a particle model in some situations, as in the Compton effect, and a wave model in other situations, as in the diffraction of x-rays. Perhaps more striking is the fact that this same wave–particle duality applies to matter as well as to radiation. The charge-to-mass ratio of the electron and its ionization trail in matter (a sequence of localized collisions) suggest a particle model, but electron diffraction suggests a wave model. We have been compelled to use both models for the same entity. It is important to note, however, that in any given measurement only one model applies—we do not use both models under the same circumstances.

Niels Bohr [12] summarized the situation in his *principle of complementarity:*

> If an experiment demonstrates the wave nature of either radiation or matter then it proves impossible to demonstrate the particle nature in the same experiment, and conversely. The description that is appropriate depends entirely on the nature of the experiment performed.

The essence of this principle is that, even though the wave and particle descriptions seem to be mutually exclusive, we are never forced to choose between them because they cannot be simultaneously revealed. The two descriptions—wave and particle—are complementary. Furthermore, our description of radiation (or of matter) is not complete unless we take into account experiments that reveal both of these aspects. Hence, radiation and matter are not simply waves nor simply particles, and it seems to be true that no classical model is available to help us probe any deeper into the subject.

An Instructive "Thought Experiment." To make the concept of complementarity more concrete, consider the "thought experiment" shown in Fig. 6-9. A beam of electrons falls on a double-slit arrangement in screen A and produces interference fringes on a screen at B. We have accepted this as convincing proof of the wave nature of the incident electrons.

Suppose, however, that we replace screen B with a small electron detector, designed to generate a "click" every time an electron strikes it. By moving the detector up and down in Fig. 6-9 we shall, by plotting the click rate as a function of position, be able to trace out the interference pattern. The clicks suggest particles (electrons) falling like raindrops on the detector; the fringes suggest waves. Have we not simultaneously shown that electrons are both waves and particles?

The answer to this question is "No." A mere click is not enough evidence to establish an entity as a particle. The concept of "particle" is tied closely to the concept of "trajectory"; our mental image is of a point following a path. As a minimum, we would like to be able to say which of the two slits in screen A the electron passed through on its way to generate a "click" at screen B. Can we obtain this information?

We can, in principle, put a very thin detector in front of each slit, designed so that if an electron passes through it, an electronic signal will be generated. We can then try to correlate each click or "screen-arrival signal" with a "slit-passage signal" (allowing, of course, for the slit-screen travel time) and in this way estab-

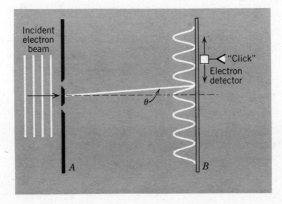

Figure 6-9. An electron beam falling on a double-slit arrangement in screen A produces interference fringes on screen B. Screen B can be replaced by an electron detector, which can be moved up and down as shown.

lish the electron trajectories. If we succeed in modifying the apparatus to do this, we shall discover a surprising thing. The fringes—which were our evidence for the wave character of the electrons—have disappeared! In passing through the slit-passage detectors the electrons were deflected in ways that totally destroyed the interference pattern.

The converse to our thought experiment is also true. If we start with an experiment that shows that electrons are particles and modify it so as to bring out the electron's wave aspect, we always destroy the evidence for particles. In addition, our thought experiment could equally well be carried out using a beam of light rather than a beam of electrons. The wave–particle duality problem and the complementarity principle apply both to radiation and to matter. We cannot develop the wave and particle aspects of either simultaneously in the same experiment, any more than an ordinary tossed coin can display simultaneously both its "head" and its "tail" aspects.

Wave–Particle Duality and Probability.   The link between the wave model and the particle model is provided by a probabilistic interpretation of the wave–particle duality. In the case of radiation, it was Einstein who united the wave and particle theories; subsequently Max Born applied a similar argument to unite wave and particle theories of matter.

On the wave picture the intensity of radiation I, is proportional to $\overline{\mathscr{E}^2}$, where $\overline{\mathscr{E}^2}$ is the average value over one cycle of the square of the electric field strength of the wave, (*Physics*, Part II, Sec. 41-9). (I is the average value of the Poynting vector and we use the symbol $\mathscr{E}$ instead of E for electric field to avoid confusion with the total energy E.) On the photon, or particle, picture the intensity is written as $I = Rh\nu$, where R is the average number of photons per unit time crossing unit area perpendicular to the direction of propagation. It was Einstein who suggested that $\overline{\mathscr{E}^2}$, which in electromagnetic theory is proportional to the radiant energy in a unit volume, could be interpreted as a measure of the average number of photons per unit volume.

Recall that Einstein introduced a granularity to radiation, abandoning the continuum interpretation of Maxwell. This leads to a statistical view of intensity. In this view, a point source of radiation emits photons randomly in all directions. The average number of photons crossing a unit area will decrease with the distance the area is from the source. This is because the photons spread out over a larger cross-sectional area the farther they are from the source. Since this area is proportional to the distance squared, we obtain, on the average, an inverse square law of intensity just as we did on the wave picture. In the wave picture we imagined spherical waves to spread out from the source, the intensity dropping inversely as the square of the distance from the source. Here, these waves, whose strength can be measured by $\overline{\mathscr{E}^2}$, can be regarded as guiding waves for the photons; the waves themselves have no energy—there are only photons—but they are a construct whose intensity measures the average number of photons per unit volume.

We use the word "average" because the emission processes are statistical in nature. We don't specify exactly how many photons cross unit area in unit time, only their average number; the exact number can fluctuate in time and space, just as in the kinetic theory of gases there are fluctuations about an average value for many quantities. We can say quite definitely, however, that the probability of having a photon cross a unit area 3 m from the source is exactly one-ninth the probability that a photon will cross a unit area 1 m from the source sometime. In the formula $I = Rh\nu$, therefore, R is an average value and is a measure of the probability of finding a photon crossing unit area in that time. If we equate the wave expression to the particle expression, we have

$$I = \frac{1}{\mu_0 c} \overline{\mathcal{E}^2} = h\nu R,$$

so that $\overline{\mathcal{E}^2}$ is proportional to R. Einstein's interpretation of $\overline{\mathcal{E}^2}$ as a probability measure of photon density then becomes clear. We expect that, as in kinetic theory, fluctuations about an average will become more noticeable at low intensities than at high intensities, so that the granular quantum phenomena would contradict the continuum classical view more dramatically there (as, for example, in the photoelectric effect discussed in Chapter 5).

In analogy to Einstein's view of radiation, Born proposed a similar uniting of the wave–particle duality for matter. Although this came after a wave theory for material particles, called wave mechanics, had been developed by Schrödinger, it was Born's interpretation that conceptualized the formal theory. Let us associate with matter waves not only a wavelength but also an amplitude. The function representing the de Broglie wave is called a *wave function*, signified by ψ. For particles moving in the x direction with constant linear momentum, for example, the wave function can be described by simple harmonic functions, such as *

$$\psi(x, t) = \psi_{\max} \cos 2\pi(\kappa x - \nu t) \qquad \text{(matter wave).} \qquad (6\text{-}5a)$$

This is analogous to

$$\mathcal{E}(x, t) = \mathcal{E}_{\max} \cos 2\pi(\kappa x - \nu t) \qquad \text{(electromagnetic wave)} \qquad (6\text{-}5b)$$

for the electric field of a harmonic electromagnetic wave of frequency ν traveling in the positive x direction. The quantity $\kappa(= 1/\lambda)$ is the *wave number* of the traveling wave.** Then $\overline{\psi^2}$ will play a role for matter waves analogous to that played by $\overline{\mathcal{E}^2}$ for waves of radiation. $\overline{\psi^2}$, the average of the square of the wave function of matter waves, is a measure of the average number of particles per unit volume; or, put another way, it is proportional to the probability of finding a particle in unit volume at a given place and time. Just as $\mathcal{E}$ is a function of space and time, so is ψ. And just as $\mathcal{E}$ satisfies a wave equation, it turns out that so does ψ (Schrödinger's equation). The quantity $\mathcal{E}$ is a (radiation) wave associated with a photon, and ψ is a (matter) wave associated with a mass particle; neither $\mathcal{E}$ nor ψ gives the path of the motion but instead are associated waves that measure probability densities.

As Born [13] says:

> According to this view, the whole course of events is determined by the laws of probability; to a state in space there corresponds a definite probability, which is given by the de Broglie wave associated with the state. A mechanical process is therefore accompanied by a wave process, the guiding wave, described by Schrödinger's equation, the significance of which is that it gives the probability of a definite course of the mechanical process. If, for example, the amplitude of the guiding wave is zero at a certain point in space, this means that the probability of finding the electron at this point is vanishingly small.

---

* Actually, the wave function describing the motion of a free particle turns out to be a complex quantity, in the mathematical sense that it involves the complex parameter $\sqrt{-1}$. The function $\psi(x, t)$ is the real component only of that complex quantity. Because we do not deal quantitatively with wave functions in this chapter, this formulation will serve quite well to guide our thinking.

** Often the quantity $k(= 2\pi/\lambda)$ is called the wave number. It might better be called the *angular wave number*. These two definitions are in direct analogy with the definitions of *frequency* ν and of *angular frequency* $\omega(= 2\pi\nu)$.

Just as in the Einstein view of radiation we do not specify the exact location of a photon at a given time but specify instead by $\overline{\mathscr{E}^2}$ *the probability* of finding a photon at a certain space interval at a given time, so here in Born's view we do not specify the exact location of a particle at a given time but specify instead by $\overline{\psi^2}$ the *probability* of finding a particle in a certain space interval at a given time. And, just as we are accustomed to adding wave functions $(\mathscr{E}_1 + \mathscr{E}_2 = \mathscr{E})$ for two superposed electromagnetic waves whose resultant intensity is given by $\mathscr{E}^2$, so we add wave functions for two superposed matter waves $(\psi_1 + \psi_2 = \psi)$ whose resultant intensity is given by ψ^2. That is, a *principle of superposition* applies to matter as well as to radiation. This is in accordance with the striking experimental fact that matter exhibits interference and diffraction properties, a fact that simply cannot be understood on the basis of ideas in classical mechanics. Because waves can be superposed either constructively (in phase) or destructively (out of phase), two waves can combine either to yield a resultant wave of large intensity or to cancel. But two classical particles of matter cannot combine in such a way as to cancel.

You might accept the logic of this fusion of wave and particle concepts but nevertheless ask whether a probabilistic or statistical interpretation is necessary. It was Heisenberg and Bohr who, in 1927, first showed how essential the concept of probability is to the union of wave and particle descriptions of matter and radiation. We investigate these matters in succeeding sections.

6-5 THE UNCERTAINTY PRINCIPLE

The use of probability considerations is not strange to classical physics. However, in classical physics the basic laws (such as Newton's laws) are deterministic, and statistical analysis is simply a practical device for treating very complicated systems. According to Heisenberg and Bohr, however, the probabilistic view is the fundamental one in quantum physics, and determinism must be discarded. Let us see how this conclusion is reached.

In classical mechanics the equations of motion of a system with given forces can be solved to give us the position and momentum of a particle at all values of time. All we need to know are the precise position and momentum of the particle at some value of time $t = 0$ (the initial conditions) and the future motion is determined exactly. This mechanics has been used with great success in the macroscopic world—for example, in astronomy and in space navigation, to predict the subsequent motions of comets and space probes in terms of their initial motions, in agreement with our subsequent observations. Note, however, that in the process of making observations, the observer interacts with the system. An example from astronomy is the precise measurement of the position of the moon by bouncing radar pulses from it. The motion of the moon is disturbed by the measurement, but because of the very large mass of the moon, the disturbance can be ignored. On a somewhat smaller scale, as in a very well-designed macroscopic experiment on earth, such disturbances are also usually small, or at least controllable, and they can be taken into account accurately ahead of time by suitable calculations. Hence, it was naturally assumed by classical physicists that in the realm of microscopic systems the position and momentum of an object, such as an electron, could be determined precisely by observations in a similar way. Heisenberg and Bohr questioned this assumption.

The situation is somewhat similar to that existing at the birth of relativity theory. Physicists spoke of length intervals and time intervals without asking critically how we actually measured them. For example, they spoke of the simultaneity of two separated events without even asking how one would physically go about establishing simultaneity. In fact, Einstein showed that simultaneity was not an absolute concept at all, as had been assumed previously, but that two separated events that are simultaneous to one inertial observer occur at different times to another inertial observer. Simultaneity is a relative concept. Similarly, then, we must ask ourselves critically how we actually measure position and momentum.

Can we determine by actual experiment at the same instant both the position and momentum of matter or radiation? The answer given by quantum theory, is "not more accurately than is allowed by the Heisenberg uncertainty principle." This principle [14,15] states that it is not possible to measure simultaneously the exact component of momentum, p_x, say, of a particle and its exact corresponding coordinate position, x. Instead, our precision of measurement is limited in a fundamental way by the measuring process itself such that

$$\Delta p_x \cdot \Delta x \gtrsim h \qquad \text{(Heisenberg's uncertainty principle)}, \qquad (6\text{-}6a)$$

where Δp_x is the uncertainty in our measurement of the momentum p_x, Δx is the uncertainty in our measurement of the position x, and the mathematical symbol means "approximately equal to or greater than." The quantity h is the Planck constant, which (once again!) turns up in a context of fundamental significance.* There are corresponding relations for other components of momentum and position, namely, $\Delta p_y \cdot \Delta y \gtrsim h$ and $\Delta p_z \cdot \Delta z \gtrsim h$.

It is important to realize that the uncertainty principle has nothing to do with improvements in instrumentation leading to better simultaneous determinations of p_x and x. Rather, the principle says that even with ideal instruments we can never do better than Eq. 6-6a; in practice we shall always do worse. Note also that the *product* of the uncertainties is involved so that, for example, the more we modify an experiment to improve our measurement of p_x, the more we give up our ability to measure x precisely. If we know p_x exactly, then we know nothing at all about x; that is, as $\Delta p_x \to 0$, then $\Delta x \to \infty$, and conversely. Hence, the restriction is not on the accuracy to which x or p_x can be measured, but on the product $\Delta p_x \cdot \Delta x$ in a simultaneous measurement of both. Finally, we must not think that x and p_x *really exist* to infinite precision, but that the measuring process somehow interferes with our ability to unearth these hidden values. Equation 6-6a represents, rather, a limitation beyond which the assignment of position and momentum values simply has no justification; it represents a fundamental breakdown in the concept of the notion of "particle."

The uncertainty principle can also be written in an equivalent form, namely,

$$\Delta E \cdot \Delta t \gtrsim h \qquad \text{(Heisenberg's uncertainty principle)}. \qquad (6.6b)$$

Here ΔE is the uncertainty in our knowledge of the energy of a system (an atom in an excited state, say) and Δt is the time interval available for the measurement of E (the mean life of the atom in that state, say). This relation also sets an ideal limit to the measurement of two corresponding quantities.

---

* We have not given a formal definition of just what we mean by Δp_x or by Δx in Eq. 6-6a, nor is it important that we do so in this introductory treatment. The essential point to grasp is that the product of these two quantities, *in whatever reasonable way they may be defined*, cannot be reduced to zero, no matter how hard we may try. For some formal definitions of the uncertainties the quantity $h/2\pi$ or $h/4\pi$ appears in place of h on the right side of Eq. 6-6a. In all the examples that we shall give, however, our definitions of the uncertainties will be reasonable and their product will always work out to be very close to the Planck constant h.

Heisenberg's relations can be shown to follow from wave (or quantum) mechanics, but that mechanics was invented, after all, to explain the experiments we have already discussed (and others). Hence, the principle is grounded in experiment. We shall show shortly examples of the consistency of Eqs. 6-6 with experiment. Notice first, however, that it is the Planck constant h that again distinguishes the quantum results from the classical ones. If h in Eqs. 6-6 were zero there would be no basic limitation on our measurement at all, which is the classical view. And again it is the smallness of h that takes the principle out of the range of our ordinary experiences. This is analogous to the smallness of β ($= v/c$) in the macroscopic situations taking relativity out of the range of ordinary experience. In principle, therefore, classical physics is of limited validity and in the microscopic domain it will lead to contradictions with experiments. For if we cannot determine x and p_x simultaneously, then we cannot specify the initial conditions of motion exactly; therefore, we cannot precisely determine the future behavior of a system. Instead of making deterministic predictions, we can only state the possible results of an observation, giving the relative probabilities of their occurrence. Indeed, since the act of observing a system disturbs it in a manner that is not completely predictable, the observation changes the previous motion of the system to a new state of motion that cannot be completely known.

Let us now illustrate the physical origin of the uncertainty principle. First, we use a thought experiment due to Bohr to derive Eq. 6-6a. Let us say that we wish to measure as accurately as possible the position of a "point" particle, such as an electron. For greatest precision we use a microscope to view the electron (Fig. 6-10a). But to see the electron we must illuminate it, for it is really the light quanta scattered by the electron that the observer sees. At this point, even before any calculations are made, we can see the uncertainty principle emerge. The very act of observing the electron disturbs it. The moment we illuminate the electron, it recoils because of the Compton effect, in a way that cannot be completely determined (see Question 26). But if we don't illuminate the electron, we don't see

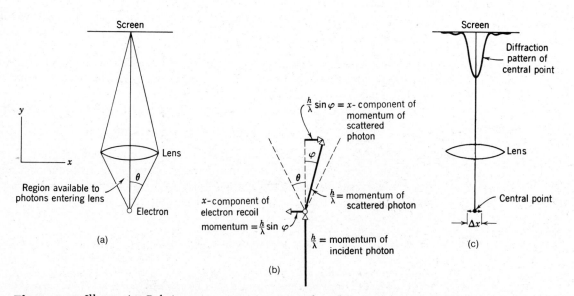

Figure 6-10. Illustrating Bohr's gamma-ray microscope thought experiment. (*a*) The geometry and the coordinate axes. (*b*) The scattering of the gamma ray from the recoiling electron. (*c*) The separation Δx of points barely resolvable from the central point, whose diffraction pattern image is shown.

(detect) it. Hence the uncertainty principle refers to the measuring process itself, and expresses the fact that there is always an undetermined interaction between observer and observed; there is nothing we can do to avoid the interaction or to allow for it ahead of time. Let us try to reduce the disturbance to the electron as much as possible by using a very weak source of light. The very weakest we can get it is to assume that we can see the electron if only *one* scattered photon enters the lens of the microscope (see Fig. 6-10*b*). The magnitude of the momentum of the photon is $p = h/\lambda$. But the photon may have been scattered *anywhere* within the angular range 2θ subtended by the lens at the electron. This is why the interaction cannot be allowed for. The conservation equations in the Compton effect can be satisfied for *any* angle of scattering within this angular range. Hence, we see that the x component of the momentum of the photon can vary from $+p \sin \theta$ to $-p \sin \theta$ and is uncertain after the scattering by an amount

$$\Delta p_x \cong 2p \sin \theta = \frac{2h}{\lambda} \sin \theta.$$

Because of conservation of momentum the electron must receive a recoil momentum in the x direction that is equal to the x momentum change in the photon and, therefore the x momentum of the electron is uncertain by this same amount. Notice that to reduce Δp_x we should use light of longer wavelength, or use a microscope with a lens subtending a smaller angle θ.

What about the location along x of the electron? Recall that the image of a point object illuminated by light (photons) is not a point, but a diffraction pattern (see Fig. 6-10*c*). The image of the electron is "fuzzy." It is the resolving power of a microscope that determines the highest accuracy to which the electron can be located. Let us take the uncertainty Δx to be the linear separation of points in the object barely resolvable in the image. Then (see Eq. 6-4) we have

$$\Delta x \cong \frac{\lambda}{\sin \theta}.$$

The one scattered photon at our disposal must have originated then *somewhere* within this range of the axis of the microscope, so the uncertainty in the electron's location is Δx. (We can't be sure exactly where any one photon originates, even though a large number of photons will show the statistical regularity of the diffraction pattern shown in the figure.) Notice that to reduce Δx we should use light of shorter wavelength, or a microscope with larger θ.

If now we take the product of the uncertainties, we obtain

$$\Delta p_x \cdot \Delta x = \left(\frac{2h}{\lambda} \sin \theta\right)\left(\frac{\lambda}{\sin \theta}\right) = 2h. \tag{6-7}$$

The quantity on the right is of the order of magnitude of h. Also, because we have given only somewhat arbitrary definitions of the uncertainties and because we have assumed absolutely ideal instruments and measuring techniques, we are justified in replacing the "equal" sign in this equation by an "approximately equal to or greater than" sign. Thus we may properly say that we have derived Eq. 6-6*a*, the Heisenberg uncertainty principle.

We cannot *simultaneously* make Δp_x and Δx in Eq. 6-7 as small as we wish, for the procedure that makes one small makes the other large. For instance, if we use light of short wavelength (for example, gamma rays) to reduce Δx (better resolution), we increase the Compton recoil and increase Δp_x, and conversely. Indeed, the wavelength λ doesn't even appear in the result. In practice an experiment might do much worse than Eq. 6-7 suggests, for that result represents the very ideal possible. We arrive at it, however, from genuinely measurable physical results, namely, the Compton effect and the resolving power of a lens.

There really should be no mystery in your mind about our result; it is a necessary consequence of the quantization of radiation. For to have light we had to have at least one quantum of it. It is this single scattered light quantum, carrying a momentum of magnitude $p = h/\lambda$, that gives rise to an interaction between the microscope and the electron. The electron is disturbed in a manner that cannot be exactly predicted or controlled. Consequently we cannot know exactly what the coordinates and momentum of the electron will be after the interaction. If classical physics were valid, and since radiation is regarded there as continuous rather than granular, we could reduce the illumination to arbitrarily small levels and deliver arbitrarily small momentum while using arbitrarily small wavelengths for "perfect" resolution. In principle there would be no simultaneous lower limit to resolution or momentum recoil and there would be no uncertainty principle. But we cannot do this; the single photon is indivisible. Again we see, from $\Delta p_x \cdot \Delta x > h$ that the value of the Planck constant determines the minimum uncontrollable disturbance that distinguishes quantum physics from classical physics.

Now let us consider Eq. 6-6b relating energy and time. For the case of a free particle we can obtain Eq. 6-6b from Eq. 6-6a, which relates position and momentum, as follows. Consider again an electron moving along the x axis whose energy we can write as $E = p_x^2/2m$. If p_x is uncertain by Δp_x, then the uncertainty in E is given by differentiation as $\Delta E = (2p_x/2m)\,\Delta p_x = v_x\,\Delta p_x$. Here v_x can be interpreted as the recoil velocity along x of the electron that is illuminated with light. If the time interval required for the observation of the electron is Δt, then the uncertainty in its x position is $\Delta x = v_x\,\Delta t$. Combining $\Delta t = \Delta x/v_x$ and $\Delta E = v_x\,\Delta p_x$, we obtain $\Delta E \cdot \Delta t = \Delta p_x\,\Delta x$. But $\Delta p_x \gtrsim h$. Hence, we obtain the result

$$\Delta E \cdot \Delta t \gtrsim h.$$

EXAMPLE 7.

The Uncertainty Principle—Bullets and Electrons. The speed of a bullet ($m = 50$ g) and the speed of an electron ($m =9.11 \times 10^{-28}$ g) are measured to be the same, namely, 300 m/s, with an uncertainty of 0.01 percent. With what fundamental accuracy could we have located the position of each, if the position is measured simultaneously with the speed in the same experiment?

For the electron,

$$p = mv = (9.11 \times 10^{-31}\ \text{kg})(300\ \text{m/s})$$
$$= 2.73 \times 10^{-28}\ \text{kg} \cdot \text{m/s}$$

and

$$\Delta p = 0.01\%\ \text{of}\ p = (0.0001)(2.7 \times 10^{-28}\ \text{kg} \cdot \text{m/s})$$
$$= 2.73 \times 10^{-32}\ \text{kg} \cdot \text{m/s}$$

so that

$$\Delta x \gtrsim \frac{h}{\Delta p} = \frac{6.63 \times 10^{-34}\ \text{J} \cdot \text{s}}{2.73 \times 10^{-32}\ \text{kg} \cdot \text{m/s}} \cong 2 \times 10^{-2}\ \text{m} = 2\ \text{cm}.$$

For the bullet,

$$p = mv = (0.05\ \text{kg})(300\ \text{m/s}) = 15\ \text{kg} \cdot \text{m/s}$$

and

$$\Delta p = (0.0001)(15\ \text{kg} \cdot \text{m/s}) = 1.5 \times 10^{-3}\ \text{kg} \cdot \text{m/s}$$

so that

$$\Delta x \gtrsim \frac{h}{\Delta p} = \frac{6.63 \times 10^{-34}\ \text{J} \cdot \text{s}}{1.5 \times 10^{-3}\ \text{kg} \cdot \text{m/s}} = 4 \times 10^{-31}\ \text{m}.$$

Hence, for macroscopic objects such as bullets the uncertainty principle sets no practical limit to our measuring procedure, Δx in this example being about 10^{-16} times the diameter of a nucleus. But, for microscopic objects such as electrons, there are practical limits, Δx in this example being about 10^8 times the diameter of an atom.

EXAMPLE 8.

The Uncertainty Principle Demonstrated. You wish to localize the y coordinate of an electron by having it pass through a narrow slit. By considering the diffraction of the associated wave, show that as a result you introduce an uncertainty in the momentum of the electron such that $\Delta p_y \, \Delta y \gtrsim h$.

In Fig. 6-11 we show the diffraction pattern formed on a screen by a parallel beam of monoenergetic electrons that first passed through a single slit placed in the path of the beam. From the wave point of view we can regard this as the passage of a plane monochromatic wave of wavelength λ through a single slit of width a. From physical optics (*Physics*, Part II, Sec. 46-2), we know that the angle θ to the first diffraction minimum is given by $a \sin \theta = \lambda$. From the particle point of view, the diffraction pattern gives the statistical distribution on the screen of a large number of electrons of incident momentum p that may be deflected up or down on passing through the slit.

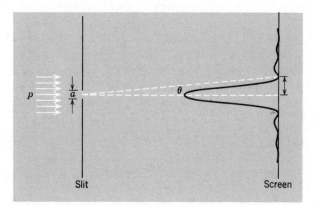

Figure 6-11. Example 8. A parallel beam of electrons, each of momentum p, is diffracted at a slit of width a, forming a diffraction pattern on a screen. The deflection angle θ to the first minimum is shown.

Each electron is regarded as being diffracted independently. The probability that an electron hits some point on the screen is determined by ψ^2, the square of the amplitude of the guiding matter wave. Where ψ^2 is zero, no electrons are observed; where ψ^2 is large, many electrons are observed. At low intensities, a short exposure will give an average pattern like the high-intensity one but with statistical fluctuations, whereas a long exposure will clearly reveal the high-intensity pattern.

For a *single* electron arriving at the screen, therefore, we do not know exactly where it passes through the slit, only that it did pass through. Hence, the uncertainty in its y coordinate at the slit is of the order of the slit width, that is,

$$\Delta y \cong a.$$

As for the y component of the momentum of a single electron at the slit, again we are uncertain of its value, but we know from the existence of the diffraction pattern that electrons do acquire vertical momentum on being deflected there. If we consider θ to be an *average* deflection angle (we don't know exactly where a single electron will hit the screen), then the uncertainty Δp_y in the y component of momentum of a single electron is of the order of $p \sin \theta$, that is,

$$\Delta p_y \cong p \sin \theta.$$

Hence,

$$\Delta y \cdot \Delta p_y \cong ap \sin \theta.$$

But $p = h/\lambda$ and $a \sin \theta = \lambda$, so

$$\Delta y \cdot \Delta p_y \cong \frac{\lambda h}{\lambda} = h.$$

Allowing for the approximations we have made and the ideal conditions we have assumed (see footnote on page 213), we may conclude that we have demonstrated the Heisenberg uncertainty principle.

EXAMPLE 9.

Putting Particles into Boxes. (*a*) Consider an electron confined to an atom of (say) neon, whose diameter we may take to be 3×10^{-10} m. What, approximately, must be the electron's momentum? Its kinetic energy?

Because we are seeking only an order-of-magnitude answer, let us treat the problem as one-dimensional and imagine the electron's motion always to be parallel to some arbitrary direction, which we may take as an x axis. The uncertainty Δx in our knowledge of the position of the electron along this axis is roughly equal to the diame-

ter of the atom. Thus, from Eq. 6-6a but dropping the subscript, the uncertainty in its momentum must be

$$\Delta p > \frac{h}{\Delta x} = \frac{6.63 \times 10^{-34} \text{ J} \cdot \text{s}}{3 \times 10^{-10} \text{ m}} = 2.2 \times 10^{-24} \text{ kg} \cdot \text{m/s}.$$

If the electron has a momentum p inside the atom, we have no idea in which direction along our axis this momentum vector points. Thus Δp must be roughly equal to $p - (-p)$ or $2p$. This gives us $p \cong \Delta p/2 = 1.1 \times 10^{-24}$ kg·m/s. The kinetic energy follows from

$$K = \frac{p^2}{2m} = \frac{(1.1 \times 10^{-24}\text{ kg·m/s})^2}{2(9.11 \times 10^{-31}\text{ kg})(1.60 \times 10^{-19}\text{ J})} \cong 4\text{ eV}.$$

The attractive Coulomb force is easily strong enough to confine an electron with this much kinetic energy to an atom. Our conclusion: The presence of electrons in atoms is totally consistent with the Heisenberg uncertainty principle.

The entire point of this example is to point out that, strictly as a consequence of the uncertainty principle, confining an electron to a restricted region of space means that it must have at least a certain minimum kinetic energy. The electron cannot, for example, be at rest within the atom, because in that case there would be no uncertainty as to its momentum ($\Delta p = 0$), which in turn would require that $\Delta x \to \infty$. An electron that fulfills this condition is no longer confined!

(b) Consider now the nucleus of (say) a silver atom, whose diameter may be taken to be 1×10^{-14} m. If this nucleus contains an electron, what must its kinetic energy be?

Just as above, we find that

$$p = \frac{\Delta p}{2} > \frac{h}{2\,\Delta x} = \frac{6.63 \times 10^{-34}\text{ J·s}}{2(1 \times 10^{-14}\text{ m})}$$

$$= 3.3 \times 10^{-20}\text{ kg·m/s}.$$

This is so much higher than the momentum that we found in (a) that we had better calculate the kinetic energy from the relativistic formula. From Eq. 3-13a, then

$$K = \sqrt{(pc)^2 + (m_0 c^2)^2} - m_0 c^2$$

$$= \sqrt{\left[\frac{(3.3 \times 10^{-20}\text{ kg·m/s})}{\times (3.00 \times 10^{8}\text{ m/s})}}{(1.60 \times 10^{-13}\text{ J/MeV})}\right]^2 + (0.511\text{ MeV})^2} - 0.511\text{ MeV}$$

$$\cong 60\text{ MeV}.$$

No known force could hold an electron with anything like this amount of kinetic energy confined to a nucleus. Our conclusion: The presence of electrons in the nucleus is *not* consistent with the Heisenberg uncertainty principle. The evidence from other sources that there are indeed no electrons in the nucleus is overwhelming.

(c) Repeat part (b), but consider the possibility of a neutron confined to the nucleus.

The neutron momentum is just that calculated in part (b), namely, 3.3×10^{-20} kg·m/s. However the mass of a neutron ($= 1.68 \times 10^{-27}$ kg) is so much greater than that of an electron that we can safely revert to the classical formula for finding the kinetic energy. Proceeding as in part (a), then,

$$K = \frac{p^2}{2m} = \frac{(3.3 \times 10^{-20}\text{ kg·m/s})^2}{2(1.68 \times 10^{-27}\text{ kg})(1.6 \times 10^{-13}\text{ J/MeV})}$$

$$\cong 2\text{ MeV}.$$

The strong nuclear binding forces can easily confine a neutron (or a proton) with this much kinetic energy to the nucleus. Our conclusion: The presence of neutrons in the nucleus is totally consistent with the Heisenberg uncertainty principle. The evidence from other sources that neutrons are indeed nuclear constituents is overwhelming.

6-6 A DERIVATION OF THE UNCERTAINTY PRINCIPLE

We can derive the Heisenberg uncertainty principle in a somewhat formal way by combining the de Broglie-Einstein relations (Eqs. 6-1) with certain properties that are universal to all waves. Let us first examine those properties.

Consider an infinitely long sinusoidal wave of frequency ν and wave number κ ($= 1/\lambda$), whose equation is

$$y(x,\, t) = y_{\max} \cos 2\pi(\kappa x - \nu t).$$

If we view this wave at one instant of time, say, $t = 0$, we can picture it as extending in one dimension over all values of x, as Fig. 6-12a suggests. Its wave number κ has a perfectly definite value, subject to no uncertainty, as Fig. 6-12b suggests. On the other hand, the infinite wave of Fig. 6-12a is featureless, its amplitude being the same for all values of x. If, for the moment, we were to regard this wave as a matter wave representing an electron, it is clear that the wave gives no indication whatever of the position of the electron along the x axis. The uncertainty in x, then, is infinitely great.

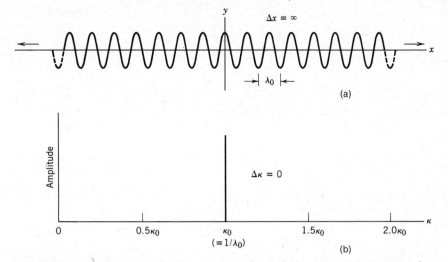

Figure 6-12. (a) A "snapshot" of an infinitely long sinusoidal wave at the instant $t = 0$. (b) Such a wave has only a single, precisely fixed, wave number κ_0 $(= 1/\lambda_0)$.

If we replace the x axis of Fig. 6-12a by a t axis, we can then view this figure as the time variation of the wave at a particular position, say, $x = 0$. As we watch the wave go endlessly by, we note that its frequency ν has a perfectly definite value, subject to no uncertainty. On the other hand, to be absolutely sure of this we would have to make measurements for an infinitely long time so that the uncertainty in our knowledge of time for this featureless wave is infinitely great. Summarizing what we have said, then, about the uncertainties in various measured quantities of the infinite wave of Fig. 6-12, we put

$$\Delta\kappa = 0, \Delta x = \infty \quad \text{and} \quad \Delta\nu = 0, \Delta t = \infty.$$

Consider now Fig. 6-13a, which represents a *wave train* or a *pulse*, consisting, in this case, of nine half-periods (that is, 4.5 full periods) of the wave of Fig. 6-12a; its amplitude is zero outside this region. In order to have such a pulse we must superimpose a number of infinitely long sinusoidal waves of different wave numbers. In the simplest case of two such wave numbers we obtain the familiar phenomenon of beats, the amplitude varying in this case in a regular way along the wave (see *Physics*, Part I, Sec. 20-6). To construct a pulse having a definite extent in space, that is, having a beginning and an end like the pulse of Fig. 6-13a, we must synthesize waves having a continuous spectrum of wave numbers within a roughly defined range $\Delta\kappa$.

There is a well-known procedure, using Fourier theory [16], by which a synthesis of the kind we have described above can be carried out. The component waves, because of their phase differences and amplitudes, interfere destructively to give zero resultant amplitude everywhere outside a limited region, whereas within this region they combine constructively to form the pulse. For the particular pulse shown in Fig. 13a, whose extension in space is Δx $(= 4.5 \lambda_0)$, Fourier theory requires that we choose the particular distribution of wave numbers shown in Fig. 6-13b; we define the quantity $\Delta\kappa$ shown in that figure as the uncertainty of the distribution in wave number.

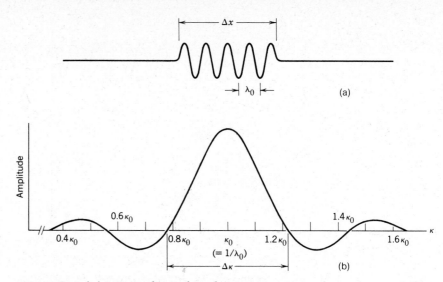

Figure 6-13. (*a*) A "snapshot" of a pulse or wave train at the instant $t = 0$. The uncertainty Δx in x is shown. (*b*) Such a pulse can be synthesized by combining a large number of infinite waves, whose wave numbers are distributed as shown. The amplitude is in arbitrary units, a negative amplitude signifying a phase shift of 180°. The uncertainty $\Delta\kappa$ in the wave number is (somewhat arbitrarily) defined as indicated.

In general, the relationship between Δx and $\Delta\kappa$ proves to be[*] [16]

$$\Delta x \cdot \Delta\kappa \cong 1. \tag{6-8a}$$

If, for example, a light pulse of spatial extent Δx is sent through an optical system containing lenses or other transparent elements, the system will behave exactly as if it had received a superposition of sinusoidal waves whose wave numbers were distributed throughout a range $\Delta\kappa$ in a way determined by Fourier theory.

In the case illustrated in Fig. 6-13*a* the uncertainty in position, Δx, is given by $4.5\lambda_0 = 4.5/\kappa_0$. The uncertainty in wave number, $\Delta\kappa$, is seen from Fig. 6-13*b* to be $(1.22 - 0.78)\kappa_0 = 0.44\kappa_0$. The product of these two uncertainties is $(4.5/\kappa_0)(0.44\kappa_0) \cong 2$, which is consistent with Eq. 6-8*a* when the arbitrariness in the definition of $\Delta\kappa$ is taken into account.

A group of waves that has a finite extension in space also has a finite duration at a given observation point. If Δt is the duration of the pulse, then its time of arrival can be determined within this precision. To construct a wave train having a finite duration in time, we must synthesize sinusoidal waves having a continuous spectrum of frequencies within a range $\Delta\nu$. The relationship between Δt and $\Delta\nu$ proves to be [16]

$$\Delta\nu \cdot \Delta t \cong 1. \tag{6-8b}$$

An electromagnetic pulse of duration Δt has no periodicity, but if such a pulse is sent through a circuit whose behavior is frequency-dependent, the circuit will

---

* The approximation signs appear in Eqs. 6-8*a* and 6.8*b* because, although there would be no argument about our definition of Δx in Fig. 6-13*a*, we have been a little arbitrary about our definition of $\Delta\kappa$ in Fig. 6-13*b*. We might, for example, have defined it to be the half-width of the central maximum in that figure.

respond precisely the same as it would if it had received a superposition of infinitely long sinusoidal waves whose frequencies were distributed throughout a range $\Delta\nu$ in a way determined by Fourier theory.

So much for the generalized properties of waves. We now turn to the derivation of the Heisenberg uncertainty principle. First, we write down the de Broglie-Einstein relations (Eqs. 6-1)

$$p = \frac{h}{\lambda} = h\kappa \quad \text{and} \quad E = h\nu,$$

expressing the former in terms of the wave number κ. Combining these relations with Eqs. 6-8 leads at once to

$$\Delta p \cdot \Delta x \gtrsim h \quad \text{and} \quad \Delta E \cdot \Delta t \gtrsim h,$$

which are the statements of the Heisenberg uncertainty principle given in Eqs. 6-6. Note that we have included the approximation symbol in these equations because it appears in Eqs. 6-8, and we have included the "greater than" symbol because our derivation deals strictly with the case of ideal measurements although we know that in practice our uncertainties will always be greater; compare the footnotes on pages 213 and 220.

As another illustration of the Heisenberg uncertainty principle, consider the excitation of an atom. Suppose that we excite an atom by having it absorb a photon from incident monochromatic light, bearing in mind that for true monochromatic light there is no uncertainty in the frequency ν or in the photon energy E. Let us ask: At what time does the atom absorb a photon and become excited? From Eq. 6-6b the answer is $\Delta t > h/\Delta E$ or $\Delta t \to \infty$, so that the time of absorption is completely indeterminate. Experimentally, to find the moment of absorption we would have to mark the original, featureless, wave train somehow in order to follow it; in practice we might use a shutter mechanism to interrupt it and "chop" the light beam. But this act of measurement disturbs the assumed monochromatic nature of the light; we are now dealing with a finite pulse, which necessarily has a frequency spread $\Delta\nu$ and a corresponding uncertainty ΔE in the energy. The narrower the pulse and the sharper Δt, the greater the spread in energy ΔE.

EXAMPLE 10.

Designing a Radar Receiver. A radar transmitter emits pulses 0.50 μs long at a wavelength of 1.5 cm. (a) To what frequency should the radar receiver be tuned? (b) What should be its bandwidth? That is, to what range of frequencies should it be able to respond?

(a) The receiver should be tuned to a frequency given by

$$\nu_0 = \frac{c}{\lambda_0} = \frac{3.0 \times 10^8 \text{ m/s}}{0.015 \text{ m}} = 2.0 \times 10^{10} \text{ Hz} = 20 \text{ GHz}.$$

The wave number κ_0 corresponding to this central frequency is simply $1/\lambda_0 = 1/(0.015 \text{ m})$ or 67m^{-1}.

(b) The receiver's bandwidth is given approximately by

$$\Delta\nu = \frac{1}{\Delta t} = \frac{1}{0.50 \times 10^{-6} \text{ s}} = 2.0 \times 10^6 \text{ Hz} = 2.0 \text{ MHz}.$$

If the receiver cannot respond to frequencies throughout this range, it will be unable to reproduce faithfully the shape of the transmitted radar pulse. If the receiver's bandwidth is substantially greater than this value, random "noise" frequencies, not significantly present in the transmitted pulse, will make it more difficult to detect weak signals.

EXAMPLE 11.

An Excited Atom Emits a Photon. An atom can radiate at any time after it is excited. It is found that in a typical case the average excited atom has a lifetime of about 10^{-8} s. That is, during this period it emits a photon and is deexcited.

(a) What is the minimum uncertainty $\Delta\nu$ in the frequency of the photon?

From Eq. 6-8b, we have

$$\Delta\nu\,\Delta t \gtrsim 1$$

or

$$\Delta\nu \gtrsim \frac{1}{\Delta t}.$$

With $\Delta t = 10^{-8}$ s we obtain $\Delta\nu \gtrsim 10^8$ s^{-1}.

(b) Such photons from sodium atoms appear in a spectral line centered at $\lambda = 589$ nm. What is the fractional width of the line, $\Delta\nu/\nu$?

For $\lambda = 589$ nm, we obtain

$$\nu = \frac{c}{\lambda} = \frac{3 \times 10^8 \text{ m/s}}{589 \times 10^{-9} \text{ m}} = 5 \times 10^{14} \text{ s}^{-1}.$$

Hence

$$\frac{\Delta\nu}{\nu} = \frac{1 \times 10^8 \text{ s}^{-1}}{5 \times 10^{14} \text{ s}^{-1}} = 2 \times 10^{-7},$$

or about one part in five million.

This is the so-called *natural width* of the spectral line. The line is much broader in practice because of the Doppler broadening and pressure broadening due to the motions and collisions of atoms in the source.

(c) Calculate the uncertainty ΔE in the energy of the excited state of the atom.

The energy of the excited state is not precisely measurable because only a finite time is available to make the measurement. That is, the atom does not stay in an excited state for an indefinite time but decays to its lowest energy state, emitting a photon in the process. The uncertainty in energy of the photon equals the uncertainty in energy of the excited state of the atom, in accordance with the energy conservation principle. From Eq. 6-6b, with Δt equal to the mean lifetime of the excited state, we have

$$\Delta E \gtrsim \frac{h}{\Delta t} = \frac{h}{\Delta t} = \frac{6.63 \times 10^{-34} \text{ J}\cdot\text{s}}{(10^{-8}\text{ s})(1.50 \times 10^{-19}\text{ J/eV})} = 4 \times 10^{-7}\text{ eV}.$$

This agrees, of course, with the value obtained from part (a) by multiplying the uncertainty in photon frequency $\Delta\nu$ by h to obtain $\Delta E = h\,\Delta\nu$.

The uncertainty in energy of an excited state is usually called the *energy-level width* of the state.

(d) From the previous results determine, to within an accuracy ΔE, the energy E of the excited state of a sodium atom, relative to its lowest energy state, that emits a photon whose wavelength is centered at 589 nm.

We have

$$\frac{\Delta\nu}{\nu} = \frac{h\,\Delta\nu}{h\nu} = \frac{\Delta E}{E}.$$

Hence,

$$E = \frac{\Delta E}{(\Delta\nu/\nu)} = \frac{4 \times 10^{-7}\text{ eV}}{2 \times 10^{-7}} = 2\text{ eV},$$

in which we have used the results of the calculations in parts (b) and (c).

6-7 INTERPRETATION OF THE UNCERTAINTY PRINCIPLE

If we take the uncertainty principle as fundamental, then we can understand why it is possible for both light and matter to have a dual, wave–particle nature. Max Born [13] has stated the connection as follows:

> It is just the limited feasibility of measurements that define the boundaries between our concepts of a particle and a wave. The corpuscular [particlelike] description means at bottom that we carry out the measurements with the object of getting exact information about momentum and energy relations [for example, in the Compton effect], while experiments which amount to determination of place and time we can always picture to ourselves in terms of the wave representation [for example, passage of electrons through thin foils and observations of deflected beam]. . . . [But] just as every determination of position carries with it an uncertainty in the momentum, and every determination of time an uncertainty in the energy, so the converse is also true. The more accurately we determine momentum and energy, the more latitude we introduce into the position of the particle and time of an event.

If we try to determine experimentally whether radiation is a wave or a particle, for example, we find that an experiment that forces radiation to reveal its wave character strongly suppresses its particle character. If we modify the experiment to bring out the particle character, its wave character is suppressed. We can never bring the wave and the particle view face to face in the same experimental situation. Matter and radiation are like coins that can be made to display either face at will but not both simultaneously. This, of course, is the essence of Bohr's principle of complementarity, the ideas of wave and of particle complementing rather than contradicting one another.

We have also seen that the probability idea emerges directly from the uncertainty principle. In classical mechanics, if at any instant we know exactly the position and momentum of each particle in an isolated system, then we can predict the exact behavior of the particles of the system for all future time. In quantum (or wave) mechanics, however, the uncertainty principle shows us that it is impossible to do this for systems involving small distances and momenta because it is impossible to know, with the required accuracy, the instantaneous positions and momenta of the particles of such a system. As a result, we are able to make predictions only of the *probable* behavior of these particles.

The connection between probability and the wave function becomes apparent in single- or double-slit interference experiments. Consider a parallel beam of electrons incident on a double slit. The motion of the electrons is governed by the de Broglie waves associated with them. The original wavefront is split into two coherent wavefronts by the slits, and these overlapping wavefronts produce the interference fringes on the screen. The intensity along the screen follows the usual interference pattern and corresponds to ψ^2, where ψ is the wave function. But the same intensity variation along the screen can be obtained by counting the numbers of electrons arriving at different parts of the screen in a given time. The probability that a single electron will arrive at a given place is just the ratio of the number that arrives there to the total number. In other words, the statistical distribution of a large number of electrons yields the interference pattern predicted by the wave function. The intensity ψ^2 of the de Broglie wave at a particular position is then a measure of the probability that the electron is located at that position.

Quantum theory, therefore, abandons strict causality in individual events in favor of a fundamentally statistical interpretation of their collective regularity. The commonly accepted statistical view (called the Copenhagen interpretation) is given by Born [13], whom we can paraphrase as follows. The law of causation, according to which the course of events in an isolated system is completely determined by the state of the system at $t = 0$, loses its validity in quantum theory. If we view processes from a wave–particle picture, then because this duality essentially involves indeterminacy, we are forced to abandon a deterministic theory. Likewise, if we examine processes analytically in quantum theory, we must describe the instantaneous state of a system by a wave function ψ. It is true that ψ satisfies a differential equation and therefore changes with time from its form at $t = 0$ in a causal way. But physical significance is confined to ψ^2, the square of the amplitude, and this only partially defines ψ. That is, the initial value of the wave function ψ is necessarily not completely definable. If we cannot in principle know the initial state exactly, it is empty to assert that events develop in a strictly causal way, for physics is then in the nature of the case indeterminate and statistical.

Heisenberg stated the matter succinctly as follows:

We have not assumed that the quantum theory, as opposed to the classical theory, is essentially a statistical theory, in the sense that only statistical conclusions can be drawn from exact data. . . . In the formulation of the causal law, namely, "If we know the present exactly, we can predict the future," it is not the conclusion, but rather the premise which is false. We *cannot* know, as a matter of principle, the present in all its details.

It is interesting to note that Einstein, who contributed so much to the development of quantum theory, actively rejected the philosophical interpretation of that theory that is held today by most physicists. That viewpoint, called the *Copenhagen interpretation* of quantum mechanics because so much of it was worked out by Niels Bohr and his collaborators, centers on the question of the ultimate reality of individual events and on the related question of whether or not quantum mechanics gives a *complete* description of events.

As for the quantum theory itself, Einstein had no doubts as to its accuracy and validity. He wrote:

This theory is, until now, the only one which unites the corpuscular and undulatory dual character of matter in a logically satisfactory fashion; and the testable relations which are contained in it are, within the natural limits fixed by the indeterminacy relation, complete. The formal relations which are given in this theory—that is, its entire mathematical formalism—will probably have to be contained in every useful theory.

But, he goes on,

What does not satisfy me in that theory is its attitude toward that which appears to me to be the programmatic aim of all physics: the complete description of any individual real situation as it supposedly exists, irrespective of any act of observation and substantiation.

In a lighter mood he once inquired:

If a mouse looks at the moon, does that change the Universe?

Einstein regarded quantum theory as incomplete, being a limiting case of the whole truth. He believed in what philosophers called realism, that an objective world exists independent of any (subjective) observation process; whereas the Copenhagen view forbids, as being outside the framework of science, an inquiry about something that exists independently of whether it is observed—the experimental conditions of observation being held to be inseparable from the phenomenon described. According to Einstein, though the uncertainty principle is valid and the predicted probabilities are statistically correct, the unpredictability of individual events is due to the incompleteness of the theory. The theory is not incomplete just because it is a statistical theory, and so does not predict individual events, but chiefly because it does not even *describe* the individual events [17]. "I still believe," said Einstein, "in the possibility of giving a model of reality which shall represent events themselves and not merely the probability of their occurrence."

The situation today, more than five decades after the development of the Copenhagen view, is that although a number of distinguished physicists align themselves with Einstein's views, the more comprehensive theory that Einstein envisioned has not yet come to light. Furthermore, during the last decade there has been a renewed interest in exploring the foundations of quantum mechanics, both theoretically and in the laboratory. All laboratory experiments to date have verified the accuracy of even the most subtle predictions of quantum mechanics and have reinforced the belief, held by many, that quantum mechanics gives a description of events that is not only accurate but also complete. Einstein, of course, may still have the last word; he very often did.

questions

1. Why is the wave nature of matter not more apparent to us in our daily observations?

2. How many experiments can you recall whose explanation calls for the wave theory of light? For the particle theory of light? For the wave theory of matter? For the particle theory of matter?

3. Discuss the analogy between (a) physical optics → geometrical optics and (b) wave mechanics → classical mechanics.

4. Discuss similarities and differences between a matter wave and an electromagnetic wave.

5. Discuss similarities and differences between an electron and a photon, both viewed as particles.

6. We have seen evidence that an electron has a particle aspect and also that it has a wave aspect. When all is said and done, just what is the electron anyhow, a particle or a wave? Discuss.

7. Why is it permissible for the speed of a matter wave, calculated as in Example 3, to exceed the speed of light?

8. An electron and a neutron each have the same kinetic energy. Which particle has the shorter de Broglie wavelength?

9. A de Broglie wavelength can be assigned to the electron, a true elementary particle with no known internal structure. We can also assign a de Broglie wavelength to the neutron; does this mean that the neutron is also a true elementary particle? Can a de Broglie wavelength be assigned to the nucleus of a uranium atom? To an ammonia molecule? To a drifting speck of dust?

10. If, in the de Broglie formula $(\lambda = h/mv)$, we let $m \to \infty$, do we get a result consistent with classical physics?

11. Can the de Broglie wavelength of a small spherical particle be smaller than the diameter of the particle? Larger? Is there any necessary relationship between the wavelength and the diameter?

12. Does the de Broglie wavelength of a particle depend on the reference frame of the observer? [See P. C. Peters, "Consistency of the de Broglie Relations with Special Relativity," *Am. J. Phys.* **38**, 7, 933 (1970).]

13. Could studies of crystal structure be carried out with beams of protons or heavier ions? Explain.

14. Why is electron diffraction more suitable than x-ray diffraction for studying the surfaces of solids?

15. A beam of electrons, each with kinetic energy K, falls on the surface of a crystal at right angles. Can you make a strong reflected beam appear at any arbitrary angle by a suitable choice of K? For an arbitrary choice of K, will there always be an angle at which a strong reflected beam will appear?

16. Do electron diffraction experiments give different information about crystals than can be obtained from x-ray diffraction experiments? From neutron diffraction experiments? Give examples.

17. In the Davisson-Germer electron diffraction experiment, why do no diffracted peaks appear if the target C in Fig. 6-1 is made up of many small crystallites rather than a single large crystal?

18. In Fig. 6-4b (made with x-rays) the diffraction circles are speckled, but in Fig. 6-4c (made with electrons) they are smooth. Can you explain why?

19. Explain the Laue and the Debye-Scherrer techniques for studying x-ray and electron diffraction. Why does a Laue pattern display spots, whereas a Debye-Scherrer pattern displays circles?

20. What is the inherent advantage of an electron microscope over an optical microscope?

21. How does the wave nature of matter enter into the design and operation of an electron microscope?

22. Explain qualitatively how a magnetic lens works to focus an incident electron beam in an electron microscope.

23. Why is an x-ray microscope difficult to construct? What about a proton microscope? A neutron microscope?

24. Give examples of how the process of measurement disturbs the system being measured. In each of your examples, can the disturbance be taken into account ahead of time by suitable calculations?

25. Why is the Heisenberg uncertainty principle not more readily apparent in our daily observations?

26. Show the relation between the unpredictable nature of the Compton recoil in Bohr's gamma-ray microscope and the fact that there are four unknowns and only three conservation equations in the Compton effect.

27. You measure the pressure in a tire, using a pressure gauge. The gauge, however, bleeds a little air from the tire in the process, so that the act of measuring changes the property that you are trying to measure. Is this an example of the Heisenberg uncertainty principle? Explain.

28. Argue from the Heisenberg uncertainty principle that the lowest energy of an oscillator cannot be zero. (See Problem 42.)

29. In Fig. 6-12, the infinite wave shown has $\Delta x = \infty$ and $\Delta \nu = 0$. Are these values consistent with the Heisenberg uncertainty principle?

30. How does the bandwidth of a radar receiver, designed to accept microwave pulses of a fixed time duration, vary with the wavelength of the transmitted signal?

31. Is there a contradiction between the statements that the predictions of wave mechanics are exact and the fact that the information derived is of a statistical character? Explain.

32. Games of chance contain events that are ruled by statistics. Do such games violate the strict determination of individual events? Do they violate cause and effect?

33. The quantity $\psi(x, t)$, the amplitude of a matter wave, is called a *wave function*. What is the relationship between this quantity and the particles that form the matter wave?

34. Explain in your own words just how the experiment described in connection with Fig. 6-9 illustrates the complementarity principle.

35. In the text we have discussed (a) Bohr's complementarity principle, (b) the Heisenberg uncertainty principle, (c) the wave function ψ and its statistical interpretation, and (d) the relationship $\Delta \nu \cdot \Delta x = 1$. Describe each of these in your own words and say what light they throw on the wave–particle duality question.

problems

$$hc = 1.99 \times 10^{-25} \text{ J} \cdot \text{m} = 1240 \text{ eV} \cdot \text{nm}$$
$$h = 6.63 \times 10^{-34} \text{ J} \cdot \text{s} = 4.14 \times 10^{-15} \text{ eV} \cdot \text{s}$$
$$k = 1.38 \times 10^{-23} \text{ J/K} = 8.62 \times 10^{-5} \text{ eV/K}$$
$$1 \text{ J} = 1.60 \times 10^{-19} \text{ eV}$$
$$\text{Electron mass} = 9.11 \times 10^{-31} \text{ kg} = 0.511 \text{ MeV}/c^2$$
$$\text{Neutron mass} = 1.68 \times 10^{-27} \text{ kg} = 940 \text{ MeV}/c^2$$

1. The de Broglie wavelength of the electron. (a) Show that, using the classical relation between momentum and kinetic energy, the de Broglie wavelength of an electron can be written as

$$\lambda_c = \frac{1.226 \text{ nm}}{\sqrt{K}}, \qquad (K \text{ in eV})$$

in which K is the kinetic energy in electron volts. (b) Show further that, using the relativistic relation, the de Broglie wavelength can be written as

$$\lambda_r = \frac{1.226 \text{ nm}}{\sqrt{K + 9.78 \times 10^{-7} K^2}}, \qquad (K \text{ in eV}).$$

(*Hint:* See Example 2.)

2. A good approximate formula. (a) Derive the commonly memorized (nonrelativistic) formula for the de Broglie wavelength of an electron (in Å) and the corresponding accelerating potential V (in volts):

$$\lambda = \sqrt{\frac{150}{\sqrt{V}}}$$

(b) For $V = 50.0$ V, compare the predictions of this formula, the proper nonrelativistic formula and the (correct) relativistic formula. (c) Repeat for $V = 100$ kV. (*Hint:* See Problem 1.)

3. Calculating the kinetic energy. (a) Show that, in terms of its de Broglie wavelength λ and its rest energy E_0 ($= m_0 c^2$), a particle's kinetic energy is given by

$$K = \sqrt{E_0^2 + (hc/\lambda)^2} - E_0.$$

(b) Calculate the kinetic energy for a positron with a de Broglie wavelength of 1.000 pm.

4. The extreme relativistic case. The linear electron accelerator at the Stanford Linear Accelerator Center (SLAC) can accelerate electrons to 20 GeV ($= 20 \times 10^9$ eV). At such high energies the electron beam has a very short de Broglie wavelength, suitable for probing the fine details of nuclear structure by scattering experiments. What is this wavelength and how does it compare with the size of an average middle-mass nucleus (radius = 5 fm = 5×10^{-15} m)? (*Hint:* At these energies it is simpler to use the extreme relativistic relationship between momentum and energy, namely $p = E/c$. This is the same relationship used for photons and is justified whenever, as in this case, the kinetic energy of the particle is very much greater than its rest energy.)

5. Is relativity needed?   At what kinetic energy will the nonrelativistic expression for the de Broglie wavelength be in error by 1.0% for (*a*) an electron and (*b*) a neutron? (*Hint:* See Example 2.)

6. The wavelength of a bullet.   A bullet of mass 40 g travels at 450 m/s. (*a*) What wavelength can we associate with it? (*b*) Why does the wavelength of the bullet not reveal itself through diffraction effects?

7. Thermal neutrons.   Thermal neutrons have an average kinetic energy of $(3/2)kT$ where $T (= 300$ K) is room temperature. Such neutrons are in thermal equilibrium with normal surroundings. (*a*) What is the average energy of such a thermal neutron? (*b*) What is the corresponding de Broglie wavelength?

8. Photons and electrons (I).   The wavelength of the yellow spectral emission line of sodium vapor is 590 nm. At what kinetic energy would an electron have the same de Broglie wavelength?

9. Photons and electrons (II).   (*a*) Photons and electrons travel in free space with wavelengths of 1.00 nm. What are the energy of the photon and the kinetic energy of the electron? (*b*) Repeat for a wavelength of 1.00 fm.

10. Photons and electrons (III).   (*a*) A photon in free space has an energy of 1.00 eV and an electron, also in free space, has a kinetic energy of that same amount. What are their wavelengths? (*b*) Repeat for an energy of 1.00 GeV.

11. Photons and electrons (IV).   A photon and an electron each have a wavelength of 1.00 nm. What are (*a*) their momenta (in keV/*c*) and (*b*) their relativistic total energies (in keV)? (*c*) Compare the kinetic energy of the electron to the energy of the photon.

12. Electrons and neutrons.   An electron and a neutron each have a kinetic energy of 5.00 MeV. What percent error is made in calculating their de Broglie wavelengths using the classical formula?

13. A graphical comparison.   Make a plot of the de Broglie wavelength against kinetic energy for (*a*) electrons and (*b*) neutrons. Restrict the range of energy values to those in which classical mechanics applies reasonably well. Example 2 supplies a convenient criterion; it is shown there

that if the range of kinetic energy is restricted to 4% of the rest energy of the particle involved an error no greater than 1% will result.

14. Electrons, neutrons and photons.   Compare the wavelengths of a 1.00-MeV (kinetic energy) electron, a 1.00-MeV (kinetic energy) neutron and a 1.00-MeV photon.

15. Locating a uranium nucleus.   We wish a uranium-238 nucleus to have enough energy so that its de Broglie wavelength is equal to its nuclear radius, which is 6.8 fm. How much energy is required? Take the nuclear mass to be 238 u.

16. de Broglie waves and Rutherford scattering.   As we shall see in Chapter 7, the existence of the atomic nucleus was uncovered in 1911 by Ernest Rutherford, who properly interpreted some experiments in which a beam of alpha particles was scattered from a foil of atoms such as gold. (*a*) If the alpha particles had a kinetic energy of 7.5 MeV, what was their de Broglie wavelength? (*b*) Should the wave nature of the incident alpha particles have been taken into account in interpreting these experiments? The mass of an alpha particle is 4.00 u and its distance of closest approach to the nuclear center in these experiments was about 30 fm. (The wave nature of matter was not postulated until more than a decade after these crucial experiments were first performed.)

17. Can we treat gas molecules as small particles?   Consider a balloon filled with (monatomic) helium gas at room temperature and pressure. (*a*) How does the average de Broglie wavelength of the helium atoms compare with the average distance between atoms under these conditions? The average kinetic energy of an atom is equal to $(3/2)kT$. (*b*) What is the answer to the question posed by the title to this problem?

18. An unknown particle (I).   A non-relativistic particle is moving three times as fast as an electron. The ratio of their de Broglie wavelengths, particle to electron, is 1.813×10^{-4}. Identify the particle. (*Hint:* See table of rest masses in the Appendix.)

19. An unknown particle (II).   A particle moving with kinetic energy equal to its rest energy has a de Broglie wavelength of 0.1920 fm. (*a*) If the kinetic energy of the particle doubles, what is the new de Broglie wavelength? (*b*) Identify the particle. (*Hint:* See table of rest masses in the Appendix.)

20. The phase speed of a matter wave.   In Example 3 we defined the *phase speed* v_{ph} of a matter wave representing a free particle from $v_{ph} = \lambda \nu$. (*a*) Show that this is consistent with the definition $v_{ph} = E/p$ where E is the relativistic energy of the particle and p is its momentum. (*b*) Show further that, for a free particle, v_{ph} = c/β in which β is the speed parameter. Thus the phase speed of the matter wave for a free particle always exceeds the speed of light. See Supplementary Topic G.

21. The group speed of a matter wave. In Problem 20 we saw that the phase speed v_{ph} of a matter wave is given by E/p, in which E is the relativistic total energy of the particle and p is its momentum. The *group speed* v_{gr} of matter waves is given by dE/dp. Use the relativistic relation between energy and momentum to show that the group speed of a matter wave representing a free particle is equal to the actual speed of the particle. See Supplementary Topic G.

22. The wavelengths of thermal neutrons. The de Broglie wavelengths in a collimated beam of thermal neutrons emerging from a reactor wall can be shown to be distributed according to

$$n(\lambda) = \frac{C}{\lambda^5} e^{-h^2/2mkT\lambda^2}$$

in which $n(\lambda)\,d\lambda$ gives the flux of neutrons whose de Broglie wavelengths lie between λ and $\lambda + d\lambda$. T is the temperature at which the neutrons are in thermal equilibrium, m is the mass of the neutron, k is the Boltzmann constant and C is an instrumental constant. (a) Show that the most probable de Broglie wavelength present in the beam is given by

$$\lambda_p = \frac{h}{\sqrt{5mkT}}.$$

(b) For T = 300 K, what is this most probable wavelength?

★ **23. The energies of thermal neutrons.** An expression for the distribution in de Broglie wavelength of thermal neutrons in a collimated beam is given in Problem 22. Starting from this relationship show that the neutrons in the collimated beam are distributed in kinetic energy according to

$$n(K) = C'Ke^{-K/kT}.$$

Here $n(K)\,dK$ gives the flux of neutrons whose kinetic energies lie between K and $K + dK$ and C' is an instrumental constant.

24. Higher-order diffraction peaks. In the experiment of Davisson and Germer (a) show that the second- and third-order diffraction peaks corresponding to the strong maximum in Fig. 6-2c cannot occur; these peaks correspond to m = 2, 3 etc. in Eq. 6-3. (b) Find the angle at which the first-order diffraction peak would occur if the accelerating potential were changed from 54 V to 60 V. (c) What minimum accelerating potential is needed to produce a second-order diffraction peak?

25. Bragg reflections for electrons and x rays. A potassium chloride crystal is cut so that the layers of atomic planes parallel to its surface have an interplanar spacing of 0.314 nm. An incident beam makes an angle ϕ with the surface and a first-order diffracted peak is generated. Find ϕ if the incident beam is (a) a 40.0-keV x-ray beam and if it is (b) a beam of electrons whose kinetic energy is 40.0 keV. (*Hint:* At these relatively high energies the bending of the electron beam at the crystal surface can be ignored and the electron diffraction treated in strict analogy with x-ray diffraction; see *Physics,* Part II, Sec. 47-5.)

26. Bragg reflections of thermal neutrons. A beam of thermal neutrons from a nuclear reactor falls on a crystal of calcium fluoride, the beam direction making an angle θ with the surface of the crystal. The atomic planes parallel to the crystal surface have an interplanar spacing of 5.464 Å. The de Broglie wavelength corresponding to the maximum intensity of the incident beam is 1.100 Å. For what values of θ will the first three orders of Bragg-reflected neutron beams occur? (*Hint:* Neutrons, which carry no charge and are thus not subject to electrical forces, are not refracted as they pass through a crystal surface. Thus neutron diffraction can be treated in strict analogy with x-ray diffraction; see Example 5 and also *Physics,* Part II, Sec. 47-5.)

27. Two microscopes compared. What accelerating voltage would be required for electrons in an electron microscope to obtain the same resolving power as that which could be obtained from a "gamma-ray microscope" using 200-keV gamma rays? Assume that the resolving power is limited only by the wavelength for each instrument.

28. The electron microscope and the optical microscope. The highest achievable resolving power of a microscope is limited only by the wavelength used. Suppose one wishes to "see" inside an atom. Assuming the atom to have a diameter of about 1.0 Å. This means that we wish to resolve detail of separation of about 0.1 Å. (a) If an electron microscope is used, what minimum kinetic energy of electrons is needed? (b) If a light microscope is used, what minimum photon energy is needed? In what region of the electromagnetic spectrum are these photons? (c) Which microscope seems more practical for the purpose?

29. Locating an electron. A microscope using photons is used to pin down the location of an electron in an atom to an uncertainty of about 0.2 Å. What is the uncertainty of the velocity of the electron located in this way? A nonrelativistic treatment is justified.

30. The uncertainty in velocity. Show that if the uncertainty in the location of a particle is about equal to its de Broglie wavelength then the uncertainty in its velocity is about equal to its velocity.

31. A problem in retrospect. In a repetition of J. J. Thomson's experiment for measuring e/m for the electron a beam of 10-keV electrons is collimated by passage through a slit whose width is 0.50 mm. Explain quantitatively why the beamlike character of the emerging electrons is not destroyed by diffraction of the electron waves as they pass through the slit.

32. The uncertainty principle. Show that for a free particle the Heisenberg uncertainty principle can be written as

$$\Delta\lambda \cdot \Delta x \gtrsim \lambda^2$$

where Δx is the uncertainty in the location of the particle and $\Delta\lambda$ is the simultaneous uncertainty in the corresponding de Broglie wavelength.

33. Locating a photon. If $\Delta\lambda/\lambda = 1.0 \times 10^{-7}$ for a photon, what is the simultaneous positional uncertainty Δx for (a) A gamma ray ($\lambda = 50$ fm)? (b) An x ray ($\lambda = 500$ pm)? (c) A visible light ray ($\lambda = 500$ nm)? (d) An infrared ray ($\lambda = 1.0$ μm)? (*Hint:* See Problem 32.)

34. An excited atomic state. What is the approximate lifetime of an excited atomic state whose emission wavelength $\lambda = 500$ nm is known to a precision $\Delta\lambda/\lambda = 10^{-7}$?

35. An excited nuclear state. A nucleus in an excited state will return to its ground state, emitting a gamma ray in the process. If its mean lifetime in a particular excited state is about 10^{-12} s, what is the uncertainty in the energy of the corresponding emitted gamma ray photon?

36. An atomic transition between two excited states. An atom in an excited state has a lifetime of 1.2×10^{-8} s; in a second excited state the lifetime is 2.3×10^{-8} s. What is the uncertainty in energy for a photon emitted when an electron makes a transition between these two states?

37. The decay of a pion. A neutral pion (symbol: π^0) decays with a meanlife of 8.4×10^{-17} s. (a) If repeated measurements of its rest energy are made, what precision can be expected? (b) The rest energy listed in the Appendix for this particle is 134.965 MeV with an uncertainty of ± 0.007 MeV. Does this reported precision exceed the limits set by the uncertainty principle?

38. The decay of a ϕ particle. The elementary particle represented by ϕ has a measured rest energy of 1019 MeV, the experimental uncertainty of this quantity being ± 4 MeV. What is the approximate minimum value of the meanlife for this unstable particle?

39. "Viewing" an orbiting atomic electron. Suppose we wish to test the possibility that electrons in atoms move in classical orbits by "viewing" them with photons of sufficiently short wavelength, say 0.1 Å. (a) What would be the energy of such a photon? (b) How much energy would such a photon transfer to a free electron in a head-on Compton collision? (c) What does this tell you about the possibility of verifying orbital motion by "viewing" an atomic electron at two or more points along its path?

★ **40. Dropping marbles—a quantum viewpoint.** A child on top of a ladder of height H is dropping marbles of mass m to the floor and trying to hit a crack in the floor. Aiming equipment of the highest possible precision is available. (a) Show that the marbles will miss the crack by an average distance of the order of $(h/m)^{1/2}(H/g)^{1/4}$, where g is the acceleration due to gravity. (b) Find this distance for $H \cong 10$ m and $m \cong 1$ g. (c) In practice can you use this experiment to prove the uncertainty principle? Can it be used to disprove the uncertainty principle?

★ **41. An electron leaves its track.** A 1.0-MeV electron leaves a visible track in a cloud chamber. The track is a series of water droplets each about 10^{-5} m in diameter. Show, from the ratio of the uncertainty in the transverse momentum to the actual momentum of the electron, that the electron path should not noticeably differ from a straight line.

★ **42. The linear harmonic oscillator.** The total mechanical energy of a linear harmonic oscillator is given by

$$E = \frac{p_x^2}{2m} + \tfrac{1}{2}kx^2.$$

(a) Show, using the uncertainty relation, that this can be written as

$$E \cong \frac{p_x^2}{2m} + \frac{kh^2}{2p_x^2}.$$

(b) Then show that the minimum energy of the oscillator is given by $h\nu$, where

$$\nu \cong \frac{1}{2}\sqrt{\frac{k}{m}}.$$

(*Hint:* Classically the minimum energy is zero. This requires that x and p_x simultaneously be zero, but our knowledge of these quantities is limited by the uncertainty principle. Very approximately we can equate Δx with x and Δp_x with p_x, as in part (a). Then, in part (b) we must minimize E with respect to p_x. The correct quantum answer for the minimum energy differs from that given in part (b) by a factor of 2π; however, the answer given here is close enough to show that the classical answer (zero) is not correct.)

43C. Calculating the de Broglie wavelength. Write a program for your handheld programmable calculator that will allow you to select as a particle of interest either the electron, the neutron, or any other particle whose rest mass you choose to enter. As a second input enter the kinetic energy of the particle. As outputs display successively the de Broglie wavelength and the energy of a photon that has this same wavelength. Use the relativistic formula for the particle momentum. See the Appendix for the rest masses of some selected particles.

44C. Checking it out. Check out the program you have written in Problem 43C by (a) verifying the calculations of Examples 1*a*, 1*b*, and 4*b*. (b) Find the wavelength and the equivalent photon energy for an electron whose kinetic energy is 1.00 keV; 1.00 MeV; 1.00 GeV. (c) Find the wavelength and the equivalent photon energy for a neutron whose kinetic energy is 0.01 eV; 0.10 eV; 1.00 eV. (d) Find the de Broglie wavelength of a 1.00-MeV alpha particle, its mass being 6.65×10^{-27} kg.

45C. Resolving a pulse into its harmonic components. Figure 6-13*a* shows a pulse formed by adding harmonic waves whose amplitudes are distributed with wave number as in Fig. 6-13*b*. Let $\kappa = \rho \kappa_0$ in which κ_0 is the wave number corresponding to the fundamental pulse wave-

length λ_0. The pulse amplitude $A(\kappa)$ is then given [see reference 16] by

$$A(\kappa) = \frac{nE_0}{4\pi\epsilon_0}\left[\frac{\sin(n\pi/2)(1+\rho)}{(n\pi/2)(1+\rho)} + \frac{\sin(n\pi/2)(1-\rho)}{(n\pi/2)(1-\rho)}\right]$$

in which n (a constant; see Fig. 6-13*a*) is the number of half waves in the pulse and E_0 is the amplitude of the pulse. Write a program for your handheld programmable calculator that will accept n and the variable ρ as inputs and will deliver $A(\kappa)$ as an output. For simplicity imagine that E_0 has been chosen so that the quantity outside the square brackets above is equal to unity.

46C. Checking it out. Check out the program that you have written in Problem 45C by (*a*) verifying the plot of Fig. 6-13*b* and (*b*) plotting $A(\kappa)$ for a microwave pulse of duration 1.05 ns and of fundamental wavelength $\lambda_0 = 3.00$ cm.

references

1. Louis de Broglie, "Investigations on Quantum Theory" (the text of his doctoral thesis), Gunther Ludwig, Ed., *Selected Readings in Physics—Wave Mechanics* (Pergamon Press, Elmsford, N.Y., 1968).

2. Louis de Broglie, "The Undulatory Aspects of the Electron" (Nobel Prize address, Stockholm, 1929), in Henry A. Boorse and Lloyd Motz, Eds., *The World of the Atom* (Basic Books, New York, 1966), p. 1048.

3. Clinton J. Davisson, "Are Electrons Waves?," *J. Franklin Inst.* **205,** 597 (1928); reprinted in Henry A. Boorse and Lloyd Motz, Eds., *The World of the Atom* (Basic Books, New York, 1966), p. 1144.

4. Karl K. Darrow, "Davisson and Germer," *Sci. Am.* (May 1948).

5. George P. Thomson, "The Early History of Electron Diffraction," *Contemp. Phys.* (January 1968).

6. Max Jammer, *The Philosophy of Quantum Mechanics* (Wiley, New York, 1974), p. 254.

7. Samuel A. Werner, "Neutron Interferometry," *Phys. Today* (December 1980).

8. G. E. Bacon, *Neutron Diffraction*, 3rd ed. (Clarendon Press, Oxford, 1975), sec. 1.2.

9. Eugene Hecht and Alfred Zajac, *Optics* (Addison-Wesley, Reading, Mass., 1974), sec. 5.7.5.

10. John M. Cowley and Sumio Iijima, "Electron Microscopy of Atoms in Crystals," *Phys. Today* (March 1977).

11. Thomas E. Everhart and Thomas L. Hayes, "The Scanning Electron Microscope," *Sci. Am.* (January 1972).

12. Niels Bohr, "Discussion with Einstein on Epistemological Problems in Atomic Physics," in *Library of Living Philosophers,*" vol. VII (1949). This volume is dedicated to Einstein on his 70th birthday. Bohr's long essay is reprinted in Henry A. Boorse and Lloyd Motz, Eds., *The World of the Atom* (Basic Books, New York, 1966), p. 1223.

13. Max Born, *Atomic Physics*, 7th ed. (Hafner, New York, 1962).

14. Henry A. Boorse and Lloyd Motz, Eds., *The World of the Atom* (Basic Books, New York, 1966), p. 1094.

15. George Gamow, "The Principle of Uncertainty," *Sci. Am.* (January 1958).

16. Eugene Hecht and Alfred Zajac, *Optics* (Addison-Wesley, Reading, Mass., 1974), sec. 7-9. See also various mathematics texts dealing with Fourier transforms.

17. L. E. Ballentine, "Einstein's Interpretation of Quantum Mechanics," *Am. J. Phys.* (December 1972).

CHAPTER 7

early quantum theory of the atom

. . . we shall consider a case in which the positive electricity is distributed . . . as a sphere of uniform density, throughout which the corpuscles [electrons] are distributed.

J. J. Thomson (1907)

In comparing the theory outlined in this paper with the experimental results, it has been supposed that the atom consists of a central charge supposed concentrated at a point. . . .

Ernest Rutherford (1911)

. . . it seems necessary to introduce . . . a quantity foreign to the classical electrodynamics, i.e., Planck's constant. . . .

Niels Bohr (1913)

7-1 J. J. THOMSON'S MODEL OF THE ATOM [1,2]

The *atomic theory of the elements*, advanced by John Dalton about 1800, is the notion that each element is made up of atoms that are characteristic of it [3,4]. By 1900 this viewpoint, which seems so transparently true to us today, had been almost universally adopted. However, almost nothing was then known about the structure of the atoms of the various elements. We know now that before 1897 no such insights were possible because the principal constituent of all atoms, the electron, was only discovered in that year.

It was indeed in 1897 that J. J. Thomson, Cavendish Professor at Cambridge University, first identified in electrical discharges in gases streams of what he called "corpuscles," each of which proved to have a mass lighter than that of the hydrogen atom by a factor of almost 2000 and a negative electrical charge—to give its modern value—of 1.602×10^{-19} C. Thomson saw these corpuscles—or electrons as they came to be called—as a fundamental constituent of atoms, and he developed an atomic model based on this assumption.

In their normal state atoms are electrically neutral, so if an atom contains (negatively charged) electrons, it must also contain an equal charge of positive electricity. Thomson knew nothing about the form taken by this positive electricity, nor did he have any idea of the number of electrons in an atom of a given element. (The concept of atomic number had not yet been developed and Mendeleev's periodic table was still a listing of the elements by atomic *weight*, with a few adjustments for compelling chemical reasons.)

232   EARLY QUANTUM THEORY OF THE ATOM

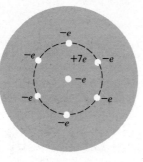

Figure 7-1. A stable configuration of seven electrons in a "plum pudding" atom. The attractive force on each electron due to the positive charge of the "pudding" is balanced by the repulsive force exerted on it by the other electrons.

Thomson assumed that the positive electricity in an atom is in the form of a sphere of "electrical fluid" of uniform density whose radius is that of the atom, a quantity known from the density of solid matter to be about 10^{-10} m. He made this assumption deliberately, "in default of exact knowledge of the nature of the way in which positive electricity is distributed in the atom," and because it was an arrangement "most amenable to mathematical calculation." As for the electrons, he assumed that they were embedded in this electrical fluid, the whole arrangement becoming known as the "plum pudding" atom model; see Fig. 7-1. With great ingenuity Thomson devised concentric ringlike arrangements for the electrons that were dynamically stable, and he was able to make plausible a loose connection between these configurations and the periodicity of the chemical properties of the elements.

Every species of atom can emit or absorb radiation at certain sharply defined characteristic frequencies, the spectrum of these frequencies providing a "signature" for the element. Thomson tried—with little success—to associate the emission of spectrum lines with the vibrations of the electrons about their stable configurations. Thomson's model was never meant to be more than a guide to thought and a framework around which experiments could be planned. In this respect it was a successful first step in understanding atomic structure. Soon, however, its central provisional assumption—that the positive charge of the atom is distributed throughout a sphere of atomic dimensions—was shown by experiment to be incorrect.

EXAMPLE 1.

The Simplest Thomson Atom. (a) The simplest atom possible on the basis of the Thomson model is a spherical positive charge of radius a and uniform charge density ρ containing a single electron. Show that if this electron is initially at rest a distance r from the center ($r < a$) and is then released, its motion will be simple harmonic.

From arguments based on Gauss's law (see *Physics*, Part II, Sec. 28-8), we know that we can calculate the force on the electron from Coulomb's law. We obtain

$$F(r) = -\frac{1}{4\pi\varepsilon_0}\left(\frac{4}{3}\pi r^3\rho\right)\frac{e}{r^2} = -\left(\frac{\rho e}{3\varepsilon_0}\right)r,$$

where $\frac{4}{3}\pi r^3\rho$ is the net positive charge in a sphere of radius r. Hence, we can write $F(r) = -kr$ where the constant $k = \rho e/3\varepsilon_0$. This is the condition on a force responsible for

simple harmonic motion, namely, that it is proportional to the displacement but oppositely directed (see *Physics*, Part I, Sec. 15-2).

(b) Let the total positive charge equal that of the electron and be distributed in a spherical region of radius $a = 1.0 \times 10^{-10}$ m. Find the force constant k and the frequency of the motion of the electron.

We have

$$\rho = \frac{e}{\frac{4}{3}\pi a^3},$$

so that

$$k = \frac{\rho e}{3\varepsilon_0} = \frac{e}{\frac{4}{3}\pi a^3}\frac{e}{3\varepsilon_0} = \frac{e^2}{4\pi\varepsilon_0 a^3} = 230 \text{ N/m}.$$

The frequency of the simple harmonic motion is given by

$$\nu = \frac{1}{2\pi} \sqrt{\frac{k}{m}} = \frac{1}{2\pi} \sqrt{\frac{230 \text{ N/m}}{9.11 \times 10^{-31} \text{ kg}}} = 2.5 \times 10^{15} \text{ s}^{-1}.$$

Assuming (in analogy to radiation emitted by electrons oscillating in an antenna) that the radiation emitted by the atom has this same frequency, it will correspond to a wavelength

$$\lambda = \frac{c}{\nu} = \frac{3.0 \times 10^8 \text{ m/s}}{2.5 \times 10^{15} \text{ s}^{-1}} = 1.2 \times 10^{-7} \text{ m} = 120 \text{ nm},$$

in the far ultraviolet portion of the electromagnetic spectrum. It is easy to show that an electron moving in a stable circular orbit of any radius inside this Thomson atom has this same frequency.

A different assumed radius of the sphere of positive charge would give a different emission frequency. However, the fact that this model has only a single characteristic frequency conflicts with the large number of frequencies observed in the spectrum of even the simplest atom.

7-2 THE NUCLEAR ATOM [5]

In 1909 Ernest Rutherford, then at the University of Manchester in England, proposed a research problem to Ernest Marsden, a 20-year-old student who had not yet taken his bachelor's degree. He suggested that Marsden, under the direction of Rutherford's assistant Hans Geiger, allow a beam of alpha particles to fall on a thin metallic foil and that he measure the extent to which the alpha particles are reflected from the foil, that is, the extent to which they are "scattered" in a backward direction. This suggestion set in motion a train of events that lead to an entirely new view of the structure of the atom. In 1911 Rutherford, interpreting the results of these scattering experiments, demonstrated that the positive charge of the atom did *not* fill the entire atomic volume as Thomson had postulated, but was concentrated in a tiny volume at the atomic center. Rutherford had discovered the atomic nucleus.

Rutherford had already been awarded the Nobel Prize in 1908 for his "investigations in regard to the decay of elements and . . . the chemistry of radioactive substances." He was a talented, hard-working physicist with enormous drive and self-confidence. In a letter written later in life, the then Lord Rutherford wrote "I've just been reading some of my early papers and, you know, when I'd finished, I said to myself, 'Rutherford, my boy, you used to be a damned clever fellow.'" Although pleased at winning a Nobel Prize, he was not happy that it was a prize in chemistry rather than one in physics (any research in the elements was then considered chemistry). In his speech accepting the Prize, he noted that he had observed many transformations in his work with radioactivity but never had seen one as rapid as his own, from physicist to chemist.

Alpha particles were already known to Rutherford to be doubly ionized helium atoms (that is, helium atoms with two electrons removed), emitted spontaneously from several radioactive materials at high speed. Rutherford recognized that he could use these particles as a probe to examine the structure of matter. In Fig. 7-2 we show schematically a later (1913) arrangement used by Geiger and Marsden to study the scattering of alpha particles that fall on thin foils of various substances. A lead block containing a radioactive source of alpha particles emits a narrow beam of alphas into a vacuum chamber at the center of which is a very thin metallic target foil. The alphas pass through the foil with little loss of energy, but may be deflected from their incident path by the electrical forces exerted on them by the atoms in the foil. A single scattered alpha particle can be detected by observing with a microscope the flash of light it produces on a fluorescent screen placed in its path. The detecting screen and microscope can be rotated to observe different angles ϕ of scattering and, during an experiment, the observer records the rate at which light flashes are produced for each of different

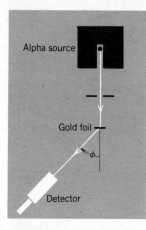

Figure 7-2. An arrangement (top view) used in Rutherford's laboratory to study the scattering of alpha particles by thin metal foils. The detector can be rotated to various values of the scattering angle ϕ.

angular positions of the detector. Careful measurements of the scattering should reveal information about the nature of the forces encountered by the alphas in passing through the foil, and this, in turn, might reveal the actual arrangement of the positive and negative charge in the atoms making up the foil.

A single Thomson model atom is expected to produce a very small deflection of an alpha particle passing through it. Because of its net zero charge, such an atom has no effect on the alpha particle until the alpha is inside the atom. Once inside, the alpha interacts with the electrons and the positive charge. The electrons, however, have such a small mass compared to the alphas that the alphas are hardly deflected at all by them—much as a bowling ball is unaffected by an inflated rubber beach ball in its path. And the positive charge, because it is distributed over the volume of the Thomson atom, cannot produce a large deflection on an alpha particle either. One can make an estimate (see Problem 8) that the average deflection of an alpha caused by one Thomson atom (see Fig. 7-3*a*) is not more than about 0.005°. After passage through many atoms in the foil, the cumulative effect of the deflections is still not large because of the randomness of the encounters—the alpha is deflected a small amount this way and then a small amount that way (see Fig. 7-3*b*). One can use statistical theory to show that, for a particular thin gold foil used, only ~0.01 percent of the alphas should be scattered at angles greater than 3° and that the chance of a scattering of 90° or more (a backward scattering) was only about one in 10^{3500}. The Thomson model atom involves small angle scattering from *many* atoms.

The results of the initial experiments confirmed the predominance of small angle scattering, but the percentage of particles scattered at large angles was in disagreement with the predictions of the Thomson model. Indeed, for the foil referred to above, one alpha in ten thousand (1 in 10^4, *not* 1 in 10^{3500}) came off backwards! To Rutherford, accustomed to thinking in terms of Thomson's model, it came as a great surprise that some alpha particles were deflected through very large angles, up to 180°. In his words

> It was quite the most incredible event that ever happened to me in my life. It was as incredible as if you fired a 15-inch shell at a piece of tissue paper and it came back and hit you. On consideration I realized that the scattering backwards must be the result of a single collision, and when I made the calculation I saw that it was impossible to get anything of that order of magnitude unless you took a system in which the greater part of the mass of the atom was concentrated in a minute nucleus. It was then that I had the idea of an atom with a minute massive center carrying a charge.

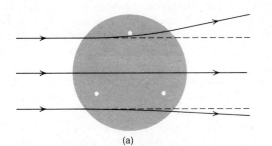

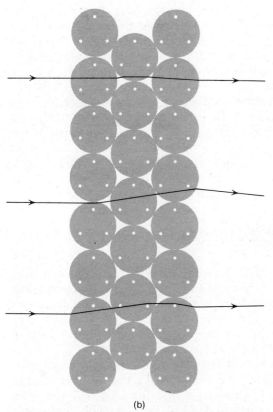

(b)

Figure 7-3. (*a*) Showing the deflection (greatly exaggerated) of various alpha particles in passing through the single Thomson atom. (*b*) Showing the deflections of various alpha particles in passing through a foil containing many layers of Thomson atoms. The Thomson model atom involves small angle scattering from many atoms.

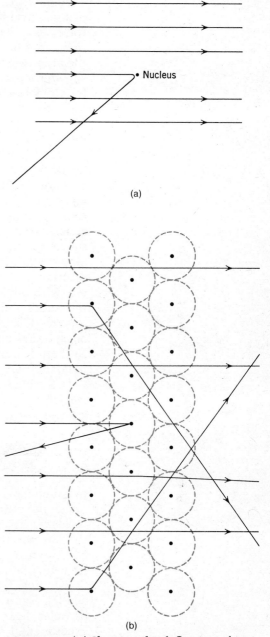

(b)

Figure 7-4. (*a*) Showing the deflection of various alpha particles in passing the nucleus of a single Rutherford atom. (*b*) Showing the deflection of various alpha particles in passing through a foil containing many layers of Rutherford atoms. The Rutherford model atom involves large angle scattering from a single atom.

Alpha particles traverse a different number of atoms N in passing through foils of different thickness. In Thomson's multiple scattering model, the alpha particle traverses a Thomson atom in each layer of the foil (see Fig. 7-3*b*) but its deflections are not cumulative, the alpha being randomly scattered—now this way, then that way. The number of alpha particles scattered through any angle can be shown to be proportional to the square root of N (and thus to $\sqrt{t}$, where t is the foil thickness) in Thomson's model. However, subsequent experiments in which foils of various thicknesses were used showed this number to be proportional to t rather than to $\sqrt{t}$. This is consistent with the scattering picture in Rutherford's model (see Fig. 7-4). The scattering of alphas due to atomic electrons can be ignored for scattering angles greater than a few degrees, so Rutherford concentrated on the effect of the nucleus in seeking to explain large angle scatterings. Because the nucleus is very small, Rutherford's atom is mostly empty space, so alpha particles don't often come near the nucleus in passing through the foil. But those that do come near feel very large forces, for all the positive charge is concentrated in a small region and the Coulomb force varies inversely as the square of the separation distance. Furthermore, because the nucleus contains almost all the mass of the atom, it would be the (lighter) alpha particle that would recoil now. In other words, the large-scale scatterings are caused by near encounters with nuclei, whereas the predominance of small angle scatterings is due to the fact that most encounters are distant ones. The thicker the foil, the larger the number of layers of atoms and the greater the chance of a near encounter between an alpha and an atom. Hence, for thin foils, the scattering beyond a few degrees in Rutherford's model is expected to be proportional to t, consistent with experimental observations.

In Supplementary Topic I we consider quantitatively the scattering of alpha particles by a nuclear atom. The probability that, in the apparatus of Fig. 7-2, an alpha particle will be scattered into the detector is shown there to be

$$P(\phi) = \left(\frac{e^2}{4\pi\varepsilon_0}\right)^2\left(\frac{\mathscr{A}\rho t n^2}{4R^2K^2}\right)\frac{1}{\sin^4(\phi/2)} \qquad \text{(Rutherford scattering)}, \qquad (7\text{-}1)$$

in which ne gives the charge of the scattering nucleus, ρ is the number of target nuclei per unit volume in the foil, t is the foil thickness, K is the energy of the incident alpha particles, $\mathscr{A}$ is the effective area of the detector, R is the distance of the detector from the scattering foil, and ϕ is the scattering angle. On the basis of the Thomson atom model $P(\phi)$ should vary as $e^{-\phi^2}$, a variation vastly different from that predicted by Eq. 7-1, particularly at large angles.

Equation 7-1 was verified experimentally in every detail by Geiger and Marsden [7]. Figure 7-5, prepared from their data, shows $N(\phi)$, the number of alpha particles scattered from a gold foil for various scattering angles. For a given experimental setup, $N(\phi)$ is proportional to $P(\phi)$ of Eq. 7-1. The constant of proportionality was evaluated by requiring that experiment and theory agree at one of the fixed data points, shown marked by a double circle in the figure. The remaining experimental points are in excellent agreement with the theoretical curve.

Geiger and Marsden also investigated the variation of the scattering with foil thickness. As we mentioned earlier, the number of scattered alpha particles for a fixed position of the detector proved to be directly proportional to t and not to $\sqrt{t}$, thus supporting the single scattering concept characteristic of the Rutherford atom model (see Fig. 7-4) rather than the multiple scattering concept characteristic of the Thomson atom model (see Fig. 7-3).

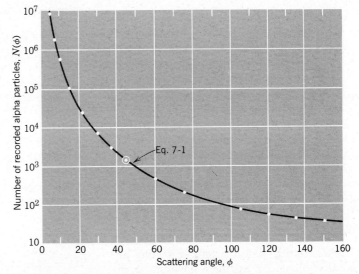

Figure 7-5. Rutherford scattering of alpha particles from a gold foil. The circles are the experimental points of Geiger and Marsden [7]. The solid curve is a plot of Eq. 7-1. The curve has been adjusted arbitrarily to fit the single data point, marked with a double circle, at $\phi = 45°$. Note that the vertical scale is logarithmic, so as to encompass the wide range of values involved.

From Geiger and Marsden's data, Rutherford was able to conclude that n, the number of electrons in the atom, was about equal to $A/2$, where A is the atomic weight. On this basis a gold atom would contain about 100 electrons. In 1913, however, two years after Rutherford proposed the nuclear atom, the speculation was first put forward (by A. van den Broek, a Dutch lawyer and an amateur physicist) that n is identical with Z, the position number of the atom in Mendeleev's table. In 1920 James Chadwick, in Rutherford's laboratory, remeasured the scattering of alpha particles from foils of copper ($Z = 29$), silver ($Z = 47$), and platinum ($Z = 78$) with improved precision. For the nuclear charges he obtained, respectively, $29.3e$, $46.3e$, and $77.4e$, in agreement to about 1 percent with the supposition that $n = Z$.

EXAMPLE 2.

Boring in Close to the Nucleus. An alpha particle with initial kinetic energy K happens to be moving directly toward the nucleus of a gold atom ($Z = 79$). It will slow down and, we assume, come momentarily to rest at a certain distance r_0 from the nuclear center, and then reverse its direction, picking up speed as it recedes. (The scattering angle ϕ for such a particle is 180°.) This *distance of closest approach* proves to be a convenient parameter in analyzing alpha particle scattering. Find r_0 if $K = 8.00$ MeV.

We assume that, because of its large momentum and

its closeness to the nucleus at closest approach, the motion of the alpha particle is not affected by the external atomic electrons. The alpha particle (charge = $2e$) will then be slowed down and eventually stopped by the repulsive Coulomb force exerted on it by the gold nucleus (charge = $79e$). From the energy point of view its initial kinetic energy will, at the moment of closest approach, have been turned entirely into electric potential energy. We have then

$$K = \frac{(2e)(Ze)}{4\pi\varepsilon_0 r_0}$$

or

$$r_0 = \left(\frac{e^2}{4\pi\varepsilon_0}\right)\left(\frac{2Z}{K}\right)$$

$$= (1.44 \text{ MeV} \cdot \text{fm})\left(\frac{2 \times 79}{8.00 \text{ MeV}}\right) = 28.4 \text{ fm}.$$

Note that we have arranged the equation for r_0 above to display the frequently occurring quantity $e^2/4\pi\varepsilon_0$, whose value can be shown to be 1.440 MeV·fm; see the list of constants at the beginning of the problem set.

The radius of the gold nucleus is known to be about 6.4 fm so that the incoming alpha particle above does not come close to the nuclear surface. Expressed otherwise, the incoming alpha particle does not have enough kinetic energy to penetrate the Coulomb barrier surrounding the nucleus.

The distribution of scattered alpha particles at large scattering angles will deviate from the prediction of Eq. 7-1 as the incident alpha particle energy is increased to large values. This happens because very energetic alpha particles can penetrate the nucleus so that the scattering results are affected not only by the Coulomb force between alpha and nucleus, which Rutherford took into account, but also by specifically nuclear forces, which were then not known. Thus the nuclear radius can be found by measuring the distance of closest approach at that incident alpha energy at which deviations from Rutherford scattering set in.

EXAMPLE 3.

Calculating the scattering. A Rutherford scattering arrangement like that of Fig. 7-2 has the following characteristics: The scattering foil is of gold ($A = 197$ g/mol; $n = Z = 79$) and is 2.0 μm thick. The detector has a sensitive collecting area of 120 mm², located 7.0 cm from the scattering foil. The kinetic energy of the alpha particles is 8.0 MeV.

(a) Using Eq. 7-1 and this information, evaluate $P(\phi)$.

Let us work on Eq. 7-1 piecemeal. The first quantity in parentheses (see the list of constants at the beginning of the problem set) has the value $(2.307 \times 10^{-28} \text{ J}\cdot\text{m})^2$ or $5.32 \times 10^{-56} \text{ J}^2\cdot\text{m}^2$. The quantity ρ, the number of atoms (and thus of nuclei) per unit volume in the foil can be found from $\rho = d N_A/A$, in which d (= 19.3 g/cm³) is the density of gold, A is its atomic weight and N_A is the Avogadro constant. Thus

$$\rho = \frac{(19,300 \text{ kg/m}^3)(6.02 \times 10^{23} \text{ mol}^{-1})}{(0.197 \text{ kg/m}^3)} = 5.90 \times 10^{28} \text{ m}^{-3}.$$

We have then, for the second parentheses in Eq. 7-1,

$$\left(\frac{\mathscr{A}\rho t Z^2}{4R^2K^2}\right)$$

$$= \frac{(120 \times 10^{-6} \text{ m}^2)(5.90 \times 10^{28} \text{ m}^{-3})(2.0 \times 10^{-6} \text{ m})(79)^2}{(4)(7.0 \times 10^{-2} \text{ m})^2(8.00 \text{ MeV})^2(1.60 \times 10^{-13} \text{ J/MeV})^2}$$

$$= 2.8 \times 10^{48} \text{ J}^{-2}\cdot\text{m}^{-2}.$$

Combining our results yields, for Eq. 7-1,

$$P(\phi) = \frac{(5.32 \times 10^{-56} \text{ J}^2\cdot\text{m}^2)(2.8 \times 10^{48} \text{ J}^{-2}\cdot\text{m}^{-2})}{\sin^4(\phi/2)}$$

$$= \frac{1.5 \times 10^{-7}}{\sin^4(\phi/2)}.$$

(b) If 10^8 alpha particles are allowed to fall on the foil, how many will enter the detector at angles of 10°, 30°, 90°, and 180°?

At 10° we have, using the above formula,

$$N(\phi) = 10^8 P(\phi) = \frac{15}{\sin^4(10°/2)} = 260,000.$$

For the angles in sequence we find then: 260,000, 3250, 60, and (about) 15 scattered particles. The rapid fall-off as the scattering angle increases is clear.

7-3 THE STABILITY OF THE NUCLEAR ATOM

When Rutherford did away with Thomson's atom-sized sphere of positive charge and replaced it by the tiny nucleus, Thomson's stable arrangements of electrons within this "pudding"—so carefully worked out—had to go with it. It can be shown that in free space no stable stationary arrangement of point charges is possible (see Problem 5). The electrons in Rutherford's nuclear atom, Z in number, must exist somewhere outside the nucleus, within a sphere whose radius is roughly 10,000 times greater than the nuclear radius. The whole question of the arrangement and the stability of the atomic electrons was thus reopened by Rutherford's discovery.

The electrons cannot be at rest outside the nucleus because the attractive Coulomb force would pull them toward the nucleus and thus, according to classical ideas, we would end up with a nuclear-sized atom. One is led then to consider a planetary model in which electrons revolve in circular (or elliptical) orbits of atomic dimensions about a nuclear "sun." A serious problem arises at once. According to Maxwell's theory of electromagnetism, accelerated charged particles radiate electromagnetic waves. Electrons in planetary orbits are continually accelerating and should therefore be continually radiating energy. Hence, electrons would lose energy and spiral in toward the nucleus, just as an earth satellite in low orbit, losing energy by atmospheric friction, spirals in toward the earth.

According to Einstein's general theory of relativity, it is also true that an accelerating *mass* should radiate energy in the form of (gravitational) waves; see Supplementary Topic C, page 299. Thus an orbiting planet (the earth, say) should also spiral in toward its attracting center (the sun). Gravitational waves, however, are so weak that they have not so far been directly detected. The rate at which the earth should lose energy by this process can be calculated and proves to be utterly negligible. The electromagnetic situation for accelerating *charge*, as for orbiting electrons about a nuclear center, however, is quite different; the time during which an atom of diameter 10^{-10} m would collapse to nuclear size can be computed to be about 10^{-12} s! Apart from contradicting the stability of atoms, this picture requires that the spiraling electrons emit radiation in a continuous spectrum of wavelengths. The most characteristic thing about radiations from atoms, however, is that they do *not* form a continuous spectrum but rather a spectrum of discrete lines at sharply defined wavelengths.

It was Niels Bohr who, in 1913 at age 28, advanced a theory that preserved the stability of the nuclear atom and predicted the discrete atomic emission spectra. Bohr saw the instability of the planetary electron model not as a difficulty but as an important clue that some new element, outside the framework of classical physics, must be introduced. This he proceeded to do. We now examine what was then known about atomic spectra, preparatory to discussing the Bohr theory.

7-4 THE SPECTRA OF ATOMS

It had been known for some time that every atom has its own characteristic spectrum. Whereas solids or liquids emit a continuous spectrum, which is easily observed when they are at high temperatures, individual free atoms are observed to emit a spectrum consisting of discrete wavelengths. The spectrum can be produced by energizing atoms in a gas container, such as by passing an electrical discharge through it. The atoms thereafter give up the energy they absorbed through collisions by emitting radiation and thereby return to their lowest energy state. In Fig. 7-6 we show schematically how the radiation from a source is made to reveal its spectrum. After being collimated by a slit system, the radiation passes through a prism and each wavelength in the spectrum is separately recorded on a photographic plate. The plate reveals a series of lines, each one an image of the line slit formed by a different wavelength.

In Fig. 7-7 we show the spectral lines emitted by hydrogen that span the visible and the near ultraviolet regions of the spectrum. The hydrogen atom is the simplest of all atoms and certainly must be understood if we are to understand the more complicated ones. Furthermore, most of the universe consists of isolated hydrogen atoms, so its spectrum is of much practical interest, too. Note, in Fig. 7-7, that the hydrogen lines seem to form a converging series, the spacing between lines decreasing steadily toward a limit in the ultraviolet.

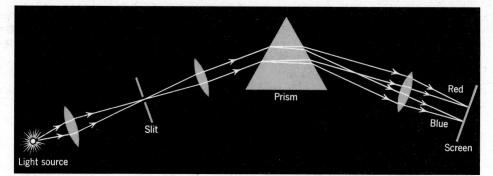

Figure 7-6. A prism spectrometer. For precise wavelength measurements the prism is usually replaced by a diffraction grating.

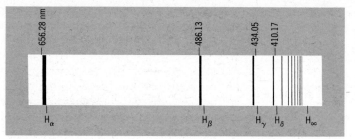

Figure 7-7. The emission spectrum of atomic hydrogen in the visible and near ultraviolet region. The series shown is the Balmer series, the position marked H_∞ indicating the series limit. From Gerhard Herzberg, *Atomic Spectra and Atomic Structure* (Dover, New York, 1944).

In 1885 Johann Balmer [8] found that he could express by an empirical formula the frequencies of the 14 then-known hydrogen lines. Put later by J. R. Rydberg in terms of reciprocal wavelengths (frequency divided by c, the speed of light), Balmer's formula became

$$\frac{1}{\lambda} = R_H \left(\frac{1}{2^2} - \frac{1}{n^2} \right) \qquad n = 3, 4, 5, \ldots \qquad (7\text{-}2)$$

Here R_H is a constant, now called the Rydberg constant, and n takes on all consecutive integral values greater than 2. The Rydberg constant for hydrogen is known today to have the value $R_H = 10,967,758.5 \pm 0.8 \text{ m}^{-1}$, which suggests how great the precision of spectral measurements can be. The set of lines given by Balmer's formula is now called the Balmer series. By the time Rutherford's model was accepted, more than 30 lines of the Balmer series had been found, all having wavelengths predicted by his formula. Balmer correctly concluded that there was a spectral *series limit*, the shortest wavelength (363.8 nm), corresponding to $n = \infty$, and that the first visible hydrogen line (656.3 nm) was the longest possible of the series, with $n = 3$.

Balmer's formula suggested to him that other series of hydrogen lines might exist. He correctly inferred that his original formula could be generalized to

$$\frac{1}{\lambda} = R_H \left(\frac{1}{m^2} - \frac{1}{n^2} \right), \qquad (7\text{-}3)$$

in which m takes on a fixed integral value and n takes on all integral values with $n \geq m + 1$. With $m = 2$ and $n \geq 3$, for example, we obtain Balmer's original formula. In the early 1900s new hydrogen lines were discovered in the far ultraviolet and in the infrared and subsequently Balmer's generalized formula was confirmed over a wide spectral range. In addition to the Balmer series, there is

> the Lyman series, with $m = 1$ and $n \geq 2$, in the ultraviolet;
> the Paschen series, with $m = 3$ and $n \geq 4$, in the near infrared;
> the Brackett series, with $m = 4$ and $n \geq 5$, in the infrared; and
> the Pfund series, with $m = 5$ and $n \geq 6$, in the far infrared.

The emission spectra of elements other than hydrogen were also observed to be series of lines, and in the 1890s Rydberg found approximate empirical laws to describe some of them, such as the alkali elements. His formulas for reciprocal wavelengths contained for each element a constant and a difference of squares. These Rydberg constants, however, had very slightly larger values than the hydrogen one, the deviation from hydrogen increasing the heavier the element. Absorption spectra, too, were known, that is, the spectrum of lines that are absorbed by atoms from a continuous spectrum in a beam incident upon them. We shall discuss their characteristics later. It is clear, however, that by 1913 a large amount of accurate spectroscopic data of various kinds was available, with much of it organized into empirical formulas, but for which no theoretical explanation existed. It was at this time that Neils Bohr was trying to construct an atomic theory by synthesizing Rutherford's model of the atom and Planck's quantization of the energy of oscillators. He later reported: "As soon as I saw Balmer's formula, everything became clear to me."

EXAMPLE 4.

Five Series of Lines in the Spectrum of Hydrogen. From the generalized Balmer formula (Eq. 7-3), calculate the wavelength range within which each of the five observed discrete series of hydrogen lines falls. Indicate also the region of the electromagnetic spectrum within which each of the five series falls.

With $R_H = 10{,}967{,}758.5 \text{ m}^{-1}$, we have

Lyman series	$\dfrac{1}{\lambda_1} = R_H\left(\dfrac{1}{1^2} - \dfrac{1}{2^2}\right)$	$\lambda_1 = 121.6 \text{ nm}$	Ultraviolet
	$\dfrac{1}{\lambda_\infty} = R_H\left(\dfrac{1}{1^2} - \dfrac{1}{\infty}\right)$	$\lambda_\infty = 91.2 \text{ nm}$	
Balmer series	$\dfrac{1}{\lambda_1} = R_H\left(\dfrac{1}{2^2} - \dfrac{1}{3^2}\right)$	$\lambda_1 = 656.5 \text{ nm}$	Visible and ultraviolet
	$\dfrac{1}{\lambda_\infty} = R_H\left(\dfrac{1}{2^2} - \dfrac{1}{\infty}\right)$	$\lambda_\infty = 364.7 \text{ nm}$	
Paschen series	$\dfrac{1}{\lambda_1} = R_H\left(\dfrac{1}{3^2} - \dfrac{1}{4^2}\right)$	$\lambda_1 = 1.876 \text{ } \mu\text{m}$	Near infrared
	$\dfrac{1}{\lambda_\infty} = R_H\left(\dfrac{1}{3^2} - \dfrac{1}{\infty}\right)$	$\lambda_\infty = 0.822 \text{ } \mu\text{m}$	
Brackett series	$\dfrac{1}{\lambda_1} = R_H\left(\dfrac{1}{4^2} - \dfrac{1}{5^2}\right)$	$\lambda_1 = 4.052 \text{ } \mu\text{m}$	Infrared
	$\dfrac{1}{\lambda_\infty} = R_H\left(\dfrac{1}{4^2} - \dfrac{1}{\infty}\right)$	$\lambda_\infty = 1.459 \text{ } \mu\text{m}$	
Pfund series	$\dfrac{1}{\lambda_1} = R_H\left(\dfrac{1}{5^2} - \dfrac{1}{6^2}\right)$	$\lambda_1 = 7.460 \text{ } \mu\text{m}$	Far infrared
	$\dfrac{1}{\lambda_\infty} = R_H\left(\dfrac{1}{5^2} - \dfrac{1}{\infty}\right)$	$\lambda_\infty = 2.279 \text{ } \mu\text{m}$	

In each case λ_∞ is called the *series limit*.

7-5 THE BOHR ATOM

In 1913 Niels Bohr, then working in Rutherford's laboratory at the University of Manchester, presented his theory of the atom [9,10]. His postulates were grounded in experimental results and, more so than Planck, he showed a willingness to give up those established classical ideas that simply contradicted observation. For example, Bohr accommodated the fact that atoms in their normal states do not radiate by simply adopting as a postulate:

1. *The postulate of stationary states.* An atom can exist without radiating in any one of a number of stable "stationary states" of definite energy.

The postulate does not identify these states further, and it does not necessarily invoke any pictorial representation of the atom, classical or otherwise. It gives no rule for discovering the energies of these states, although such a rule will emerge later. This first Bohr postulate is equivalent to the idea of the quantization of energy of atoms. If one pictures an electron orbiting a nucleus in the hydrogen atom, as Bohr did, this postulate simply contradicts—by fiat if you will—the classical electromagnetic assertion that such an electron must radiate energy.

If an atom in a stationary state does not radiate, it then becomes necessary to state the conditions under which an atom *does* radiate. Bohr handled this by means of a second postulate:

2. *The radiation postulate.* An atom emits (or absorbs) radiation only when the atom goes from one of its stationary states to another. The energy of the emitted (or absorbed) photon is equal to the energy difference between these two states.

That is, if the atom makes a transition between allowed energy states of energies E_2 and E_1, where E_2 is greater than E_1, then (a) if the transition is from (higher) E_2 to (lower) E_1, we have

$$E_2 = E_1 + h\nu_{21},$$

in which ν_{21} is the frequency of the *emitted* photon; whereas (b) if the transition is from (lower) E_1 to (higher) E_2, we have

$$E_1 + h\nu_{12} = E_2,$$

in which ν_{12} is the frequency of the *absorbed* photon. This second postulate contains the conservation of energy idea, of course, but more specifically invokes Einstein's quantum picture of radiation. This postulate, like the first, is meant to be quite general, applying to atoms and to atomic systems of all kinds. Again, it does not rely on any pictorial representation of the atom, classical or otherwise.

Finally, Bohr presented what can be called a quantization rule, that is, a way to determine the energies of the stationary states of the atom. This rule, called the correspondence principle, states:

3. *The correspondence principle.* Quantum theory must give the same results as classical theory in the limit in which classical theory is known to be correct.

This idea has already been used in relativity, where we showed that as $v/c \to 0$, the relativistic results become the same as the classical results, which are known to be valid in the region of low speeds. In the case of the hydrogen atom, the correspondence principle requires that in the limit of large systems, where the allowed energies form a continuum, the quantum radiation postulate must yield the same result as a classical calculation. If one pictures an electron orbiting about a nucleus, this means that for *very* large orbital radii, such that the atom has macroscopic size, the frequency of the radiation emitted by hydrogen should be the same as the frequency of revolution of the electron. Let us now apply Bohr's postulates.

We shall proceed much as Bohr did. First we shall use the experimentally verified Balmer formula together with Bohr's first two postulates to determine the allowed energies of the hydrogen atom. After getting familiar with this result, we shall then *derive* it from Bohr's third postulate, that is, from theory alone independent of experimentally determined quantities.

We start with the Balmer formula $1/\lambda = R_H(1/m^2 - 1/n^2)$ and, to get it in terms of emitted frequency ν_{nm}, multiply by c, yielding $\nu_{nm} = cR_H(1/m^2 - 1/n^2)$. The energy of the radiated photon, $h\nu_{nm}$, is then

$$h\nu_{nm} = hcR_H\left(\frac{1}{m^2} - \frac{1}{n^2}\right). \tag{7-4}$$

From Bohr's second postulate for emission, however, we have, for a transition from a stationary state of energy E_n to one of energy E_m,

$$h\nu_{nm} = E_n - E_m. \tag{7-5}$$

Comparison of these equations shows that *the allowed energies of the stationary states* must be

$$E_n = -\frac{hcR_H}{n^2}, \tag{7-6}$$

in which each integer n determines an allowed (quantized) energy. The lowest (most negative) energy corresponds to $n = 1$ and the highest $(E = 0)$ to $n = \infty$. The negative sign in Eq. 7-6 signifies that the system (hydrogen atom) is a *bound* one, whose total energy is negative. The binding is just barely broken when, on the planetary picture, enough energy is added to separate the electron and nucleus; they are then at rest an infinite distance apart, the total (kinetic plus potential) energy of the system being zero. If this energy, or more, is added, the atom is ionized.

EXAMPLE 5.

The Binding Energy of Hydrogen. Calculate the binding energy of the hydrogen atom, that is, the energy that must be added to the atom to separate the electron and the nucleus from each other.

The total energy of the highest bound state of a hydrogen atom is zero, obtained by putting $n = \infty$ into Eq. 7-6. The binding energy E_b is therefore numerically equal to the energy of the lowest state, corresponding to $n = 1$ in Eq. 7-5. That is,

$$E_b = -E_1$$

$$= -\left(-\frac{hcR_H}{1^2}\right)$$

$$= \frac{(6.63 \times 10^{-34}\ \text{J}\cdot\text{s})(3.00 \times 10^8\ \text{m/s})(1.097 \times 10^7\ \text{m}^{-1})}{1}$$

$$= 2.18 \times 10^{-18}\ \text{J} = 13.6\ \text{eV},$$

which agrees with the experimentally observed binding energy for hydrogen.

In Fig. 7-8 we show an energy-level diagram for hydrogen, with the quantum number n for each level. The energies shown are computed from Eq. 7-6. Notice that the levels become closer together as n increases and, from the fact that there are an infinite number of levels, we see that the classical continuum of energy is approached at high values of n. Normally the atom is in the state of lowest energy, $n = 1$, called the ground state ("ground" state means "fundamental" state, the term originating from the German word "grund," meaning fundamen-

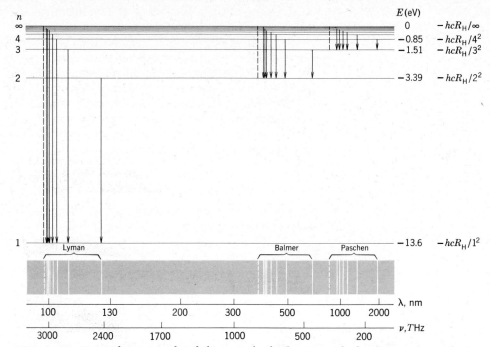

Figure 7-8. *Top:* The energy-level diagram for hydrogen with the quantum number n for each level, the energy of the atom in each stationary (or quantum) state, and some of the transitions between states. An infinite number of levels is crowded in between the levels marked $n = 4$ and $n = \infty$. *Bottom:* The corresponding spectral lines for the three series indicated. Within each series the spectral lines follow a regular pattern, approaching the series limit at the short-wave end of the series. Note that neither the wavelength nor the frequency scale is linear. The Brackett and Pfund series (not shown) are in the far infrared.

tal). When the atom absorbs energy from some excitation process, it makes a transition to some higher allowed energy state, called an excited state, for which $n > 1$. The atom may emit this energy thereafter in a series of transitions to allowed states of lower energy and return to its ground state. For each transition a photon is emitted. In a spectral discharge tube containing a large number of atoms, all possible transitions occur and the entire spectrum is emitted. The spectrum can then be analyzed as consisting of distinct series of lines. The transitions between allowed stationary states of the hydrogen atom that give rise to the observed spectral series of lines are shown in the figure, the Balmer series, for example, arising from transitions that end on the second stationary state, $n = 2$. Actually, Bohr had predicted the existence of a stationary state corresponding to $n = 1$ before the Lyman series of lines, for transitions ending in this ground state, had been observed.

Up to now we have been examining and interpreting Eq. 7-6. But that equation is an empirical relation that combines Bohr's first two postulates—the quantization of energy and the radiation postulate—with the generalized Balmer formula. In that formula the Rydberg constant is determined by experiment. Bohr used his third postulate to *derive* a relation giving the Rydberg constant in terms of other known fundamental constants. His theoretically determined value agreed with the precisely determined experimental value, which was a triumph for his theory.

Let us now use the third postulate—the correspondence principle—to determine the Rydberg constant.

A review of the section so far will convince you that up to this point we have not invoked the classical picture of an electron orbiting a central nucleus. We do so now, with the full realization that only in the classical limit of large quantum numbers can we expect such a model to be correct. The frequency ν_{orb} at which an electron orbits the nucleus can readily be shown to be

$$\nu_{\text{orb}} = \frac{4\varepsilon_0}{e^2}\sqrt{\frac{2}{m}}\,|E|^{3/2}, \tag{7-7}$$

in which E is the total mechanical energy (kinetic plus potential) of the system; the vertical bars denote an absolute value, E being a negative quantity. (The algebraic proof needed to get this result is diverting, so we leave the proof—which is quite simple—to a guided problem, Problem 40.)

If we substitute for E from Eq. 7-6 we find, as a result that should hold true at large quantum numbers,

$$\nu_{\text{orb}} = \frac{4\varepsilon_0}{e^2}\sqrt{\frac{2}{m}}\,\frac{(hcR_{\text{H}})^{3/2}}{n^3} \qquad (n \gg 1). \tag{7-8}$$

The accelerating electron in this classical picture should radiate continuously at a fundamental frequency equal to its frequency of revolution; as it loses energy and spirals inward, it radiates at continuously higher frequencies. In the quantum picture, however, the atom—starting in a state of high energy—jumps from one stationary state to another lower one, emitting a photon in each transition, which gives rise to a discrete set of spectral lines. At high energies, however, the allowed states are very close together in energy (see Fig. 7-8), so the successively emitted photons should be very close together in frequency. This is the region in which the quantum levels approach a classical continuum and in which, according to the correspondence principle, the quantum radiation postulate should yield the known classical result.

Imagine an atom passing successively through every stationary state. For a transition from state n to state $n - 1$, the quantum radiation postulate gives us, for the frequency,

$$\nu_{\text{rad}} = cR_{\text{H}}\left[\frac{1}{(n-1)^2} - \frac{1}{n^2}\right]$$

$$= cR_{\text{H}}\left[\frac{2n-1}{(n-1)^2 n^2}\right]. \tag{7-9a}$$

At very large values of n we can neglect 1 in comparison to n, so this formula becomes

$$\nu_{\text{rad}} = cR_{\text{H}}\frac{2n}{n^2 n^2} = \frac{2cR_{\text{H}}}{n^3} \qquad (n \gg 1). \tag{7-9b}$$

This is the quantum result in the "classical" region of large quantum numbers—the region where the energy levels are so close together as to form a continuum for practical purposes—and, according to the correspondence principle, it should agree with the classical result, which assumed that the allowed energies *do* form a continuum.

Equating Eqs. 7-8 and 7-9b and solving for the Rydberg constant for hydrogen, we obtain

$$R_\mathrm{H} = \frac{e^4 m}{8\varepsilon_0^2 h^3 c}. \qquad (7\text{-}10)$$

Hence, we now have a *theoretically predicted value* for the Rydberg constant in terms of other fundamental constants—the charge e and mass m of an electron, the speed c of light, and the Planck constant h.

Bohr, using data available in his time for these fundamental constants, obtained good agreement with experiment, the agreement today being within the extremely narrow limits of experimental error. Therefore, we can now regard the constant R_H as theoretically determined and write the allowed energies of the hydrogen atom, Eq. 7-6, in terms of other constants as

$$E_n = -\left(\frac{me^4}{8\varepsilon_0^2 h^2}\right)\frac{1}{n^2}. \qquad (7\text{-}11)$$

In summary, then, Bohr's postulates enabled him to deduce a theoretical formula for the allowed energies of the hydrogen atom and to establish rules for emission of radiation that agreed with the observed wavelengths in the hydrogen emission spectrum. The only place that a classical picture was invoked—the picture of a point electron orbiting in a definite circular path about a nucleus—was in the classical limit where such a picture is valid according to the correspondence principle.*

Table 7-1 shows the operation of the correspondence principle in detail. We see clearly that as the quantum number n increases, the frequency of revolution of the electron in its orbit, calculated from Eq. 7-8, comes to agree with the frequency of transition to the next lower state, calculated from Eq. 7-9a.

Table 7-1
THE CORRESPONDENCE PRINCIPLE AS APPLIED TO THE HYDROGEN ATOM

Quantum Number, n	Frequency of Revolution in Orbit (Eq. 7-8) Hz	Frequency of Transition to Next Lowest State (Eq. 7-9a), Hz	Difference, %
2	8.22×10^{14}	24.7×10^{14}	67
5	5.26×10^{13}	7.40×10^{13}	29
10	6.58×10^{12}	7.72×10^{12}	15
50	5.26×10^{10}	5.43×10^{10}	3.1
100	6.580×10^{9}	6.680×10^{9}	1.5
1,000	6.5797×10^{6}	6.5896×10^{6}	0.15
10,000	6.5797×10^{3}	6.5807×10^{3}	0.015
25,000	4.2110×10^{2}	4.2113×10^{2}	0.007
100,000	6.5798	6.5799	0.0007

* In the classical portion of our calculations we have made a tacit assumption, namely, that the mass of the nucleus of the hydrogen atom may be assumed infinitely great in comparison to the mass of the electron. In Section 7-8 we shall take the finite nuclear mass fully into account, which will require a small correction in Eqs. 7-10 and 7-11.

The Bohr postulates met with other successes. It was possible, for example, to understand absorption spectra in terms of the Bohr atom. A Bohr atom can absorb only certain energies from a beam of incident radiation. If a continuous spectrum of wavelengths is incident on a container of gas, the radiation transmitted is found to be missing a set of discrete wavelengths that must have been absorbed by the atoms in the container. Each so-called absorption line has the same wavelength as a line in the emission spectrum of the atom. However, not every emission line appears in the absorption spectrum. This is explained by noting that the atoms in a gas are normally in the ground state. Hence in hydrogen, for example, only those lines corresponding to exciting the atom from the ground state ($n = 1$) will appear in the absorption spectrum; indeed, only the Lyman series is normally observed. If the gas is at a high temperature, many of the atoms may be in excited states to begin with, so absorption lines corresponding to the Balmer series, for example, may appear. This, in fact, is observed to be the case in stellar spectra (see Question 17).

The nature of the absorption spectrum, it should be noted, cannot be at all understood on the basis of any atom model in which the emission spectrum is identified with the frequency of revolution of an electron in its orbit or with the frequency of oscillation of an electron about a fixed equilibrium position. In such a model every emission line should also appear in the absorption spectrum.

7-6 THE BOHR SEMICLASSICAL PLANETARY MODEL OF THE ONE-ELECTRON ATOM

With his postulates confirmed, Bohr then ventured further. Rather than limiting himself to correspondence between classical and quantum motion at large values of n, he constructed a model for the one-electron atom that pictured the electron as moving in a classical orbit for *all* values of n. This model was meant to be suggestive only. Bohr, in fact, in his first paper on atomic structure [7], wrote about such a classical representation of the atom: "While there obviously can be no question of a mechanical foundation of the calculations given in this paper, it is, however, possible to give a very simple interpretation of the results of the calculation . . . by help of symbols taken from ordinary mechanics." The specifics of Bohr's planetary atom model did not survive in the final quantum mechanical picture that eventually emerged. Nevertheless, Bohr's semiclassical planetary model made plausible many observed properties of atoms and introduced key new ideas, such as the quantization of angular momentum, that did survive and that helped to develop the new theory.

Consider an electron, charge e, moving about a nucleus of charge Ze in a circular orbit. From $F = ma$, we have

$$\frac{1}{4\pi\varepsilon_0} \frac{Ze^2}{r^2} = \frac{mv^2}{r},$$ (7-12)

so that the kinetic energy K is

$$K = \frac{1}{2} mv^2 = \frac{1}{8\pi\varepsilon_0} \frac{Ze^2}{r}.$$

The mutual electric potential energy U, (*Physics*, Part II, Sec. 29-6) is

$$U = -\frac{1}{4\pi\varepsilon_0} \frac{Ze^2}{r},$$

so that the total energy $E = K + U$ is

$$E = -\frac{1}{4\pi\varepsilon_0}\frac{Ze^2}{2r}.$$ (7-13)

Bohr adopted this classical planetary picture as the starting point in his model of the one-electron atom. Such a formula would correspond to a hydrogen atom if we set $Z = 1$. It would apply to a singly ionized helium atom, if we set $Z = 2$, and to a doubly ionized lithium atom with $Z = 3$, and so forth. It is the general one-electron atom classical relation.

The generalized Balmer-Rydberg empirical formula for one-electron atoms was assumed to be $1/\lambda = Z^2 R_H(1/m^2 - 1/n^2)$. This was confirmed for singly ionized helium lines observed in 1896 by Pickering (see Problem 32), and in subsequent laboratory measurements on other one-electron atoms. For example, singly ionized helium atoms He^+ or doubly ionized lithium atoms Li^{2+} can be formed by sending a high-voltage discharge through a container of the normal gas. The so-called spark spectrum emitted by these ions is much simpler than the "arc" spectrum emitted by the normal atom and is easily identified. It is found that each frequency of He^+, for example, is almost precisely four times that of each corresponding hydrogen frequency, consistent with the formula above in which $Z^2 = 4$.

Bohr found that his entire earlier interpretation and analysis for hydrogen was consistent with a generalization to one-electron atoms of nuclear charge greater than one, if, in his hydrogen analysis, he simply replaced e^2 by the more general $(Ze)e$, to account for the force between a nucleus of charge Ze and a single electron. Hence, in Eq. 7-11 for the energy, e^4 is replaced by $Z^2 e^4$, and our one-electron energy formula becomes

$$E_n = -Z^2 \left(\frac{me^4}{8\varepsilon_0^2 h^2}\right)\frac{1}{n^2} \qquad \text{(energy).}$$ (7-14)

Since, according to his earlier postulates, the atom can exist only in these energy states, Bohr then modified the classical planetary picture so that the radius of the electron's orbit in Eq. 7-13 is allowed to take on only certain values. If we now combine Eqs. 7-13 and 7-14, we can solve for these allowed orbital radii, obtaining

$$r_n = \frac{\varepsilon_0 h^2}{\pi m Z e^2}n^2 = a_0 n^2 \qquad n = 1, 2, 3, \ldots \qquad \text{(orbital radius).}$$ (7-15)

Here $a_0 \ (= \varepsilon_0 h^2/\pi m Z e^2)$ is the radius of the Bohr orbit of lowest energy, corresponding to $n = 1$. For hydrogen, with $Z = 1$, we obtain

$$a_0 = \frac{\varepsilon_0 h^2}{\pi m e^2} = 5.292 \times 10^{-11} \text{ m} = 0.05292 \text{ nm} \qquad \text{(Bohr radius).}$$

This agrees in magnitude with what was known about the size of the hydrogen atom in its normal state. Although we no longer use the orbit concept, the Bohr radius given above proves to be a useful parameter in which to express lengths on the atomic scale. The orbits of the successively higher excited states in this picture are then $4a_0$, $9a_0$, $16a_0$, and so on, so that a highly excited atom with a very large value of n can be regarded as approaching a classical macroscopic atom. In interstellar space, where the density of atoms is low and the mean free path between collisions is large, it is possible for hydrogen atoms to exist in such highly excited states of large orbital radius with little disturbance and emissions from states of high n are characteristically observed there rather than in high-density discharge tubes.

Corresponding to each allowed orbital radius is an allowed orbital speed. From Eq. 7-12, we have $v^2 = (1/4\pi\varepsilon_0)(Ze^2/mr)$ and, on substituting for r the allowed orbital radii r_n of Eq. 7-15, we obtain the allowed orbital speeds

$$v_n = \frac{Ze^2}{2\varepsilon_0 h}\frac{1}{n} \qquad n = 1, 2, 3, \ldots \qquad \text{(orbital speeds)}. \tag{7-16}$$

We find the highest speed to be in the ground state, $n = 1$, and for hydrogen, with $Z = 1$, its value in terms of the speed of light is

$$\beta = \frac{v_1}{c} = \frac{e^2}{2\varepsilon_0 hc} = 0.0073.$$

Since, in this picture, the speed of the electron decreases as the orbital radius increases ($v_n \propto 1/n$ in Eq. 7-16), this result justifies the use of nonrelativistic mechanics for the hydrogen orbits. Of course, in one-electron atoms with large nuclear charge Z, relativistic considerations become more important, the electron moving at higher speed in the stronger Coulomb field of such nuclei. Even in hydrogen there are small relativistic effects, and such effects are taken into account in more detailed theories to explain the fine structure of spectral lines found in all atomic spectra.

The Bohr picture helps us to understand yet another feature of atomic spectra. We have seen (Example 5) that the binding energy, that is, the energy difference between the ground state of an atom and its state of zero energy, is the minimum energy required to ionize a normal atom. In Bohr's orbiting planetary picture of a one-electron atom, the electron in an ionized atom would no longer be bound to the nucleus. The highest discrete bound energy state of the atom is $E_\infty = 0$. For positive total energies, the atom would be ionized, the electron no longer being bound and becoming a free particle. The electron's "orbit" is one of infinite radius, and the correspondence principle tells us that this is the classical region. Therefore, the energy of a totally free particle is not quantized, so that a continuum of energy exists above the highest quantized state at $E = 0$. If an energy greater than the binding energy is supplied to a one-electron atom, the electron will be free in this energy continuum. When a photon supplies this energy to the atom, we simply have the photoelectric effect. Correspondingly, an ionized atom can capture a free electron, the neutral atom thereafter being in an allowed quantized energy state. Radiation of frequency greater than that of the series limit for that state will be emitted in such a process. Since the free electron can have any energy $E > 0$ initially, there ought to be a continuum in the spectrum of the atom beyond each series limit. This, in fact, is observed experimentally under the appropriate circumstances.

EXAMPLE 6.

Some Properties of Bohr Orbits. (*a*) Find the allowed frequencies of revolution of an electron in the Bohr model one-electron atom. (*Note:* In this example we represent the electron mass by m_e, reserving m to represent a quantum number.)

If ω represents angular velocity, we can write Eq. 7-12 as $m_e\omega^2 r = Ze^2/4\pi\varepsilon_0 r^2$ or

$$\omega^2 = \frac{1}{4\pi\varepsilon_0}\frac{Ze^2}{m_e r^3}.$$

From Eq. 7-15, however, only certain radii, $r_n = (\varepsilon_0 h^2/\pi m_e Ze^2)n^2$, are allowed. Substituting these values for r into the above equation and solving for ω yields

$$\omega_n = \left(\frac{Z^2 m_e e^4}{8\varepsilon_0^2 h^3}\right)\frac{4\pi}{n^3},$$

as you can verify. But $\omega = 2\pi\nu$, so the allowed frequencies ν_n are

$$\nu_n = \left(\frac{Z^2 m_e e^4}{8\varepsilon_0^2 h^3}\right)\frac{2}{n^3}.$$

(b) Compare the frequencies of revolution of an electron in an upper and lower state of the Bohr model atom to the frequency of radiation emitted in a transition between states.

We can write the frequencies of revolution (above) on Bohr's model as

$$\nu_n = \left(\frac{Z^2 m_e e^4}{8\varepsilon_0^2 h^3}\right)\frac{2}{n^3} \quad \text{and} \quad \nu_m = \left(\frac{Z^2 m_e e^4}{8\varepsilon_0^2 h^3}\right)\frac{2}{m^3}$$

for the upper nth and lower mth states. However, the frequency of radiation ν_{nm} emitted in a transition $n \to m$ is given by the radiation condition as

$$\nu_{nm} = \frac{E_n - E_m}{h} = \left(\frac{Z^2 m_e e^4}{8\varepsilon_0^2 h^3}\right)\left(\frac{1}{m^2} - \frac{1}{n^2}\right).$$

This observed frequency of radiation does *not* correspond to *either* of the previous frequencies. Indeed, one can show (see Problem 36) that

$$\frac{2}{m^3} \geq \left(\frac{1}{m^2} - \frac{1}{n^2}\right) \geq \frac{2}{n^3},$$

so that the frequency of the emitted radiation lies *between* the frequencies of revolution of the electron in the orbits between which, in Bohr's picture, the transition occurs.

We would expect the Bohr picture to be strictly correct in the classical limit, and, indeed, if $n = m + 1$ and n is very large, the radiation frequency equals the rotational frequency—the very condition Bohr originally imposed in his theory. But here again we see that in the quantum region, the classical picture must be discarded and the Bohr postulates used instead.

EXAMPLE 7.

The Sizes of Atoms. Consider a discharge tube filled with hydrogen at a pressure of 1.00 atm $(=1.01 \times 10^5 \text{ Pa})$ and a temperature T of 300 K. Assume, for simplicity, that the gas is ideal and that the hydrogen is present in atomic rather than molecular form.

(a) On the average, how far apart are the hydrogen atoms?

If V is the volume per mole and N_A is the Avogadro constant then, using the ideal gas law, the volume per atom of hydrogen is given by

$$v = \frac{V}{N_A} = \frac{RT}{N_A p}$$

$$= \frac{(8.31 \text{ J/mol} \cdot \text{K})(300 \text{ K})}{(6.02 \times 10^{23} \text{ atoms/mol})(1.01 \times 10^5 \text{ Pa})}$$

$$= 4.10 \times 10^{-26} \text{ m}^3/\text{atom}.$$

The mean separation of atoms is of the order of the cube root of this quantity, or

$$l \cong v^{1/3} = (4.1 \times 10^{-26} \text{ m}^3)^{1/3}$$
$$= 3.5 \text{ nm} = 65a_0,$$

in which a_0 $(=0.0529 \text{ nm})$ is the Bohr radius.

(b) For what Bohr orbit would the orbit diameter $(= 2r_n)$ be equal to the quantity calculated above?

From Eq. 7-15 we can write

$$n = \sqrt{\frac{r_n}{a_0}} = \sqrt{\frac{l}{2a_0}} = \sqrt{\frac{65a_0}{2a_0}} \cong 6.$$

It will be difficult to excite hydrogen atoms under such circumstances to states as high as $n = 6$; the atoms will be too readily deexcited by collision. Show that, if the pressure in the discharge tube is reduced to 100 Pa $(= 0.001$ atm$)$, the quantum number corresponding to that calculated above becomes ~ 18; higher orders of excitation should now be possible.

7-7 *THE QUANTIZATION OF ANGULAR MOMENTUM*

We have seen that the allowed stationary states of one-electron atoms are completely specified by a single integral quantum number n. Once we have selected a value for this quantity we can, for example, compute the total energies E of these states from Eq. 7-14. These values have meaning whether or not we choose to view the electron in such an atom as circulating in a classical orbit. If we do choose the orbit point of view, however, specifying n also fixes the orbit radius (Eq. 7-15), the electron's linear speed in this orbit (Eq. 7-16), and also the electron's angular speed (Example 6). Bohr drew attention* to another property of the stationary state, namely, its characteristic angular momentum L, that is also specified once the quantum number n is known.

The angular momentum is the moment of momentum, that is, the linear momentum times the moment arm. Thus, from Eqs. 7-15 and 7-16,

$$L_n = m_e v_n r_n = m_e \left(\frac{Ze^2}{2\varepsilon_0 h n} \right) \left(\frac{\varepsilon_0 h^2 n^2}{\pi m_e Ze^2} \right)$$

or

$$L_n = n \frac{h}{2\pi} = n\hbar \qquad n = 1, 2, 3, \ldots , \tag{7-17}$$

in which $\hbar$ is a convenient abbreviation for $h/2\pi$. *The orbital angular momentum of the electron is quantized, taking on only integral multiples of $\hbar$.* This result was so simple that Bohr felt compelled to change his third postulate. As Bohr arrived at it, the quantization of the angular momentum was a result of his earlier postulates and his planetary model. But, with his intuition for simple central principles, Bohr made the quantization of angular momentum, rather than the correspondence principle, the central postulate for determining the quantized energy states of the atom. Starting from Eq. 7-17, one can derive all the results of the Bohr atom and the Bohr model of the one-electron atom (see Problem 37) and can show thereafter that the results are consistent with the correspondence principle.

We see that there are two approaches to Bohr's theory of the one-electron atom. Both make use of Bohr's first and second postulates as presented in Section 7-5. The first approach that we discussed also makes use of the correspondence principle. Many prefer this approach because it does not require the visualization of planetary orbits except in the limit of large quantum numbers where such classical pictures are valid. The second approach to Bohr theory does not lean on the correspondence principle but postulates instead the quantization of angular momentum. This approach *does* involve a planetary orbit representation (for all quantum numbers, not only for large ones). However, some prefer this version of Bohr's theory—in spite of its drawbacks—because the quantization of angular momentum plays such a large role in modern quantum theory.

In 1924 (eleven years after Bohr had presented his theory) de Broglie gave a physical interpretation of the Bohr quantization rule for angular momentum. If p represents the linear momentum of an electron moving in an allowed circular orbit, we can write the angular momentum as pr and Eq. 7-17 becomes

$$pr = n \frac{h}{2\pi} \qquad n = 1, 2, 3, \ldots .$$

But, in terms of the de Broglie wavelength, we can write $p = h/\lambda$, and the equation for angular momentum becomes

$$\frac{h}{\lambda} r = n \frac{h}{2\pi}$$

or

$$2\pi r = n\lambda \qquad n = 1, 2, 3, \ldots . \tag{7-18}$$

---

* J. W. Nicholson had earlier emphasized the possible importance of angular momentum in the discussion of atomic systems.

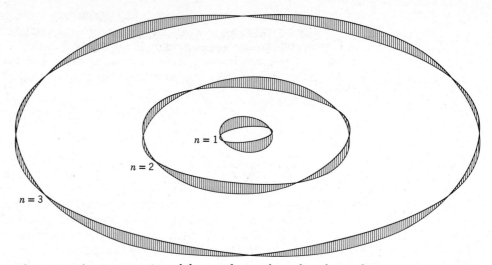

Figure 7-9. A representation of the envelopes of standing de Broglie waves set up in the first three Bohr orbits. The location of the nodes is, of course, arbitrary.

Therefore, only those orbits are allowed in which the circumference contains an integral number of de Broglie wavelengths.

Imagine the electron to be moving in a circular orbit with constant speed and to have a de Broglie wave associated with it. The wave, of wavelength λ, is then wrapped repeatedly around the circular orbit. The resultant wave that is produced will have zero intensity at any point unless the wave at each passage is exactly in phase at that point with the wave in other passages. If the waves in each passage are exactly in phase, then the orbits must contain an integral number of de Broglie wavelengths, as required by Eq. 7-18. If this requirement were not met, then in a large number of passages the waves would interfere with one another such that their average intensity would be zero. However, the average intensity of the waves, $\bar{\Psi}^2$, gives the probability of locating the particle, so that an electron has no chance of being in an orbit in which $\bar{\Psi}^2$ is zero.

This wave picture gives no suggestion of progressive motion. Rather, it suggests standing waves, as in a stretched string of a given length. In a stretched string only certain wavelengths, or frequencies of vibration, are permitted. Once such modes are excited, the vibration goes on indefinitely if there is no damping. To get standing waves, however, we need oppositely directed traveling waves of equal amplitude. For atoms this requirement is presumably satisfied by the fact that the electron can traverse an orbit in either direction and still have the magnitude of angular momentum required by Bohr. The de Broglie standing wave interpretation, illustrated in Fig. 7-9, therefore provides a satisfying basis for Bohr's quantization condition.

7-8 *CORRECTION FOR THE NUCLEAR MASS*

So far in our analysis of the Bohr model we have assumed that the nucleus remains at rest as the electron revolves about it. This is equivalent mechanically to regarding the nucleus as having an infinite mass compared to the electron. Because the mass of the hydrogen atom nucleus is nearly 2000 times larger than the mass of the electron, our procedure has been approximately correct. However, just as two bodies of finite mass that are under the influence of each other's gravitational attraction move in circular orbits about their common center of mass with the same angular frequency (see *Physics*, Part I, Sec. 16-7), so here the electron and nucleus move similarly under the influence of each other's electrical attraction. Because spectral data can be determined to such high accuracy, it turns out to be necessary, in order to get agreement with the data, to take into account in our formulas the actual finite mass of the nucleus and its effect on the motion. This can be done rather easily because in such a planetarylike system the electron moves relative to the nucleus as though the nucleus were fixed and the mass of the electron m were reduced to μ, the *reduced mass* of the system. The equations of motion of the system are the same as those we have considered if we simply substitute μ for m, where

$$\mu = m\left(\frac{M}{m + M}\right) \qquad \text{(reduced mass)}, \qquad (7\text{-}19)$$

in which M is the mass of the nucleus (see *Physics*, Part I, Sec. 15.8). Notice that μ is *less* than m by a factor $1/(1 + m/M)$.

In Bohr's planetary model treatment of the one-electron atom he made the necessary correction by taking into account the angular momentum of the nucleus as well as that of the electron. He simply postulated that *the total orbital angular momentum of the whole atom* (not just the electron) is an integral multiple of h, so that Eq. 7-17 is generalized to

$$L_n = \mu v_n r_n = nh \qquad n = 1, 2, 3, \ldots \qquad (7\text{-}20)$$

If now we were to proceed with the Bohr analysis, we would find the equations to be the same as before except that we must replace the electronic mass by the reduced mass, (see Problem 44). In particular, the Rydberg constant for finite nuclear mass R_M is related to the one derived by Bohr (Eq. 7-10) for infinite nuclear mass, $R_\infty \equiv me^4/8\varepsilon_0^2 h^3 c$, by

$$R_M = \frac{\mu e^4}{8\varepsilon_0^2 h^3 c} = \frac{\mu}{m} R_\infty = \left(\frac{M}{m + M}\right) R_\infty. \qquad (7\text{-}21)$$

The experimental value of R_H cited in Sec. 7-4 should be compared to this value R_M of the theory, rather than to the value $R_\infty = 10{,}973{,}731.8 \text{ m}^{-1}$, which incorrectly assumed infinite nuclear mass for hydrogen. When this is done, it is found that the values for hydrogen,

$$R_H = \left(\frac{M}{M + m}\right) R_\infty = \left(\frac{1836.15}{1836.15 + 1}\right) (10{,}973{,}731.8 \text{ m}^{-1})$$

$$= 10{,}967{,}758.5 \text{ m}^{-1},$$

agree to at least six significant figures! We see also in Eq. 7-21 confirmation of Rydberg's empirical findings that the effective Rydberg constant increases slightly the heavier the atom. For example, $R_H = 10{,}967{,}758.5 \text{ m}^{-1}$, $R_D = 10{,}970{,}742.8 \text{ m}^{-1}$, and $R_{He^+} = 10{,}972{,}227.8 \text{ m}^{-1}$ give the Rydberg constants for hydrogen, deuterium (see Example 8), and singly ionized helium, respectively. Finally, with the correct Rydberg constant, the formula for the reciprocal wavelength of the spectral lines becomes

$$\frac{1}{\lambda} = R_M Z^2 \left(\frac{1}{m^2} - \frac{1}{n^2}\right), \qquad (7\text{-}22)$$

with R_M, for an atom with a nucleus of mass M, being given by Eq. 7-21.

EXAMPLE 8.

Discovering Deuterium. Ordinary hydrogen contains about one part in 6000 of *deuterium*, often called heavy hydrogen. This is a hydrogen isotope whose nucleus (containing a proton *and* a neutron) has a mass nearly twice that of the nucleus of ordinary hydrogen, which is a single proton.

(*a*) Compare the emission spectra of hydrogen and deuterium.

The spectra would be identical if it were not for the correction for finite nuclear mass. The Rydberg constants for hydrogen and for deuterium have been calculated in the text above to be

$$R_{\mathrm{H}} = 10{,}967{,}758.5 \ \mathrm{m}^{-1}$$

and

$$R_{\mathrm{D}} = 10{,}970{,}742.8 \ \mathrm{m}^{-1}.$$

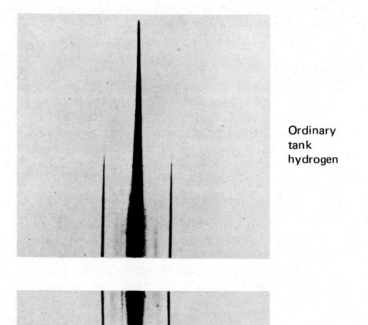

Ordinary
tank
hydrogen

Evaporated
hydrogen

Figure 7-10. The H-β lines for ordinary hydrogen (above) and for a sample of hydrogen treated in such a way as to enhance its deuterium content (below). The outer lines in each case are instrumental artifacts (called "ghosts"). The main line in the center is the H-β for ordinary hydrogen and has about the same intensity in both exposures. The faint line to the left of the main line (see arrow) is the deuterium H-β line. It is considerably more enhanced in the lower spectrum because of the increased concentration of deuterium in that sample. (From Urey, Brickwedde, and Murphy.)

These quantities differ by only 2,984.3 m^{-1} or 0.03 percent, but such is the precision of spectroscopic measurements that such a difference generates quite a detectable wavelength shift (called an *isotope shift*) in the spectrum lines. Because $R_D > R_H$, analysis of Eq. 7-22 shows that the wavelengths of the spectrum lines for deuterium should be shifted slightly toward *shorter* wavelengths as compared to the corresponding lines in the normal hydrogen spectrum.

(*b*) What wavelength shift do you expect between hydrogen and deuterium for the second line of the Balmer series (called the H-β line), corresponding to $m = 2$ and $n = 4$ in Eq. 7-22? (See also Problem 46.)

For hydrogen that equation becomes

$$\frac{1}{\lambda_H} = R_H Z^2 \left(\frac{1}{m^2} - \frac{1}{n^2} \right) = R_H (1)^2 \left(\frac{1}{2^2} - \frac{1}{4^2} \right)$$

$$= \frac{3 R_H}{16}$$

Similarly, for deuterium we have

$$\frac{1}{\lambda_D} = \frac{3 R_D}{16}.$$

The wavelength difference is then

$$\lambda_H - \lambda_D = \frac{16}{3 R_H} - \frac{16}{3 R_D} = \frac{16}{3} \frac{R_D - R_H}{R_D R_H}$$

$$= \frac{(16)(2984.3 \ \text{m}^{-1})}{(3)(1.097 \times 10^7 \ \text{m}^{-1})}$$

$$= 1.32 \times 10^{-10} \ \text{m} = 0.132 \ \text{nm}.$$

Deuterium was indeed first detected in this way in 1932, by H. C. Urey and his collaborators. By increasing the concentration of the heavy isotope (deuterium) above its normal value using an electrochemical technique, they were able to enhance the intensity of the deuterium lines in a hydrogen discharge tube, which, ordinarily, are too weak to detect. Figure 7-10 shows their results, the faint shifted line due to deuterium being plainly visible and in just the expected wavelength position. For this work Urey received the 1934 Nobel Prize in Chemistry.

EXAMPLE 9.

Muonic Atoms [11]. A muonic atom consists of a nucleus of charge Ze with a negative muon circulating about it. A (negative) muon, symbol μ^-, is a particle of charge $-e$ and a mass that is 207 times as large as the electron mass. Such an atom is formed when a proton, or some other nucleus, captures a negative muon.

(*a*) Calculate the radius of the first Bohr orbit of a muonic atom with $Z = 1$.

The reduced mass of the system, with $m_\mu = 207 m_e$ and $M = 1836 m_e$, is, from Eq. 7-19,

$$\mu = \frac{(207 m_e)(1836 m_e)}{207 m_e + 1836 m_e} = 186 m_e.$$

Then, from Eq. 7-15, with $n = 1$, $Z = 1$, and m replaced by $\mu = 186 m_e$, we obtain

$$r_1 = \frac{\varepsilon_0 h^2}{\pi (186 m_e) Z e^2} n^2 = \frac{a_0}{186}$$

$$= \frac{5.29 \times 10^{-11} \ \text{m}}{186} = 2.84 \times 10^{-4} \ \text{nm}.$$

The muon then is much closer to the nuclear (proton) surface than is the electron in a hydrogen atom. It is this feature that makes such muonic atoms interesting, information about nuclear properties being revealed from their study.

(*b*) Calculate the binding energy of a muonic atom with $Z = 1$.

From Eq. 7-14, with $Z = 1$, $n = 1$, and $\mu (=m) = 186 m_e$, we have

$$E_1 = -186 \frac{m_e e^4}{8 \varepsilon_0^2 h^2} = -(186)(13.6 \ \text{eV}) = -2530 \ \text{eV}$$

as the ground-state energy. Hence, the binding energy is 2530 eV.

(*c*) What is the wavelength of the first line in the Lyman series for such an atom?

From Eq. 7-22, with $Z = 1$, we have

$$\frac{1}{\lambda} = R_M \left(\frac{1}{m^2} - \frac{1}{n^2} \right).$$

For the first Lyman line, $n = 2$ and $m = 1$. In this case, $R_M = (\mu / m_e) R_\infty = 186 R_\infty$. Hence,

$$\frac{1}{\lambda} = 186 R_\infty \left(1 - \frac{1}{4} \right) = 139.5 R_\infty.$$

With $R_\infty = 1.097 \times 10^7 \ \text{m}^{-1}$, we obtain

$$\lambda \simeq 0.653 \ \text{nm}.$$

so that the Lyman lines lie in the x-ray part of the spectrum. x-Ray techniques are necessary therefore to study the spectrum of muonic atoms.

7-9 BOHR THEORY—THE HIGH-WATER MARK

In discussing the one-electron planetary Bohr atom, so far we have dealt with circular orbits only. Beginning in 1916 Arnold Sommerfeld, working in Munich, extended Bohr's analysis to elliptical orbits. Figure 7-11 shows the family of such ellipses for $n = 3$, all with the same major axis. It turns out that the energy of the system depends only on the major axis, so all three orbits in the figure have the same energy. However, they have different angular momenta, the angular momentum being $3h$ (see Eq. 7-17) for the circular orbit and decreasing by units of $\hbar$ for the remaining two orbits.

Sommerfeld also showed that if relativistic effects are taken into account, then the energy of the state no longer depends on the length of the major axis of the ellipse alone but also depends to a small extent on the length of the minor axis. Put another way, the energy of the state is no longer a function of the principal quantum number n alone, but also depends slightly on the value of the quantum number used to specify the angular momentum of the state. In this way Sommerfeld was able to explain the *fine structure* of spectrum lines, that is, the fact that many such lines, when examined under high resolution, are shown to consist of several components.

Meanwhile Bohr, drawing on the work of Sommerfeld and extending his planetary model to the limit, sought to develop a theoretical basis for understanding the arrangement of the elements in the periodic table. In this he had at best a modest success. Figure 7-12 shows Bohr's 1922 representaton of the krypton atom, with its 36 electrons. As Heilbron [8] puts it, Bohr ". . . commissioned a set of commemorative portraits [of the atoms]." The reference is to the demise of such models that was to come just a few years later with the emergence of quantum mechanics.

Bohr's crowning achievement in extending his theory to many-electron atoms was his prediction that element 72, then only a blank space in the periodic table, should resemble zirconium in its chemical properties. This flew in the face of a claim by French chemists that this element (which they had named "celtium") was to be found among the rare earth elements. In 1922, however, a successful search for element 72 in zirconium ores was carried out in Bohr's Institute in Copenhagen. Coster and Hevesy, who carried out the search, named the element *hafnium*, after the early neo-Latin name for Copenhagen. Word of this discovery came dramatically to Bohr in Stockholm, just hours before he was scheduled to receive the Nobel Prize.

In 1913, when Bohr first presented his theory, (1) the nuclear atom was only two years old and not yet widely accepted; (2) the concept of atomic number was just being developed and it was not known how many electrons there were in the atom of any given element; (3) Einstein's photon concept had yet to be confirmed by Millikan and was not accepted by many leading physicists, including Planck; (4) the wave nature of matter was unknown; (5) the fact that the electron has an

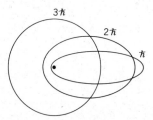

Figure 7-11. The three allowed Bohr orbits corresponding to $n = 3$; their predicted angular momenta are indicated. If relativity is taken into account, Sommerfeld's analysis predicts that the three states will differ slightly in energy.

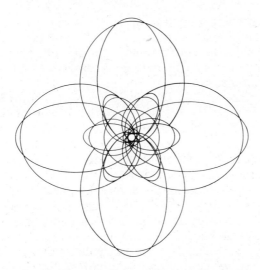

Figure 7-12. Bohr's 1922 representation of the krypton atom ($Z = 36$) in terms of classical orbits. Such pictures represent the high-water mark of the use of classical orbit concepts.

intrinsic angular momentum of its own—apart from any orbital angular momentum it may also possess—was unknown; (6) the uncertainty principle had yet to be developed; (7) the Pauli exclusion principle—vital for any concept of atom building—had not yet been put forward. Under these circumstances Bohr's achievement in devising his theory must be viewed with amazement.

In spite of its impressive successes, the scope of Bohr theory was clearly limited. It could not cope quantitatively with the experimental observations on the neutral helium atom, let alone more complicated atoms. Even in the case of one-electron atoms, the theory had no prescription for calculating the observed *intensities* of the spectrum lines and had no way to account for the fact that some lines that were energetically possible did not appear at all. Fundamentally more serious was the fact that neither Bohr's theory nor Planck's theory was a coherent explanation of the physics of microscopic systems. Rather, they resembled a patchwork in which some classical ideas that did not seem to work were simply declared invalid and were replaced in certain special circumstances. What was needed was a reformulation and a generalization of the laws of physics that would give the correct results for *all* systems—microscopic and macroscopic—reducing to the classical laws in the latter domain. Such a reformulation was indeed made, starting with the work of Schrödinger and Heisenberg in the mid-1920s, and the modern theory of quantum mechanics was born.

7-10 QUANTUM MECHANICS—A PREVIEW

Quantum mechanics is rather formal and abstract in comparison with earlier theories in physics. It is conceptually difficult for the mind trained to think at the level of macroscopic experience. Although familiarity with it soon enables one to feel at home with the theory, this is partly so because of the knowledge and experience gained with early quantum theory. The breakdown of classical ideas in the numerous experiments we have cited in the past four chapters, and the emergence of key ideas that explain our observations, not only motivate us to seek and accept a new theory but give us an intuitive feeling and a conceptual basis for the theory that emerged. Indeed, every major idea we have discussed in these chapters remains valid and is incorporated into the new theory. This includes the quantization of energy of bound systems, the wave properties of matter and the particle properties of radiation, the wave–particle duality, the uncertainty principle and the probabilistic interpretation, the nuclear model of the atom, the correspondence principle, the quantization of the angular momentum, and the role of Planck's fundamental constant h. And throughout, experiment is the guide and test for theory.

It is our plan in this section to provide a glimpse into the world of wave mechanics* by giving a brief description of the hydrogen atom from this new point of view. This will serve as a bridge between the early quantum treatment that we have given so far and the more complete treatment that can only be provided by a full course in the subject.

We start by pointing out that an orbit of the kind shown in Fig. 7-12 is not permitted in wave mechanics because, by its very existence, an orbit implies that you know the position and the momentum of the electron at all times. This, as we have seen, violates the uncertainty principle, one of the foundation stones of quantum physics. We describe the electron instead by a *wave function* $\psi(\mathbf{r})$ whose square gives the probability per unit volume that the electron will be at a specified position $\mathbf{r}$. Wave functions are derived by solving a wave equation (called *Schrödinger's equation*) under boundary conditions appropriate to the problem at hand. This equation, which we do not present here because it is not our purpose to examine it in detail or to solve it [see, however, Reference 12], is the basic postulate on which the entire structure of wave mechanics rests.

It turns out that the energies of the allowed states of the hydrogen atom are given in wave mechanics by precisely the same formula (Eq. 7-11) that we derived in Bohr theory. In both theories, then, the energy of the ground state of the hydrogen atom, found by putting $n = 1$ into that equation, is -13.6 eV. The wave function that describes the ground state proves to be

$$\psi(r) = \frac{1}{\sqrt{\pi a_0^3}} e^{-r/a_0} \qquad (n = 1),$$

in which a_0 is the familiar Bohr radius. As we promised in Section 7-6, this quantity turns out to be a useful parameter in wave mechanics. We see that the wave function is spherically symmetric, by which we mean that its value depends only of the magnitude of the position vector $\mathbf{r}$ and not on its direction.

In describing the location of the electron it is useful to take as a volume element dV the volume lying between two concentric spheres, centered on the origin and of radii r and $r + dr$. Thus we define the *radial probability density*, a quantity more directly informative than the wave function itself, from

$$P(r)\, dr = \psi^2(r)\, dV = \psi^2(r)(4\pi r^2)\, dr.$$

For the ground state this becomes

$$P(r) = \left(\frac{4}{a_0^3}\right) r^2 e^{-2r/a_0} \qquad (n = 1). \tag{7-23}$$

Figure 7-13 is a plot of the radial distribution function $P(r)$ for this state. We see that it has a maximum value precisely at $r = a_0$, which is the radius of the ground state orbit in Bohr theory. Figure 7-14 contrasts further the two representations.

It can readily be shown (see Problem 49) that

$$\int_0^\infty P(r)\, dr = 1.$$

---

* Wave mechanics is the name given to a particular mathematical formulation of the theory known more generally as quantum mechanics.

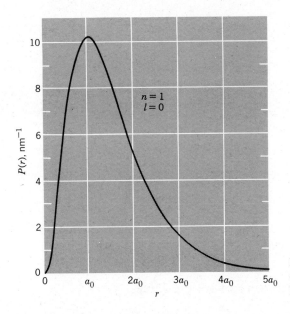

Figure 7-13. The radial probability density describing the electron in the hydrogen atom in its ground state.

A probability of unity corresponds to a certainty. The expression above simply asserts that, even though the position of the electron can be specified only statistically, the electron must certainly be *somewhere* outside the nucleus.

In wave mechanics the magnitude of the angular momentum associated with a stationary state is given by the relation

$$L = \sqrt{l(l + 1)}\hbar \qquad l = 0, 1, 2, \ldots, (n - 1), \qquad (7\text{-}24)$$

in which l, called the *orbital quantum number,* can have only the values shown. For the ground state then, defined by $n = 1$, we can have only $l = 0$, so this state has no angular momentum. Here we have both a similarity to and a departure from Bohr theory. In that theory angular momentum *is* quantized (and is described in Sommerfeld's extension of Bohr theory by a second quantum number), but the angular momentum of the ground state (see Eq. 7-17) is taken as $\hbar$ rather than zero. In this respect Bohr theory is simply incorrect.

We turn now to the second excited state of the hydrogen atom, for which $n = 2$. Equation 7-24 above reveals that for $n = 2$ we may have $l = 0$ or $l = 1$ for the orbital quantum number. Figure 7-15 shows the radial probability functions for these two cases. We note that the most probable location for the state described by $n = 2$, $l = 0$ is $r = 4a_0$. This (see Eq. 7-15) is just the radius of the second Bohr orbit.

The orbital quantum number l specifies, as we said above, only the *magnitude* of the angular momentum vector. According to the predictions of wave mechanics, however, its direction in space is also quantized, the relationship being

$$L_z = m_l\hbar,$$

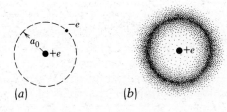

(a)   (b)

Figure 7-14. (*a*) The ground state of the hydrogen atom according to Bohr theory. (*b*) The ground state of the hydrogen atom according to wave mechanics. The dots represent the electron's radial probability density of Fig. 7-13.

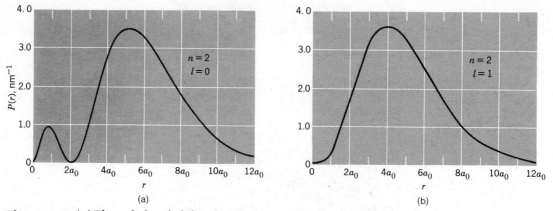

Figure 7-15. (*a*) The radial probability density for a hydrogen atom state with $n = 2$ and $l = 0$. (*b*) The same, for a state with $n = 2$ and $l = 1$.

in which m_l can have the values

$$-l, \ldots, -2, -1, 0, +1, +2, \ldots, +l.$$

Thus, for $l = 1$ we can have the values $m_l = +1$, $m_l = 0$, and $m_l = -1$, corresponding to three allowed orientations of the angular momentum vector. For $l = 0$, of course, the magnitude of the angular momentum is zero, so $m_l = 0$ is the only possibility. The table below summarizes the quantum numbers of the four states identified with $n = 2$.

n	*l*	m_l
2	0	0
2	1	−1
2	1	0
2	1	+1

The energy of a hydrogen atom state depends only on the principal quantum number n, so all four states listed above would be expected to have the same energy, namely, −3.4 eV. This seems reasonable because, even though the states with different values of m_l correspond to different orientations of the angular momentum vector, for an atom in free space one direction is as good as another and there is no physical reason that these states should have different energies. The states are said to be *degenerate*, which is simply a convenient way of saying that, though described by a different set of quantum numbers, they have the same energy.

Matters change, however, if the hydrogen atom is placed in a magnetic field. Here there *is* a unique direction, namely, that of the field. The orbiting electrons (if we permit a relapse into Bohr language) have a magnetic moment associated with them, so we expect that states with different orientations of their angular momentum vector (that is, with different values of m_l) will have different energies. All this is born out by experiment. It is a well-known phenomenon (called the *Zeeman effect*) that when a discharge tube, say, is immersed in a magnetic field, spectrum lines that are single in the absence of a field become split into several components, reflecting the splittings of the atomic levels.

The system of quantum numbers worked out to describe the states of the hydrogen atom can also be used to describe individual electrons in multielectron atoms. The energies of these states, however, are no longer given by Eq. 7-11 and depend not only on the principal quantum number n but also on the orbital quantum number l. We state without proof that, guided by wave mechanics and by certain reasonable principles, it is possible to assign chemical properties to the elements and thus to reconstruct the entire periodic table, a task far beyond the scope of Bohr theory.

Wave mechanics can account for the stability of stationary states—which was simply postulated in Bohr theory. It does so by demonstrating that the elements of radiation, emitted on the classical picture by the individual moving elements of the electron cloud, annul each other by interference. Wave mechanics can also be used to calculate various properties of atoms, such as their electric dipole moments. It can further be used to calculate the intensities of spectrum lines. This is of special interest when these intensities turn out to be zero; not every pair of levels has a transition between them, and before the advent of wave mechanics, such "missing" spectrum lines could simply be classified as "forbidden" by certain empirical "selection rules." Wave mechanics provides a theoretical basis for these rules. Finally, wave mechanics forms the basis for understanding the mechanisms by which atoms bind together to form molecules or solids. Thus it underlies, in principle at least, most of chemistry and the vast domain called solid-state physics.

questions

1. What assumptions did J. J. Thomson make in setting up his model of the atom? What were the successes of his model? The failures?

2. Explain why Rutherford could safely neglect the effect of the extranuclear electrons in computing the distance of closest approach of an alpha particle to a nucleus.

3. How does the force acting on an electron vary with distance from the atomic center for an electron in a Thomson "plum pudding" atom? In a Rutherford nuclear atom?

4. (a) Explain why scattering of alphas due to the atomic electrons can be ignored for scattering angles greater than a few degrees. (b) The scattering of alpha particles for very small angles disagrees with the predictions of the Rutherford formula. Can you explain why?

5. The Rutherford scattering formula (Eq. 7-1) yields an infinite value for $P(\phi)$ for a scattering angle of zero degrees. What is the significance of this? (*Hint:* How far from a target nucleus would the initial track of an incident alpha particle have to be if its deflection were to be truly zero?)

6. Why does a head-on collision ($\phi = 180°$) give the distance of closest approach in alpha particle scattering?

7. How does the alpha particle scattering probability vary with foil thickness on the assumption of single scattering? Of multiple scattering? Why do we specify that the foil be "thin" in experiments intended to check the Rutherford scattering formula?

8. In verifying the variation of the alpha particle scattering probability with angle (Eq. 7-1), Geiger and Marsden used a source that emitted alpha particles with several different energies. Would this affect their results? Explain.

9. We have neglected the recoil of the target nucleus in calculating the distance of closest approach in alpha particle scattering. In principle, how could this be taken into account? In practice, does it make much difference for heavy target nuclei?

10. The Rutherford scattering formula (Eq. 7-1) is said to explain the scattering results equally well no matter whether the nucleus carries a positive or a negative charge. How does this assertion manifest itself in the structure of Eq. 7-1? If the nuclear charge were negative, how would Fig. I-1 in Supplementary Topic I have to be changed? (Assume that the charge on the alpha particle is positive and that its trajectory remains as drawn in that figure.)

11. Why was the Balmer series, rather than the Lyman or the Paschen series (see Fig. 7-8), the first to be detected and analyzed in the hydrogen spectrum?

12. Upon emitting a photon, the hydrogen atom recoils to conserve momentum. Explain the fact that the energy of the emitted photon is less than the energy difference between the levels involved in the emission process.

13. If only lines in the absorption spectrum of hydrogen need to be calculated, how would you modify Eq. 7-4 to obtain them?

14. (*a*) Can a hydrogen atom absorb a photon whose energy exceeds its binding energy (=13.6 eV)? (*b*) What minimum energy must a photon have to initiate the photoelectric effect in hydrogen gas? (Careful!)

15. Would you expect to observe all the lines of atomic hydrogen if such a gas were excited by 13.6-eV electrons?

16. How would you estimate the temperature of hydrogen gas at which atomic collisions cause significant ionization of the atoms?

17. Only a relatively small number of Balmer lines can be observed from laboratory discharge tubes, whereas a large number are observed in stellar spectra. Explain this in terms of the small density, high temperature, and large volume of gases in stellar atmospheres.

18. In what two ways does the Balmer formula for He$^+$ differ from that for neutral hydrogen? What effect does each difference have on the He$^+$ spectrum compared to the spectrum of hydrogen?

19. Is the ionization energy of deuterium different from that of hydrogen? Explain.

20. What were the assumptions made by Bohr in setting up his model of the atom? What were the successes of his model? The failures?

21. Discuss the analogy between the Kepler-Newton relationship in the development of Newton's law of gravitation and the Balmer-Bohr relationship in developing the Bohr theory of atomic structure.

22. For the Bohr hydrogen atom planetary orbits, the potential energy is negative and greater in magnitude than the kinetic energy. What does this imply?

23. The nuclear atom as introduced by Rutherford, in contrast to the Thomson "plum pudding" atom, no longer contained a characteristic length that could be identified even approximately with the dimensions of an atom. What, in fact, prevents the electrons from simply falling directly into the nucleus, thus forming a nuclear-sized atom? How did Bohr introduce a characteristic length into Rutherford's model? What is this characteristic length?

24. Why couldn't Bohr allow the quantum number n to take the value $n = 0$, as it does in Planck's quantization condition?

25. Keeping in mind the correspondence principle, what physical significance do you ascribe to Bohr orbits having very large values of n, say, $n > 1000$?

26. Bohr's postulate of stationary states asserts that a hydrogen atom can exist (without radiating) in a number of discrete stationary states. Why must these states form a "discrete" set? What would be the nature of the emission spectrum if these states were distributed continuously in energy?

27. Does the Bohr radiation postulate (Eq. 7-5) apply to multielectron atoms? Molecules?

28. According to classical mechanics, an electron moving in an orbit should be able to do so with any angular momentum whatsoever. According to Bohr's theory of the hydrogen atom, however, the angular momentum is quantized at values $L = n\hbar$. Can you use the correspondence principle to reconcile these two statements?

29. To bring out the notion that an atom is mostly empty space, it has been said that the nucleus takes up about as much space in the atom as does a fly in a cathedral. Guess at the dimensions of a fly and a cathedral and check out this analogy. An atom is about 100 pm in linear dimension and a nucleus about 10 fm, 10,000 times smaller.

30. Exactly how is the notion of a Bohr orbit inconsistent with the uncertainty principle?

31. Even though an atom is mostly empty space, it is still pretty incompressible. You do not fall through the floor when you step on it. Can you make up an argument based on the uncertainty principle to account for the fact that an atom cannot be readily squeezed down to a geometrical point?

32. It has been said that the modern wave mechanical view of the atom simply inverts the original view of J. J. Thomson in that he envisaged tiny negative charges (the electrons) embedded in a cloud of positive charge, whereas the modern view envisages a tiny positive charge (the nucleus) embedded in a cloud of negative charge. Is this view valid? What other differences between the two models occur to you?

33. What is the relationship between the wave function $\psi(r)$ and the probability density function $P(r)$ as applied to the hydrogen atom? In what units are each of these quantities expressed? What value does each of these quantities have at the center of the atom in its ground state?

34. "The energy of the ground state of an atomic system can be precisely known, but the energies of its excited states are always subject to some uncertainty." Can you explain this statement on the basis of the uncertainty principle?

problems

$$e^2/4\pi\varepsilon_0 = 2.31 \times 10^{-28} \text{ J} \cdot \text{m} = 1.44 \text{ eV} \cdot \text{nm}$$
$$= 1.44 \text{ MeV} \cdot \text{fm}$$
$$h = 6.63 \times 10^{-34} \text{ J} \cdot \text{s} = 4.14 \times 10^{-15} \text{ eV} \cdot \text{s}$$
$$N_A = 6.02 \times 10^{23} \text{ mol}^{-1} \qquad a_0 = 0.0529 \text{ nm}$$
$$R_\infty = R_H = 1.097 \times 10^7 \text{ m}^{-1} \qquad c = 3.00 \times 10^8 \text{ m/s}$$
$$1 \text{ MeV} = 10^6 \text{ eV} = 1.60 \times 10^{-13} \text{ J}$$
$$1 \text{ m} = 10^6 \ \mu\text{m} = 10^9 \text{ nm} = 10^{10} \text{ Å} = 10^{12} \text{ pm} = 10^{15} \text{ fm}$$
$$\text{Electron mass} = 9.11 \times 10^{-31} \text{ kg} = 0.511 \text{ MeV}/c^2$$
$$\text{Hydrogen atom mass} = 1.67 \times 10^{-27} \text{ kg} = 939 \text{ MeV}/c^2$$

1. How big is an atom? The density and atomic weight of solid copper are 8.94 g/cm³ and 63.6 g/mol respectively. (a) What volume may be assigned to a copper atom in solid copper? (b) To what effective atomic radius does this volume (assumed to be spherical) correspond?

2. An electron in the Thomson atom. (a) Show, for a Thomson atom, that an electron moving in a stable circular orbit does so with the same angular frequency with which it would oscillate in moving through the center along a diameter. (b) Does this frequency depend on the radius of the circular orbit?

3. A Thomson atom radiates. What radius must the Thomson model of the one-electron atom have if it is to radiate a spectral line of wavelength 600 nm? Comment on the deficiency of the model to account for such monochromatic radiation.

4. A stable two-electron Thomson atom. Figure 7-16 shows two electrons at rest inside a Thomson atom. (a) Show that the configuration will be in static equilibrium if the electrons are separated by a distance equal to the radius R of the sphere of positive charge. (b) Show that the equilibrium is stable by imagining the separation of the electrons to be slightly increased or slightly decreased and verifying that there is a restoring force.

5. A two-electron nuclear atom. Figure 7-17 shows a nuclear atom version of the simple Thomson atom of Fig. 7-16. (a) Show that, if the electrons are at rest, the arrangement cannot be stable. (b) If the electrons are separated by

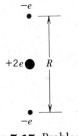

Figure 7-17. Problem 5.

a distance R and are rotating, show that their rotation frequency must be given by

$$\nu = \frac{1}{2\pi} \sqrt{\frac{7e^2}{2\pi\varepsilon_0 mR^3}}$$

if the arrangement is to be dynamically stable. (Assume that the rotating system does not radiate.)

★6. The classical electron radius. It is possible to calculate a radius r_e for the electron if it is assumed that the rest energy of the electron is equal to the electrostatic energy stored outside the electron; see *Physics*, Part II, Sec. 30-3. (a) Show that this assumption leads to the prediction that $r_e = (\tfrac{1}{8}\pi\varepsilon_0)(e^2/m_0 c^2)$. (b) Evaluate this quantity. (As a physical concept the classical electron radius is now of

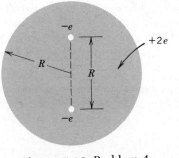

Figure 7-16. Problem 4.

historical interest only. The present view is that the electron is essentially a point particle, with no measurable radius.)

★7. An alpha particle hits an electron. An alpha particle of mass M and initial speed V collides with a free electron of mass m at rest. Show that the maximum deflection of the alpha particle is about 10^{-4} radians. (*Hint*: Apply the conservation equations; solve for the deflection of the alpha particle; maximize this quantity; make approximations justified by the fact that $m \ll M$; treat nonrelativistically throughout.)

★8. An alpha particle hits a Thomson atom. (*a*) An alpha particle of mass M and kinetic energy K collides with a Thomson atom of radius R. Assume the atom to be gold ($Z = 79$) and take $R = 1.0$ Å; assume $K = 5.0$ MeV. Show that the maximum deflection of the alpha particle caused by its interaction with the positive charge of the gold atom is about 10^{-4} radians. (*Hint*: Estimate the maximum force acting on the alpha particle and also the time of interaction. Multiply to find the maximum momentum transfer; see *Physics*, Part I, Sec. 10-2.) (*b*) In view of this result and of the result of Problem 7 argue that the maximum deflection of an alpha particle in an encounter with a single Thomson atom is about 10^{-4} radians.

9. The Thomson and the Rutherford atoms compared. In each atom model the positive charge of the atom is assumed to be distributed uniformly throughout a sphere of radius R. For the Thomson model $R = 10^{-10}$ m and for the Rutherford model $R = 10^{-14}$ m. For each model calculate and plot the electric field, due to the positive charge alone, from $r = 0$ to $r = 2R$. In comparing the results can you see why the Rutherford model accounts for the scattering of alpha particles through large angles and the Thomson model does not?

10. Boring in towards a copper nucleus. What is the distance of closest approach for a head-on collision between a 5.30-MeV alpha particle and the nucleus of a copper atom?

11. Penetrating the nucleus. Assume that the gold nucleus has a radius of 6.4 fm and the alpha particle has a radius of 1.8 fm. What minimum energy must an incident alpha particle have to experience non-Coulombic nuclear forces, that is, to penetrate the nucleus?

12. The nucleus recoils (I). When an alpha particle collides elastically with a nucleus the nucleus recoils. A 5.00-MeV alpha has a head-on elastic collision with a gold nucleus, initially at rest. What is the kinetic energy (*a*) of the recoiling nucleus? (*b*) of the rebounding alpha particle?

13. The nucleus recoils (II). When an alpha particle collides elastically with a nucleus the nucleus recoils. A 5.00-MeV alpha particle has a head-on elastic collision with a gold nucleus, initially at rest. (*a*) What is the distance of closest approach of the alpha particle to the recoiling nucleus? (*Hint*: At closest approach the alpha particle and the recoiling nucleus are moving with the same velocity in the laboratory reference frame.) (*b*) How does this compare with the result one gets if, as in Example 2, we assume the target nucleus to remain at rest?

14. A single Rutherford scattering event. If an alpha particle approaches a heavy target nucleus the *impact parameter* b describing the encounter is the distance by which the alpha particle would have missed its target had it not been deflected; see Fig. I-1 in Supplementary Topic I. The impact parameter can be shown to be related to the scattering angle and the distance of closest approach r_0 by (see Supplementary Topic I)

$$\tan (\phi/2) = \frac{r_0}{2b} = \left(\frac{e^2}{4\pi\varepsilon_0}\right)\frac{Z}{bK}$$

in which K is the kinetic energy of the incident alpha particle and Z is the atomic number of the target nucleus. (*a*) A 5.30-MeV alpha particle is scattered through an angle of 60° in an encounter with a gold nucleus ($Z = 79$). What is the impact parameter? (*b*) If an incident alpha particle has an impact parameter equal to r_0, its distance of closest approach, through what angle will it be scattered?

15. Scattering from gold and from copper. A 5.30-MeV alpha particle is scattered through 60° in passing through a thin foil. Calculate the distance of closest approach (for an assumed head-on collision) and the impact parameter (for the actual collision) if the foil is (*a*) gold or (*b*) copper. The atomic numbers of gold and copper are 79 and 29. See Problem 14. Neglect the recoil of the nucleus.

16. A Rutherford scattering experiment (I). A beam of alpha particles, of kinetic energy 5.30 MeV and intensity 1.0×10^4 particles/s, falls normally on a lead foil of atomic number 82, density 11.4 g/cm³, atomic weight 207 g/mol, and thickness 1.0 μm. A detector of area 1.0 cm² is placed at a distance of 10 cm from the foil. What is the counting rate for a scattering angle (*a*) of 10°? (*b*) of 45°?

17. A Rutherford scattering experiment (II). In Problem 16 a copper foil 5.0 μm thick is substituted for the lead foil. At a scattering angle of 10° the counting rate is 688 min⁻¹. What atomic number for copper may be deduced from these data? The atomic weight and the density of copper are 63.6 g/mol and 8.94 g/cm³.

★18. Rutherford scattering—another view. Show that the probability $P_\phi(\phi)$ that an alpha particle will be scattered through an angle greater than ϕ is given by

$$P_\phi(\phi) = \left(\frac{e^4\rho t Z^2}{16\pi\varepsilon_0^2 K^2}\right)\cot^2(\phi/2).$$

(*Hint*: In Eq. 7-1 consider a detector whose sensitive area $\mathscr{A}$ is a circular strip of width $R\,d\phi$ and circumference $2\pi R \sin \phi$ on the surface of a sphere of radius R. Integrate from $\phi = \phi$ to $\phi = 180°$.)

19. Calculating the scattering. A 5.30-MeV alpha particle is incident on a gold foil $(Z = 79)$ of thickness 1.00 μm. What is the probability that it will be scattered through an angle greater than (a) 30° (b) 90° (c) 120°? Use the formula derived in Problem 18. The density of gold is 19.3 g/cm^3.

★20. The trajectory in Rutherford scattering. The equation for the trajectory of an alpha particle, incident with kinetic energy K and impact parameter b on a target nucleus of atomic number Z, is shown in Supplementary Topic I to be

$$r(\theta) = \frac{a}{d \cos \theta - 1},$$

in which

$$a = 2b^2/r_0 \quad \text{and} \quad d = \sqrt{1 + (2b/r_0)^2}.$$

The trajectory (see Fig. I-1 in Supplementary Topic I) is given in polar coordinates, the origin of the coordinate system being the target nucleus. Assume a lead target $(Z = 82)$, a 5.30-MeV alpha particle and an impact parameter of 40.0 fm. (a) What is the scattering angle ϕ? (b) What is r_0, the distance of closest approach for an assumed head-on collision? (c) What is the actual distance of closest approach, corresponding to $\theta = 0$ in Fig. I-1? (d) What is the kinetic energy of the alpha particle at its actual distance of closest approach? (*Hint*: Consider energy conservation.)

21. Balmer's formula. Using Balmer's formula (Eq. 7-2) (a) calculate the wavelengths of the three longest lines of the Balmer series. (b) Calculate the series limit.

22. Three series compared. (a) What are the wavelength intervals over which the Lyman, the Balmer and the Paschen series extend? (The intervals extend from the longest wavelength to the series limit.) (b) What are the corresponding frequency intervals?

23. Name something after yourself. Pick a series of lines from the hydrogen spectrum and name it after yourself. (Why should Lyman, Balmer, Paschen, etc. have all the glory?) Try $m = 158,257$ for example. (a) Calculate the range of wavelengths (and also of frequencies) within which "your" series lies and identify the region of the electromagnetic spectrum in which these lines occur. (b) What is the radius of a hydrogen atom with "your" value of m? (c) What are the possibilities that "your" series will actually be observed?

24. An emitted photon. What are the energy, momentum, wavelength and frequency of a photon that is emitted by a hydrogen atom making a direct transition from an excited state with $n = 10$ to the ground state?

25. Exciting a hydrogen atom. A hydrogen atom is excited from its ground state $(n = 1)$ to a state with $n = 4$. (a) Calculate the energy that must be absorbed by the atom. (b) Calculate and display on an energy-level diagram the different photon energies that may be emitted if the atom returns spontaneously to its ground state. (c) Calculate the recoil speed and kinetic energy of the hydrogen atom, assumed initially at rest, if it makes the transition from $n = 4$ to its ground state in a single quantum jump.

26. An energy-level puzzle. A hydrogen atom emits a photon of wavelength 486.3 nm. (a) What are the quantum numbers of the atom both before and after it emitted this photon? (b) To what series does the transition belong? (*Hint*: Calculate the photon energy and see if you can identify the transition on Fig. 7-8.)

27. Another energy-level puzzle. A hydrogen atom in a state having a binding energy (this is the energy required to remove an electron) of 0.85 eV makes a transition to a state with an excitation energy (this is the difference in energy between the state and the ground state) of 10.2 eV. (a) What is the energy of the emitted photon? (b) Show the transition on an energy-level diagram for hydrogen, labeling the appropriate quantum numbers.

28. The Ritz combination principle. From the energy-level diagram for hydrogen explain the observation that the frequency of the second Lyman series line is the sum of the frequencies of the first Lyman series line and the first Balmer series line. This is an example of the empirically discovered *Ritz combination principle*. Use the diagram to find some other valid combinations. Verify the combinations using the generalized Balmer formula, Eq. 7-3.

29. Interpreting the Franck-Hertz experiment. In a Franck-Hertz experiment atomic hydrogen is bombarded with electrons and excitation potentials are found at 10.21 V and 12.10 V. (a) Explain the observation that three different spectrum lines are observed to accompany these excitations. (*Hint*: Draw an energy level diagram for hydrogen.) (b) Calculate the wavelengths of these three lines and identify the series to which each belongs.

30. Removing the electron. (a) How much energy is required to remove the electron from a hydrogen atom in a state with $n = 8$? With $n = 25$? (b) From a He$^+$ ion in its ground state? (c) From a Li^{++} ion in a state with $n = 3$?

★31. Two Bohr atoms compared. Radiation from a helium ion He$^+$ is nearly equal in wavelength to the H$_\alpha$ line from hydrogen (the first line of the Balmer series). (a) Between what states (values of n) does the helium transition occur? (b) Is the wavelength greater or smaller than that of the H$_\alpha$ line? (c) Compute the wavelength difference.

32. Bohr atoms in the stars. In stars the Pickering series is found in the He$^+$ spectrum. It is emitted when the electron in He$^+$ jumps from higher levels to the level with $n = 4$. (a) Show that the wavelengths in this series are given by

$$\lambda = \frac{1}{R_\infty}\left(\frac{2n}{n-2}\right)^2\left(1 + \frac{me}{M_{\mathrm{He}}}\right)$$

in which $n = 5, 6, 7, \ldots$ and M_{He} is the mass of the helium nucleus. (b) Calculate the wavelength of the first line of this series and of the series limit. (c) In what region of the spectrum does this series occur?

33. A busy gas discharge tube. A gas discharge tube contains H^1, H^2, He^3, He^4, Li^6 and Li^7 single-electron ions and atoms (the superscript is the atomic mass number). (a) As the potential across the tube is raised from zero, which spectral line should appear first? (b) Give, in order of increasing frequency, the source of the lines corresponding to the first line of the Lyman series for H^1.

34. Checking the correspondence principle. According to the correspondence principle, as $n \to \infty$ we expect classical results in the Bohr atom. Hence, the de Broglie wavelength associated with the electron (a quantum result) should get smaller compared to the radius of the orbit as n increases. Indeed, we expect that $\lambda/r_n \to 0$ as $n \to \infty$. Show that this is the case.

35. Radio waves from a hydrogen atom. (a) Show that the smallest quantum number of the levels in hydrogen between which transitions giving rise to radio waves are possible is given by $n = (2R_H\lambda)^{1/3}$ where λ is the wavelength of the radio wave. (b) If the 21-cm radio emission from interstellar hydrogen were due to such a transition (it isn't) what would be the value of n?

36. An interesting inequality. Prove that the frequency of radiation emitted in a transition between levels in a Bohr atom is intermediate between the frequencies of revolution of the electron in the initial and the final levels; see Example 6. (*Hint*: Note that n exceeds m by at least unity.)

37. The Bohr atom based on angular momentum quantization. Starting (in Bohr's model) from the classical mechanics of an electron moving about the nucleus in a circular orbit, show that, taking the quantization of angular momentum (Eq. 7-17) as a postulate, it is possible to derive Eq. 7-14, the expression for the energy of the stationary states. This relation was derived earlier on the basis of the correspondence principle.

38. The hydrogen atom in its ground state. In the ground state of the hydrogen atom, according to Bohr's model, what are (a) the quantum number, (b) the orbit radius, (c) the angular momentum, (d) the linear momentum, (e) the angular velocity, (f) the linear speed, (g) the force on the electron, (h) the acceleration of the electron, (i) the kinetic energy, (j) the potential energy and (k) the total energy?

39. An exercise in dimensions. (a) Prove that the dimensions of the Planck constant h are those of angular momentum. (b) Prove that the dimension of the Bohr radius, $a_0 = \varepsilon_0 h^2/\pi m e^2$, is that of length. (c) Prove that the so-called *fine structure constant*, $\alpha = 2\varepsilon_0 hc/e^2$, is dimensionless. (Note: The speed parameter β_n for electrons in Bohr orbits is given by $Z\alpha/n$; see Eq. 7-16.) (d) Prove that the dimension of the Rydberg constant, $R_\infty = me^4/8\varepsilon_0^2 h^3 c$, is that of inverse length.

40. An orbiting electron. An electron moves in a classical circular orbit of radius r about a proton. Show from classical analysis that the frequency of revolution is given by $\nu_{\mathrm{orb}}^2 = e^2/(16\pi^3\varepsilon_0)mr^3$. (b) Show that the total energy of the system, kinetic plus potential, is given by $E = -(e^2/8\pi\varepsilon_0 r)$. (c) Combine these two results to obtain Eq. 7-7, a purely classical result

$$\nu_{\mathrm{orb}} = \frac{4\varepsilon_0}{e^2}\sqrt{\frac{2}{m}}\,|E|^{3/2}.$$

41. Taking the recoil into account. (a) Show that when the recoil kinetic energy of the atom, $p^2/2M$, is taken into account the frequency of the photon emitted in a transition between two atomic levels of energy difference ΔE is reduced by a factor of, approximately, $(1 - \Delta E/2Mc^2)$. (Hint: The recoil momentum of the atom is $h\nu/c$.) (b) Compare the wavelength of the light emitted by a hydrogen atom in the $3 \to 1$ transition when the recoil is taken into account to the wavelength without accounting for recoil; express as $\Delta\lambda/\lambda$.

42. Two atoms collide. An atom, at rest and in its ground state, is struck by another atom of the same kind which is also in its ground state but has a kinetic energy K. Show, from conservation principles, that the collision must be elastic if $K < 2E$, where E is the first excitation energy of the atom (that is, the energy difference between the ground state of the atom and its first excited state).

43. Excitation by electron collision. (a) An atom can be raised to an excited state by means of a collision with an energetic electron. Let E be the difference between the ground state and the first excited state of the atom (mass M) and let K be the energy of the incident electron (mass m). Show that K must be at least

$$K = \left(1 + \frac{m}{M}\right)E$$

if the atom is to be excited. (*b*) Show that this result reduces to the result of Problem 42 under the appropriate conditions.

44. Quantized energy and reduced mass. In a one-electron Bohr atom the electron and the nucleus revolve about their common center of mass with an angular velocity ω (see Fig. 7-18). (*a*) Show, from the definition of the center of mass, that $R = (m/M)r$, where m, M, r and R are defined in the figure. (*b*) Show that the quantization of total angular momentum gives the relation $n(h/2\pi) = m\omega r^2(1 + m/M)$. (*c*) Show, if the nuclear charge is Ze, that the equation of motion can be written as

$$m\omega^2 r = \frac{1}{4\pi\varepsilon_0} \frac{Ze^2}{(R + r)^2} = \frac{1}{4\pi\varepsilon_0} \frac{Ze^2}{r^2} \left(\frac{M}{M + m}\right)^2.$$

(*d*) Then show, finally, that the quantized energies are given by

$$E_n = -\frac{\mu Z^2 e^4}{8\varepsilon_0^2 h^2} \frac{1}{n^2}.$$

Compare Eq. 7-14.

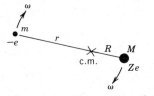

Figure 7-18. Problems 44 and 45.

★45. Reduced mass and the orbit radius. Assume that an electron of mass m and charge $-e$ and a heavy nucleus of mass M and charge Ze revolve as a Bohr atom around their common center of mass as shown in Fig. 7-18. (*a*) Show that the orbit radius (r in Fig. 7-18) of the electron is given exactly by Eq. 7-15. (*b*) Apply this conclusion to solve Problem 38*b*. (*c*) Show that if m in Eq. 7-15 is replaced by the reduced mass μ then the quantity r_n calculated from that equation is not the orbit radius r but the distance $(=r + R)$ in Fig. 7-18) between the electron and the central nucleus. Alternatively, the quantity calculated is the orbit radius of the electron from the point of view of an observer stationed on the nucleus, which is itself orbiting in a circle of radius R. (*Hint*: See Problem 44.)

46. Tritium. Hydrogen of mass number three (tritium) is added to a gas discharge tube containing ordinary atomic hydrogen. What wavelength separation would be observed for the first line of the Balmer series for each gas? (*Hint*: See Example 8, noting that the *second* line of the Balmer series is used in that case.)

47. Positronium. In Chapter 5 we spoke of the positronium atom, consisting of a positron and an electron revolving around their common center of mass, which lies halfway between them. (*a*) If such a system were a normal atom, show that the wavelengths of the emitted lines would be double those emitted by a hydrogen atom (assuming an infinitely heavy nucleus). (*b*) What would be the radius of the ground state orbit of positronium (that is, the distance between either particle and the common center of mass)? (*c*) Assume that electron-positron annihilation occurs from the ground state of positronium. How, if at all, does this alter the annihilation photon energies of the two-photon decay discussed in Chapter 5, where we ignored the bound system.

48. A Muonic atom emits a photon. What is the wavelength of the most energetic photon that can be emitted from a muonic atom with $Z = 1$? (*Hint*: See Example 9.)

49. The radial distribution function. The radial distribution function for the ground state of the hydrogen atom is given by Eq. 7-23, or

$$P(r) = (4/a_0^3)\, r^2 e^{-2r/a_0},$$

in which a_0 is the Bohr radius ($=0.0529$ nm). (*a*) Show that $P(r)$ has its maximum value for $r = a_0$. (*b*) Prove that

$$\int_0^\infty P(r)\, dr = 1.$$

★50. Where is the electron? According to wave mechanics, what is the probability that the electron in the ground state of the hydrogen atom will be found outside a sphere whose radius is the Bohr radius a_0?

51. The wave function and the radial probability density. In the ground state of the hydrogen atom evaluate the wave function $\psi(r)$, the square of this quantity, $\psi^2(r)$, and the radial probability density $P(r)$ for the positions (*a*) $r = 0$ and (*b*) $r = a_0$. Explain what these quantities mean.

52C. Rutherford scattering. Write a program for your handheld programmable calculator that will accept as inputs the atomic number, the atomic weight, the density and the thickness of the target; the kinetic energy of the incident alpha particle; and the effective area $\mathcal{A}$, the distance R and the angular position ϕ (the scattering angle) of the detector. Let the outputs be $P(\phi)$, the probability that the incident alpha particle will enter the detector (see Eq. 7-1), and $P_\phi(\phi)$, the probability that the incident alpha particle will be scattered through an angle greater than ϕ (see Problem 18). (*a*) Use this program to verify the calculation of Example 3. (*b*) Solve that Example if an aluminum foil whose thickness is 5.0 μm is substituted for the gold foil. (c) Plot the function $P_\phi(\phi)$ over the range $\phi = 0$ to $\phi = 180°$ for the conditions of Example 3.

53C. The hydrogen spectrum.   Write a program for your handheld programmable calculator that will accept as inputs the quantum numbers m and n in Eq. 7-3 and will deliver as successive outputs the wavelength associated with the transition between these levels, the series limit corresponding to the value of m used, and the energy difference between the levels. (a) Evaluate these quantities for $m = 1$ and $n = 2$. (b) Evaluate these quantities in the classical region, for $m = 200$ and $n = 201$. (c) Use your

program to write down the successive wavelengths of the Paschen series, for which $m = 3$.

54C. The Bohr one-electron atom.   Write a program for your handheld programmable calculator that will accept as inputs the nuclear atomic number Z and the quantum number n and will deliver as successive outputs the energy of the state, the orbit radius, the orbital frequency, and the frequency of a transition to the next lowest energy state. (a) Evaluate these quantities for hydrogen ($Z = 1$) for $n = 2$ and for $n = 200$. (b) Evaluate these quantities for a single electron orbiting a uranium nucleus ($Z = 92$) for $n = 2$.

references

1. John L. Heilbron, "J. J. Thomson and the Bohr Atom," *Phys. Today* (April 1977).

2. J. J. Thomson, *The Corpuscular Theory of Matter* (Scribner's, New York, 1907).

3. Frank Greenaway, *John Dalton and the Atom* (Cornell University Press, Ithaca, N.Y., 1966).

4. Henry A. Boorse and Lloyd Motz, *The World of the Atom* (Basic Books, New York, 1966), chap. 10.

5. Ernest Rutherford, "The Scattering of α and β Particles by Matter and the Structure of the Atom," *Phil. Mag.* **21,** 669 (1911). Reprinted in Henry A. Boorse and Lloyd Motz, Eds., *The World of the Atom* (Basic Books, New York, 1966), chap. 44.

6. E. N. de C. Andrade, "The Birth of the Nuclear Atom," *Sci. Am.* (November 1956).

7. Hans Geiger and Ernest Marsden, "The Laws of Deflexion of α Particles Through Large Angles," *Phil. Mag.* **25,** 604 (1913). Reprinted in Henry A. Boorse and Lloyd Motz, Eds., *The World of the Atom* (Basic Books, New York, 1966), chap. 44.

8. Leo Banet, "Balmer's Manuscripts and the Construction of his Series," *Am. J. Phys.* (July 1970).

9. Niels Bohr, "On the Constitution of Atoms and Molecules," *Phil. Mag.* **26,** 1 (1913). Reprinted in Henry A. Boorse and Lloyd Motz, Eds., *The World of the Atom* (Basic Books, New York, 1966), chap. 45.

10. J. L. Heilbron, "Rutherford-Bohr Atom," *Am. J. Phys.* **49,** 223 (1981).

11. E. H. S. Burhop, "Exotic Atoms," *Contemp. Phys.* **11,** No. 4 (1970).

12. Robert Eisberg and Robert Resnick, *Quantum Physics* (Wiley, New York, Second Edition, 1985), chap. 5.

SUPPLEMENTARY TOPIC A

the geometric representation of spacetime

Oh, that Einstein, always cutting lectures—I really would not have believed him capable of it.

Hermann Minkowski (ca. 1908)

A-1 SPACETIME DIAGRAMS

We have seen that in classical physics it is proper to treat the space and time coordinates separately. In relativity, however, it is natural to treat them together, their intimate interconnection being clearly displayed in the Lorentz transformation equations; see Tables 2-2 and 2-3. The common use of the single word "spacetime" (without a hyphen) to represent the coordinate description of events is symbolic of the general acceptance of this view.

As we have learned, it was Einstein [1] who first set forth, in his special theory of relativity, the physical basis for the proper description of events in space and time. Shortly afterwards the mathematician Hermann Minkowski (who, incidentally, had formerly been Einstein's mathematics professor in Zurich) [2] presented a simple and symmetrical geometric representation of these ideas, a representation that permits a ready understanding in geometric terms of such matters as the relativity of simultaneity, the length contraction, and the time dilation, including their reciprocal nature.

In what follows, we shall consider only one space axis, the x axis, and shall ignore the y and z axes. We lose no generality by this algebraic simplification, and this procedure will enable us to focus more clearly on the interdependence of space and time and its geometric representation. The coordinates of an event are given, then, by x and t. All possible spacetime coordinates can be represented on a spacetime diagram in which the space axis is horizontal and the time axis is vertical. It is convenient to keep the dimensions of the coordinates the same; this is easily done by multiplying the time t by the universal constant c, the velocity of light. Let ct be represented by the symbol w. Then, the Lorentz transformation equations (see Table 2-2 and Problem 11 of Chapter 2) can be written as follows:

$$
\begin{aligned}
&(a)\ \ x' = \gamma(x - \beta w) \qquad &(a')\ \ x = \gamma(x' + \beta w') \\
&(b)\ \ w' = \gamma(w - \beta x) \qquad &(b')\ \ w = \gamma(w' + \beta x')
\end{aligned}
\tag{A-1}
$$

Notice the symmetry of this form of the equations.

To represent the situation geometrically, we begin by drawing the x and w axes of frame S at right angles to one another, as in Fig. A-1. If we want to represent a moving particle in this frame, we draw a curve, called the *world line* of the particle, which gives the loci of spacetime points corresponding to the motion.

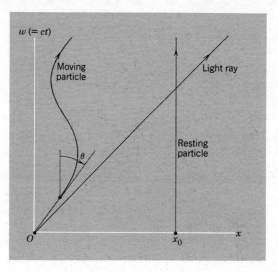

Figure A-1. The *world lines* of light and some particles.

The tangent to the world line at any point makes an angle θ with the direction of the time axis that is given by $\tan \theta = dx/dw = (dx/dt)(1/c) = u/c$. Because we must have $u < c$ for a material particle, the angle θ at any point on its world line must always be less than 45°. If the particle is at rest, say, at position x_0 on the x axis of Fig. A-1, its world line is parallel to the w axis, with θ ($= \tan^{-1} u/c$) = 0 at all points. For a light ray traveling along the x axis we have $u = c$, so its world line is a straight line making an angle of 45° with the axes.

Consider now the primed frame (S'), which moves relative to S with a velocity **v** along the common x-x' axis. The equation of motion of the origin of S' relative to S can be obtained by setting $x' = 0$; from Eq. A-1a, we see that this corresponds to $x = \beta w$. We draw the line $x' = 0$ (that is, $x = \beta w$) on our diagram (Fig. A-2) and note that since $v < c$ and $\beta < 1$, the angle this line makes with the w axis, ϕ($= \tan^{-1} \beta$), is less than 45°. Just as the w axis corresponds to $x = 0$ and is the time axis in frame S, so the line $x' = 0$ gives the time axis w' in S'. Now, if we draw the line $w' = 0$ (giving the location of clocks that read $t' = 0$ in S'), we shall have the space axis x'. That is, just as the x axis corresponds to $w = 0$, so the x' axis corresponds to $w' = 0$. But, from Eq. A-1b, $w' = 0$ gives us $w = \beta x$ as the equation of this axis on our w-x diagram (Fig. A-2). The angle between the space axes is the same as that between the time axes. Note that, for simplicity, we have shown in Fig. A-2 only the quadrant in which both x and w are positive.

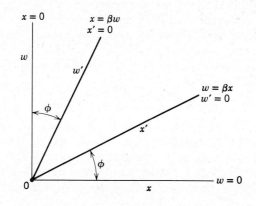

Figure A-2. The *Minkowski diagram* for frames S and S'.

You should compare Fig. A-2 carefully with the standard representation of Fig. 1-1, which we have used exclusively in the main body of the text. A point in the coordinate reference frames of Fig. 1-1 shows only the space coordinates of the event to which it corresponds; the time of occurrence of the event must be given separately. A point on the Minkowski diagram of Fig. A-2, however, shows both the space and the time coordinates of the event in a single geometric representation.

A-2 CALIBRATING THE SPACETIME AXES

Before we can make practical use of the spacetime diagram we must establish scales on its x, w and its x', w' axes. We can use the Lorentz transformation equations of Eq. A-1 for this purpose. Consider first point O, located at the common origin of the two pairs of axes in Fig. A-3. It has coordinates $x = w = 0$ and $x' = w' = 0$, and the event to which it corresponds is the coincidence in time of the origins of the S and S' reference frames.

Point P_1 on the x' axis of Fig. A-3 has been chosen as a point to which we wish to assign the value $x' = 1$, representing a unit of length on this axis. As for all points on the x' axis, the time coordinate w' of P_1 is zero. Putting $x' = 1$ and $w' = 0$ into Eq. A-1a' yields, by simple inspection, $x = \gamma$ for the x coordinate of P_1. With this information we can easily construct numerical scales for both the x and the x' axes, based on our initially assumed unit length.

Consider now point P_2 on the w' axis of Fig. A-3, to which we wish to assign the value $w' = 1$, representing a unit of time (measured in terms of ct', to be sure) on that axis. We wish the scales on both the x' and the w' axes to be based on the same unit length, so we choose to locate P_2 so that the line segment OP_2 is equal in length to the segment OP_1. As for all points on the w' axis, the space coordinate x' of P_2 is zero. Putting $w' = 1$ and $x' = 0$ into Eq. A-1b' yields, again by simple inspection, $w = \gamma$ for the w coordinate of P_2. We are now able to construct numerical scales for both the w and the w' axes, based on the same unit length as we assumed in calibrating the space axes.

To gain some physical familiarity with the Minkowski diagram, let us consider a clock at rest at the origin of the S' frame. For that clock we have $x' = 0$ (always), so events involving it must correspond to points along the w' axis of Fig.

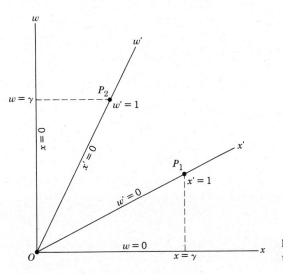

Figure A-3. Establishing the scales on the spacetime axes.

A-3. Point O, which is on that line, could represent the coincidence of the clock hand with a fiducial marker on the clock face, corresponding to zero time. Point P_2, whose time coordinate in the S' frame gives unit time ($w' = 1$) on that resting clock, is also on that line. The event represented by P_2 might correspond to a second coincidence of the clock hand with the fiducial marker. In frame S, however, the clock would be seen as a moving clock. We have seen above that $w' = 1$ in the S' frame corresponds to $w = \gamma$ in the S frame. Thus, by S-frame clocks, the unit time interval of the S' clock would be recorded as γ, corresponding exactly to the time dilation effect described by Eq. 2-14b.

In Fig. A-4 we show the calibration of the axes of the frames S and S', the unit time interval along w' being a longer line segment than the unit time interval along w and the unit length interval along x' being a longer line segment than the unit length interval along x. The first thing we must be able to do is to determine the spacetime coordinates of an event such as P directly from the Minkowski diagram. To find the space coordinate of the event, we simply draw a line parallel to the time axis from P to the space axis. The time coordinate is given similarly by a line parallel to the space axis from P to the time axis. The rules hold equally well for the primed frame as for the unprimed frame. In Fig. A-4, for example, the event P has the spacetime coordinates $x = 3.0$ and $w = 2.5$ in S (long dashed lines) and spacetime coordinates $x' = 2.0$ and $w' = 1.2$ in S' (short dashed lines). Figure A-4 was drawn assuming that $\beta = 0.50$, which yields $\gamma = 1.15$. Using these values for β and γ, you can readily derive the S-frame coordinates from the S'-frame coordinates—or conversely—by means of the Lorentz transformation equations (Eq. A-1), thus verifying the graphical relationships displayed in the Minkowski diagram.

In using the Minkowski diagram it is almost as if the rectangular grid of coordinate lines of S (Fig. A-5a) became squashed toward the 45° bisecting line when the coordinate lines of S' are put on the same graph (Fig. A-5b). In more formal language, we say that the Lorentz transformation equations transform an orthogonal (perpendicular) reference frame into a nonorthogonal one. Note that as $\beta \to 1$, corresponding to $v \to c$, the angle ϕ in Fig. A-5b ($= \tan^{-1}\beta$) approaches

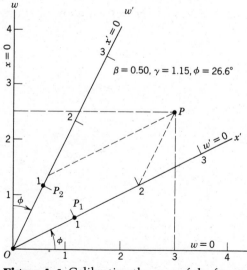

Figure A-4. Calibrating the axes of the frames S and S'.

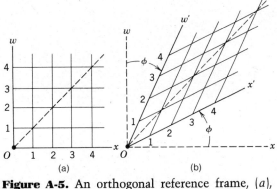

Figure A-5. An orthogonal reference frame, (a), transforms into a nonorthogonal one, (b).

45°, thus compressing the S'-frame coordinate space into a thinner and thinner wedge of the S-frame coordinate space. Alternatively, as $\beta \rightarrow 0$, corresponding to an approach to classical conditions, the angle ϕ between corresponding S and S' axes becomes very small. Even for a speed as high as that of a typical earth satellite ($\sim$17,000 mi/h), we note that $\beta = 2.5 \times 10^{-5}$, which yields a value of only 0.0015° for ϕ; relativistic mechanics is not much different from classical mechanics in these circumstances.

A-3 SIMULTANEITY, CONTRACTION, AND DILATION

Now we can easily show the relativity of simultaneity. As measured in S', two events will be simultaneous if they have the same time coordinate w'. Hence, if the events lie on a line parallel to the x' axis, they are simultaneous to S'. In Fig. A-6, for example, events Q_1 and Q_2 are simultaneous in S'; they obviously are not simultaneous in S, occurring at different times w_1 and w_2 there. Similarly, two events R_1 and R_2, which are simultaneous in S, are separated in time in S'.

As for the space contraction, consider Fig. A-7a. Let a meter stick be at rest in the S frame, its end points being at $x = 3$ and $x = 4$, for example. As time goes on, the world line of each end point traces out a vertical line parallel to the w axis.

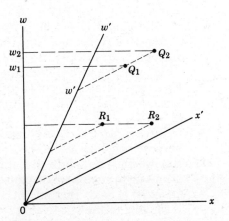

Figure A-6. Showing the relativity of simultaneity.

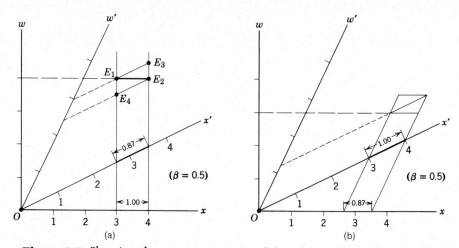

Figure A-7. Showing the space contraction, (*a*), and its reciprocal nature, (*b*).

The length of the stick is defined as the distance between the end points measured simultaneously. In *S*, the rest frame, the length is the distance in *S* between the intersections of the world lines with the *x* axis, or any line parallel to the *x* axis, for these intersecting points represent simultaneous events in *S*. The rest length is one meter. To get the length of the stick in *S'*, where the stick moves, we must obtain the distance in *S'* between end points measured simultaneously. This will be the separation in *S'* of the intersections of the world lines with the *x'* axis, or any line parallel to the *x'* axis, for these intersecting points represent simultaneous events in *S'*. The length of the (moving) stick is clearly less than one meter in *S'* (see Fig. A-7*a*).

Notice how very clearly Fig. A-7*a* reveals that it is a disagreement about the simultaneity of events that leads to different measured lengths. Indeed, the two observers do not measure the same pair of events in determining the length of a body (for example, the *S* observer uses E_1 and E_2, say, whereas the *S'* observer would use E_1 and E_3, or E_2 and E_4) for events that are simultaneous to one inertial observer are *not* simultaneous to the other. We should also note that the *x'* coordinate of each end point decreases as time goes on (simply project from successive world-line points parallel to *w'* onto the *x'* axis), consistent with the fact that the stick that is at rest in *S* moves towards the left in *S'*.

The reciprocal nature of this result is shown in Fig. A-7*b*. Here, we have a meter stick at rest in *S'*, and the world lines of its end points are parallel to *w'* (the end points are always at *x'* = 3 and *x'* = 4, say). The rest length is one meter. In *S*, where the stick moves to the right, the measured length is the distance in *S* between intersections of these world lines with the *x* axis, or any line parallel to the *x* axis. The length of the (moving) stick is clearly less than one meter in *S* (Fig. A-7*b*).

It remains now to demonstrate the time-dilation result geometrically. For this purpose consider Fig. A-8. Let a clock be at rest in frame *S*, ticking off units of time there. The solid vertical line in Fig. A-8, at *x* = 2.3, is the world line corresponding to such a single clock. T_1 and T_2 are the events of ticking at *w* (= *ct*) = 2 and *w* (= *ct*) = 3, the time interval in *S* between ticks being unity. In *S'*, this clock is moving to the left so that it is at a different place there each time it ticks. To measure the time interval between events T_1 and T_2 in *S'*, we use two different clocks, one at the location of event T_1 and the other at the location of event T_2.

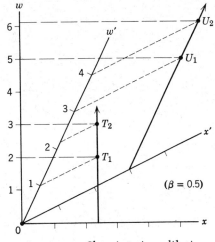

Figure A-8. Showing time dilation.

The difference in reading of these clocks in S' is the difference in times between T_1 and T_2 as measured in S'. From the graph, we see that this interval is greater than unity. Hence, from the point of view of S', the moving S clock appears slowed down. During the interval that the S clock registered unit time, the S' clock registered a time greater than one unit.

The reciprocal nature of the time-dilation result is also shown in Fig. A-8. You should construct the detailed argument. Here a clock at rest in S' emits ticks U_1 and U_2 separated by unit proper time. As measured in S, the corresponding time interval exceeds one unit.

A-4 THE TIME ORDER AND SPACE SEPARATION OF EVENTS

We can also use the geometric representation of spacetime to gain further insight into the concepts of simultaneity and the time order of events that we discussed in Chapter 2. Consider the shaded area in Fig. A-9, for example. Through any point P in this shaded area, bounded by the world lines of light waves, we can draw a w' axis from the origin; that is, we can find an inertial frame S' in which the events O and P occur at the same place $(x' = 0)$ and are separated only in time.* As shown in Fig. A-9, event P follows event O in time (it comes later on S' clocks), as is true wherever event P is in the upper half of the shaded area. Hence, events in the upper half (region 1 on Fig. A-10) are absolutely in the future relative to O, and this region is called the Absolute Future. If event P is at a spacetime point in the lower half of the shaded area (region 2 on Fig. A-10), then P will precede event O in time. Events in the lower half are absolutely in the past relative to O, and this region is called the Absolute Past. In the shaded regions, therefore, there is a definite time order of events relative to O, for we can always find a frame in which O and P occur at the same place; a single clock will determine absolutely the time order of the event at this place.

* We cannot draw an x' axis through points such as P in Fig. A-9 because the angle ϕ in Fig. A-2 would then exceed 45°, which requires that $\beta > 1$ (or, equivalently, that $v > c$). For the same reason, we cannot draw a w' axis through points such as Q in Fig. A-9.

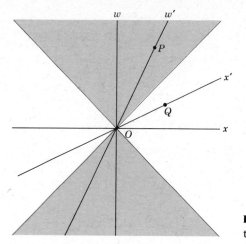

Figure A-9. The time order and space separation of events.

Consider now the unshaded regions of Fig. A-9. Through any point Q we can draw an x' axis from the origin; that is, we can find an inertial frame S' in which the events O and Q occur at the same time $(w' = ct' = 0)$ and are separated only in space. We can always find an inertial frame in which events O and Q appear to be simultaneous for spacetime points Q that are in the unshaded regions (region 3 of Fig. A-10), so that this region is called the Present. In other inertial frames, of course, O and Q are not simultaneous, and there is no absolute time order of these events but a relative time order, instead.

If we ask about the space separation of events, rather than their time order, we see that events in the present are absolutely separated from O, whereas those in the absolute future or absolute past have no definite space order relative to O. Indeed, region 3 (present) is said to be "spacelike" whereas regions 1 and 2 (absolute past or future) are said to be "timelike." That is, a world interval such as OQ is spacelike and a world interval such as OP is timelike.

The geometric considerations that we have presented are connected with the invariant nature of the spacetime *interval*, described in Section 2.3. As presented there, the interval involves a pair of events. For our purposes we can choose as one universal member of this pair the standard reference event represented by point O in Fig. A-9. It corresponds to the coincidence in time of the

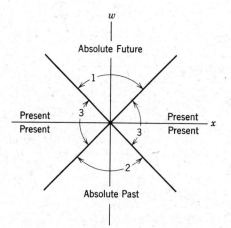

Figure A-10. Location in time of events relative to the origin.

origins of the two reference frames, S and S', and has the spacetime coordinates $x = w = 0$ and $x' = w' = 0$. The other member of the event pair can then be a generalized event represented by points such as P or Q in Fig. A-9. In this way we can associate the spacetime interval with P and Q alone, and can write (from Eq. 2-16, recalling that $w = ct$),

$$s^2 = w^2 - x^2 = w'^2 - x'^2. \tag{A-2}$$

We have seen that s^2, which has the same numerical value in all reference frames, can be either positive, negative, or zero, depending on the relative magnitudes of w and x (or of w' and x'). If $w > x$, as it is for points such as P in Fig. A-9, then s^2 is positive and s is a real quantity; we write it as $c\tau$, where τ is the *proper time interval* associated with the event pairs such as OP; see Eq. 2-17. If $w < x$, as it is for points such as Q in Fig. A-9, then $-s^2$ is a positive quantity; we call its square root σ, the *proper distance interval* for the event pairs such as OQ. We have then two relations,

$$c^2\tau^2 = w^2 - x^2 \tag{A-3a}$$

and

$$\sigma^2 = x^2 - w^2. \tag{A-3b}$$

Now consider Fig. A-10. In regions 1 and 2 we have spacetime points for which $w > x$, so the proper time is a real quantity, $c^2\tau^2$ being positive; see Eq. A-3a. In regions 3 we have spacetime points for which $x > w$, so the proper distance σ is a real quantity; see Eq. A-3b. Hence either τ or σ is real for any two events (that is, the event at the origin and the event elsewhere in spacetime) and either τ or σ may be called the spacetime interval between the two events. When τ is real the interval is called "timelike"; when σ is real the interval is called "spacelike." Because σ and τ are invariant properties of two events, it does not depend at all on what inertial frame is used to specify the events whether the interval between them is spacelike or timelike.

In the spacelike region we can always find a frame S' in which the two events are simultaneous, so that σ can be thought of as the spatial interval between the events in that frame. (That is, $\sigma^2 = x^2 - w^2 = x'^2 - w'^2$. But $w' = 0$ in S', so $\sigma = x'$.) In the timelike region we can always find a frame S' in which the two events occur at the same place, so that τ can be thought of as the time interval between the events in that frame. [That is, $\tau^2 = t^2 - (x^2/c^2) = t'^2 - (x'^2/c^2)$. But $x' = 0$ in S', so $\tau = t'$.]

What can we say about points on the 45° lines? For such points, $x = w$. Therefore, the proper time interval between two events on these lines vanishes, for $c^2\tau^2 = w^2 - x^2 = 0$ if $x = w$. We have seen that such lines represent the world lines of light rays and give the limiting velocity $(v = c)$ of relativity. On one side of these 45° lines (shaded regions in Fig. A-9), the proper time interval is real; on the other side (unshaded regions), it is imaginary. An imaginary value of τ would correspond to a velocity in excess of c. But no signals can travel faster than c. All this is relevant to an interesting question that can be posed about the unshaded regions.

In this region, which we have called the Present, there is no absolute time order of events; event O may precede event Q in one frame but follow event Q in another frame. What does this do to our deep-seated notions of cause and effect? Does relativity theory negate the causality principle? To test cause and effect, we would have to examine the events at the same place so that we could say absolutely that Q followed O, or that O followed Q, in each instance. But in the Present, or spacelike, region these two events occur in such rapid succession that the time difference is less than the time needed by a light ray to traverse the spatial distance between two events. We cannot fix the time order of such events absolutely, for no signal can travel from one event to the other faster than *c*. In other words, no frame of reference exists with respect to which the two events occur at the same place; thus, we simply cannot test causality for such events even in principle. Therefore, there is no violation of the law of causality implied by the relative time order of *O* and events in the spacelike region. We can arrive at this same result by an argument other than this operational one. If the two events, *O* and *Q*, are related causally, then they must be capable of interacting physically. But no physical signal can travel faster than *c*, so events *O* and *Q* cannot interact physically. Hence, their time order is immaterial, for they cannot be related causally. Events that can interact physically with *O* are in regions other than the Present. For such events, *O* and *P*, relativity gives an unambiguous time order. Therefore, relativity is completely consistent with the causality principle.

questions and problems

1. Interpreting events on a spacetime diagram (I). Draw a spacetime diagram and on it locate an event *P* whose coordinates are $x = 450$ m and $t = 1.00$ μs ($w = ct = 300$ m). With respect to the standard reference event *O* at the origin, (*a*) does *P* represent an event in the future? The present? The past? (*b*) Is the interval *OP* spacelike? Timelike? Lightlike? (*c*) What proper time interval is associated with *OP*? (*d*) What proper space interval? (*e*) Can you find another frame *S'* for which the events *OP* would occur at the same time? If so, draw the spacetime axes of that frame on your diagram and give the speed parameter β of the frame. (*f*) Can you find another frame *S"* in which the events *OP* would occur at the same place? If so, draw the spacetime axes of *that* frame on your diagram and give *its* speed parameter β.

2. Interpreting events on a spacetime diagram (II). Solve Problem 1 for an event whose coordinates are $x = 450$ m and $t = 1.50$ μs, and for an event whose coordinates are $x = 450$ m and $t = 1.00$ μs. Plot both events on the spacetime diagram of Fig. A-10 and compare.

3. Present or Absolute Future? Consider two events, both of which have $x = 1$ in, say, Fig. A-4. Event *A* has $w = 0.9$ and event *B* has $w = 1.1$. Comparison with Fig. A-10 shows that, with respect to the reference event *O* at the origin, the first of these events would be classified as Absolute Future and the second as Present. However, they both seem to occur in the future from the point of view of observer *S*. If Present means "right now," then neither of these events seems to qualify. It also seems clear that these two events could differ as slightly as you please; one, for example, could have $w = 0.99999$ and the other $w = 1.00001$, and a fundamental difference would still remain between them. Can you identify and clarify this difference?

★4. Calibrating the axes. (*a*) Let *d* represent the length (measured with a ruler) of a unit interval on the *x* (or the *w*) axis of Fig. A-4 and let *d'* represent the corresponding quantity for the *x'* (or the *w'*) axis of that figure. Show that

$$\frac{d'}{d} = \frac{\gamma}{\cos\phi} = \sqrt{\frac{1+\beta^2}{1-\beta^2}}.$$

(*b*) Evaluate this ratio for $\beta = 0.50$ (for which Fig. A-4 is drawn) and verify (using a ruler) that the axes in that figure are calibrated correctly.

5. Changing the scales of a spacetime diagram. (*a*) Redraw the spacetime diagram of Fig. A-4 but, in place of the dimensionless scale factor of unity, take 200 m as the unit scale distance. On this diagram, locate an event *P* whose spacetime coordinates, as determined by observer *S*, are $x = 800$ m and $t = 1.00$ μs. Determine the spacetime coordinates of this event as determined by observer *S'* (*b*) directly from your diagram and (*c*) using the Lorentz transformation equations.

6. Learning about the time axes. In the spacetime (or Minkowski) diagram of Fig. A-2, the *w* axis, from the point of view of observer *S*, represents the world line of a particle resting at the origin of the *S* frame. Identify on the diagram the world line (*a*) of the *S* origin from the point of view of *S'*; (*b*) of the *S'* origin from the point of view of *S*; (*c*) of the *S'* origin from the point of view of *S'*. (*d*) Write down the equations of all four of these world lines, in coordinates appropriate to the observer, using the Lorentz transformation equations (see Eqs. A-1) as needed.

7. Learning about the position axes. In the spacetime diagram of Fig. A-2, the *x* axis, from the point of view of observer *S*, is made up of points at each of which there is a clock, fixed in the *S* frame, that reads $w = 0$. (*a*) What times would observer *S'* read on these same *S* clocks? (*b*) Identify the locus of points each of which contains a clock fixed in the *S'* frame that reads *w'* = 0. (*c*) What times would the *S* observer read on those *S'* clocks?

8. S and S' watch a clock. In Fig. A-4, consider a clock at rest at the origin of the *S'* frame and consider an event corresponding to a reading of "3" (as seen by observer *S'*) on that clock. What reading will observer *S* (who uses his own clocks) record for this event? Solve by direct measurement from the spacetime diagram and also by use of the Lorentz transformation equations. Recall that β = 0.50 for the conditions of Fig. A-4.

9. S and S' measure a rod. In Fig. A-4, consider a rod 2.00 units long at rest along the *x'* axis of the *S'* reference frame, one end of the rod being at the origin of that frame. Both observers, *S* and *S'*, measure the length of the rod. What values do they find? Solve by direct measurement from the spacetime diagram and also by use of the Lorentz transformation equations. Show on the spacetime diagram the two events that are used by each observer in the measuring process.

10. You can't get there that fast. Let the departure of a plane from Boston be an event whose coordinates are *x* = 0 and *w* = *t* = 0. Let a second event be the arrival of that plane in Seattle. Plot these two events qualitatively on the spacetime diagram of Fig. A-10. (*a*) Can you find a second frame *S'* in which these two events are simultaneous? If so, describe that frame. (*b*) Can you find a frame in which these two events occur at the same place? If so, describe *that* frame. (*c*) Is the interval associated with these two events spacelike or timelike?

11. Three reference frames. A system *S'* moves to the right relative to *S* at a speed of 0.60*c* and another system *S"* moves to the right relative to *S* at a speed 0.35*c*. (*a*) Using the Minkowski diagram, find the velocity of *S"* relative to frame *S'*. (*b*) Repeat, with *S"* moving to the right at a speed of 0.50*c*. (*Hint*: Construct lines of constant *x'* and *t'* on the diagram. Using this gridwork of lines, find the slope of the world line for β = 0.35 and for β = 0.50.)

12. An event viewed from three reference frames. (*a*) Redraw the spacetime diagram of Fig. A-4 and superimpose on it a set of axes corresponding to frame *S"*, which is moving in the positive *x* direction with a speed 0.80*c*. Calibrate the *x"-w"* axes described in Problem 4. (*b*) Consider an event, represented by *P* in Fig. A-4, whose coordinates in *S* are *x* = 3.0 and *w* = 2.5. Find the coordinates of this event in frame *S'* and in frame *S"*, both directly from the diagram and by use of the Lorentz transformation equations.

13. A collision on a spacetime diagram. A particle of mass *m* is at rest at *x* = 3m along the *x* axis of a coordi-

nate system. A second particle, of mass 2*m*, passes through the origin at *t* = 0 and moves toward the first particle with a speed of 1 m/s. The two particles then undergo a head-on collision, both particles moving forward along the *x* axis after the collision. (*a*) Show that, according to classical physics, after the collision the initially moving particle moves forward with speed of $\frac{1}{3}$m/s and the initially resting particle does so with a speed of $\frac{4}{3}$m/s. (*b*) Draw the world lines for these particles on a spacetime diagram, for an interval encompassing the collision. (*c*) Draw on the same diagram the world line representing the motion of the center of mass of the colliding particles.

14. Using spacetime diagrams (I). Read again Problems 31 and 32 of Chapter 2. (*a*) Draw a world diagram for the problem, including on it the world lines for observers *A*, *B*, *C*, *D*, and *E*. Label the points *AD*, *BD*, *AC*, *BC*, and *EC*. (*b*) Show, by means of the diagram, that the clock at *A* records a shorter time interval for the events *AD*, *AC* than do clocks at *D* and *C*. (*c*) Show that, if observers on the cart try to measure the length *DC* by making simultaneous markings on a measuring stick in their frame, they will measure a length shorter than the rest length *DC*. Explain this result in terms of simultaneity, using the diagram. For convenience, take *v* = $\frac{1}{2}$*c*.

15. Using spacetime diagrams (II). Do Example 4, Chapter 2, by means of a Minkowski diagram. Check your results by calculating the invariants $c^2\tau^2$ or σ^2.

16. Using spacetime diagrams (III). Do Example 5, Chapter 2, by means of a Minkowski diagram. Check your results by calculating the invariants $c^2\tau^2$ or σ^2.

17. Calibration by hyperbolas. Equations A-3 represent the equations of hyperbolas, each of which has two branches. If we choose both σ and $c\tau$ equal to one scale unit on the *x* and *w* axes of a spacetime diagram, we can write, for the equations of these hyperbolas,

$$w^2 - x^2 = 1 \quad \text{(timelike region)}$$

and

$$x^2 - w^2 = 1 \quad \text{(spacelike region)}.$$

Figure A-11 shows these hyperbolas; they represent but one typical family of an infinite set of hyperbolas, corresponding to different choices for σ and for $c\tau$ in Eqs. A-3. Figure A-12 shows the upper right quadrant of Fig. A-11, with the spacetime axes of two reference frames *S'* (β = 0.50) and *S"* (β = 0.80) drawn in.

(*a*) Prove that the hyperbola branches in Fig. A-11 approach the 45° lightlike lines asymptotically. (*b*) Sketch in roughly the curves that would correspond to a choice of σ = 2 in Eqs. A-3 and to a choice of *c*τ = 2. (*c*) With respect to the standard reference event *O*, what proper time interval (in terms of *w*) can you assign to events such as *J*,

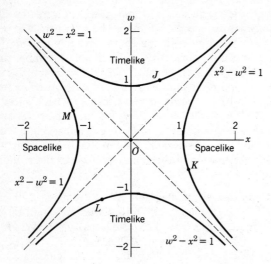

Figure A-11. Problem 17; calibration by hyperbolas.

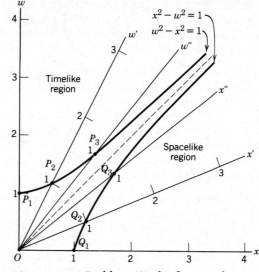

Figure A-12. Problem 17; the first quadrant in detail.

K, L, and M in Fig. A-11? What proper space interval? (d) In Fig. A-11, convince yourself that the point representing the intersection of the w axis and the upper branch of the timelike hyperbola corresponds to a clock at rest at the origin of the S frame and reading "1." Other points on this hyperbola branch read times greater than "1"; where are these clocks located? Are they fixed in the S frame, or are they moving? What does it mean when we say that, though one of these clocks may read, say, "1.5," its *proper* time is still "1"? (e) In Fig. A-12 we have seen that P_1 represents a clock at the origin of the S frame reading "1."

Can you see that P_2 represents a clock resting at the origin of the S' frame and also reading "1"? What does P_3 represent? Can you see how the hyperbolas of Figs. A-11 and A-12 can be used to establish scales for the axes of a given reference frame? (The hyperbolas are often called "calibration curves," for this reason.) (f) Analyze the spacelike hyperbola branch shown in Fig. A-12 in physical terms, following the pattern outlined above for the timelike branch. In particular, to what do events Q_1, Q_2, and Q_3 correspond physically?

references

1. A. Einstein, H. Minkowski, H. A. Lorentz, and H. Weyl, *The Principle of Relativity: A Collection of Original Memoirs on the Special and General Theory of Relativity,* notes by A. Sommerfeld, Dover, New York, 1952.

2. Hermann Minkowski, "Space and Time" (a translation of an address given September 21, 1908), in *The Principle of Relativity* (Dover, New York).

SUPPLEMENTARY TOPIC B
the twin paradox

If we placed a living organism in a box . . . one could arrange that the organism, after any arbitrary lengthy flight, could be returned to its original spot in a scarcely altered condition, while corresponding organisms which had remained in their original positions had already long since given way to new generations. For the moving organism the lengthy time of the journey was a mere instant, provided the motion took place with approximately the speed of light.

Albert Einstein (1911)

In the above statement Einstein describes what has come to be called the twin paradox or the clock paradox [1]. If the stationary organism is a man and the traveling one is his twin, then the traveler returns home to find his twin brother much aged compared to himself. The paradox centers around the contention that, in relativity, either twin could regard the other as the traveler, in which case each should find the other younger—a logical contradiction. This contention assumes that the twins' situations are symmetrical and interchangeable, an assumption that is not correct. Furthermore, the accessible experiments have been done and support Einstein's prediction. In succeeding sections, we look with some care into the many aspects of this problem.

B-1 THE ELAPSED PROPER TIME DEPENDS ON THE ROUTE

Figure B-1a shows the world lines of three particles, their motions being confined to the x axis of an inertial reference frame. Particle 1 is at rest on the x axis, its world line being a simple vertical line. Particle 2 is moving along this axis in the direction of increasing x, its (constant) speed v being $dx/dt = c\, dx/(c\, dt) = c \tan \theta$, where θ is the angle made by the world line of this particle with the vertical. A reference frame moving with particle 2 would thus be an inertial frame. Particle 3 is in accelerated motion along the x axis, its speed v being, in general, different for every position of the particle. An inertial frame moving with *this* particle would *not* be an inertial frame.

Let us assume that particle 3 in Fig. B-1a is, in fact, a clock, and let us consider the problem of calculating the elapsed proper time on this clock as it travels from one point to another. We can assume that this traveling clock is in an inertial frame only for differential elements of its path. We can compute the elapsed proper time $d\tau$ for such an element from Eq. 2-17, which we write in differential form as

$$(c\, d\tau)^2 = (c\, dt)^2 - (dx)^2$$

or

$$d\tau = \sqrt{(dt)^2 - \left(\frac{dx}{c}\right)^2}. \tag{B-1}$$

The total elapsed proper time between any two points would then be the integral of this quantity between those points.

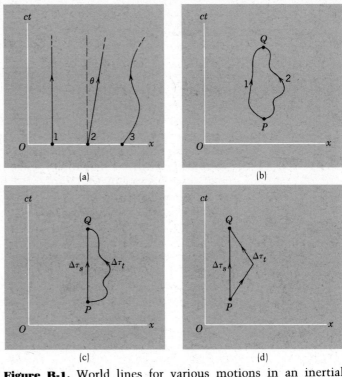

Figure B-1. World lines for various motions in an inertial frame.

Consider now Fig. B-1*b*, in which the world lines of two particles, each with a clock attached, start from point P and reconvene at Q. For either path the proper time interval—that is, the elapsed time on the particle's clock—between P and Q is given by (see Eq. B-1)

$$\Delta\tau = \int_P^Q d\tau = \int_P^Q \sqrt{(dt)^2 - \left(\frac{dx}{c}\right)^2}.$$ (B-2)

Both dt and dx in the above are differential spacetime path elements as measured by the observer in the inertial reference frame of Fig. B-1*b*. We are not surprised that the two paths shown in this figure differ as far as x is concerned (odometer readings), and we have learned not to be surprised that clock readings vary in much the same way. Simple inspection of Eq. B-2 shows that the quantity depends not only on the initial and final points but also on the path taken between them.

In Fig. B-1*c* we let one of these paths be a straight line, corresponding to the simple passage of time for a stationary particle; the other path remains arbitrary. From Eq. B-2 we have, for the straight path,

$$\Delta\tau_s = \int_P^Q \sqrt{(dt)^2 - \left(\frac{dx}{c}\right)^2} = \int_P^Q dt = t_Q - t_P,$$

in which the subscript on $\Delta\tau$ refers to the stationary clock. In such a case dx is zero along the path, and the proper time coincides with the time interval, $t_Q - t_P$, recorded by the stationary clocks of the inertial reference frame. Along the second world line, however, the elapsed power time is

$$\Delta\tau_t = \int_P^Q \sqrt{(dt)^2 - \left(\frac{dx}{c}\right)^2},$$

in which the subscript refers to the traveling clock; we see that $\Delta\tau_t$ will *not* equal $\Delta\tau_s$. In fact, since $(dx)^2$ is always positive, we find that

$$\Delta\tau_t < \Delta\tau_s. \tag{B-3}$$

The clocks will read different times when brought back together, the traveling clock running behind (recording a smaller time difference than) the stay-at-home clock. Figure B-1*d* is a special case of Fig. B-1*c* in that the traveling clock moves with constant velocity over most of its path, its motion being accelerated only near its "turnaround point." Note that, although the turnaround may occupy only a small fraction of the total travel time, it is vitally necessary to the motion if the two clocks are to reconvene.

We have noted that the reference frame whose axes are drawn in Fig. B-1 is an inertial frame. The motion of the traveling clock is represented in this frame by a curved world line, for this clock undergoes accelerated motion rather than motion with uniform velocity. It could not return to the stationary clock, for example, without reversing its velocity. The special theory of relativity can predict the behavior of accelerated objects as long as, in the formulation of the physical laws, we take the view of the inertial (unaccelerated) observer. This is what we have done so far. A frame attached to the clock traveling along its round-trip path would not be an inertial frame. We could reformulate the laws of physics so that they have the same form for accelerated (noninertial) observers—this is the program of general relativity theory—but it is unnecessary to do so to explain the twin paradox. All we wish to point out here is that the situation is *not* symmetrical with respect to the clocks (or twins); one is always in a single inertial frame and the other is not.

B-2 SPACETIME DIAGRAM OF THE TWIN PARADOX

In our earlier discussions of time dilation, we spoke of "moving clocks running slow." What is meant by that phrase is that a clock moving at a constant velocity **u** relative to an inertial frame containing synchronized clocks will be found to run slow by the factor $\sqrt{1 - u^2/c^2}$ *when timed by those clocks*. That is, to time a clock moving at constant velocity relative to an inertial frame, we need at least *two* synchronized clocks in that frame. We found this result to be reciprocal in that a single S' clock is timed as running slow by the many S clocks, and a single S clock is timed as running slow by the many S' clocks.

The situation in the twin paradox is different. If the traveling twin traveled always at a constant speed in a straight line, he would never get back home. And each twin would indeed claim that the other's clock runs slow compared to the synchronized clocks in his own frame. To get back home—that is, to make a round trip—the traveling twin would have to change his velocity. What we wish to compare in the case of the twin paradox is a single moving clock with a *single* clock at rest. To do this we must bring the clocks into coincidence twice—they must come back together again. It is not the idea that we regard one clock as moving and the other at rest that leads to the different clock readings, for if each of two observers seems to the other to be moving at constant speed in a straight line, they cannot absolutely assert who is moving and who is not. Instead, it is because one clock has *changed* its velocity and the other has not that makes the situation unsymmetrical.

Now you may ask how the twins can tell who has changed his velocity. This is clearcut. Each twin can carry an accelerometer. If he changes his speed or the direction of his motion, the acceleration will be detected. We may not be aware of an airplane's motion, or a train's motion, if it is one of uniform velocity; but let it move in a curve, rise and fall, speed up or slow down, and we are our own accelerometer as we get thrown around. Our twin on the ground watching us does not experience these feelings—his accelerometer registers nothing. Hence, we can tell the twins apart by the fact that the one who makes the round-trip experiences and records accelerations whereas the stay-at-home does not.

A numerical example, suggested by C. G. Darwin [2], is helpful in fixing the ideas. We imagine that, on New Year's Day, Bob leaves his twin brother Dave, who is at rest on a spaceship, fires rockets that get him moving at a speed of $0.8c$ relative to Dave, and by his own clock travels away at this constant speed toward a distant star, which he reaches after three years of travel. He then fires more powerful rockets that exactly reverse his motion and gets back to Dave after another three years by his clock. By firing rockets a third time, he comes to rest beside Dave and compares clock readings. Bob's clock says he has been away for six years (the $\Delta\tau_t$ of Eq. B-3), but Dave's clock says that ten years have elapsed (the $\Delta\tau_s$ of Eq. B-3). Let us see how this comes about.

First, we can simplify matters by ignoring the effect of the accelerations on the traveling clock. Bob can turn off his clock during the three acceleration periods, for example. The error thereby introduced can be made very small compared to the total time of the trip, for we can make the trip as far and as long as we wish without changing the acceleration intervals. It is the total time that is at issue here in any case.* We do not destroy the asymmetry, for even in the ideal simplification of Fig. B-2 (where the world lines are straight lines rather than curved ones), Dave is always in one inertial frame whereas Bob is definitely in two different inertial frames—one going out $(0.8c)$ and another coming in $(-0.8c)$.

Let the spaceships be equipped with identical clocks that send out light signals at one-year intervals. Dave receives the signals arriving from Bob's clock and records them against the annual signals of his own clock; likewise, Bob receives the signals from Dave's clock and records them against the annual signals of his clock.

In Fig. B-2, Dave's world line is straight along the ct-axis; he is at $x = 0$ and we mark off ten years (in terms of ct), a dot corresponding to the annual New Year's Day signal of his clock. Bob's world line at first is a straight line inclined to the ct axis, corresponding to a ct' axis of a frame moving at $+0.8c$ relative to Dave's frame. We mark off three years (in terms of ct'), a dot corresponding to the annual New Year's Day signal of his clock. After three of Bob's years, he switches to another inertial frame whose world line is a straight line inclined to the ct axis, corresponding to the ct'' axis of a frame moving at $-0.8c$ relative to Dave's frame. We mark off three years (in terms of ct''), a dot corresponding to the annual New Year's Day signal of his clock. Note the dilation of the time interval of Bob's clock compared to Dave's.

Now let us draw the light signals from Bob's clock on the spacetime diagram of Fig. B-2. Recall (see Fig. A-1) that such signals are drawn at 45° to the spacetime axes, corresponding to their speed of c. Thus from each dot on Bob's world line we draw such a 45°-line headed back to Dave on the line at $x = 0$. There are six signals, the last one emitted when Bob returns home to Dave. Likewise, the

---

* An analogy is that the total distance traveled by two drivers between the same two points, one along the hypotenuse of a right triangle and the other along the other two sides of the triangle, can be quite different. One driver always moves along a straight line, whereas the other makes a right turn to travel along two straight lines. We can make the distance between the two points as long as we wish without altering the fact that only one turn must be made. The difference in mileage traveled by the drivers certainly is not acquired at the turn that one of them makes.

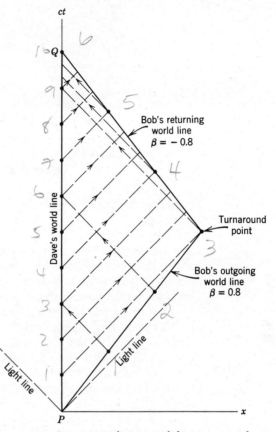

Figure B-2. Spacetime diagram of the twin paradox.

signals from Dave's clock are straight lines, from each dot on Dave's world line, inclined 45° to the axes and headed out to Bob's spaceship. We see that there are ten signals, the last one emitted when Bob returns home to Dave.

How can we confirm this spacetime diagram numerically? Simply by the Doppler effect. As the clocks recede from each other, the frequency ν of their signals is reduced from the proper frequency ν_0 by the Doppler effect. From Eq. 2-30*b* we thus have

$$\frac{\nu}{\nu_0} = \sqrt{\frac{1 - \beta}{1 + \beta}} = \sqrt{\frac{1 - 0.8}{1 + 0.8}} = \frac{1}{3}.$$

Hence, Bob receives the first signal from Dave after three of his years, just as he is turning back. Similarly, Dave receives messages from Bob on the way out once every three of his years, receiving three signals in nine years. As the clocks approach one another, the frequency ν of their signals is increased from the proper frequency ν_0 by the Doppler effect. In this case (see Eq. 2-30*a*) we have

$$\frac{\nu}{\nu_0} = \sqrt{\frac{1 + \beta}{1 - \beta}} = \sqrt{\frac{1 + 0.8}{1 - 0.8}} = 3.$$

Thus, Bob receives nine signals from Dave in his three-year return journey. Altogether, Bob receives ten signals from Dave. Similarly, Dave receives three signals

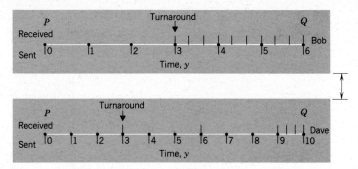

Figure B-3. The signal logs for the twins.

from Bob in the last year before Bob is home. Altogether, Dave receives six signals from Bob.

Figure B-3 shows the signal logs for Dave's and Bob's spaceships. Signals sent are indicated below the time axis in each case and signals received are shown above that axis. There is no disagreement about the signals: Bob sends six and Dave receives six; Dave sends ten and Bob receives ten. Everything works out, each seeing the correct Doppler shift of the other's clock and each agreeing to the number of signals that the other sent. The different total times recorded by the twins corresponds to the fact that Dave sees Bob recede for nine years and return in one year, although Bob both receded for three of his years and returned for three of his years. Dave's records will show that he received signals at a slow rate for nine years and at a rapid rate for one year. Bob's records will show that he received signals at a slow rate for three years and at a rapid rate for another three years. The essential asymmetry is thereby revealed by a Doppler effect analysis. When Bob and Dave compare records, they will agree that Dave's clock recorded ten years and Bob's recorded only six. Ten years have passed for Dave during Bob's six-year round trip.

B-3 SOME OTHER CONSIDERATIONS

Will Bob really be four years younger than his twin brother? Since for the word "clock" we could have substituted any periodic natural phenomena, such as heartbeat or pulse rate, the answer is yes. We might say that Bob lived at a slower rate than Dave during his trip, his bodily functions proceeding at the same slower rate as his physical clock. Biological clocks behave in this respect the same as physical clocks. There is no evidence that there is any difference in the physics of organic processes and the physics of the inorganic materials involved in these processes. If motion affects the rate of a physical clock, we expect it to effect a biological clock in the same way.

It is of interest to note the public acceptance of the idea that human life processes can be slowed down by refrigeration, so that a corresponding different aging of twins can be achieved by temperature differences. What is paradoxical about the relativistic case, in which the different aging is due to the difference in motion, is that since (uniform) motion is relative, the situation appears (incorrectly) to be symmetrical. But, just as the temperature differences are real, measurable, and agreed upon by the twins in the foregoing example, so are the differences in motion real, measurable, and agreed upon in the relativistic case—the changing of inertial frames, that is, the accelerations, are not symmetrical. The results are absolutely agreed upon.

Although there is no need to invoke general relativity theory in explaining the twin paradox, the student may wonder what the outcome of the analysis would be if we knew how to deal with accelerated reference frames. We could then use Bob's spaceship as our reference frame, so that Bob is the stay-at-home, and it would be Dave who, in this frame, makes the round-trip space journey. We would find that we must have a gravitational field in this frame to account for the accelerations that Bob feels and the fact that Dave feels no accelerations even though he makes a round trip. If, as required in general relativity, we then compute the frequency shifts of light in this gravitational field, we come to the same conclusion as in special relativity [3].

B-4 EXPERIMENTAL TESTS

Testing the conclusions we have reached with actual twins and clocks in spaceships moving with speeds close to the speed of light is, of course, more than can be managed at present. However, totally equivalent high-speed tests can be carried out using as clocks unstable elementary particles such as muons or pions or one of the hundreds of varieties of radioactive atoms available to us. At lower speeds—those of jet planes, for example—tests can be carried out with macroscopic atomic clocks, thanks to impressive improvements in the stability and time-keeping ability of such clocks.

In Section 2-7 we described the precise measurements of Kundig [Ref. 9, Chapter 2] on the transverse Doppler effect. This effect, as we noted in that section, is a direct measure of the time dilation, and we can use it to illustrate the twin paradox. From the point of view of the observer on the rotor axis (the stay-at-home twin), the absorbing foil on the perimeter of the rotor (the traveling twin) has a characteristic resonant absorption frequency that matches the source frequency only when the rotor is *not* turning. When the rotor *is* turning, the resonant frequency of the moving foil drops, just as predicted by Eqs. 2-32. Put another way, the round-trip twin ages less than his stay-at-home brother and (to within 1 percent) by exactly the amount predicted by relativity theory.

In 1968 a careful measurement of time dilation was reported from CERN (the European Nuclear Research Center, located near Geneva) in which laboratory-generated 1.18-GeV muons, for which the corresponding speed is $0.9966c$, served as high-speed traveling clocks [4]. These muons were constrained to circulate in an orbit 5.0 m in diameter in the muon storage ring in that laboratory. Thus, like the traveling twin (and also like the resonant absorbing foil in the experiment described above), they traverse a closed path and undergo (centripetal) acceleration during their journey. Their mean life for decay in flight can then be compared with the mean life observed when muons are brought to rest in an absorbing block. Many experiments give the accepted value of 2.200 ± 0.0015 μs for the decay of resting muons (the stay-at-home twin); the CERN experimenters measured 26.15 ± 0.03 μs for the mean decay time for the traveling muons (the traveling twin). This agrees within about 2 percent with the lifetime predicted for these traveling muons because of the time dilation, namely, 26.72 μs. The time dilation phenomenon is universally accepted and is, in fact, turned to specific advantage in the design of certain high-energy particle experiments. As one high-energy physicist has written [5]: "We frequently transport beams of unstable particles over long distances such that no particles would be left without the help of Einstein's factor."

Refinements of atomic clocks have so improved the accuracy of timekeeping that time-dilation effects can be detected at speeds as low as those of jet planes. In

Table B-1
ROUND-THE-WORLD ATOMIC CLOCKS [6]
(The numbers shown are time differences, in nanoseconds, with respect to reference clocks at the U.S. Naval Observatory.)

	Eastward		Westward	
Predicted:				
Special relativity	-184 ± 18		96 ± 10	
General relativity	144 ± 14		179 ± 18	
Net predicted:		-40 ± 23		275 ± 21
Observed:		-59 ± 10		273 ± 7

October 1977, Joseph Hafele and Richard Keating [6] carried four cesium-beam atomic clocks around the world on commercial airline flights, ". . . to test Einstein's theory of relativity with macroscopic clocks." They took their clocks around once each way, that is, once eastward and once westward, comparing the traveling clocks to those that stayed at home at the Naval Observatory on the Earth, rotating (eastward) below them. The calculations must take into account not only the kinematic time-dilation effect (which is related only to the speed of the traveling clocks), but also relativistic frequency shifts associated with changes encountered in the strength of the earth's gravitational field (see Supplementary Topic C, Section C-2). Table B-1 shows the predictions of relativity theory along with the experimental findings. Hafele and Keating conclude: "There seems to be little basis for further arguments about whether clocks will indicate the same time after a round trip, for we find that they do not."

Today, when precision clocks move from one location to another, cumulative time corrections with respect to a stay-at-home clock are made routinely [7]. Such considerations enter, for example, when precision clocks are moved for comparison purposes between Washington, D.C., and the National Bureau of Standards Laboratory at Boulder, Colorado. Similarly, relativistic time-dilation effects must be considered in the design and operation of the Global Positioning System (GPS/NAVSTAR), a precision navigation system in which it is planned to employ 24 orbiting satellites.

questions and problems

1. The shortest distance between two points is a straight line (?). Comparison of Fig. B-1c and Eq. B-3 shows us that, in terms of elapsed proper time in units of ct on a spacetime diagram, a straight line is not the shortest distance between two points but the *longest*. Is this statement still true if one of the two particles involved is not stationary (as in Fig. B-1c) but moves with constant speed? Draw a spacetime diagram to represent this situation.

2. Einstein on the clock "paradox." Einstein, in his first paper on the special theory of relativity, wrote the following: "If one of two synchronous clocks at A is moved in a closed curve with constant velocity until it returns to A, the journey lasting t seconds, then by the clock that has remained at rest the travelled clock on its arrival at A will be $tv^2/2c^2$ seconds slow." Prove this statement. (*Note:* Elsewhere in his paper Einstein indicated that this result is an approximation, valid only for $v \ll c$.)

3. Do you really want to do it? You wish to make a round trip from earth in a spaceship, traveling at constant speed in a straight line for six months and then returning at the same constant speed. You wish further, on your return, to find the earth as it will be a thousand years in the future. (*a*) How fast must you travel? (*b*) Does it matter whether or not you travel in a straight line on your journey? If, for example, you traveled in a circle for one year, would you still find that a thousand years had elapsed by earth clocks when you returned?

4. Synchronizing clocks. Consider two clocks fixed along the x axis of an inertial reference frame, one at $x = x_1$ and the other at $x = x_2$. In Section 2-1 we saw how to synchronize such clocks, using light signals. Here is another proposed method that, at first glance, may seem quite reasonable: Let a traveler move out along the x axis with constant speed v, wearing a wristwatch. Let the traveler then set each of the two x axis clocks to agree with the wristwatch as she passes them. What is wrong with this method of synchronization?

5. Bob and Dave. (*a*) In the spacetime diagram of Fig. B-2, how far apart are Bob and Dave when Bob turns around? (*b*) Suppose that Dave did not know beforehand when Bob was planning to turn around. When (by his own clocks and calendars) would Dave find out that Bob had done so? (*c*) If Bob's clock runs slow on the outbound trip (as it does), then why does it not run fast on the inbound trip, for which his velocity is reversed in sign?

6. Bob changes his mind. Suppose that Bob, after noting the passage of three years by his on-board clock, decides not to return to Dave but simply stops. He compares his on-board clock with one of the local clocks belonging to the synchronized array of stationary clocks fixed in Dave's inertial frame. (*a*) What will this local clock read? (*b*) Draw Bob's world line for this new situation on a spacetime diagram.

★7. Bob is older than Dave this time. Bob, once started on his outward journey from Dave, keeps on going at his original uniform speed of 0.8c. Dave, knowing that Bob was planning to do this, decides, after waiting for three years, to catch up with Bob and to do so in three additional years. (*a*) To what speed must Dave accelerate to do so? (*b*) What will be the elapsed time by Bob's clock when they meet? (*c*) How far will they each have traveled when they meet, measured in Dave's original inertial reference

frame? (*d*) Draw the world lines for Bob and Dave on a spacetime diagram and compare it with Fig. B-2. Notice that the present scenario is the mirror image of the one discussed in connection with that figure; there Dave turned out to be four years older than Bob when they reconvened; here Bob will turn out to be four years older than Dave [8].

8. Bob and Dave are twins again. Suppose that Bob and Dave each agree to follow the scenario described by Fig. B-2 for three years, each counting the years by his own on-board clock. Then Bob will come to rest and Dave will accelerate to 0.8c and eventually catch up with Bob. (*a*) What will be the total elapsed times on each of their clocks when they meet? (*b*) Draw a spacetime diagram and compare it carefully with that of Fig. B-2. Note the total symmetry of the situation. In the scenario as given originally, Dave turned out to be four years older than Bob when they met; in Problem 7, the reverse turned out to be true; in this case they turn out to be the same age at the end of their journeys [8].

9. The twins talk it over. Explain (in terms of heartbeats, physical and mental activities, and so on) why the younger returning twin has not lived any longer than his own proper time even though his stay-at-home brother may say that he has. Hence, explain the remark: "You age according to your own proper time."

10. The twin paradox and time dilation. Time dilation is a symmetric (reciprocal) effect. The twin-paradox result is asymmetric (nonreciprocal). Discuss these two effects from this point of view and explain how they are related.

11. Asymmetric aging and acceleration. Is asymmetric aging always associated with acceleration? Can you have acceleration (of one or both twins) without asymmetric aging?

12. Getting younger. Can you think of any way to use space travel to reverse the aging process, that is, to get younger? Could you send your parents out on a long space voyage and have them be younger than you are when they get back?

references

1. Many articles on this topic are reproduced in Gerald Holton, Ed., *Special Relativity—Selected Reprints* (American Institute of Physics, New York, 1963). See also a number of Letters to the Editor in the September (1971) and the January (1972) issues of *Physics Today*.

2. C. G. Darwin, "The Clock Paradox in Relativity," *Nature*, **180**, 976 (1957).

3. O. R. Frisch, "Time and Relativity: Part I," *Contemp. Phys.*, **3**, 16 (1961), and O. R. Frisch, "Time and Relativity: Part II," *Contemp. Phys.*, **3**, 194 (1962).

4. F. J. M. Farley, J. Bailey, and E. Picasso, "Experimental Verifications of the Theory of Relativity," *Nature*, **217**, 17 (1968). The test of time dilation using the muon storage ring—important in its own right—was an incidental feature of another investigation.

5. R. W. Williams, *Phys. Today* (January 1972), p. 11.

6. J. C. Hafele and Richard E. Keating, "Around-the-World Atomic Clocks: Predicted Relativistic Time Gains," *Science,* **177,** 166 (1972), and J. C. Hafele and Richard E. Keating, "Around-the-World Atomic Clocks: Observed Relativistic Time Gains," *Science,* **177,** 168 (1972).

7. See Edward M. Purcell, in Harry Woolf, Ed., *Some Strangeness in the Proportion: A Centennial Symposium to Celebrate the Achievements of Albert Einstein,* (Addison-Wesley, Reading, Mass. 1980), p. 106.

8. Problems 7 and 8 are adapted from Frank S. Crawford, "Symmetrization of C. G. Darwin's Clock Paradox Scenario," *Am. J. Phys.* **51,** 1145 (1983).

SUPPLEMENTARY TOPIC C

the principle of equivalence and the general theory of relativity

> *I was sitting in a chair in the patent office at Bern when all of a sudden a thought occurred to me: "If a person falls freely he will not feel his own weight." I was startled. This simple thought made a deep impression on me. It impelled me toward a theory of gravitation.*
>
> Albert Einstein (1922)

We have seen that special relativity requires us to modify the classical laws of motion. However, the classical laws of electromagnetism, including the Lorentz force law, remain valid relativistically. What about the gravitational force, that is, Newton's law of gravitation—does relativity require us to modify that? Despite its great success in harmonizing the experimental observations, Newton's theory of gravitation is suspect conceptually if for no other reason than that it is an action-at-a-distance theory. The gravitational force of interaction between bodies is assumed to be transmitted instantaneously, that is, with infinite speed, in contradiction to the relativistic requirement that the limiting speed of a signal is c, the velocity of light. And there are worrisome features about the interpretation of the masses in the law of gravitation. For one thing, there is the equality of inertial and gravitational mass, which in the classical theory is apparently an accident (see *Physics*, Part I, Sec. 16-4). Surely there must be some physical significance to this equality. For another thing, the relativistic concept of mass-energy suggests that even particles of zero rest mass will exhibit masslike properties (for example, inertia and weight). But such particles are excluded from the classical theory. If gravity acts on them, we must find how to incorporate this fact in a theory of gravitation.

In 1911 Einstein advanced his principle of equivalence, which became the starting point for a new theory of gravitation. In 1916, he published his theory of general relativity, in which gravitational effects propagate with the speed of light and the laws of physics are reformulated so as to be invariant with respect to accelerated (noninertial) observers. The equivalence principle is strongly confirmed by experiment. Let us examine this first.

C-1 THE PRINCIPLE OF EQUIVALENCE

Consider two reference frames: (1) a nonaccelerating (inertial) reference frame S in which there is a uniform gravitational field and (2) a reference frame S', which is accelerating uniformly with respect to the inertial frame but in which there is no gravitational field. Two such frames are physically equivalent. That is, experiments carried out under otherwise identical conditions in these two frames should give the same results. This is the *principle of equivalence*. If the principle is restricted to mechanical experiments alone, it may be taken as a consequence of Newtonian mechanics. Einstein, however, broadened the principle to include *all* physical experiments, including optical (that is, electromagnetic) experiments, and used this principle as the basis for his general theory of relativity.

To explore this principle further, let us imagine a spaceship to be at rest in an inertial reference frame S in which there is a uniform gravitational field, say at the surface of the earth. Inside the spaceship, objects that are released will fall with an acceleration, g, in the gravitational field; an object that is at rest, such as an astronaut sitting on the floor, will experience a force opposing its weight. Now let the spaceship proceed to a region of outer space where there is no gravitational field. Its rockets are accelerating the spaceship, our new frame S', with $\mathbf{a} = -\mathbf{g}$ with respect to the inertial frame S. In other words, the ship is accelerating away from the earth beyond the region where the earth's field (or any other gravitational field) is appreciable. The conditions in the spaceship will now be like those in the spaceship when it was at rest on the surface of the earth. Inside the ship an object released by the astronaut will accelerate downward relative to the spaceship with an acceleration $\mathbf{g}$. And an object at rest relative to the spaceship, for instance, the astronaut sitting on the floor, will experience a force indistinguishable from that which balanced its weight before. From observations made in his own frame, the astronaut could not tell the difference between a situation in which his ship was accelerating relative to an inertial frame in a region having no gravitational field and a situation in which the spaceship was unaccelerated in an inertial frame in which a uniform gravitational field existed. The two situations are exactly equivalent.

Indeed, it follows that if a body is in a uniform gravitational field—such as an elevator in a building on earth—and is at the same time accelerating in the direction of the field with an acceleration whose magnitude equals that due to the field—such as the same elevator in free fall—then particles in such a body will behave as though they are in an inertial reference frame with no gravitational field. They will be free of acceleration unless a force is impressed on them. This is the situation inside an orbiting space laboratory, in which objects released by the astronaut will not fall relative to the orbiting vehicle (they appear to float in space) and the astronaut himself will be free of the force that countered the pull of gravity before launching (he feels weightless).

Einstein pointed out that, from the principle of equivalence, it follows that we cannot speak of the absolute acceleration of a reference frame, only a relative one, just as it followed from the special theory of relativity that we cannot speak of the absolute velocity of a reference frame, only a relative one. This analogy to special relativity is a formal one, for there is no absolute acceleration provided that we grant that there is also no absolute gravitational field. It also follows from the principle of equivalence (it is *not* an accident) that inertial mass and gravitational mass are equal (see Question 1).

C-2 THE GRAVITATIONAL RED SHIFT

Now let us apply the principle of equivalence to see what gravitational effects there might be that are not accounted for in the classical theory. Consider a pulse of radiation (a photon) emitted by an atom A at rest in frame S (a spaceship at rest on the earth's surface, for example). A uniform gravitational field $\mathbf{g}$ is directed

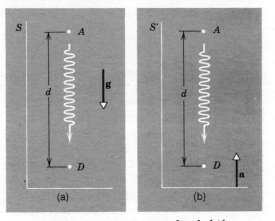

Figure C-1. Gravitational red shift.

downward in S, the photon falling down a distance d through this field before it is absorbed by the detector D (see Fig. C-1a). To analyze what effect gravity has on the photon, let us consider the equivalent situation, shown in Fig. C-1b. Here we have an atom and a detector separated by a distance d in a frame S' in which there is no gravitational field, the frame S' (a spaceship in outer space, for example) accelerating uniformly upward relative to S with $a = g$. When the photon is emitted, the atom has some speed u in frame S'. The speed of the detector, when the photon reaches it, is $u + at$, where t is the time of flight of the photon. But (see Question 6) t is (approximately) d/c and $a = g$, so the detector's speed on absorption is $u + g\,d/c$. In effect, the detector has an approach velocity relative to the emitter of $v = g\,d/c$, independent of u. This corresponds to a speed parameter $\beta(= v/c)$ of $g\,d/c^2$. Hence ν', the frequency received, is greater than ν, the frequency emitted, the Doppler formula (see Eq. 2-31a) giving us

$$\nu' = \nu(1 + \beta + \tfrac{1}{2}\beta^2 + \cdots).$$

Because $g\,d/c^2 \ll 1$, we need keep only the first two terms in this equation and can write, for the fractional shift in frequency,

$$\frac{\Delta\nu}{\nu} = \frac{\nu' - \nu}{\nu} \cong \beta = g\,\frac{d}{c^2}. \tag{C-1}$$

By the principle of equivalence, we should obtain this same result in frame S. In this frame, however, A and D are at rest and there is no Doppler effect to explain the increase in frequency. There is a gravitational field, however, and the result in S' suggests that this field might act on the photon. Let us explore this possibility by ascribing to the photon a *gravitational mass* equal to its inertial mass, E/c^2. Then, in falling a distance d in a gravitational field of strength g, the photon gains energy $(E/c^2)g\,d$. How can we connect the energy E to the frequency ν? In the quantum theory, the connection is $E = h\nu$, where h is a constant called the Planck constant. For the moment, let us use this relation so that the energy of the photon on absorption at D is its initial emission energy plus the energy gained in falling from A to D, or $h\nu + (h\nu/c^2)g\,d$. If we call this absorption energy $E' = h\nu'$, then we have

$$h\nu' = h\nu + \frac{h\nu g\,d}{c^2},$$

or, for the fractional frequency shift,

$$\frac{\Delta \nu}{\nu} = \frac{\nu' - \nu}{\nu} = g\frac{d}{c^2},$$ (C-2)

the same result obtained in frame S' (Eq. C-1).

Actually, it is not necessary to use quantum theory in deriving this result, a fact that we may suspect to be true because the Planck constant h, which is the characteristic constant of quantum theory, cancels out and does not appear in Eq. C-2. We can show in relativity itself that E is proportional to ν, because it follows, from the relativistic transformation of energy and momentum, that the energy in an electromagnetic pulse changes by the same factor as its frequency when observed in a different reference frame. The conclusion then is that, in falling through a gravitational field, light gains energy and frequency (its wavelength decreases and we say that it is shifted toward the blue). Clearly, had we reversed emitter and detector, we would have concluded that in rising against a gravitational field light loses energy and frequency (its wavelength increases and we say that it is shifted toward the red).

Even with d in Eq. C-2 being the distance from sea level to the top of the highest mountain, the predicted value of $\Delta \nu / \nu$, the fractional frequency shift, is only about 10^{-12}. Nevertheless, Pound and Rebka [1] in 1960 were able to confirm the prediction, using the 74-ft-high Jefferson tower at Harvard! For this small distance we have

$$\frac{\Delta \nu}{\nu} = \frac{g\,d}{c^2} = \frac{(9.8 \text{ m/s}^2)(22.5 \text{ m})}{(3.0 \times 10^8 \text{ m/s})^2} = 2.5 \times 10^{-15},$$

an incredibly small effect. By using the Mössbauer effect (which permits a highly sensitive measurement of frequency shifts) with a gamma-ray source, and taking admirable care to control the competing variables, Pound and Rebka observed this gravitational effect on gamma-ray photons and confirmed the quantitative prediction with a precision of about 10 percent. In a subsequent refinement of the original experiment, Pound and Snider [2] in 1965 found, comparing experimental observation with theoretical prediction, that

$$\frac{(\Delta \nu / \nu)_{\text{exp}}}{(\Delta \nu / \nu)_{\text{theory}}} = 0.9990 \pm 0.0076.$$

The most precise verification of the gravitational frequency shift, however, was reported in 1980 by Vessot and his co-workers [3]. They fired a space probe vertically upward to a height of 10,000 km, the probe containing a microwave transmitter whose frequency was accurately controlled by a hydrogen maser. By comparing the frequency received at the ground station with that of a similar ground-based maser, these workers were able to measure, among other quantities, the gravitational frequency shift of the space-borne "clock." They concluded that the predictions of relativity theory of the effect of changes in gravitational potential on this clock during this experiment can be relied on to a precision of one part in 14,000.

We can easily generalize our result (Eqs. C-1 and C-2) to photons emitted from the surface of stars and observed on earth. Here we assume that the gravitational field need not be uniform and that the result depends only on the difference in gravitational potential between the source and the observer. Then, in place of $g\,d$ we have GM_s/R_s, where M_s is the mass of the star of radius R_s, and because the photon *loses* energy in rising through the gravitational field of the star, we obtain

$$\frac{\Delta\nu}{\nu} = -\frac{GM_s}{R_s c^2}. \tag{C-3}$$

This effect is known as the *gravitational red shift*, for light in the visible part of the spectrum will be shifted in frequency toward the red end. This effect is distinct from the Doppler red shift from receding stars. Indeed, because the Doppler shift is much larger, the gravitational red shift is difficult to verify, in part because the masses and radii of stars other than the sun are not well known. The effect has been confirmed with certainty, however, for the white dwarf star Sirius B and verified to a precision of about 5 percent for light from the sun.

C-3 GENERAL RELATIVITY THEORY

Features and Foundations A full treatment of the general theory of relativity is beyond the scope of a book at this level [4]. We therefore limit ourselves to a qualitative description of some of its important characteristics. Let us return, then, to Fig. C-1a, in which detector D measures a greater frequency than source A emits. It may appear strange that a frequency can increase with no relative motion of source and detector. After all, D surely receives the same *number* of vibrations that A sent out in its light pulse (photon). Even the distance of separation remains constant between A and D, so how do we interpret the measured frequency increase? The answer, once again, is that there is a disagreement about time. That is, frequency is the number of vibrations per unit time, so the frequency difference must be due to a time difference; the rate of A's clock must differ from the rate of D's clock.

We conclude that clocks in a region of higher gravitational potential (that is, higher above the earth's surface) run faster than those in a region of lower gravitational potential. Let the emitting atom A be the clock, for example, its rate being the frequency of its emitted radiation. Then the higher up in the gravitational field the atom is, the higher its frequency appears to be to D (that is, compared to the same atom radiating at D). Similarly, if we reverse A and D, then the lower down in the gravitational field the atom is, the lower its frequency appears to be to D (that is, compared to the same atom radiating at D).

We can use these considerations to throw some light on the twin paradox of Supplementary Topic B. Consider the experiment (see pages 75 and 287) in which a "clock" is placed on the rim of a rotor, a radial distance R from the axis. The central fact is that this clock (the traveling twin) runs slow compared to a clock located on the rotor axis (the stay-at-home twin). In our earlier discussion we took the point of view of an (inertial) observer on the rotor axis and explained the phenomenon as the familiar time dilation of special relativity, associated with the speed v of the traveling clock. We can equally well take the point of view of a (noninertial) observer on the rotor rim. This observer, using an argument from general relativity, would explain the same fact by asserting that the clock is in an effective gravitational field of magnitude v^2/R that acts on the clock and slows it. Both approaches yield the same numerical result for the rate difference between the axis clock and the rim clock, as indeed the principle of relativity requires.

All of the above conclusions about the effect of a gravitational field on the rate of a clock follow from the principle of equivalence, first enunciated by Einstein. It was also Einstein who called attention to the gravitational red shift required from theory and to the need to ascribe a gravitational mass m ($= E/c^2$) to an energy E. Still another of his results was that the direction of the velocity of light is not constant in a gravitational field; indeed, light rays *bend* in such a field because of their gravitational mass, and Einstein predicted that this bending would cause a displacement in the apparent positions of fixed stars that are seen near the edge of the sun.

When we discussed the special theory of relativity in Chapters 1, 2, and 3, we deliberately excluded gravitational fields from consideration, without even mentioning that we were doing so; thus the possibility that such fields might affect clock rates never arose. Their presence would certainly have complicated, for example, our discussion of simultaneity (see Section 2-1) on which our derivation of the Lorentz transformation equations (Section 2-2) was based. Thus it becomes clear that a more general theory* is needed, which takes into account the principle of equivalence and which generalizes even that principle to include nonuniform gravitational fields as well. Furthermore, gravitational effects themselves must be treated by a theory in which the propagation speed is finite. In a series of papers [5], Einstein formulated just such a general theory of relativity.

We have seen that the special theory of relativity asserts that physics is the same for all *inertial* observers, assuming only that gravitational forces are excluded. The general theory goes further and asserts that physics is the same for *all* observers, whether or not they are in inertial (unaccelerated) reference frames and whether or not gravitational forces are present. Thus the special theory appears as a limiting case of the general theory.

Einstein was lead to the general theory by the realization of the deep significance of the principle of equivalence. By use of this principle it is possible to "transform away"—as the technical phrase goes—inhomogeneous gravitational fields by having at *each* point a *different* accelerated reference frame that replaces the local (infinitesimal) gravitational field there. In such local frames, the special theory of relativity *is* valid, so the invariance of the laws of physics under a Lorentz transformation applies to infinitesimal regions. Second, through an invariant spacetime metric that follows from this, we can link geometry to gravitation and geometry becomes non-Euclidean. That is, the presence of a large body of matter causes spacetime to warp in the region near it so that spacetime becomes non-Euclidean. This warping is equivalent to the gravitational field. The curvature of spacetime in general relativity replaces the gravitational field of classical theory. Hence, the geometry of spacetime is determined by the presence of matter. In this sense, geometry becomes a branch of physics. The fact that special relativity is valid in small regions corresponds to the fact that Euclidean geometry is valid over small parts of a curved surface. In large regions, special relativity and Euclidean geometry need not apply, so the world lines of light rays and of inertial motion need not be straight; instead, they are geodetic, that is, as straight as possible.

The relationship between matter and geometry has been aptly summarized by the statement: "Geometry tells matter how to move, and matter tells geometry how to curve." General relativity has, in fact, been described as *geometrodynamics*, to reflect this interaction [6]. Figure C-2 suggests the interaction between matter and geometry by what has been called the rubber sheet analogy. In Fig. C-2*a* a stretched rubber sheet is shown, marked with coordinate lines. No matter is present, the geometry is Euclidean, special relativity holds throughout, and particles not acted on by forces move in straight lines as suggested by the

---

* Just as Galilean relativity is a limiting case of special relativity, so special relativity itself is a limiting case of general relativity. The gravitational field near the earth is so weak that there are no readily measurable discrepancies between special and general relativity. We usually operate in the special relativistic limit of general relativity.

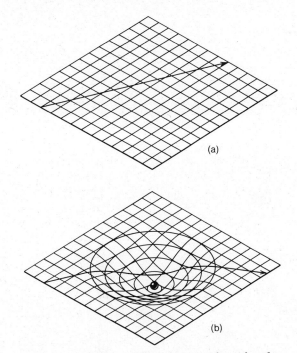

Figure C-2. The rubber sheet analogy for the interaction between matter and geometry.

trajectory shown. In Fig. C-2*b*, however, a heavy ball bearing has been placed on the sheet, deforming it and its coordinate lines. Now matter *is* present and is affecting the geometry, which becomes non-Euclidean; general relativity must be used, and the trajectory shown suggests the motion of a particle deflected by a centrally directed gravitational force.

Experimental Tests The gravitational red shift, discussed in the previous section, is a test of the equivalence principle, as formulated by Einstein, rather than a specific test of the theory itself. A second critical test of this principle is the verification of the observation that, in free fall, all bodies fall with the same acceleration, regardless of their mass or composition. This is equivalent to the assertion that the gravitational mass of a body is equal to its inertial mass. Measurement has confirmed that these quantities are indeed equal, to within one part in 10^{12}. Thus the foundations of the general theory of relativity seem to be firmly rooted in experiment. Note that Newton's theory of gravity has no explanation at all for the gravitational red shift and can only view the equality of gravitational and inertial mass as a coincidence of astonishing proportions.

As for specific predictions of the theory itself, there are three that have been subject to careful experimental scrutiny. (1) The precession of the perhelion of the planet Mercury (Fig. C-3*a*) should differ from the classical prediction, the difference being about 43 seconds of arc per century. (2) The positions of stars whose light passes near the edge of the sun (observed, say, during a total eclipse) should be displaced (due to deflection in the sun's gravitational field) by 1.75 seconds of arc from their positions observed at night (Fig. C-3*b*). (3) There should be a time delay in radio or radar signals (sent, say, from earth to one of the other planets and back again) if the light path passes near the sun (Fig. C-3*c*). We consider each prediction in turn.

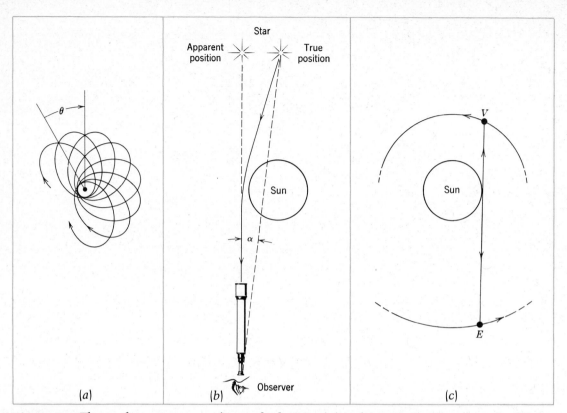

Figure C-3. *Three solar system tests of general relativity.* (*a*) A planet moving in a precessing ellipti-cal orbit about the sun. The perhelion, or point of closest approach, shifts by an angle θ after each excursion. (*b*) Light passing the sun from a star to earth is deflected, making the star appear displaced away from the sun by an angle (= 1.75 arc seconds). (*c*) A radar pulse sent from earth to Venus and back is delayed slightly if its path passes near the sun. All drawings are schematic and clearly not to scale.

1. The observed rate of precession of the perhelion of Mercury about the sun is 5599.74 ± 0.41 seconds of arc per century. Of this, all but 43.11 ± 0.45 seconds of arc per century can be accounted for by the fact that the earth is not an inertial reference frame and by the Newtonian gravitational effects of other planets. The prediction of the general theory of relativity is 43.03 seconds of arc per century, so the ratio of the observed to the predicted value is 1.00 ± 0.01, a striking confir-mation of the theory. Perhelion shift rates have also been measured for other planets, and for the asteroid Icarus; the agreement is again within the error of the measurements, but the precision of those measurements is not nearly as good as it is for the planet Mercury.

2. The bending of star light as the rays pass near the rim of the sun has been measured many times during solar eclipses. The angular deflection predicted by theory is 1.75 arc seconds. The observed deflections, when the possibility of systematic errors is taken into account, are judged to agree with this to a preci-sion within the range of 10 to 20 percent. More recently, radio emissions from distant quasars, passing near the sun's rim on their path to earth, have been observed to be deflected by amounts that agree with the prediction of the general theory to within 1 percent or so. Fomalont and Sramek, for example, measured a solar-rim deflection of 1.775 ± 0.019 arc seconds by this method [7].

3. The delay in the echo times of radio signals reflected from Venus, when that planet is on the far side of the sun from us, have been measured by Shapiro and his co-workers [8] and found to agree with the predictions of the general theory to within about 2 percent.

See Ref. 6 (part IX; see Summary on p. 1129) for a detailed analysis of the experimental tests to which Einstein's general theory has been subjected as of 1972. These authors, commenting on the success of Einstein's theory with respect to competing theories that have been advanced from time to time, conclude: "As experiment after experiment has been performed, and one theory of gravity after another has fallen by the wayside a victim of the observations, Einstein's theory has stood firm. No purported inconsistency between Einstein's laws of gravity has ever surmounted the test of time."

Predictions and Consequences There are several other areas now under active investigation that relate to Einstein's general theory of relativity:

1. *Gravitational lenses.* In 1979 two "twin quasars" were discovered, after intensive investigation at both optical and radio wavelengths and after much theoretical analysis, not to be two distinct objects at all but rather to be images of a single distant quasar, formed by the bending of light rays by the gravitational action of an intervening galaxy [9]. Since that time several other gravitationally lensed quasar images have been found. See Problem 15.

2. *Black holes.* The term originates from the possibility that a body of a given size might be massive enough for its strong gravitational field to prevent the escape of light from it. In general relativity, theory predicts that, under certain conditions, a sufficiently massive body can indeed undergo gravitational collapse to form a black hole. Such an object can be detected only by secondary effects associated with its strong gravitational field. Thus, if a black hole forms a binary system with an ordinary star, it is possible for matter to be transferred from the star to the black hole. As the matter spirals in toward the black hole, the particles of matter collide with each other and may generate sufficient heat that they emit x-rays. On this basis, many astrophysicists believe that one component of the x-ray binary identified as Cygnus X-1 is a black hole. Others believe that there is strong evidence for the existence of a black hole at the center of our own galaxy.

3. *Gravity waves.* Just as an accelerated charge emits electromagnetic radiation traveling at the speed of light, so implicit in the general theory of relativity is the prediction that an accelerated mass can emit gravitational waves, also traveling at the speed of light. Much effort is now being expended to detect such radiation unambiguously. The best evidence for the existence of such waves, however, is indirect but nevertheless convincing. The binary pulsar identified as PSR 1913 + 16 is thought to emit gravitational waves, because of the acceleration associated with the orbital motion of the pulsar and its companion star about their center of mass [10]. The energy presumably carried away by gravitational radiation results in a slow decrease in the period of this orbital motion, amounting to 67 ns per orbit. Fortunately, this slow decrease has measurable effects on the arrival times of the pulses from the pulsar. In particular, the times of periastron (that is, the times when the components of the binary are closest together in their orbital motions) can be measured with some precision. The general theory of relativity predicts that loss of energy by gravitational radiation will cause the periastron times to be cumulatively displaced—as time goes on—when compared with the periastron times expected for a hypothetical system that does not radiate. Figure C-4 shows the excellent agreement between theory (solid curve) and observation.

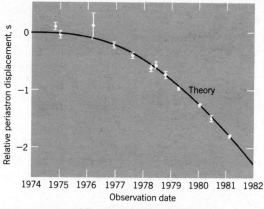

Figure C-4. The points show the deviations in time of periastron passage for the binary pulsar PSR 1913 + 16 compared to a hypothetical binary that does not emit gravity radiation. The solid curve is the prediction of the general theory of relativity [10].

4. *Cosmological predictions.* The general theory of relativity is one of the greatest intellectual achievements of all time. Its originality and unorthodox approach exceed that of special relativity. And far more than special relativity, it was almost completely the work of a single man, Albert Einstein. The philosophic impact of relativity theory on the thinking of man has been profound and the vistas of science opened by it are literally endless. To quote Max Born, writing in 1962: "The idea first expressed by Ernst Mach, that the inertial forces are due to the total system of the fixed stars, suggests the application of the theory of general relativity to the whole universe. This step was actually made by Einstein in 1917, and from that time dates the modern development of cosmology and cosmogony, the sciences of the structure and genesis of the cosmos. This development is still in full swing and rich in important results, though far from final conclusions." In 1972 Wheeler wrote: "Einstein's description of gravitation as curvature of spacetime led directly to that greatest of all predictions of his theory, that the universe itself is dynamic. Physics still has far to go to come to terms with this amazing fact and what it means for man and his relation with the universe."

questions and problems

1. Free fall, the same for all. Starting from the fact that all bodies that are free of forces move with uniform velocity relative to an inertial reference frame, then considering the motion of those bodies in an accelerated frame, and finally using the principle of equivalence, show that all bodies fall with the same acceleration in a uniform gravitational field. Hence the equality of gravitational and inertial mass.

2. Transforming away gravity. Can gravity be regarded as a "fictitious" force, arising from the acceleration of one's reference frame relative to an inertial reference frame, rather than a "real" force? Discuss an analogy with the forces encountered on a rotating (that is, accelerating) reference frame such as a merry-go-round.

3. The equivalence principle and the bending of light. (a) Consider the path of a light ray to be straight in an unaccelerated reference frame where there is no gravitational field. What will the path look like if the frame accelerates upward with an acceleration **a**? (b) Use the principle of equivalence to show that we should expect the path to look just the same if a homogeneous gravitational field $\mathbf{g}$ ($= -\mathbf{a}$) is present.

4. The gravitational mass of light. A photon may be viewed as a particle of light, its energy being given by $h\nu$, where ν is the frequency and h ($= 6.63 \times 10^{-34}$ J·s) is the Planck constant. In terms of m_e, the electron rest mass, what is the gravitational mass of a 600-nm visible-light photon?

5. Rest mass or total mass? The mass of a moving particle can be considered to be either its rest mass or its total relativistic mass. Which mass is appropriate for gravitation, that is, for the concept of mass as causing gravitation or being responsive to gravitation? Consider the special case of light, which has zero rest mass.

6. Clearing up a point. Why is the relation $t = d/c$, used for the light pulse's (that is, the photon's) time of flight in frame S' of Fig. C-1b, only approximate, rather than exact?

7. The clock on the top floor is running fast. An atomic clock is placed in the basement of the World Trade Center in New York, a second similar clock being placed on the 110th floor, 1,350 ft higher. How long will the clocks have to run before (because of the difference in gravitational potential between their locations) the upper clock gains 1.0 μs by comparison with the lower clock? Before the upper clock gains 1.0 ns?

8. Comparing clocks. Occasionally, an atomic clock is flown from Washington, D.C. to Boulder, Colorado, a distance of 2,400 km, for intercomparison with standard clocks kept at the Boulder Laboratory of the National Bureau of Standards. (a) How much time has been lost by the transported clock upon its arrival in Boulder, by virtue of its trip from Washington at an assumed average speed of 650 mi/h? Note that this effect can be accounted for by special relativity alone; see Supplementary Topic B. (b) If the transported clock stays in Boulder for ten days before being returned to Washington, what correction must be subtracted from its reading, by virtue of the fact that Boulder is 1,600 m higher than Washington? Note that this effect is a general-relativistic effect.

9. The gravitational red shift for light from a white dwarf star. (a) Show that, in terms of wavelength, the gravitational red shift (see Eq. C-3) can be written as

$$\lambda = \lambda_0 \left(1 + \frac{GM_s}{R_s c^2} \right).$$

(b) Light of wavelength 600.0 nm is emitted by a white dwarf star. Due to the gravitational red shift, what is the wavelength received on earth? (c) What relative speed between the white dwarf and the earth would produce this same wavelength shift? Assume that the white dwarf has a mass ($= 2.0 \times 10^{30}$ kg) equal to that of the sun but a

radius ($= 8.0 \times 10^6$ m) only about 1.1 percent as great as the solar radius.

10. Gravitational red shift for sunlight. The solar spectrum contains the sodium D line, which is a doublet, one of its components having a wavelength of 588.997 nm. Calculate the wavelength shift expected for this component because of (a) the gravitational red shift and (b) the Doppler effect associated with the sun's rotation, assuming in this latter case that the light originates at the rim of the sun at its equator. The sun's mass and radius are 2.0×10^{30} kg and 7.0×10^8 m, respectively. The sun's period of rotation, measured at the equator, is 26 d. (*Hint:* Use the result of Problem 9a.)

11. A gravitational blue shift. A satellite in orbit at an altitude of 200 km sends a radio signal of frequency 1020 MHz to the ground station directly below on earth. What is the frequency shift of the signal, as received by the ground station, associated with the earth's gravitational field?

12. Deflection of light à la Newton. It can be shown by classical mechanics that the deflection φ of a particle of initial speed v_0 whose extended initial trajectory passes within a distance b (the so-called impact parameter) of a body of mass M is given by (see Fig. C-5)

$$\tan \frac{\varphi}{2} = \frac{GM}{bv_0^2}.$$

In deriving this expression it has been assumed that $m << M$, where m is the mass of the deflected particle; note that m does not appear in the above expression. Let us now apply this purely Newtonian result to a "particle" of light whose extended trajectory grazes the rim of the sun. Putting $v_0 = c$, $b = R_s$, and $M = M_s$ in the above expression, and assuming further that, because φ is small, $\tan \varphi/2 \cong \varphi/2$, leads to

$$\varphi = \frac{2GM_s}{R_s c^2}.$$

Evaluate this quantity and show that the classically predicted value is just half of the value predicted by general relativity ($= 1.75$ seconds) for this quantity.

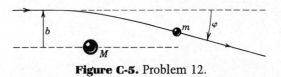

Figure C-5. Problem 12.

13. How big is a black hole? If an object of mass M has a radius less than a certain critical value, it will undergo gravitational collapse and become a black hole. This critical radius—known as the *Schwarzschild radius*—is given by

$$R_s = \frac{2GM}{c^2}.$$

(a) "Derive" this result from classical mechanics in this way: Assume (incorrectly) that the kinetic energy of light is $\frac{1}{2}mc^2$ and (also incorrectly) use Newton's law of gravitation to find the radius of a mass M from which the escape velocity is c. (These two errors happen to cancel, giving a correct result!) (b) Calculate the Schwarzschild radius for an object having a mass equal to that of the earth (= 6×10^{24} kg), of the sun (= 2×10^{30} kg), and of our Milky Way galaxy ($\cong 3 \times 10^{41}$ kg).

14. The density of black holes. (a) Show that the average density ρ of a black hole, that is, of the material inside the Schwarzschild radius (see Problem 13), is given by

$$\rho = \frac{3c^6}{32\pi G^3 M^2}.$$

(b) Evaluate this density numerically for the three objects described in Problem 13b. (c) Note that, if M is large enough, the average density of a black hole can be quite low. For what mass M would the average density be equal to that of water?

★15. A gravitational lens. Mass M in Fig. C-6 (a galaxy) is along the line of sight from earth E to a distant quasar Q. It deflects the light from the quasar so that an earth observer sees two images A and B of the quasar, each displaced from the quasar's actual position. The angle α by which each light ray is bent is, from the general theory

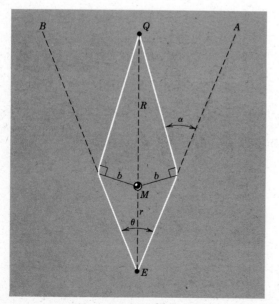

Figure C-6. Problem 15.

of relativity,

$$\alpha = \frac{4GM}{bc^2},$$

where b is the so-called impact parameter of the incident light ray. The twin quasars identified as 0957 + 561, at a distance R from earth of about 6×10^{24} m, appear in the sky in virtually the same direction as a galaxy whose distance r from earth is about 2×10^{24} m. Assume that the mass of this galaxy is about the same as that of the Milky Way galaxy (= 3×10^{41} kg). Assuming that the quasars are two images of a single quasar, calculate the expected angular separation θ between the images. (*Note:* The angles α and θ are very small; the calculated answer agrees well with observation.)

★16. Precession of the perhelion. According to the general theory of relativity, the angular advance per revolution of a planet's orbit is given by

$$\theta = \frac{A}{r},$$

where A is a constant, the same for all planets, and r is the mean radius of the planet's orbit. P_M, the rate of precession of the perhelion of the orbit of Mercury due to general relativistic effects, is 43.11 arc seconds/century. What is P_E, the corresponding rate of precession of the perhelion of the earth's orbit? The ratio of the mean radius of Mercury's orbit to that of the earth is 0.386. (*Hint:* Use Kepler's third law.)

17. Gravitational waves versus electromagnetic waves. What similarities might you expect in the production, detection, and nature of gravitational and electromagnetic waves?

18. Life on the rim of a centrifuge. A clock is located in an inertial frame on the axis of a centrifuge. A similar clock is located on the centrifuge rim, a radial distance R from the axis. Show that the fractional difference in the rates of the clocks is given, in the case of $v \ll c$, by $v^2/2c^2$, in which v is the speed of the rim clock. Derive this result from the point of view of (a) an axis observer, using special relativity, and also (b) a rim observer, using the equivalence principle. [*Hint:* Replace the centripetal acceleration by an equivalent gravitational field. The fractional difference in clock rates is given by $(V_r - V_a)/c^2$, where the quantity in parentheses is the difference in gravitational potential between the two clock sites.]

19. The equivalence principle in Fantasy Land. Figure C-7 suggests the design of a spaceship that uses the principle of equivalence to permit high accelerations while subjecting passengers in the life capsule to a standard acceleration of 1 "gee". M is a sphere whose radius is 12.0 m,

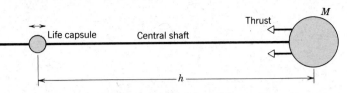

Figure C-7. Problem 19.

constructed by "compressed matter, electromagnetically stabilized" to a density of 1.20×10^{12} kg/m³. The position of the life capsule along the central shaft can be adjusted in flight. (a) If the ship is coasting in free space at constant velocity, what must be the separation h between the center of the life capsule and the center of sphere M if the passengers are to experience a 1.0-gee environment? (b) At an acceleration of 35 gee for the spaceship as a whole, what must this separation be to maintain the 1.0-gee environment? What is the so-called tidal force (measured in gees per meter, along the central shaft) in the life capsule under these conditions? (Adapted from Charles Sheffield, "Moment of Inertia," *Analog Science Fiction/Science Fact*, October 1980. In the story the life capsule gets stuck in position on the central shaft while the spaceship is in high-gee acceleration with the consequence that, if the acceleration is reduced, the passengers will be flattened by the gravitational forces due to sphere M. Read the story to see how they escape, again using the equivalence principle.)

references

1. R. V. Pound and G. A. Rebka, "Apparent Weight of Photons," *Phys. Rev. Lett.*, **4,** 337 (1960).

2. R. V. Pound and J. L. Snider, "Effect of Gravity on Gamma Radiation," *Phys. Rev. B*, **140,** 788 (1965).

3. R. F. C. Vessot, M. W. Levine, E. M. Mattison, E. L. Blomberg, T. E. Hoffman, G. U. Nystrom, and B. F. Farrel; R. Decher, P. B. Eby, C. R. Baugher, J. W. Watts, D. L. Teuber, and F. D. Wills, "Test of Relativistic Gravitation with a Space-Borne Hydrogen Maser," *Phys. Rev. Lett.*, **45,** 2081 (1980).

4. Max Born, *Einstein's Theory of Relativity* (Dover, New York, 1965), chap. VII; Albert Einstein, *Relativity, The Special and the General Theory* (Crown, New York, 1952).

5. A. Einstein, H. Minkowski, H. A. Lorentz, and H. Weyl, *The Principle of Relativity: A Collection of Original Memoirs on the Special and General Theory of Relativity*, notes by A. Sommerfeld, Dover, N.Y. 1952.

6. Charles W. Misner, Kip S. Thorne, and John Archibald Wheeler, *Gravitation* (W. H. Freeman, San Francisco, 1973), chap. 1.

7. E. B. Fomalont and R. A. Sramek, "A Confirmation of Einstein's General Theory of Relativity by Measuring the Bending of Microwave Radiation in the Gravitational Field of the Sun," *Astrophys. J.*, **199,** 749 (1975).

8. Irwin I. Shapiro, Michael E. Ash, Richard P. Ingalls, and William B. Smith; Donald B. Campbell, Rolf B. Dyce, Raymond F. Jurgens, and Gordon H. Pettingill, "Fourth Test of General Relativity: New Radar Result," *Phys. Rev. Lett.*, **26,** 1132 (1971).

9. Frederic H. Chaffee, Jr., "The Discovery of a Gravitational Lens," *Sci. Am.* (November 1980).

10. Joel M. Weisberg, Joseph H. Taylor, and Lee A. Fowler, "Gravitational Waves from an Orbiting Pulsar," *Sci. Am.* (October 1981).

SUPPLEMENTARY TOPIC D

the radiancy and the energy density

Be sure of it; give me the ocular proof.

William Shakespeare

Our purpose here is to derive Eqs. 4-4a and 4-4b, which express the relationship between the spectral radiancy $\mathcal{R}(\lambda)$ and the spectral energy density $\rho(\nu)$ for radiation from a cavity whose walls are held at a temperature T. Figure D-1a shows a small area σ on the inner wall of such a cavity; because the nature of the cavity radiation does not depend on the material of which the walls are made, we may safely assume, for simplicity, that the walls are perfectly black. As we watch area σ for a time Δt, an amount of energy (per unit wavelength interval) given by $\mathcal{R}\sigma \Delta t$ is emitted into the forward hemisphere. Of this, however, only a differential amount $d\mathcal{R}\,\sigma\,\Delta t$ travels outward parallel to the axis of the cylinder shown in Fig. D-1a, the rest going in other directions. The cylinder axis makes an angle θ with the normal to the wall.

We can describe this outward flow of energy in another, totally equivalent way. Let the energy density (again, per unit wavelength interval) immediately in front of the area σ be $\rho(\lambda)$. Consider the energy contained in the cylindrical volume whose cross-sectional area is $\sigma \cos \theta$ and whose length is $\Delta x (= c\,\Delta t)$. Of this energy we discard half, which corresponds to radiation moving *inward* rather than outward. Of the remaining half, only a fraction $d\Omega/2\pi$ moves outward parallel to the cylinder axis, the rest moving outward in other directions. In this expression $d\Omega$ is the solid angle corresponding to the cylinder and 2π is the solid angle corresponding to the entire forward hemisphere.

The two expressions we have derived for the energy traveling outward parallel to the cylinder axis of Fig. D-1a must be equal. Thus

$$d\mathcal{R}\,\sigma\,\Delta t = \left(\frac{\rho}{2}\right)\left(\frac{d\Omega}{2\pi}\right)(\sigma \cos \theta)(c\,\Delta t)$$

or

$$d\mathcal{R} = \left(\frac{c}{4\pi}\right)\rho \cos \theta \, d\Omega. \tag{D-1}$$

It remains to evaluate the solid angle element $d\Omega$ corresponding to the cylinder and then to integrate the above expression to find the radiancy $\mathcal{R}$.

Figure D-1b shows the spot of area σ from a larger perspective. We have constructed a hypothetical hemisphere of arbitrary radius r centered on this spot, r being chosen large compared with any linear dimension of σ. On this scale the cylinder at angle θ is now best viewed as the solid angle between two cones, whose angles with the normal are θ and $\theta + d\theta$. The quantity $d\Omega$ is, then, by definition of solid angle, dA/r^2, where $dA\ [=(2\pi r \sin \theta)(r\,d\theta)]$ is the area of the shaded strip shown on the hemisphere of Fig. D-1b. Thus $d\Omega = 2\pi \sin d\theta$. Substi-

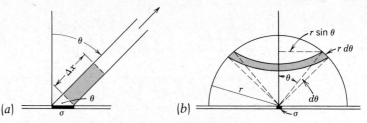

Figure D-1. (a) Radiation is emitted from a small area σ on the inner wall of a cavity held at a temperature T. (b) The spot of area σ viewed from a larger perspective.

tuting this value into Eq. D-1 yields

$$d\mathcal{R} = \left(\frac{c}{4\pi}\right) (\rho \cos \theta)(2\pi \sin \theta \, d\theta)$$

$$= \left(\frac{c}{2}\right) \rho \sin \theta \cos \theta \, d\theta.$$

Integrating the above expression yields

$$\mathcal{R} = \int d\mathcal{R} = \left(\frac{c}{2}\right) \rho \int_0^{90°} \sin \theta \cos \theta \, d\theta$$

or, finally,

$$\mathcal{R}(\lambda) = \left(\frac{c}{4}\right) \rho(\lambda). \tag{D-2}$$

We now wish to express the spectral energy density $\rho(\lambda)$ in terms of the frequency rather than the wavelength, that being the form used in deriving theoretical formulations for this quantity. We can do so by considering that to every wavelength interval $d\lambda$ there is a corresponding frequency interval $d\nu$, the energy contained in this interval being expressed either by $\rho(\lambda) \, d\lambda$ or by $\rho(\nu) \, d\nu$. Thus

$$\rho(\lambda) \, d\lambda = -\rho(\nu) \, d\nu,$$

in which the minus sign expressed the fact that $d\lambda$ and $d\nu$ are intrinsically opposite in sign. Bearing in mind that $\nu = c/\lambda$, we can then write

$$\rho(\lambda) = -\rho(\nu) \left(\frac{d\nu}{d\lambda}\right) = \left(\frac{c}{\lambda^2}\right) \rho(\nu).$$

Substituting this expression into Eq. D-2 yields

$$\mathcal{R}(\lambda) = \left(\frac{c}{2\lambda}\right)^2 \rho(\nu), \tag{4-4b}$$

which is the relationship we seek. The reciprocal relation (Eq. 4-4a) readily follows.

SUPPLEMENTARY TOPIC E

Einstein's derivation of Planck's radiation law

A splendid light has dawned on me about the absorption and emission of radiation.

Albert Einstein (1916)
(Letter to Michele Besso)

Planck's derivation of the radiation law that bears his name (Eq. 4-8a) was based on the assumption that the walls of a cavity resonator were populated with linear, electronic, harmonic oscillators, covering a wide range of frequencies. We now know that ordinary matter is not made up of such oscillators. In 1900, however, when Planck advanced his theory, the electron had been known for but three years, the atomic nucleus was not to reveal itself for another decade and, indeed, the very existence of atoms was much in dispute.

It is well known that the properties of cavity radiation depend only on the temperature of the cavity walls and are independent of the shape and size of the cavity and of the materials of which it is made. This suggests that it should be possible to derive Planck's radiation law without making detailed assumptions about the nature of the entities that make up the cavity walls. In 1917 Einstein, in a fundamental study of the nature of the thermal equilibrium between matter and radiation, did just that [1]. Along the way he proposed a new form of interaction between matter and radiation, the *stimulated emission* of radiation by atoms immersed in a radiation field. This concept was eventually to lead to the development of the laser [2].

Rather than consider the interaction between the radiation and the cavity walls, let us simplify matters and introduce a rarefied gas into the cavity. We then examine the thermal equilibrium that exists between the cavity radiation and the molecules of this gas. The nature of these "atoms" does not matter, and we assume about them *only* that they can exist in quantized states, each with a characteristic energy. We focus on two of these states only, with energies E_m and E_n, in which $E_m > E_n$. Boltzmann showed that the number of atoms in each of these states is given by

$$n_m = pe^{-E_m/kT} \qquad \text{and} \qquad n_n = pe^{-E_n/kT}, \tag{E-1}$$

in which p is a constant, assumed to be the same for each state. At thermal equilibrium individual atoms will be transferring from one state to another, but the average populations of the two states will remain constant. We consider three mechanisms by which such transfers can occur.

1. *Spontaneous emission.* An atom in the upper state might move spontaneously to the lower state, emitting an energy $E_m - E_n$ in the form of radiation. The rate at which transfers of this kind occur is proportional to the population of the

307

upper state and is given by

$$\left(\frac{dn_m}{dt}\right)_1 = An_m = Ape^{-E_m/kT},$$ (E-2)

in which A is a constant coefficient. The radiation field in the cavity plays no role in this process.

2. *Stimulated emission.* It might also happen (Einstein proposed) that an atom in the upper state might be *stimulated* to emit because of its interaction with the radiation field. In this case the transfer rate is proportional to both the population of the state and to the spectral energy density ρ, or

$$\left(\frac{dn_m}{dt}\right)_2 = Bn_m\rho = Bp\rho e^{-E_m/kT},$$ (E-3)

B being a second coefficient.

3. *Absorption.* An atom in the lower state might absorb energy from the radiation field and move to the upper state. The transfer rate is proportional to both the population of the lower state and to the energy density, or

$$\frac{dn_n}{dt} = B'n_n\rho = B'p\rho e^{-E_n/kT},$$ (E-4)

with B' as a third coefficient. Note that an atom in the lower state cannot move *spontaneously* to the upper state without violating the principle of energy conservation.

At equilibrium the rate at which atoms transfer from the lower to the upper state must equal the rate at which they move in the opposite direction, or

$$e^{-E_m/kT}(A + B\rho) = B'\rho e^{-E_n/kT}.$$

Solving for the energy density yields

$$\rho = \frac{A}{B'e^{(E_m-E_n)/kT} - B}.$$

This relation already begins to look like Planck's radiation law; all we have to do is assign values, on some logical basis, to the constants A, B, and B'.

Let us first consider the situation as $T \to \infty$. The energy density must also approach infinitely large values; inspection of the above equation shows that it will do so only if $B' = B$. Making the substitution gives us then

$$\rho = \frac{(A/B)}{e^{(E_m-E_n)/kT} - 1},$$ (E-5)

in which A/B is the single remaining constant.

Wien (see Problem 23, Chapter 4) was able to show from purely classical thermodynamics that the radiation law must be of the form

$$\rho(\nu) = \nu^3 f\left(\frac{\nu}{T}\right),$$ (E-6)

in which $f(\nu/T)$ is an unknown function. Comparing Eqs. E-5 and E-6 allows us then to put

$$E_m - E_n = h\nu$$ (E-7)

and

$$\frac{A}{B} = a\nu^3,$$

in which Eq. E-7 is the Bohr frequency condition (see Eq. 7-5)—here advanced independently—and a is an adjustable constant. Equation E-5 then becomes

$$\rho(\nu) = a\nu^3 \frac{1}{e^{h\nu/kT} - 1},$$ (E-8)

which is indeed Planck's radiation law. It is possible to go one step further and evaluate the remaining constant a above by requiring that Eq. E-8 reduce to the classical Rayleigh-Jeans law (Eq. 4-6) as $\nu \to 0$. We leave this as an exercise.

It is important to realize the simplicity of the assumptions that underlie this derivation and the fundamental character of the results obtained. Beyond the well-established classical laws it was assumed *only* that the atoms with which the radiation field was in thermal equilibrium have quantized energies. As for results we list, (1) Planck's radiation law was derived, without detailed assumptions; (2) the Bohr frequency condition was put forward independently, again without *ad hoc* assumptions or reliance on a detailed model; and (3) the concept of stimulated emission was proposed. We leave as a second exercise to show that, if stimulated emission is neglected in the above derivation, Wien's radiation law (Eq. 4-5), rather than Planck's, results.

references

1. For a translation of Einstein's 1917 article, along with explanatory commentary, see Henry A. Boorse and Lloyd Motz, *The World of the Atom* (Basic Books, New York, 1966); see article entitled "Quantum Theory of Radiation and Atomic Processes," vol. 2, p. 884. For a detailed scientific and historical commentary, see: Abraham Pais, *Sub-tle is the Lord . . . The Science and Life of Albert Einstein* (Clarendon Press, Oxford, 1982), sec. 21b, p. 405.

2. David Halliday and Robert Resnick, *Fundamentals of Physics—Extended Version*, (Wiley, New York, 1981), sec. 45-4.

SUPPLEMENTARY TOPIC F

Debye's theory of the heat capacity of solids

"Degrees of freedom should be weighted, not counted." Quantum theory showed how this is to be achieved.

Arnold Sommerfeld
(Referring, at a later date, to
a remark he made in 1911.)

F-1 THE THEORY

Classical theory predicts (see Section 4-6) that the molar heat capacity at constant volume of a solid, if we consider only the energy associated with the vibrations of its atoms about their lattice sites, should have the constant value $3R$, independent of temperature and the same for all solids. This is in sharp disagreement with experiment, as Fig. 4-9 shows. Einstein, as we have seen, showed that a theory based on quantizing the energy of the atomic oscillators agreed quite well with experiment. In aiming simply to bring out the main features of the situation, however, Einstein made a bold and over-reaching assumption, namely, that the atoms in the solid oscillate at a single frequency and that the oscillations of a given atom are not influenced by the oscillations of its neighbors. Although Einstein's formula (Eq. 4-19b) does indeed bring out the main features of the variation of c_v with temperature, the agreement at very low temperatures is not exact (see Fig. 4-10).

In 1912, about five years after Einstein had advanced his theory, the Dutch physicist Peter Debye (and, independently, Max Born and Theodore von Kármán) introduced an improved quantum theory in which the interactions between adjacent atoms were taken into account and the restriction to a single frequency of oscillation was relaxed. The resulting theory is in excellent agreement with experiment over the full range of temperature (see Fig. 4-11), thus confirming even more strongly the concept of energy quantization.

Rather than considering the motions of the individual atoms in the solid, Debye focused his attention on the assembly of elastic standing waves that can be thermally excited in the solid as a whole. For any given atom, the combined action of this assembly of standing waves represents the vibrational motion of that atom. Debye assumed that the energies of these standing waves are quantized, an approach in strict analogy to that employed in cavity radiation theory, in which the energy associated with the standing electromagnetic waves in the cavity is quantized.

We saw in Section 4-4 that the number of standing electromagnetic waves per unit volume and per unit frequency interval for cavity radiation is given by

$$n(\nu) = \frac{8\pi\nu^2}{c^3}.$$ [4-9]

Three changes need to be made in this formula if we are to apply it to elastic waves in a solid. (1) We must take into account the fact that electromagnetic waves, being entirely transverse, can exist only in two independent polarizations. If the wave is propagated along the z axis, for example, it can be polarized independently in the x and the y directions. Hence, to get the number of waves per unit frequency interval *associated with a single polarization component*, we should replace the 8 in the equation by 4. Elastic waves, on the other hand, have not only two such transverse components but also a single longitudinal component, making a total of three components in all. To take this properly into account we should thus replace the 8 in the above by 12 (= 3 × 4). (2) We must substitute v, the speed of the elastic waves in the solid, for c, the speed of electromagnetic waves in the cavity. It is true that the transverse and the longitudinal waves in the solids have different speeds, but we commit no error if we take v to be a suitably weighted average of these two speeds. (3) Finally, the above equation gives the distribution of modes on a unit volume basis and our concern is for the distribution on a unit mass basis. We obtain this new interpretation of $n(\nu)$ if we simply multiply the right-hand side of Eq. 4-9 by V_0, the molar volume of the substance.

Making the three changes described above yields

$$n(\nu) = \frac{12\pi V_0 \nu^2}{v^3}. \tag{F-1}$$

To get the molar energy per unit frequency interval we simply multiply the above quantity by $\bar{\mathcal{E}}$ (see Eq. 4-12), the average energy of an oscillator of frequency ν at temperature T, just as we did in deriving Planck's radiation formula. To get the total molar energy u, we integrate the expression so derived over all frequencies from zero up to a certain maximum ν_m, obtaining

$$u(T) = \left(\frac{12\pi V_0 h}{v^3}\right) \int_0^{\nu_m} \frac{\nu^3\, d\nu}{e^{h\nu/kT} - 1}. \tag{F-2}$$

The existence of a maximum frequency comes about because we wish the total number of elastic standing waves in the specimen to equal the total number of independent vibrational modes of the atoms that make up the specimen. On a molar basis this number is simply $3N_A$, in which N_A is the Avogadro constant. This implies that $\int_0^{\nu_m} n(\nu) d\nu = 3N_A$; with Eq. F-1, this leads to

$$u(T) = \left(\frac{9N_A h}{\nu_m^3}\right) \int_0^{\nu_m} \frac{\nu^3\, d\nu}{e^{h\nu/kT} - 1}. \tag{F-3}$$

Differentiating with respect to temperature to find the molar heat capacity (see Eq. 4-15) yields

$$c_v(T) = \left(\frac{9N_A h^2}{kT^2\nu_m^3}\right) \int_0^{\nu_m} \frac{e^{h\nu/kT}\nu^4\, d\nu}{(e^{h\nu/kT} - 1)^2}. \tag{F-4}$$

To put this relation in more manageable form, we introduce a new variable, namely,

$$x = \frac{h\nu}{kT},$$

obtaining

$$c_v(T) = 9R\left(\frac{T^3}{T_D^3}\right) \int_0^{T_D/T} \frac{e^x x^4 dx}{(e^x - 1)^2} \tag{F-5}$$

as a final result. Here T_D is the so-called *Debye temperature* of the solid, defined from $T_D = h\nu_m/k$. Rather than relating the Debye temperature specifically to the maximum frequency through this formal definition, we usually regard it as an arbitrary constant, its value chosen to maximize the agreement between theory and experiment. The Debye temperature thus plays the same role with respect to the Debye theory of heat capacity that the Einstein temperature T_E did with respect to the Einstein theory of heat capacity. These two characteristic temperatures, because they relate to different theories, do not have the same value for a given substance. For aluminum, for example, we have taken the Einstein temperature to be 290 K; its Debye temperature, however, is given as 428 K.

The integral in Eq. F-5 cannot be evaluated in closed form. However, a numerical integration can easily be carried out using a hand-held calculator. In addition, tables of the integral for various values of its upper limit T_D/T are readily available. See, for example, *The American Institute of Physics Handbook*, 3rd ed., Dwight E. Gray, Coordinating Editor (McGraw-Hill, New York, 1972), Section 4*e*, "Heat Capacities."

Figure 4-11 reminds us how excellent is the agreement between experiment and the predictions of the Debye theory of heat capacity. The agreement at low temperatures is particularly gratifying because that is the region in which the Einstein theory of heat capacity falls short.

F-2 THE HIGH-TEMPERATURE LIMIT

As $T \to \infty$, the value of the upper limit on the integral in Eq. F-5 ($= T_D/T$) approaches zero. Thus, in evaluating that integral, we can assume that $x \ll 1$ in the integrand and can use the approximation (see Appendix 5)

$$e^x \cong 1 + x + \frac{x^2}{2} + \cdots .$$

With this approximation suitably introduced, the predicted value of c_v approaches $3R$ at high temperatures, in full agreement with experiment (and also, incidentally, with the prediction of Einstein's theory of heat capacity).

F-3 THE LOW-TEMPERATURE LIMIT

As $T \to 0$, we see that the upper limit of the integral in Eq. F-5 ($= T_D/T$) becomes very large and can be taken to be infinitely great. In this limit the integral becomes a definite integral, with a specific numerical value independent of temperature. Equation F-5 shows us that the molar heat capacity c_v then varies as T^3, a prediction known as the *Debye T^3 law*. The definite integral can, in fact, be evaluated exactly and turns out to have the value of $4\pi^4/15$. With this substitution we then have, for the low-temperature behavior of the molar heat capacity,

$$c_v(T) = \left(\frac{12\pi^4 R}{5T_D^3}\right) T^3 \qquad (T \ll T_D). \tag{F-6}$$

Figure F-1 shows some data at extremely low temperatures for solid argon. Here c_v is plotted as a function of T^3 and the straight line that results is clear evidence that Eq. F-6 is obeyed extremely well. From the slope of the line in the figure, and from Eq. F-6, you can easily show that $T_D = 91$ K for argon.

It is not expected that the T^3 law of Eq. F-6 would hold for metals at low temperatures and, indeed, it does not. A metal is characterized by the presence of (free) conduction electrons, and the heat capacity associated with them—negligi-

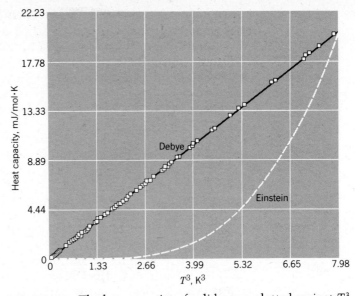

Figure F-1. The heat capacity of solid argon plotted against T^3. The straight line represents Eq. F-6 with $T_D = 24.2$ K. The dashed curve shows the (expected) failure of Einstein's heat capacity theory in this extreme low-temperature case. Adapted from Charles Kittel, *Solid State Physics*, 5th ed. (Wiley, New York, 1976), p. 139.

ble for temperatures such that $T > T_D$—reveals itself at sufficiently low temperatures because here the heat capacity associated with the lattice vibrations can become arbitrarily small. Solid argon (see Fig. F-1) has no conduction electrons, so these considerations do not apply.

The discrepancy between Einstein's and Debye's heat capacity theories should be more pronounced the lower the temperature. The dashed line in Fig. F-1, which is an attempt to fit Einstein's heat capacity theory (Eq. 4-19a) to the data for solid argon at very low temperatures, illustrates this point. If we choose $T_E = 24.2$ K, we can force agreement with Einstein's theory at 2.0 K, the highest temperature; there is no agreement at all, however, at lower temperatures. Nevertheless, as we have noted, Einstein's theory *does* fit the main features of the specific heat data over the full range of temperatures and was the first theory to do so. Its broad agreement with experiment had a marked effect in promoting general acceptance of the novel concept of the quantization of energy.

SUPPLEMENTARY TOPIC G
phase speed and group speed

[Fourier's] theorem tells us that every curve, no matter what its nature may be, or in what way it was originally obtained, can be exactly reproduced by superposing a sufficient number of simple harmonic curves—in brief, every curve can be built up by piling up waves.

Sir James Jeans (1937)

Our plan is to define the phase speed and the group speed of a wave, to examine the properties of the speeds so defined, for both matter waves and light waves, and, finally, to reintroduce the concepts in a more formal way. We shall deal throughout with the *wave number* κ $(= 1/\lambda)$ rather than with the wavelength, and our treatment will be fully relativistic.

G-1 PHASE SPEED AND GROUP SPEED—A FIRST GLANCE

Two Definitions. We can combine the wave number κ and the frequency ν to define a speed that we call the *phase speed:*

$$v_{\text{ph}} = \frac{\nu}{\kappa} \quad \text{(phase speed).} \tag{G-1}$$

We can also identify a second quantity that has the dimensions of a speed. We call it the *group speed* and define it from

$$v_{\text{gr}} = \frac{d\nu}{d\kappa} \quad \text{(group speed).} \tag{G-2}$$

A Matter Wave. We can find the phase and group speeds for a matter wave representing a free particle moving with speed v by substituting the Einstein-de Broglie relations $(\kappa = p/h$ and $\nu = E/h$; see Eqs. 6-1) into the defining equations. For the phase speed we obtain

$$v_{\text{ph}} = \frac{\nu}{\kappa} = \frac{E/h}{p/h} = \frac{E}{p} = \frac{mc^2}{mv} = \frac{c}{\beta}. \tag{G-3}$$

Because the speed parameter β $(= v/c)$ is always less than unity, we see that the phase speed exceeds the speed of light and thus we cannot identify it with the speed v of the particle.

To find the group speed of the particle, we start from the relativistic relationship between energy and momentum (Eq. 3-13*b*):

$$E^2 = (pc)^2 + (m_0c^2)^2.$$

Substituting from the Einstein-de Broglie relations gives us

$$h^2\nu^2 = h^2\kappa^2c^2 + (m_0c^2)^2. \tag{G-4}$$

Differentiating and rearranging yields

$$\frac{d\nu}{d\kappa} = \frac{c^2}{\nu/\kappa}.$$

But $d\nu/d\kappa$ is the group speed and ν/κ is the phase speed $(= c/\beta)$, so

$$v_{gr} = \frac{c^2}{v_{ph}} = \frac{c^2}{c/\beta} = c\beta = v. \tag{G-5}$$

Thus the group speed is identical with the particle speed.

A Light Wave.   We can use Eqs. G-3 and G-5 to find the phase and group speeds for a plane wave of light traveling in free space by the simple expedient of putting $\beta = 1$ in those equations. Doing so yields

$$v_{ph} = v_{gr} = c.$$

Thus we see that, for a light wave in free space, the phase and group speeds are identical and are equal to c.

G-2 PHASE SPEED AND GROUP SPEED— A SECOND LOOK

Now we examine the concepts of phase speed and group speed in a more formal manner. The simplest situation in which these two speeds emerge is in the superposition of two harmonic waves whose wave numbers and frequencies differ but slightly from each other. Let us represent the two waves by

$$y_1(x, t) = y_0 \cos 2\pi(\kappa_1 x - \nu_1 t) \tag{G-6a}$$

and

$$y_2(x, t) = y_0 \cos 2\pi(\kappa_2 x - \nu_2 t). \tag{G-6b}$$

From the trigonometric identity

$$\cos a + \cos b = 2 \cos\left(\frac{a - b}{2}\right) \cos\left(\frac{a + b}{2}\right),$$

we can write the sum of these two waves as

$$y(x, t) = 2y_0 \cos \pi[(\kappa_2 - \kappa_1)x - (\nu_2 - \nu_1)t] \cos 2\pi \left[\left(\frac{\kappa_2 + \kappa_1}{2}\right)x - \left(\frac{\nu_2 + \nu_1}{2}\right)t\right].$$

We now assume that κ_1 and κ_2 are approximately equal to each other, their average value being $\bar{\kappa}$ and their difference being $\Delta\kappa$. If we make a similar assumption for the two frequencies, we can then write the combined wave as

$$y(x, t) = [2y_0 \cos \pi(\Delta\kappa x - \Delta\nu t)] \cos 2\pi(\bar{\kappa}x - \bar{\nu}t). \tag{G-7}$$

Figure G-1a shows the variation with time (for an observer at position $x = 0$) of the two combining waves of Eqs. G-6; Fig. G-1b shows their sum (Eq. G-7). Let us see what we can learn about the nature of this combined wave.

Let us put $x = 0$ in Eq. G-7. Because $\Delta\nu << \bar{\nu}$, the cosine function inside the square brackets of that equation varies much more slowly with time than does the cosine function outside the square brackets. We are lead to view the quantity in the square brackets as the slowly varying amplitude of a wave given by $\cos 2\pi(\bar{\kappa}x - \bar{\nu}t)$ that is traveling to the right with a speed given by $\bar{\nu}/\bar{\kappa}$. The amplitude envelope, on the other hand, which is given by $\cos \pi(\Delta\kappa x - \Delta\nu t)$, travels to the right with a speed given by $\Delta\nu/\Delta\kappa$. These two speeds may or may not be equal.

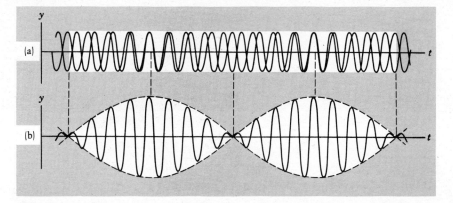

Figure G-1. (*a*) The time variation of two harmonic waves, as seen by an observer at *x* = 0; see Eqs. G-6. (*b*) The sum of these waves, showing the modulation envelope and the wavelets within that envelope; see Eq. G-7.

Consider now the limiting case in which $\kappa_1 \to \kappa_2$ and $\nu_1 \to \nu_2$ (and thus also $\Delta\kappa \to d\kappa$ and $\Delta\nu \to d\nu$). The quantity $\bar{\nu}/\bar{\kappa}$ then reduces to ν/κ, which is just the phase speed of Eq. G-1, and the second quantity reduces to $d\nu/d\kappa$, which is just the group speed of Eq. G-2.

If the phase speed and the group speed are equal to each other, then the modulation envelope of Fig. G-1*b* travels at the same speed as the wavelets within that envelope. If the phase and group speeds are *not* equal, however, then the modulation envelope and the wavelets travel at different speeds. An observer, watching the combined wave move by, would see the wavelets moving through the envelope—either forward or backward with respect to it, as the case may be—as the wave travels along.

Whether or not the phase speed and the group speed are equal in a given situation depends on whether or not the frequency ν of the wave is directly proportional to the wave number κ. If this is the case, then ν/κ and $d\nu/d\kappa$ are equal to each other, the wave and group speeds are identical, and the wave motion is said to be *nondispersive*. If, however, the relationship between the frequency and the wave number is *not* a simple direct proportionality, then the phase and group speeds will be different and the wave motion is said to be *dispersive*. For a given kind of wave traveling through a given medium, it is useful to plot the frequency ν as a function of the wave number κ. If the plot is a straight line through the origin, the wave motion is nondispersive; otherwise it is dispersive. (An equivalent statement is to say that a wave motion is nondispersive if the index of refraction of the medium through which the wave travels is independent of the wavelength; it is dispersive if this is not the case.)

G-3 MATTER WAVES AND LIGHT WAVES REVISITED

We can find the frequency as a function of wave number for the matter wave representing a free particle from Eq. G-4. Solving that equation for ν yields

$$\nu = \sqrt{\kappa^2 c^2 + \left(\frac{m_0 c^2}{h}\right)^2}.$$

We see that ν is *not* directly proportional to the wave number κ. Figure G-2*a* shows a plot of frequency versus wave number for this wave, electron kinetic energies being indicated at various points along the curve. The slope $d\nu/d\kappa$ of the

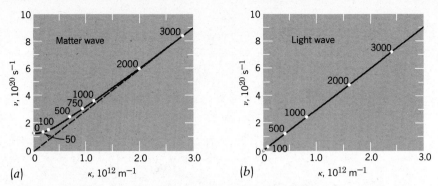

Figure G-2. (a) A plot of frequency ν versus wave number κ for a matter wave representing a free electron; the wave motion is dispersive. The labeled dots are the electron kinetic energies in keV. (b) A similar plot for a light wave in free space; the wave motion is nondispersive. The labeled dots are the photon energies in keV.

curve at any point represents the group speed of the electron at that point. We see that the slope (= group speed = particle speed) is small at low energies and increases to a limiting value of c as the energy increases.

The phase speed of the matter wave at any point along the curve of Fig. G-2a is simply the ratio of ν to κ at that point. This ratio is infinitely great at zero energy and decreases toward the limiting value of c as the energy increases. The fact that the phase speed of this matter wave exceeds the speed of light does not cause any conflict with the special theory of relativity because signals cannot be carried at the phase speed of a wave, only at its group speed. Signals carried by a wave must be encoded in its modulation envelope. We can see this in the limiting case in which the modulation envelope (in Fig. G-1b, say) is absent. We then have a featureless, infinitely long harmonic wave, with no beginning and no end. The wave maxima are all identical and there is no way we can tell one from the other. If we mark the wave in some way (by momentarily increasing the amplitude, say), that constitutes a modulation envelope and it will travel with the group speed, which is less than the speed of light.

Figure G-2b shows a plot of frequency ν versus wave number κ for a light wave in free space, a few selected photon energies being indicated. The curve is a simple straight line through the origin so that, as we have already shown, the phase speed and the group speed are equal to c for all energies.

If we wish to represent an electron or a photon by a wave train or a wave packet like that of Fig. 6-13a, we must pile up not two harmonic waves as in Fig. G-1, but an infinitely large number of them, with carefully chosen wave numbers, phases, and amplitudes. For a matter wave packet these subsidiary waves will all travel with different speeds, the wavelets will "run through" the wave packet as the packet moves, and the packet will not maintain its shape; it will *disperse* as the wave travels. If the packet represents a light wave in free space, however, the subsidiary waves all travel with the same speed c and the packet maintains its shape as it moves. If a light pulse travels not through free space but through a transparent medium, the ratio ν/κ will in general no longer be independent of frequency and the wave motion will be dispersive. High-energy laser pulses, for example, change their shape as they move through the atmosphere, as opposed to moving through the near vacuum of outer space.

SUPPLEMENTARY TOPIC H

matter waves —refraction at the crystal surface

Errors, like straws, upon the surface flow; He who would search for pearls must dive below.

John Dryden (1677)

H-1 THE PROBLEM

Figure H-1 shows an incident beam i of monoenergetic electrons falling at right angles on a single crystal. A reflected beam r is shown, making an angle ϕ with the normal to the crystal surface. This beam is generated by Bragg reflection of the incident matter wave from a family of underlying atomic planes. These planes are suggested by the dashed lines in the figure, the interplanar distance being d.

Our purpose is to measure λ, the wavelength of the reflected beam. The difficulty is that the emerging beam is refracted or bent as it passes through the surface so that the angle ϕ, which we measure above the crystal surface, is not the same as the angle 2θ that the reflected beam makes with the incident beam deep within the crystal.

H-2 THE SOLUTION

Consider first the bending of the beam at surface point s in Fig. H-1a. By analogy with the refraction of light, we can write

$$\mu \sin 2\theta = \sin \phi,$$

in which 2θ is the angle of incidence, ϕ is the angle of refraction, and μ is the index of refraction. This index, however, is by definition just λ/λ', the ratio of the wavelength of the matter wave outside the crystal to that inside the crystal. Using this relation to eliminate the index of refraction, we can write

$$\lambda' = \lambda \frac{\sin 2\theta}{\sin \phi} = \lambda \frac{2 \sin \theta \cos \theta}{\sin \phi}. \tag{H-1}$$

Now we turn attention to point p in the crystal. What happens here is governed by Bragg's law of reflection, developed initially to explain x-ray diffraction but equally applicable to the diffraction of matter waves. This law is (see *Physics*, Part II, Sec. 47–6, noting that the angles there are measured with respect to the atomic planes, whereas here they are measured with respect to the normal to those planes)

$$m\lambda' = 2d \cos \theta \qquad m = 1, 2, 3, \ldots, \tag{H-2}$$

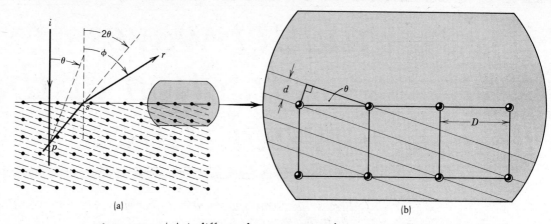

Figure H-1. (*a*) A diffracted ray emerging from a crystalline target. (*b*) An enlargement of a portion of the crystal surface. See Reference 3 in Chapter 6.

in which d is the spacing between the planes and the integer m is the so-called *order number*. Eliminating λ' between Eqs. H-1 and H-2 and solving for λ leads to

$$\lambda = \frac{1}{m}\left(\frac{d}{\sin \theta}\right)\sin \phi.$$

However (see the figure), $d/\sin \theta$ happens to be just D, the distance between parallel rows of atoms on the crystal surface. Thus we can write, as our final result,

$$m\lambda = D \sin \phi, \qquad\qquad\qquad\text{(H-3)}$$

which is just the formula defining the diffracted beams generated when light falls at right angles on a diffraction grating; see *Physics*, Part II, Sec. 47-3. Assuming that we know the order number (which we take to be unity in the examples given in Section 6-2), we can then find the de Broglie wavelength λ of the reflected electron beam in terms of the angle ϕ, measured outside the crystal, and the distance D, which may be calculated if the structure and the orientation of the crystal are known. We do not need to know which set of atomic planes is responsible for the Bragg reflections within the crystal, nor do we need to know the index of refraction of the crystal for the matter waves representing the electron beam.

SUPPLEMENTARY TOPIC I
Rutherford scattering

If circumstances lead me, I will find where truth is hid, though it were hid indeed within the center.

Hamlet

Here we derive the relationships by which Rutherford, in 1911, was able to analyze the alpha-scattering experiments of Geiger and Marsden and confirm his hypothesis that the positive charge of every atom is concentrated in a tiny massive "nucleus" at its center; see Section 7-2. In the first section we shall analyze the dynamics of a single scattering event, in which a single alpha particle is deflected by a single target nucleus. We shall then express this result in terms accessible to experiment, in which a collimated beam of many alpha particles falls on a thin target foil containing many nuclei.

I-1 A SINGLE SCATTERING EVENT

Figure I-1 shows an alpha particle, assumed by Rutherford to be essentially a point particle and to carry a positive charge of $2e$, being deflected as it passes within the range of the repulsive Coulomb force exerted on it by a heavy nucleus of charge ne.* We take this nucleus, whose mass we assume to be infinitely great, as the origin of a coordinate system, and we describe the position of the alpha particle in that reference frame by the polar coordinates r and θ. The two asymptotes to the path of the alpha particle are shown, the angle ϕ between them being the scattering angle. The *impact parameter b* shows how close to the nucleus the alpha particle would have passed if it had not been deflected.

Let v_0 be the speed of the alpha particle when it is at a large distance from the nucleus and v its speed at any arbitrary point on its path. Conservation of energy then allows us to write

$$\frac{1}{2} mv^2 + \frac{1}{4\pi\varepsilon_0} \frac{(2e)(ne)}{r} = \frac{1}{2} mv_0^2. \tag{I-1}$$

If the collision were head-on, that is, if the impact parameter b were zero, then the alpha particle would (we assume) come to rest momentarily and then reverse its direction. Its distance of closest approach under these circumstances, r_0, will prove to be a convenient parameter. We can find it by putting $v = 0$ in Eq. I-1, the result being

$$r_0 = \frac{ne^2}{\pi\varepsilon_0 mv_0^2} = \frac{ne^2}{2\pi\varepsilon_0 K}, \tag{I-2}$$

---

* The number n will turn out to be the same as the position number of the atom in the periodic table, a quantity that we now label Z and call the *atomic number*. In 1911, however, that connection had not yet been made and it was still firmly believed that the periodic table was governed by the atomic weight A.

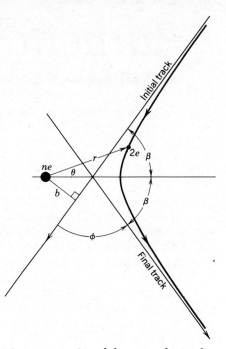

Figure I-1. An alpha particle is deflected through an angle ϕ by the repulsive Coulomb force exerted on it by a heavy nucleus of charge ne. (The trajectory is drawn to scale for a 5.00-MeV alpha particle scattered by a gold nucleus at an impact parameter b of 30 fm.)

in which K is the kinetic energy of the incident alpha particle. Also, we can express the speed v in terms of its components in polar coordinates. Thus

$$v^2 = \dot{r}^2 + (r\dot{\theta})^2,$$

in which the dot represents a differentiation with respect to time.

Using these last two results we can rewrite Eq. I-1, the energy conservation equation, as

$$\dot{r}^2 + (r\dot{\theta})^2 + \frac{v_0^2 r_0}{r} = v_0^2 \qquad \text{(conservation of energy)}. \tag{I-3}$$

Angular momentum must also be conserved in the encounter of Fig. I-1. This leads to

$$(mv_0)b = m(r\dot{\theta})r$$

or

$$\dot{\theta} = \frac{bv_0}{r^2} \qquad \text{(conservation of momentum)}. \tag{I-4a}$$

We can write for $\dot{r}$,

$$\dot{r} = \left(\frac{dr}{d\theta}\right)\dot{\theta} = \left(\frac{dr}{d\theta}\right)\frac{bv_0}{r^2}. \tag{I-4b}$$

Substituting Eqs. I-4 into Eq. I-3 leads to

$$\frac{1}{r^4}\left(\frac{dr}{d\theta}\right)^2 + \frac{1}{r^2} + \frac{1}{r}\left(\frac{r_0}{b^2}\right) = \frac{1}{b^2} \tag{I-5}$$

as the differential equation that describes the trajectory of the alpha particle. It is our task to find the function $r(\theta)$ that is a solution of this equation.

Solving Eq. I-5 is simpler if we change the variable, replacing r by $1/u$ [and $dr/d\theta$ by $-(1/u^2)(du/d\theta)$]. Making this change and then solving the resulting differential equation finally leads to, as a solution of Eq. I-5,

$$r(\theta) = \frac{a}{d\cos\theta - 1} \quad \text{(trajectory)}, \tag{I-6}$$

in which

$$a = \frac{2b^2}{r_0} \quad \text{and} \quad d = \sqrt{1 + (2b/r_0)^2}.$$

Equation I-6 can be shown to be the equation of a hyperbola.

If in Eq. I-6 we let $\theta \to \cos^{-1}(1/d)$, then $r \to \infty$ and we have found the angular position of the two asymptotes to the path of the alpha particle. In this limit $\theta \to \beta$, so (see Fig. I-1)

$$\beta = \pm\cos^{-1}(1/d).$$

Further, the scattering angle ϕ is given by

$$\phi = \pi - 2\beta.$$

We can combine these last two results and show, as a final result,

$$\tan\left(\frac{\phi}{2}\right) = \frac{r_0}{2b} = \left(\frac{e^2}{4\pi\varepsilon_0}\right)\frac{n}{bK}, \tag{I-7}$$

which gives the scattering angle ϕ in terms of the impact parameter b, the kinetic energy K of the alpha particle, and the number n that determines the charge of the target nucleus.

I-2 MANY SCATTERING EVENTS

It is not possible to test Eq. I-7 directly because we cannot aim a single alpha particle precisely at a single nucleus with a preselected impact parameter. We must deal with many alpha particles falling on a foil that contains many nuclei and we must treat the impact parameters for the individual encounters on a statistical basis.

Figure I-2 shows a portion of the target foil in the apparatus of Fig. 7-2. Let t be the thickness of the foil and ρ the number of target nuclei per unit volume of the foil. The number of target nuclei per unit area of the foil exposed to the alpha beam is then ρt. The figure shows one of these target nuclei, located at the center of an annular ring whose radius is b and whose width is db. If an alpha particle passes through this annulus it will have an impact parameter b and will be deflected through an angle ϕ, which may be calculated from Eq. I-7. The area of one such annulus is $(2\pi b)(db)$, and the fraction of the foil area covered by such annuli (assuming no overlapping) is $(2\pi b)(db)(\rho t)$. If N_0 is the number of alpha particles that fall on the foil, then the number dN that pass through one or

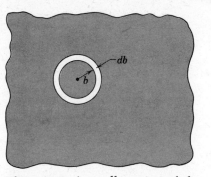

Figure I-2. A small portion of the target foil used in an alpha particle scattering experiment. One of the target nuclei is shown, centered on an annulus of radius b and thickness db.

another of such annuli is

$$dN = N_0(2\pi b)(db)(\rho t).$$

We can evaluate both b and db in this equation from Eq. I-7. Doing so and substituting the results into the above expression for dN eventually yields

$$dN = (\pi/2)r_0^2 N_0 \rho t \cot(\phi/2) \csc^2(\phi/2)\, d(\phi/2). \qquad \text{(I-8)}$$

The dN alpha particles that pass through the annuli of Fig. I-2 are scattered into the space that lies between two cones whose axis is the direction of the incident beam and whose half-angles are ϕ and $\phi + d\phi$; see Fig. I-3. Let our detector be at a distance R from the scattering foil and let us draw a sphere of this radius centered on the foil. The two cones above will intercept this sphere and will define on its surface an annular ring whose area dA is just $(2\pi R \sin\phi)(R d\phi)$ or

$$dA = 2\pi R^2 \sin\phi\, d\phi$$
$$= 8\pi R^2 \sin\left(\frac{\phi}{2}\right)\cos\left(\frac{\phi}{2}\right) d\left(\frac{\phi}{2}\right) \qquad \text{(I-9)}$$

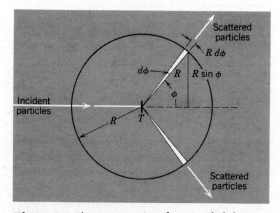

Figure I-3. A cross-sectional view of alpha particles scattered in the direction ϕ from a target foil at T.

If we divide Eq. I-8 by Eq. I-9 we get dN/dA, the number (per unit area of the sphere) of alpha particles scattered through a sphere of radius R at angles between ϕ and $\phi + d\phi$. If we further divide by N_0 we now get the *probability* per unit area that a *single* alpha particle will be so scattered. Finally, if we multiply by $\mathscr{A}$, the effective area of our detector, we get $P(\phi)$, the probability that a single incident alpha particle will be scattered into our detector. Carrying out these operations, and substituting for r_0 from Eq. I-2, leads after some rearrangement to

$$P(\phi) = \frac{\mathscr{A}}{N_0} \frac{dN}{dA}$$

$$= \left(\frac{e^2}{4\pi\varepsilon_0}\right)^2 \left(\frac{\rho t n^2}{4K^2}\right) \left(\frac{\mathscr{A}}{R^2}\right) \frac{1}{\sin^4(\phi/2)}$$

which is Eq. 7-1, the relationship that we set out to derive. As we have arranged this equation above note that the first quantity in parentheses on the right is a simple constant, the second quantity deals with the nature of the target and with the energy of the incident alpha particle. The third quantity deals with the detector, including its effective area $\mathscr{A}$, its distance R, and its angular position ϕ.

ANSWERS TO PROBLEMS

It is better to know some of the questions than all of the answers.

James Thurber (1894–1961)

CHAPTER 1

1. (a) 2.59×10^{13}. (b) 0.30. (c) 3.26.
   (d) 3.33×10^{-24}.

2. (a) 3×10^{-18}. (b) 2×10^{-12}. (c) 8.2×10^{-8}.
   (d) 6.4×10^{-6} (e) 1.1×10^{-6}.
   (f) 3.7×10^{-5}. (g) 9.9×10^{-5}. (h) 0.10.

3. (a) 30.0 cm/ns. (b) 984 ft/μs. (c) 1 ly/y.
   (d) 3.00×10^8 J$^{1/2} \cdot$ kg$^{-1/2}$.
   (e) 30.5 MeV$^{1/2} \cdot$ u$^{-1/2}$.
   (f) 3.00×10^8 m $\cdot$ F$^{-1/2} \cdot$ H$^{-1/2}$.

4. (a) Light beam in a vacuum, electron beam, light beam in air. (b) 1.1×10^{-14} s.
   (c) 9.7×10^{-9} s.

5. (a) 3.37×10^{-2} m/s^2. (b) 5.94×10^{-3} m/s^2.
   (c) 2.91×10^{-10} m/s^2.

6. (a) 6.7×10^{-10} s. (b) 2.2×10^{-18} m.

12. (a) $W_T = 48.0$ J; $W_G = 293$ J. (c) For T:
    $K_i = 0$ and $K_f = 48.0$ J; for G: $K_i = 313$ J and
    $K_f = 606$ J.

13. (b) 24.0 J.

14.

	Symbol	Steven's Data	Sally's Data
Particle masses, kg	m_1	0.107	0.107
	m_2	0.345	0.345
Particle velocities, m/s	v_1	+3.25	+0.75
	v_2	0.00	−2.50
	v_1'	−1.71	−4.21
	v_2'	+1.54	−0.96
Total system momentum, kg$\cdot$m/s	P	+0.348	−0.782
	P'	+0.348	−0.782
	$P' - P$	0.000	0.000
Total system kinetic energy J	K	0.565	1.11
	K'	0.565	1.11
	$K' - K$	0.000	0.00

15.

	Symbol	Steven's Data	Sally's Data
Particle masses, kg	m_1	0.107	0.107
	m_2	0.345	0.345
Particle velocities, m/s	v_1	+3.75	+0.75
	v_2	0.00	−2.50
	v_1'	−0.78	−3.28
	v_2'	+1.25	−1.25
Total system momentum, kg$\cdot$m/s	P	0.348	−0.782
	P'	0.348	−0.782
	$P' - P$	0.000	0.000
Total system kinetic energy J	K	0.565	1.108
	K'	0.302	0.845
	$K' - K$	−0.263	−0.263

17. (b) 0.81%; 0.073%. (c) 0.31%; 0.024%.

20. (a) 8.6° into the wind. (b) Cross-wind, by 18 s.

21. 140.

22. 6.4 cm.

25. (a) After ~2 s a ring of light appears, its plane being at right angles to the direction of the ether wind. The ring splits into two, the two separate rings moving away from each other, shrinking, and finally, after 111 ns, disappearing at diametrically opposite points. (b) After 2 s, there is a brief uniform illumination of the entire sphere, followed by darkness.

26. (a) 0.32″. (b) 0.16″. (c) Zero.

27. (a) $\sqrt{c^2 + v^2}$. (b) c.

28. 6.7×10^{-10}.

29. (b) 2.00×10^{30} kg.

31. 0.57c.

33. (a) $c - u$. (b) $c + v$. (c) $c + 2v + u$.

34. (a) c. (b) c. (c) c.

CHAPTER 2

4. (*a*) 0.140.   (*b*) 0.9950.   (*c*) 0.999950.
(*d*) 0.99999950.

7. $x' = 138$ km; $y' = 10$ km; $z' = 55$ km;
$t' = -374$ μs.

8. (*a*) $x' = 0$; $t' = 2.29$ s.   (*b*) $x' = 6.55 \times 10^8$
m; $t' = 3.16$ s.

9. 0.80 μs.

10. 0.80 m.

12. (*a*) 0.48.   (*b*) $\Delta w' = 1320$ m or $\Delta t' = 4.39$ μs.

13. (*a*) 0.866c.   (*b*) 2.000.

14. (*a*) 2.21×10^{-12}.   (*b*) 5.25 days.

15. 1.53 cm.

16. 6.4 cm.

17. 40 mi/h.

18. (*c*) 0.938 m; 32.2°.

19. 250 ns.

20. (*a*) 26.3 y.   (*b*) 52.3 y.   (*c*) 3.72 y.

21. (*b*) 0.99999943c.

22. 1.4×10^{-8} s.

23. 55 m.

24. 0.991c.

25. 4.45×10^{-13} s.

26. (*a*) 180 m.   (*b*) 750 ns.   (*c*) 0.80c.

27. (*a*) Zero; 495 m; 1360 m; 4630 m.   (*b*) Zero;
396 m; 594 m; 653 m.

28. (*a*) 26 μs.   (*b*) The red flash.

30. S' finds the flashes to be 3.46 km apart and
finds that each flash occurs 5.77 μs later than
the flash just beyond it.

33. $t'_1 = 0$; $t'_2 = -2.5$ μs.

34. (*a*) S' must move towards S, along their common axis, at a speed of 0.48c.   (*b*) The "red"
flash (suitably Doppler-shifted).   (*c*) 4.39 μs.

35. 2.40 μs.

36. (*a*) 4.00 μs. (*b*) 2.50 μs.

37. 4.0×10^{-13} s, the wavefront from BB' arriving
first.

39. (*a*) The S'-frame is seen by S to move in the
positive x-direction with a speed of
0.899c.   (*b*) Also the "red" flash (suitably
Doppler-shifted).

42. (*a*) 5.8×10^5 m². (*b*) Event 1: -225 m, 4.45
μs; Event 2: 1050 m, -0.50 μs; 5.8×10^5
m². (*c*) Event 1: 6869 m, 23.3 μs; Event 2:
5669 m, 18.6 μs; 5.8×10^5 m².

43. (*a*) -1.9×10^5 m². (*b*) 436 m. (*c*) A frame
moving in the direction of decreasing x at a
speed of 0.90c.   (*d*) No.   (*e*) Spacelike.

44. (*a*) 2.5 μs, in all three frames.   (*b*) A frame
moving in the direction of decreasing x at a
speed of 0.53c.   (*c*) No.   (*d*) Timelike.

47. (*a*) 34,000 mi/h.   (*b*) 6.4×10^{-10}.

49. 0.81c.

50. (*a*) 0.81c, in the direction of increasing x.   (*b*)
0.26c, in the direction of increasing x. The classical predictions are 1.0c and 0.20c.

51. 0.95c.

52. 0.54c.

53. 1.025 μs.

54. (*a*) 0.35c.   (*b*) 0.62c.

55. Receding at 0.59c.

56. 0.88c.

57. Seven.

59. (*a*) 0.817c; along the x-axis.   (*b*) 0.801c; 3.58°
forward from the y-axis.   (*c*) 0.801c; 3.57°
backward from the y'-axis.

61. (*a*) $v_{AB} = 0.93c$, in a direction 31° north of
west.   (*b*) $v_{BA} = 0.93c$, in a direction 31° east
of south.

63. (*a*) 0.866c.

64. 23 MHz.

66. 0.80c.

67. 0.0067 nm.

68. Yellow (551 nm).

69. (*a*) -28.8 nm; -204 nm; -393 nm.   (*b*) -29.5
nm; -236 nm; -472 nm.

70. (*b*) 0.80c.

71. (*a*) 482.903 nm.   (*b*) 489.385 nm.   (*c*) 0.011
nm.

73. $+2.97$ nm.

74. (*a*) $+2.97$ nm.   (*b*) $+0.90$ nm.   (*c*) -2.19 nm.

78. (*a*) 43.9°.   (*b*) 10.2°.

80. 10.2° in each case.

81. (*a*) Zero.   (*b*) 43°.   (*c*) 87°.

82. (*a*) 0.067.   (*b*) 10.2°; 7.0°; 2.2°.

83. (*b*) 6.8°.   (*c*) 0.996c.

85C. (c) 6.25 m. (d) 3.13 μs. (e) 1.09×10^9; timelike.

87C. (a) +157 nm. (b) −1.49 kHz.

CHAPTER 3

1. (a) V_0/γ. (b) γm_0. (c) $\gamma^2 \rho_0$; 0.10c.

2. (a) 0.13c. (b) 4.6 keV. (c) 1.2%.

3. (a) 79.1 keV. (b) 3.11 MeV. (c) 10.9 MeV.

4. (a) 0.943c. (b) 0.866c.

5. (a) 10.9 MeV. (b) 43%.

6. (a) 1.00 keV. (b) 1.05 MeV.

7. 1.13, 5.59, 25.1, 112, 504, 2254.

8. (b) 3.88 km/s. (c) 6.3 cm/s.

9. (a) 256 kV. (b) 0.746c. (c) $1.50m_0$. (d) 256 keV.

12. (a) $\sim 2 \times 10^4$. (b) 9.4 cm/s. (c) 20 GeV.

13. (a) 0.42 μm/y. (b) 1.8×10^{-16} kg. (c) 2.8 m.

14. (a) 0.0625; 1.00196. (b) 0.941; 2.96. (c) 0.999 999 87; 1960.

15. (a) 0.9988; 20.6. (b) 0.145; 1.01. (c) 0.073; 1.0027.

17. (a) 1580 km. (b) 1.2 GeV.

20. (a) 0.707c. (b) $1.414m_0$. (c) $0.414m_0c^2$.

21. (a) 126 MeV. (b) 69 keV.

24. (a) The photon. (b) The proton. (c) The proton. (d) The photon.

26. (b) 5.9 GeV/c. (c) 501 GeV/c.

27.

$\beta(= u/c)$	0.80	0.90	0.99	0.999	0.9999
E GeV	1.56	2.15	6.65	21.0	66.3
p GeV/c	1.25	1.94	6.58	21.0	66.3

28. (a) 5.71 GeV; 6.65 GeV; 6.58 GeV/c. (b) 3.11 MeV; 3.62 MeV; 3.59 MeV/c.

29. (c) $207m_e$; the particle is a muon.

30. (a) 0.948c. (b) $649m_e$. (c) 226 MeV. (d) 314 MeV/c.

32.

Proton	Frame	K(GeV)	E(GeV)	p(GeV/c)
A	S	5.711	6.649	6.583
B	S	0	0.938	0
A	S'	0.949	1.887	1.637
B	S'	0.949	1.887	−1.637

35. 11.1 ns.

36. (a) 2.04 u; 0.385c. (b) −38.4 MeV. (c) −38.4 MeV. (d) +38.4 MeV.

37. $M_0 = 2.5m_0$.

38. (a) $2.12m_0$. (b) 0.33c.

40. (a) $0.58m_0c$. (b) $0.20m_0c$. (c) $2.92m_0c^2$. (d) $2.86m_0$. (e) $0.059m_0c^2$.

42. (a) $u_1 = 0.220c$; $u_2 = 0.724c$. (b) $K_i = K_f = 0.50m_0c^2$.

43. Sam's observations are: (a) "Jim and Sue are 8 km apart." (b) "The speed of Jim's neutrons is 0.882c." (c) "I agree with Jim." (d) "One of Jim's neutrons was scattered through an angle of 12.1°." (e) "Jim is firing neutrons at the rate of 12,500 s^{-1}."

44. (a) 1.1×10^8. (b) 1.5×10^{32}. (c) 14; 40,000.

45. (a) 3.53 cm. (b) 12.0 cm. (c) 38.2 cm. (d) 121 cm.

46. 4.00 u; probably a helium-4 nucleus (alpha particle).

47. (a) 330 mT. (b) 5.89.

48. 660 km.

49. $\sim 2 \times 10^{11}$ m.

50. $10^4 - 10^5$ ly.

51. (a) 9.6×10^{15} eV = 9600 TeV = 1.5 mJ. (b) 1.0×10^7.

52. (a) 2×10^{14} eV = 200 TeV. (b) 4×10^8.

53. 92.1 MeV (7.68 MeV per constituent particle).

54. 20.6 MeV.

56. 139.5 MeV.

57. (b) 13.8 GeV. (c) 5330 GeV.

58. (b) 202 GeV. (c) 49.1 GeV.

59. (a) 4.43 MeV/c. (b) 880 eV.

61. (b) 0.511 MeV. (c) 938 MeV.

62. (a) 1.26×10^{13} J. (b) 3.5 h.

63. 2.62 mg.

64. ~ 5 μg.

65. (a) 2.8×10^{14} J. (b) 3.2 days.

66. About one part in 3×10^{11}.

67. 88 kg.

68. 6.65×10^6 mi, or 270 earth circumferences.

69. 190 tons.

70. 18 smu/y.

71. (a) 2.7×10^{14} J. (b) 1.8×10^7 kg or 18 kilotons (metric). (c) 6.0×10^6 times.

73. (a) 1430 eV/c; 45.3 keV/c; 2.46 MeV/c; 2.00 GeV/c. (b) 4.52 MeV (electron); 13.3 keV (proton). (c) 0.99882, 0.407, 0.145. (d) 2.04 MeV, 423 MeV, 3750 MeV.

75. (b) 10.4 MeV, 102 MeV.

CHAPTER 4

1. 5800 K.

2. 713 mW.

3. (a) 1630 K. (b) 1.26 cm.

4. (a) 280 K (= 45°F).

5. 198 K (= −103°F).

6. 4300 K.

7. (a) 4.4×10^9 kg/s. (b) 1.4×10^8 y; 3.1 percent.

8. (a) 0.97 mm; microwave. (b) 9.9 μm; infrared. (c) 1.6 μm; infrared. (d) 500 nm; visible. (e) 0.29 nm; x-ray. (f) 2.9×10^{-41} m; hard gamma ray.

9. 1.45 m; short radio wave.

10. 91 K.

11. 7200 K.

12. (a) 138 K. (b) 21 μm.

13. 3.2 mW.

14. 8500 K.

15. (a) Sun: 8.4×10^4 W/m²·nm; Sirius: 3.3×10^6 W/m²·nm. (b) 15. (c) 220.

16. (a) 1.45 μm. (b) 41.0 W/cm²·μm. (c) 0.89 μm and 2.63 μm.

17. (a) 7650 K. (b) 17,200 K.

18. (a) 5270 K. (b) 5.20 mW/cm²·nm. (c) 4630 K. (d) 6110 K.

19. (b) 0.655 percent K⁻¹.

21. (b) 41.0 W/cm²·μm.

27. (a) 792 nm. (b) 839 W/cm²·μm. (c) 833 W/cm²·μm. (d) 24,200 W/cm²·μm.

28. (a) $\lambda_m T = hc/5.000k$. (b) 0.7 percent lower.

29. 39 percent.

30. 27 percent below, 73 percent above.

32. (a) 3.98×10^4 W/m²·nm. (b) 4.78×10^{-8} W/m²·Hz. (c) 20 W/cm².

33. (b) 2.0×10^6 K.

37. (a) 5.83×10^{-17} J/m³·Hz. (b) 4.37×10^{-3} J/m³·μm. (c) 117 μJ/m³.

38. 1190 K.

39. (b) 12.0 mJ/m³.

41. From 0.49 mm to 2.64 mm.

42. 11.1 J/mol·K (boron) and 25.4 J/mol·K (gold).

44. (a) 6.0×10^{12} Hz. (b) 1220 J/mol. (c) 18.4 J/mol·K.

47. (a) 92 percent. (b) 58 percent.

48. (a) 8.80 T_E. (b) 0.34 T_E. (c) 232 K; 99 K.

49. (a) 1.4×10^{12} Hz; 6.0×10^{12} Hz; 14×10^{12} Hz. (b) 5.9 meV; 25 meV; 60 meV. (c) 27 N/m; 64 N/m; 120 N/m.

50. 0.015 nm or 5.2 percent.

51. (a) 1110 J. (b) 714 J.

52. 129 K.

53. 25.7 J/mol·K.

54. 26.3 J/mol·K.

55. 4.5 R.

56. (a) 40 meV. (b) ~10⁴ K.

57. 6.6×10^{-34} J·s.

58. 16,200 K.

59. (b) 103 nm; 122 nm; 658 nm.

61. 1.1×10^{-5}.

63. 6×10^{-7} m or about 2000 mercury atom diameters.

CHAPTER 5

1. (a) No. (b) 544 nm; green.

2. (a) 2.0 eV. (b) Zero. (c) 2.0 V. (d) 295 nm. (e) 2.0×10^{18} photons/m²·s.

4. (a) 382 nm. (b) 1.82 eV. (c) Cesium.

5. (a) 6.60×10^{-34} J·s. (b) 2.27 eV. (c) 545 nm.

6. Only barium and lithium will work.

7. 2.1 μm; infrared.

8. (a) 3.1 keV. (b) 14.4 keV.

9. 8.17 MeV.

11. (a) 0.48 nm. (b) X-ray region.

13. (a) 5.9×10^{-6} eV. (b) 2.05 eV.

14. (a) The infrared bulb. (b) 6×10^{20}.

15. 3.71×10^{21} photons/m²·s.

16. A's is twice B's; A's is half of B's; A's is twice B's; the same; all except the speed.

18. 3.6×10^{-17} W.

19. (a) 1.24×10^{20} Hz. (b) 2.43 pm. (c) 2.73×10^{-22} kg·m/s.

21. (a) 89 m. (b) 2.0×10^4 photons/cm³.

22. 1.3×10^{11} photons/cm³.

23. (a) 3.7×10^{23} photons/m²·s. (b) 2.5×10^{30} photons/m²·s.

24. 268 MeV (forward photon) and 17 MeV.

25. (b) 2.4×10^{-9} m/s. (c) 500 keV.

26. (*a*) 2.73 pm; 61.7 keV.   (*b*) 6.05 pm; 312 keV.

28. 300%.

29. 44°.

33. $2E^2/(m_0c^2 + 2E)$.

34. 0.682 MeV.

35. $p_x = 5.33 \times 10^{-23}$ kg·m/s; $p_y = -4.59 \times 10^{-23}$ kg·m/s.

37. 2.65×10^{-15} m = 2.65 fm.

40. (*b*) 6.67 pm.

41. (*a*) 12.4 kV.   (*b*) 511 kV.   (*c*) 1.02 MV.

42. (*a*) 5.73 keV.   (*b*) 0.0870 nm and 14.3 keV for the first photon; 0.217 nm and 5.7 keV for the second.

43. (*a*) 140 keV. (*b*) 9.2 MeV. (*c*) 0.85 − 4.5 MeV.

44. (*a*) 2.02 MeV.   (*b*) 29.7%.

46. (*a*) 5.46×10^{-22} kg·m/s.   (*b*) 2.71 eV.

48. 1.88 GeV.

49. (*a*) $c/3$.   (*b*) 1 : 2.   (*c*) 1 : 2.

50. (*d*) 0.625.

51. (*a*) 0.577.   (*b*) 417 keV and 834 keV.

52. (*b*) $E_1 = 606$ keV; $E_2 = 438$ keV; $\theta_2 = 43.8°$.

53. (*a*) 0.511 MeV.

55C. (*a*) 1.0×10^{-10} eV, 12 km; 3.3×10^{-7} eV, 3.8 m; 1.2×10^{-5} eV, 10 cm. (*b*) 2.9×10^4, 8.3×10^{-17} m; 3.9, 6.2×10^{-13} m; 5.9×10^{-6}, 4.1×10^{-7} m. (*c*) 25 MeV, 6.0×10^{21} Hz; 25 eV, 6.0×10^{15} Hz; 1.2×10^{-4} eV, 3.0×10^{10} Hz.

57C. (*a*) 2.49×10^{-10} m, 2.493×10^{-10} m, 7.11×10^{-13} m, 5.00 keV, 4.986 keV, 14.3 eV, 0.3%, 45.0°, 67.3°. (*b*) 4.972×10^{-14} m, 76.0×10^{-14} m, 71.1×10^{-14} m, 25.0 MeV, 1.64 MeV, 23.4 MeV, 93.5%, 45.0°, 2.8°.

58C. 0.30 619 453   45° 75°
0.30 463 608   100° 49°
no solution
0.30 600   471 51° 84°
0.98 300 5420 45° 2.2°
no solution

CHAPTER 6

2. (*b*) In the sequence given: 1.732 Å; 1.734 Å; 1.734 Å.   (*c*) 0.0387 Å; 0.0388 Å; 0.0370 Å.

3. 830 keV.

4. 6.2×10^{-17} m, or 1.2% of the specified nuclear radius.

5. (*a*) 21 keV.   (*b*) 37.8 MeV.

6. (*a*) 3.7×10^{-35} m.

7. (*a*) 0.0388 eV.   (*b*) 0.145 nm.

8. 4.32×10^{-6} eV.

9. (*a*) 1240 eV, 1.50 eV.   (*b*) 1.24 GeV, 1.24 GeV.

10. (*a*) 1.24 μm, 1.23 nm.   (*b*) 1.24 fm, 1.24 fm.

11. (*a*) electron: 1.24 keV/c, photon: 1.24 keV/c. (*b*) electron: 511 keV, photon: 1.24 keV. (*c*) electron: 0.0015 keV, photon: 1.24 keV.

12. 143%, 0.13%.

14. 872 fm, 28.6 fm, 1240 fm.

15. 75 keV.

16. (*a*) 5.3 fm.   (*b*) No.

17. (*a*) de Broglie wavelength = 0.073 nm; average separation = 3.4 nm.   (*b*) Yes.

18. Rest mass = 1.675×10^{-27} kg; a neutron.

19. (*a*) 0.1176 fm.   (*b*) A helium atom.

22. (*b*) 1.12 Å.

24. (*b*) 47.4°.   (*c*) 130 V at 90°.

25. (*a*) 2.83°.   (*b*) 0.55°.

26. 5.78°, 11.61°, 17.58°.

27. 37.7 kV.

28. (*a*) 15 keV. (*b*) 124 keV; gamma rays. (*c*) Electron.

29. 3.6×10^7 m/s or 0.12c.

33. (*a*) 500 nm. (*b*) 5.0 mm. (*c*) 5.0 m. (*d*) 10 m.

34. 1.7×10^{-8} s.

35. 4.1×10^{-3} eV.

36. 5.2×10^{-7} eV.

37. (*a*) 50 eV.   (*b*) No.

38. 10^{-21} s.

39. (*a*) 124 keV.   (*b*) 40.5 keV.

40. (*b*) 8×10^{-16} m. (*c*) No, no.

44C. (*b*) 38.8 pm, 0.872 pm, 0.00124 pm; 32.0 keV, 1.42 MeV, 1.00 GeV. (*c*) 286 pm, 90.4 pm, 28.6 pm; 4.34 keV, 13.7 keV, 43.4 keV. (*d*) 14.4 fm.

CHAPTER 7

1. (*a*) 1.18×10^{-29} m³.   (*b*) 1.41 Å.

2. No, as long as the orbit lies within the sphere of positive charge.

3. 0.295 nm.

6. (*b*) 1.41×10^{-15} m.

10. 15.8 fm.

11. 27.7 MeV.

12. (*a*) 0.39 MeV.   (*b*) 4.61 MeV.

13. (*a*) 46.4 fm.   (*b*) 45.5 fm.

14. (*a*) 37.2 fm.   (*b*) 53.1°.

15. (*a*) 42.9 fm; 37.2 fm.   (*b*) 15.8 fm; 13.6 fm.

16. (*a*) 430 counts/min.   (*b*) 1.15 counts/min.

17. 29.

19. (*a*) 1.2×10^{-3}. (*b*) 8.5×10^{-5}. (*c*) 2.8×10^{-5}.

20. (*a*) 58.2°.   (*b*) 44.6 fm.   (*c*) 68.1 fm.   (*d*) 1.83 MeV.

21. (*a*) 656.4 nm, 486.3 nm, 434.2 nm.   (*b*) 364.7 nm.

22. (*a*) 30.5 nm, 291.3 nm, 1053.1 nm.   (*b*) 8.250 $\times 10^{14}$ Hz, 3.654×10^{14} Hz, 2.055×10^{14} Hz.

23. (*a*) 2300 m (130 kHz) to 1.8×10^8 m (1.7 Hz); very long radio waves.   (*b*) 1.3 m.

24. 13.46 eV, 13.46 eV/*c*, 92.1 nm, 3.26×10^{15} Hz.

25. (*a*) 12.8 eV.   (*b*) 12.8 eV, 12.1 eV, 10.2 eV, 2.55 eV, 1.89 eV, 0.66 eV.   (*c*) 4.07 m/s, 8.66×10^{-8} eV.

26. (*a*) $n = 4$, $n = 2$,   (*b*) Balmer.

27. (*a*) 2.55 eV.

29. (*b*) 102.5 nm (Lyman), 121.6 nm (Lyman), 652.6 nm (Balmer).

30. (*a*) 0.213 eV, 0.0218 eV.   (*b*) 54.4 eV.   (*c*) 13.6 eV.

31. (*a*) $n = 6$ to $n = 4$.   (*b*) Smaller.   (*c*) 0.267 nm.

32. (*b*) 1.013 μm, 0.365 μm.   (*c*) Infrared and visible.

33. (*a*) $n = 2$ to $n = 1$ for H^1.   (*b*) H^1, H^2, He^3, He^4, Li^6, Li^7.

35. (*b*) 166.

38. (*a*) $n = 1$.   (*b*) 0.0529 nm.   (*c*) 1.05×10^{-34} kg·m²/s.   (*d*) 1.99×10^{-24} kg·m/s.   (*e*) 4.13×10^{16} rad/s.   (*f*) 2.19×10^6 m/s.   (*g*) 8.23×10^{-8} N.   (*h*) 9.04×10^{22} m/s².   (*i*) 13.6 eV.   (*j*) −27.2 eV.   (*k*) −13.6 eV.

41. (*b*) 6.44×10^{-9}.

46. 0.238 nm.

47. (*b*) 0.0529 nm.

48. 0.490 nm.

50. 0.677.

51. (*a*) 46.4 nm$^{-3/2}$, 2150 nm^{-3}, zero.   (*b*) 17.1 nm$^{-3/2}$, 291 nm^{-3}, 10.2 nm^{-1}.

53C. (*a*) 122 nm, 91.2 nm, 10.2 eV.   (*b*) 36.7 cm, 3.65 mm, 3.37×10^{-6} eV.   (*c*) 1.876 μm, 1.282 μm, 1.094 μm, 1.005 μm, 0.955 μm, . . .

54C. (*a*) $n = 2$: −3.40 eV, 0.212 nm, 8.22×10^{14} Hz, 24.7×10^{14} Hz. $n = 200$: -3.40×10^{-4} eV, 2.12 μm, 8.22×10^8 Hz, 8.29×10^8 Hz.   (*b*) −28.8 keV, 2.30 pm, 6.96×10^{18} Hz, 20.9×10^{18} Hz.

SUPPLEMENTARY TOPIC A

1. (*a*) The present.   (*b*) Spacelike.   (*c*) None possible.   (*d*) 335 m.   (*e*) Yes; 0.67*c*.   (*f*) No.

4. $d'/d = 1.29$.

5. (*b,c*) $x' = 751$ m; $w' = -115$ m ($t' = -0.385$ μs).

6. (*a*) The *w* axis.   (*b*) The *w'* axis.   (*c*) The *w'* axis.   (*d*) $x = 0$; $x' = -\beta w'$; $x = \beta w$; $x' = 0$.

7. (*a*) $w' = -\beta\gamma x$.   (*b*) The *x'* axis.   (*c*) $w = \beta\gamma x'$.

8. "3.46"

9. 1.73 units.

10. (*a*) No.   (*b*) Yes; the plane itself is such a frame.   (*c*) Timelike.

11. (*a*) −0.316*c*.   (*b*) −0.143*c*.

12. (*b*) In *S'*, $x' = 2.02$ and $w' = 1.15$; in *S"*, $x'' = 1.67$ and $w'' = 0.167$.

SUPPLEMENTARY TOPIC B

3. (*a*) 0.99999950*c*.

5. (*a*) 4.0 ly.   (*b*) Nine years and four months after Bob left.

6. (*a*) 5.0 y.

7. (*a*) (40/41)*c*.   (*b*) 10 y.   (*c*) (40/3) ly.

8. (*a*) 6.0 y, for each clock.

SUPPLEMENTARY TOPIC C

4. $4.0 \times 10^{-6} m_e$.

7. 258 d; 6.2 h.

8. (*a*) 3.9 ns.   (*b*) 150 ns.

9. (*b*) 600.11 nm   (*c*) 50 km/s.

10. (*a*) 1.25×10^{-3} nm.   (*b*) $\pm 3.84 \times 10^{-3}$ nm.

11. +0.022 Hz.

13. (*b*) 9 mm; 3 km; 2 light-weeks.

14. (*b*) Earth: 2.0×10^{30} kg/m³; Sun: 1.8×10^{19} kg/m³; Milky Way galaxy: 8×10^{-4} kg/m³.   (*c*) 2.7×10^{38} kg; this is about 1/1000 of the mass of the Milky Way galaxy.

15. 7 arc seconds.

16. 4.0 arc seconds/century.

19. (*a*) 243 m.   (*b*) 41.7 m; 1.6 gee/m.

APPENDIXES

1. SOME PHYSICAL CONSTANTS

Avogadro constant	N_A	6.022×10^{23} mol^{-1}
Bohr radius	a_0	5.292×10^{-11} m
Boltzmann constant	k	1.381×10^{-23} J/K
		8.617×10^{-5} eV/K
Elementary charge	e	1.602×10^{-19} C
Mass-energy equivalent	c^2	9.315×10^8 eV/u
Permeability constant	μ_0	1.257×10^{-6} H/m
Permittivity constant	ε_0	8.854×10^{-12} F/m
Planck constant	h	6.626×10^{-34} J $\cdot$ s
		4.136×10^{-15} eV $\cdot$ s
Rydberg constant	R_∞	1.097×10^7 m^{-1}
Speed of light	c	2.998×10^8 m/s
Stefan-Boltzmann constant	σ	5.670×10^{-8} W/m$^2 \cdot$ K^4
Universal gas constant	R	8.314 J/mol $\cdot$ K
Wien constant	w	2898 μm $\cdot$ K

2. SOME CONVERSION FACTORS

Mass
  1.000 kg $= 2.205$ lb (mass); 453.6 g $= 1.000$ lb (mass)

Length
  1 m $= 10^6$ μm $= 10^9$ nm $= 10^{10}$ Å (angstrom) $= 10^{12}$ pm $= 10^{15}$ fm (fermi)
  1 m $= 39.37$ in. $= 3.280$ ft; 1 in. $= 2.540$ cm; 1 mi $= 1.609$ km
  1 parsec (pc) $= 3.262$ light years (ly) $= 3.086 \times 10^{16}$ m
  1 Astronomical unit (AU) $= 1.496 \times 10^{11}$ m

Time
  1 s $= 10^6$ μs $= 10^9$ ns $= 10^{12}$ ps
  1 d $= 86,400$ s; 1 (tropical) year $= 365.24$ d $= 3.156 \times 10^7$ s

Angular measure
  1 rad $= 57.30° = 0.1592$ rev

Speed
  1 m/s $= 3.281$ ft/s $= 2.237$ mi/h; 1 mi/h $= 1.609$ km/h

Force and pressure
  1 N $= 10^5$ dyne $= 0.2248$ lb (force); 1 lb (force) $= 4.448$ N
  1 Pa $= 1$ N/m$^2 = 1.451 \times 10^{-4}$ lb/in$^2 = 9.872 \times 10^{-6}$ atm
  1 atm $= 1.013 \times 10^5$ Pa $= 760.0$ Torr $= 14.70$ lb/in^2

Energy and Power
  1 J $= 10^7$ erg $= 0.2388$ cal $= 0.7376$ ft $\cdot$ lb $= 2.778 \times 10^{-7}$ kW $\cdot$ h
  1 eV $= 1.602 \times 10^{-19}$ J
  1 horsepower (hp) $= 745.7$ W $= 550.0$ ft $\cdot$ lb/s

Magnetism
  1 T $= 1$ Wb/m$^2 = 10^4$ gauss

3. SOME MASS-ENERGY CONVERSION FACTORS*

		kg	u	MeV	J
1 kg	=	1	6.022×10^{26}	5.610×10^{29}	8.988×10^{16}
1 u	=	1.661×10^{-27}	1	931.5	1.492×10^{-10}
1 MeV	=	$1.782 = 10^{-30}$	1.074×10^{-3}	1	1.602×10^{-13}
1 J	=	1.113×10^{-17}	6.700×10^{9}	6.241×10^{12}	1

$$c^2 = 931.5016 \text{ MeV/u} = 8.987552 \times 10^{16} \text{ J/kg}$$

4. SOME REST MASSES

		kg	u	Mev/c^2
Electron	e	9.10954×10^{-31}	0.000548580	0.511003
Muon	μ	1.88357×10^{-27}	0.113429	105.660
Neutral pion	π^0	2.40598×10^{-28}	0.144889	134.965
Pion	π	2.48806×10^{-28}	0.149832	139.569
Proton	p	1.67265×10^{-27}	1.00728	938.279
Hydrogen atom	H^1	1.67356×10^{-27}	1.00783	938.791
Neutron	n	1.67495×10^{-27}	1.00867	939.573
Deuterium atom	H^2	3.34455×10^{-27}	2.01410	1876.14
Helium atom	He^4	6.64659×10^{-27}	4.00260	3728.43

$$m_\mu = 206.77 \, m_e \qquad\qquad m_\pi = 273.13 \, m_e$$
$$m_\pi^0 = 264.12 \, m_e \qquad\qquad m_p = 1836.2 \, m_e$$

5. SOME SERIES EXPANSIONS

$$(y + x)^n = x^n + \frac{n}{1!} x^{n-1}y + \frac{n(n-1)}{2!} x^{n-2}y^2 + \cdots \qquad (x^2 < y^2)$$

$$(1 + x)^n = 1 + nx + \frac{n(n-1)}{2!} x^2 + \frac{n(n-1)(n-2)}{3!} x^3 \cdots \qquad (x^2 < 1)$$

$$\sin x = x - \frac{x^3}{3!} + \frac{x^5}{5!} - \cdots \qquad\qquad \cos x = 1 - \frac{x^2}{2!} + \frac{x^4}{4!} - \cdots$$

$$e^x = 1 + x + \frac{x^2}{2!} + \frac{x^3}{3!} + \cdots$$

* Units in the shaded area are mass units; those in the unshaded area are energy units.

Index

Aberration of light, 21-22
  and refutation of ether drag, 22
  relativistic treatment, 36 (Problem 27), 72
Absolute frame of reference, 11-12
  attempts to locate, 13-19
Absolute future, 275
Absolute past, 275
Absorption of radiation by matter, 170, 176, 179
  relative probability of processes, 179, 180
Absorption spectra, 247
Acceleration, relativistic, of particle under the influence of a force, 103
  transformation of, classical, 10
Addition of velocities, 8, 66-70
Angular frequency, 211
Angular momentum, quantization of, 250-252
  Bohr quantization rule, 251, 259
    correction for finite nuclear mass, 253
  physical interpretation, 251
Angular wave number, 211
Atom, 231
  models of:
    Bohr model, 241
      correction for finite nuclear mass, 253
      critique of, 257
      one-electron atoms, 247-250
    Rutherford model, 233-237
    Thomson model, 231-233
  size of, 250
  stability of, 239
Atomic mass unit, 112
Atomic spectra, 239-241
  absorption lines, 247
  of deuterium, 254-255
  of hydrogen, 239-241
  natural line width, 222
  presence of continuum, 249
  series limit, 240
Atomic structure, internal evidence of, in gas collisions, 147
  Franck-Hertz experiment, 147-148
Atomic theory of elements, 231
Avogadro constant, 143

Balmer formula, 240
Balmer series, 240, 241
Binding energy, 113, 122 (Problem 53), 249
  of hydrogen, 243
  of muonic atom, 255
Black body, 129
  brightness of, 130
  cavity as, 129
Black-body radiation, *see* Cavity radiation

Black hole, 299, 301-302 (Problems 13, 14)
Bohr, Niels, quotation from, 209, 214, 231, 241
Bohr microscope, 214
Bohr model of atom, 241-250
  correction for finite nuclear mass, 252-253
  critique of, 257
Bohr postulates, 242, 251
Bohr radius, 248
Bondi, Herman, quotation from, 29, 79
Born, Max, quotation from, 211, 222, 300
Brackett series, 241
Bradley, J., 21
Bragg law, 319
Bremsstrahlung, 181, 182
Brightness of cavity, 130

Causality, in quantum theory, 223, 277
  and time-order of events, 87 (Problem 60)
Cavity:
  as black body, 129
  total energy in, 157 (Problem 39)
Cavity radiation, 128
  Planck formulae for, 133
  properties of, 129-130
Charge, relativistic invariance, 106
Classical electron radius, 264 (Problem 6)
Clocks, in gravitational field, 295
Complementarity, principle of, 209, 223
Compton, A. H., quotation from, 176
Compton effect, 170-176
  summary of chief features, 187
Compton line, 173
Compton shift, 170, 172
Compton wavelength, 172
Conduction electron, 169
Conservation of energy, as law of physics, 10, 33-34 (Problems 10-13), 110, 112
  equivalence with mass conservation, 112
Conservation of mass, 98, 110, 112
  connection with momentum conservation, 33 (Problem 9)
  equivalence with energy conservation, 112
Conservation of mass-energy, 112
Conservation of momentum, as law of physics, 10-11, 33 (Problems 8, 9), 98
  failure of classical expression, 95, 97
Contact potential difference, 162, 190 (Problem 10)
Copenhagen interpretation of quantum theory, 223, 224
Corpuscles, 231
Correspondence principle, 242, 245
Cosmic radiation background, 158 (Problem 4)
Cosmology, 300

Cutoff frequency, photoelectric effect, 164
  interpretation of, 165
Cutoff wavelength, x-ray production, 180
  interpretation of, 182
Cyclotron, 117 (Question 17)
Cyclotron frequency, 107

Dalton, J., 231
Darrow, K. K., quotation from, 199 (footnote)
Darwin, C. G., 284
Davisson-Germer experiment, 199
"Death dance," 185
De Broglie, Louis, 195
De Broglie, Maurice, 195
De Broglie wavelength, 196
  of electron, 227 (Problem 1)
  and quantization of angular momentum, 251
  and resolving power of electron microscope, 205
Debye, P., 170
  theory of heat capacity, 311–314
Debye-Hull-Scherrer diffraction, 201
Debye law, 313
Debye temperature, 145, 313
Degenerate state, 260
De Sitter experiment, 26, 37 (Problem 32)
Deterministic interpretation of physical
        phenomena, 212, 223
  failure of, at microscopic level, 213
Deuterium, spectrum of, 254–255
Deuteron, 112
Diffraction of matter waves, 199–201, 319–320
  neutron diffraction, 202
Dispersive wave motion, 317
Doppler effect, relativistic, 73, 186
  transverse, 74
Double star observations, 26, 37 (Problem 32)
Dulong and Petit rule, 142

Einstein, Albert, 4, 29–31, 291, 300
  quotation from, 3, 4, 27, 31, 39, 93, 113, 161, 176,
        224, 281, 291, 307
Einstein heat capacity theory, 143–145
Einstein principle of relativity, 27, 80
Einstein quantum theory of light, 165, 168
Einstein temperature, 144
Einstein theory of general relativity, 291–303
Electric and magnetic fields, interdependence of,
        116
Electromagnetic spectrum, 169
Electron:
  bound, and energy quantization, 243, 246
  De Broglie wavelength of, 227 (Problem 1)
  free, allowed energies of, 249
  radius of, classical, 264 (Problem 6)
  "refractive index" for, 199
  see also Hydrogen atom; One-electron atoms,
        Bohr model
Electron diffraction, 199–200
Electron microscope, 204–208
  resolving power, 205

scanning, 206
Emission of light from moving object, 90 (Problems
        82, 83)
Emission of radiation by atom:
  Bohr model, 242
  Thomson model, 232
Emission spectrum of hydrogen, 239–241
Emission theories of light, 25–26, 37 (Problems 32,
        33)
Energy:
  binding, 113, 249
    of hydrogen, 243
    of muonic atom, 255
  conservation of, as law of physics, 10, 33–34
        (Problems 10–13), 110, 112
    equivalence with mass conservation, 112
  conversion of, into mass, 176–180
  conversion of mass into, 184–186
  equipartition of:
    classical, 135
    quantum, 136
  of harmonic oscillator, minimum, 230 (Problem
        42)
  internal, contributions to, 111
    evidence of quantization of, for atoms,
        147–149
  kinetic, 99–101, 111
  of photon, 165
  rest-mass, 96, 110
    as internal energy, 111
  thermal, 110
  total, 100, 109
Energy density of cavity radiation, 132
  Planck radiation law, 133
    physical interpretation of, 137–139
  Rayleigh-Jeans law, 132
  "ultraviolet catastrophe," 133, 139
Energy levels, 137
  correction for finite nuclear mass, 252–253
  excited states, 244
  ground state, 243
  of harmonic oscillator, ground state, 230
        (Problem 42)
  of hydrogen, 244
    Ritz combination principle, 266 (Problem 28)
  in mercury, 149
  natural width of, 222
  of one-electron atoms, 248
  of Planck oscillator, 137
  stationary states, 243
  transitions between 242, 243, 245
Energy-mass equivalence, 111, 112
Energy quantization, 136
  Bohr model of atom, 243, 248
  bound and free systems, 249
  evidence of, in atoms, 149
  one-electron atoms, 248
  in Planck oscillator, 136
  of radiation, 165, 168
Energy-time uncertainty relation, 213

and natural width of spectral lines, 222
Equipartition of energy:
  classical, 135
  quantum, 136
Equivalence, principle of, 291-292, 296
Ether, 12, 13-14
Ether drag, 21
Event, 5

Fine structure, of spectrum, 256
Fine structure constant, 264 (Problem 44)
Fizeau experiment, 23-25
Force:
  and acceleration, relativistic, 103
  transformation of, classical, 10
Frame of reference, 5
  absolute, 11, 12, 13*ff*
  inertial, 5
    equivalence of, 11, 27, 42, 52
  proper, 52
Franck-Hertz experiment, 147
Fresnel drag, 24, 37, (Problem 30)
  relativistic treatment, 68

Galilean transformation, 6, 8, 10
  and Maxwell's equations, 12
  and Newton's laws, 10
  and velocity of light, 12
Gamow, G., 175
Gas constant, 143
General theory of relativity, 239, 291-303
  predictions of, 299-300
  tests of, 298—299
Geometric optics, analogy to classical mechanics, 197
Geometrodynamics, 296
Gravitational lens, 299, 302 (Problem 15)
Gravitational red shift, 294, 301 (Problem 9)
Gravitational waves, 239, 299
Ground state, 243
  of harmonic oscillator, 230 (Problem 42)
  of hydrogen, 243, 258
Group speed, 198, 228, (Problem 21), 315

Harmonic oscillator:
  comparison of classical and Planck oscillator, 136, 137
  energy of, 143
  energy quantization in, 136, 140
  ground state energy, 230 (Problem 42)
Headlight effect, 90 (Problems 82, 83)
Heisenberg, W., 257
  quotation from, 195, 224
Heisenberg uncertainty relations, *see* Uncertainty principle
Hertz, H., 161
High-gamma engineering, 57
Homogeneity of space and time, 44
Hydrogen atom:
  Bohr model of, 241-250

critique of, 257
Bohr radius, 248
emission spectrum of, 239-241
  comparison of, with deuterium, 254-255
energy levels, 244
  Ritz combination principle, 266 (Problem 28)
quantum theory of, 258
Rydberg constant for, 240, 246
  correction for finite nuclear mass, 253

Ideal radiator, 128
Impact parameter, 321
"Index of refraction," for electrons, 199
Inelastic collisions:
  and evidence of atomic structure, 147
  and mass-energy equivalence, 110
Inertial frame, 5
  equivalence of, 11, 27, 42, 52
Intensity:
  of cavity radiation, 129
  of radiation, 167, 210-211
Internal energy, of solid, 312
Invariant quantities and transformation laws, 11, 12, 54
  charge, 106
  for Galilean transformation, 8, 10
  for Lorentz transformation, 77-78
  proper time, 77
  rest length, 77
  spacetime interval, 78
  velocity of light, 27, 67-68
Ives-Stilwell experiment, 74

Jammer, Max, quotation from, 201
Jeans, James, quotation from, 315

Kennedy-Thorndike experiment, 20
Kinetic energy, 99-101, 111
Klein, Martin, quotation from, 30

Laue, Max von, quotation from, 127
Length:
  contraction, 50, 80 (Question 6)
    as consequence of relativity of simultaneity, 78
    as consequence of time dilation, 61
    reality of, 78
  measurement of, 50, 59, 61
    classical assumptions, 7
    geometrical interpretation, 274
    relationship of, to simultaneity, 7, 43
    relativity of, 43
    transverse, 59
  proper (rest), 52
  transverse, 59
Light:
  aberration of, 21-22, 72
  bending of, 298, 301 (Problem 12)
  effective mass, 111
  Einstein quantum theory of, 165, 168

Light (*Continued*)
  emission of, by moving object, 90 (Problems 82, 83)
  particle nature of, 161*ff*
  propagation of, 71
    emission theories, 25–26
    ether hypothesis, 12, 13–14
    in moving medium, 23–25, 68
    see also Photon; Radiation; Velocity of light; Wave-particle duality
Lightlike interval, 56
Lorentz, H. A., quotation from, 3
Lorentz factor, 47
Lorentz-Fitzgerald contraction, 19–20
Lorentz force, 99
Lorentz transformation, 46, 48, 49, 269
  consequences of, 50–58
  derivation of, 44–46
  invariant quantities, 53–54, 77–78, 106, 111, 115
Luminosity, 131
Lyman series, 241

Magnetic and electric fields, interdependence of, 116
Magnetic field, charged particle in, 104–105, 107
Mass:
  conservation of, 110
    connection with momentum conservation, 33 (Problem 9)
    equivalence with energy conservation, 111
  conversion of, into energy, 184–186
  creation of, from energy, 176–180
  effective, 111
  reduced, 253
  relativistic, 96, 105, 111
  rest, 96
    as internal energy, 110–111
    particles having zero, 113
Mass-energy conservation, 112
Mass-energy equivalence, 111–112
Matter, fusion of wave and particle models, 209, 212, 223
  interaction of, with radiation, 170, 176, 179
  wave nature of, 195*ff*
Matter waves, 196
  De Broglie wavelength, 196
  diffraction of, 199–203
  refraction of, 319–320
  speeds of, 198, 228 (Problems 20, 21), 315
  wave function, 211
    probabilistic interpretation, 211–212, 223
  see also Wave-particle duality
Maxwell, J. C., quotation from, 1
Maxwell's equations, attempts to modify, 25–26
  and Galilean transformation, 12
  and Lorentz transformation, 115
Measurement and interaction with physical system, 214–215
  indivisibility of photons, 216
Measurement and Special Relativity, 77

"Measuring" *vs.* "seeing," 66
Michelson, A. A., 14
Michelson-Morley experiment, 13–19
  generalized, 35 (Problem 18)
Millikan, R. A., 163
  quotation from, 166, 168
Minkowski, H., 269
  quotation from, 269
Minkowski diagram, *see* Spacetime diagram
Modern physics, definition, 1
Molar heat capacity, 141, 142, 144
  classical value, 143
Momentum:
  conservation of, as law of physics, 10, 11, 33 (Problems 8, 9) 98
  failure of classical expression, 95–97
  of photon, 113
  and proper time, 98
  relativistic definition, 97
Momentum-position uncertainty relation, 213
  physical origin of, 214–216
"Moving clocks run slow," 51
Muonic atoms, 255

Neutrino, 118 (Problem 12)
Neutron, thermal, 227 (Problem 7)
Neutron diffraction, 202
Newtonian relativity, 11–13, 79
Newton's laws of motion:
  deterministic character of, 212
  and Galilean transformations, 10
  need to modify, 93
  relativistic generalization, 103

Observer:
  interaction with systems, 214–215
  in relativity, 65–66
One-electron atoms, Bohr model, 247–250
  critique, 257
Orbital frequency in one-electron atom, 245
Orbital quantum number, 259
Order number, in diffraction, 320
Oscillator, *see* Harmonic oscillator

Pair-annihilation, 184–186
  summary of chief features, 187
Pair-production, 176–180
  summary of chief features, 187
Parallax, 32 (Question 18)
Particle in a box, 217–218
Paschen series, 241
Perfect absorber, 129, 130
Pfund series, 241
Phase difference in synchronization of clocks, 53, 61
Phase speed, 198, 228 (Problem 20), 315
Photoelectric effect, 161–164
  Einstein theory of, 165–166
  summary of chief features, 187
Photoelectron, 162

Photon, 161, 165, 171
  absorption of, by matter, 170, 176, 179
  gravitational mass of, 293
  indivisibility of, 216
  pair-annihilation, 184-186
  pair-production, 176-180
  production by accelerating charges, 180-183
  scattering of, by matter, 170, 173, 187
  and wave model of radiation, link between,
    210-211
  x-ray production, 180-182
  as zero-rest-mass particle, 113
  see also Wave-particle duality
Photonuclear reaction, 169
Pickering series, 266 (Problem 32)
Pion decay, 54-55
Planck, Max, quotation from, 151, 168
Planck's constant, 133, 136, 166
  and domain of observable quantum phenomena,
    139, 175, 202-203, 214
Planck oscillators, 136, 137, 140
Planck radiation law, 132, 133
  Einstein derivation of, 307-309
  physical interpretation, 137-139
"Plum pudding" model of atom, 232
Position-momentum uncertainty relation, 213
  physical origin of, 214-216
Positron, 176-177
Positron emission tomography, 185
Positronium, 184, 267 (Problem 47)
Precession, of perihelion, 298, 302 (Problem 16)
Preferred frame of reference, 11
  attempts to locate, 13-19
Probabilistic interpretation of physical phenomena,
    210-225
  as fundamental view of quantum physics,
    211-212, 223-225
  as link between wave and particle models,
    210-211
Proper distance interval, 56, 276
Proper frame of reference, 52
Proper frequency, 73
Proper length, 52
Proper mass, 97, *see also* Rest mass
Proper time, 52, 77
Proper time interval, 54, 277
  route dependence, 281-283
Pulsar, 118 (Problem 17)
Purcell, E. M., quotation from, 57

Quantization conditions, 259-260
  Bohr quantization rule, 251
  Planck rule for oscillator, 136
  see also Angular momentum, quantization of;
    Energy quantization
Quantum mechanics, 257-261
Quantum number, 136
  states of large "n," and correspondence principle,
    245
Quantum state, energy level width of, 222

Quantum theory, Copenhagen interpretation of,
  224
Quantum theory of radiation, Einstein, 165, 168
Quasar, 31 (Question 1), 123 (Problem 70)

Radial distribution function, 268 (Problem 49)
Radial probability density, 258, 268 (Problem 49)
Radiancy, 129
Radiation:
  absence of, in stationary state, 239, 242
  absorption of:
    by a gas, 247
    by matter, 170, 176, 179
      relative probability of processes, 179-180
  atomic spectra, 239-241, 254
  black-body, *see* Cavity radiation
  Einstein quantum theory of, 165, 168
  fusion of wave and particle models, 211-212
  intensity, statistical interpretation of, 211
  and loss of rest mass, 111, 114
  from moving relativistic particle, 90 (Problems
    82, 83)
  from orbiting charge, 239
    comparison of Bohr and classical models, 242
  particle nature of, 161*ff*
  scattering of, by matter, 170, 173, 187
  synchrotron, 182
  thermal, 127, 128
  x-rays, 169, 180
  see also Photon; Wave-particle duality
Radiation condition of Bohr, 242
Rayleigh-Jeans law, 132, 134-135
Reduced mass, 253
Reference frame, *see* Frame of reference
Refraction, of matter waves, 319-320
"Refractive index" for electron, 199
Relativity, practicle uses of, 57, 288
  special, *see* Special relativity
Relativity principle, 11
  of Einstein, 27, 80
  of Newton, 11, 79
Relativity of simultaneity, 39-43
  geometric interpretation, 273
  and intuitive reason, 63-64
  and length contraction, 78
Resolving power in electron microscope, 205
Rest length, 52, 77
Rest mass, 52, 96, 111
  particles having zero, 113
Rest-mass energy, 100, 110
  conversion of, into radiant energy, 184-186
  conversion of radiant energy into, 176-180
  as internal energy, 111
Rigid body, 78-79
Ritz combination principle, 266 (Problem 28)
Rocket, relativistic, 121 (Problem 44)
Rutherford, Ernest, 233
  quotation from, 231, 233, 234
Rutherford model of atom, 233-237
  stability of, 239

Rutherford scattering experiment, 233–236, 265
(Problems 14, 18, 20), 321–325
angular probability in, 236, 325
distance of closest approach in, 238
Rydberg constant, 240
correction for finite nuclear mass, 253
derivation of, 245–246
Rydberg formula, 240

Scanning electron microscope, 206
Schrödinger, E., 257
Schrödinger's equation, 258
Schwarzschild radius, 301
Series limit, 240
Shankland, R. S., quotation from, 39
Shurcliff, W. A., summary of relativity, 55
Simultaneity, 41
relationship:
to length measurement, 7, 43
to time measurement, 43
relativity of, 39–43, 53
and intuitive reason, 63–64
and length contraction, 78
Space contraction, 274
Spacelike interval, 56, 276
Spacetime, 270
Spacetime diagram, 270
of twin paradox, 285
Spacetime interval, 54, 78, 276
Special relativity:
and common sense, 76–80
domain of validity, 4
experimental basis for, 28
postulates of, 26–27
and process of measurement, 77
Specific heat capacity, 141, 159 (Problem 53)
Debye theory of, 311–314
Einstein theory of, 143–145
Spectral radiancy, 129–130, 156 (Problem 31), 305
and spectral energy density, 132
Speed parameter, 16, 47
Spontaneous emission, 307–308
Stars, temperatures of, 130–131
Stationary states, 242
transitions between, 242, 243, 245
Stefan-Boltzmann constant, 129
derivation of, 156 (Problem 22)
Stefan-Boltzmann law, 129
Stimulated emission, 307–308
Stopping potential, photoelectric effect, 163, 166
Superposition, principle of, 212
Synchronization of clocks, one frame, 40
phase difference for moving clocks, 53, 61
Synchrotron, 107–109
Synchrotron radiation, 182–183

Tachyon, 69
Thermal neutron, 227 (Problem 7)
Thermal radiation, 127
temperature dependence of, 128

see also Cavity radiation
Thomson, G. P., 200, 201
Thomson, J. J., 201
quotation from, 231
Thomson model of atom, 231–233
Time:
classical concepts, 7, 39
dilation, 52
geometric interpretation, 274
relation to length contraction, 61
and transverse Doppler effect, 75
measurement of, 39–43
order of events, 43
proper, 52, 77
Time-energy uncertainty relation, 213
and natural width of spectral lines, 222
Timelike interval, 54, 276
Total energy, 100
Transformation equations, 46, 48, 49
acceleration, classical, 10
force, classical, 10
Galilean, 6, 8, 10
Lorentz, 46, 48, 49
velocity:
classical, 8
relativistic, 70
Twin paradox, 281, 295
experimental test, 287
reality of, 286
spacetime diagram of, 285

"Ultraviolet catastrophe," 133, 139
Uncertainty principle, 212–218
and constituents of nuclei, 218
derivation of, from universal wave properties,
218–221
for free particle, 229 (Problem 32)
physical origin, 214–216
and probabilistic view of physical phenomena,
222–223
statement of, 213

Velocity addition theorem:
classical, 8
relativistic, 67
Velocity of light, 3
constancy principle of, 27
as defined standard, 3
and emission theories, 25–26
and Galilean transformation, 12
as limiting speed, 3–4, 67, 76–77, 93, 100, 291
in moving medium, 23–24
speeds in excess of, 69
and synchronization of clocks, 40
Visual appearance of moving objects, 66

Wave equation, 35 (Problem 16)
Wave function, 211, 258
probabilistic interpretation, 211–212, 223
Wave groups and uncertainty principle, 218–221

Wave mechanics, *see* Quantum mechanics
Wave motion, dispersive, 317
Wave number, 211
Wave-particle duality, 208–212
  fusion of wave and particle models, 211–212,
    223
  and measuring process, 209–210
  principle of complementarity, 209, 223
  probabilistic interpretation, 210–211
Wave pulse, 219
Wave train, 219
Weisskopf, V. F., quotation from, 66
Wheeler, J., quotation from, 300
Wien constant, 130

derivation of, 155 (Problem 20)
Wien's displacement law, 130
Wien's law, 132
Work-energy theorem, 33 (Problem 11)
Work function, 165
World line, 269

X-ray production, 180–182
  summary of chief features, 187
X-rays, 169, 180
  and microscopes, 205

Zeeman effect, 260
Zero-rest-mass particles, 113